Turbulent Flow

Turbulent Flow

Analysis, Measurement, and Prediction

Peter S. Bernard
James M. Wallace

JOHN WILEY & SONS, INC.

To the memory of Foster Arnold Bernard

Peter S. Bernard

To my wife, Barbara, and son, Jaimie, who encouraged my participation in the writing of this book in spite of the time it took away from our family, and to my friends Bob Brodkey and Helmut Eckelmann, who made my first experience in turbulence research so much fun.

James M. Wallace

Contents

Preface

To the extent that many practical turbulent engineering flows are beyond our capacity to predict, it is clear that the turbulence "problem" is not yet solved. The need to predict turbulent fluid behavior, however, is not diminished even if finding a reliably accurate solution technique is elusive. In essence, the subject of turbulence is driven by its applications: answers to difficult turbulent flow problems are required, and engineers or physicists must supply the best possible answer. Judgments must be made between competing approaches whether experimental, numerical, or analytical. Whatever decision is made, it is subject to debate and justification.

Whether the reader is an engineer faced with turbulent flow prediction or a scientist or student intending to pursue research in the field, one needs to become familiar with what might at first glance appear to be a vast literature of loosely connected theories, numerical predictions, and experimental measurements of turbulence. A principal goal of this book is to help readers see the subject as a relatively coherent whole, so that they will be able to make informed decisions as to how best to study and predict turbulent flows of interest.

The book aims at a relatively wide coverage, but without being exhaustive. Our interest is to include those areas that are essential to making sense of currently available options for measuring or predicting turbulent flows. We believe that to be most useful, the book needs to familiarize the reader with current techniques used to measure, simulate, analyze, and predict turbulence, so we have attempted to provide meaningful discussions of many such techniques. If successful, the book will put the reader in a good position to understand more advanced and specialized books in the field. As a general rule, we have striven to keep the discussion focused on those results and theories that are relatively well established. More speculative ideas are often part of the research frontier and appropriate for independent study.

After giving some essential preliminary notation and concepts in Chapter 1, in Chapter 2 we provide a natural entry to the subject of turbulence through consideration of the averaged equations of motion and the basic physical processes that they encompass. Note that this volume will be concerned exclusively with incompressible turbulent flow. The focus of Chapter 2 is to provide an overview of the principal concepts and issues in understanding turbulence, so that the basic language of the subject is available to the reader when considering the major facets of turbulence in the subsequent chapters.

In Chapter 3 we survey the main techniques used in performing physical experiments, with a view toward making clear what is feasible and what is not. Included here is a discussion of how turbulent flow can be simulated on a computer. Such algorithms, considered to be "numerical wind tunnels," have become a very large part of turbulent flow "measurement." A number of the experimental and numerical techniques mentioned in Chapter 3 have been used for many years to gain fundamental knowledge of turbulence. A review of some of the major aspects of the knowledge acquired this way is presented in Chapter 4, covering bounded flows, and in Chapter 5, covering free shear flows. The hope here is to give the reader a feeling of what turbulent flow is and something of what is presently known about it. This is an important background to have when considering the predictive methodologies developed in later chapters.

In Chapters 6 and 7 we go into greater depth in exploring the essential physics of turbulence. In view of its position as a distinguishing characteristic of turbulence, the issue of turbulent transport is considered from a number of perspectives in Chapter 6. Following this, in Chapter 7, the properties of idealized turbulent flows are considered. This permits a relatively unobstructed view of such fundamental processes as energy dissipation. Many aspects of the theoretical analysis of such flows are useful to the development of predictive theories.

The stage is then set in Chapter 8 to begin a discussion of turbulence closures, which are the predominant means by which turbulent flows are predicted in engineering work. The relationship that closure models have to our understanding of the physics of turbulent flow will become evident. Once we have discussed many features of closure schemes, in Chapter 9 we give some examples of how they perform in the practical solution of turbulent flow problems.

In Chapter 10 we consider prediction methods known as large eddy simulations (LES), which are situated between closure schemes on the one hand and direct computation of the flow on the other. First discussed are traditional grid-based LES techniques, which have long been used by meteorologists and are now finding some application in engineering work. This is followed by consideration of the application of grid-free vortex methods to LES, which is just now beginning to become a significant new methodology for predicting turbulence.

A key part of many engineering flows is the occurrence of heat or mass transfer, and in Chapter 11 we give some basic background knowledge of this subject. Finally, in Chapter 12 we introduce several of the principal trends in the theoretical analysis of turbulence. To varying degrees, these theories, which tend to be mathematically difficult, have been developed with a view toward understanding turbulence at its most fundamental level. There remains considerable controversy in this area, and our purpose here is only to hint at some of the major developments in what is a very large field in and of itself.

This book represents material that has been successfully taught by the authors in two complementary graduate-level courses on turbulent flow given at the University of Maryland over the last 20 years. One of these covers the analysis of the physics of turbulent flow through its measurement and simulation, and the other concentrates on the prediction and theoretical analysis of turbulence. As suggested by the title of this volume, we bring together here material that would be sufficient for one or both of these courses. Depending on the inclination of the instructor, more focused readings of particular research articles related to topics in the book can be included to help extend the book into current research activities. In fact, our goal for this book will be met if students are brought by it to the point where they can read classical and current research articles with comprehension and an informed critical eye.

PETER S. BERNARD
JAMES M. WALLACE

College Park, Maryland

Acknowledgments

We wish to express our appreciation to Ron Adrian, Ilias Balaras, Alexandre Chorin, Deji Demuren, Thanasis Dimas, Sharath Girimaji, Robert Hander, Ken Kiger, Jacob Krispin, Bob Rubinstein, and Petar Vukoslavčević, who have kindly answered our questions, read parts of the manuscript, and otherwise provided valuable help in completing this project. Thanks are also due to those colleagues, including Alexander Broniewicz, Michael Breuer, John Cimbala, Johan Larson, Michael Leschziner, Zi-Chao Liu, Richard Loucks, Nagi Mansour, Rajan Menon, Bob Moser, Steven Olson, Valeriy Prostokishin, Anatol Roshko, Chuck Smith, and Phillippe Spalart, who provided us with original graphics, data files, photographs, and other publications. We are also very grateful to Niem Dang, who skillfully produced many of the figures, and to Ning Li, who created data-scanning software for us and helped with some of the more complex figures.

PETER S. BERNARD
JAMES M. WALLACE

1

Preliminaries

1.1 TURBULENCE

Every airplane passenger knows firsthand that the atmosphere contains something called *turbulence*. When airplanes enter into it, they shake and vibrate unpredictably, until, to every passenger's relief, the turbulent zone is passed and the flight returns to a reassuring calm. It is evident that turbulence must consist of a collection of disordered wind gusts, which in this case are capable of pushing an airplane around.

It is also possible that the reader has experienced turbulence on a hand or back placed over the jet of fluid entering a swimming pool or jacuzzi. Far from a constant flow, one has the sensation of an ever-changing eddying motion which is particularly pronounced at the edges of the jet. Although on a very much smaller scale than in the airplane flow and in a different fluid medium, the apparent random pulsations of the turbulent flow appear to be a common characteristic.

Countless other experiences of turbulent flow accompany us in our daily lives. Examples abound of turbulent flow in technological, environmental, and biological applications. The flow around an automobile, between buildings in a downtown street, and through a diseased artery are but three commonplace occurrences.

It is natural to wonder where turbulence comes from, why it occurs, and why it so prevalent. Furthermore, it is not hard to see that there are probably many situations where one wishes to be able to predict, a priori, either the occurrence of turbulence or its behavior. Our starting point for shedding light on these questions is to be precise about exactly what is meant by turbulence.

The apparent randomness of the buffeting of an airplane as it flies through turbulence appears to be a defining characteristic of such flows. However, since it is believed that fluid flows evolve deterministically according to the Navier–Stokes equation, evidently, a useful definition of turbulence must be more precise than the statement that the flow displays random characteristics. To be explicit about the role of determinism and randomness in turbulent flow, consider the flow in a controlled setting such as a laboratory wind tunnel. Imagine that the following experiment is repeatable at will: Starting from quiescent conditions, the wind tunnel is turned on to a particular speed setting. A fixed time after it is started, the fluid velocity is recorded at arbitrary points within the test section of the tunnel. Depending on the speed of the flow, one of two possible outcomes is possible: either the measured velocities are identical, within measurement accuracy, each time the experiment is repeated, or they vary. The former case is referred to as *laminar flow*, and the second, in which the velocity field is not repeatable in either the whole or in part of the flow domain, is what will formally be referred to as *turbulent flow*, or *turbulence*. By this definition, transitional flows in which the motion is undergoing a change from laminar to turbulent conditions qualify as turbulent.

The differences in velocity from realization to realization of a turbulent flow are explained by the sensitivity of the evolving flow to small uncontrollable perturbations in the initial and boundary conditions. These originate from slight thermal currents, from changing surface roughness at a small scale, from small variations in the input power of the tunnel, perhaps even from microscopic sources such as the subcontinuum molecular motions that cause Brownian motion [4]. In no way is the determinism of the evolving flow in doubt. Rather, it is the ability of the flow to amplify slight, unpredictable changes in boundary and initial conditions to measurable scale which is the defining characteristic of turbulence. The result is a randomly appearing flow structure over a sizable physical extent, such as is encountered by airplanes in the atmosphere. This contrasts with laminar flows in that the same small perturbations experienced in turbulent flow are present, but in this case they are damped out successfully as soon as they appear, and the flow velocity, if measured, is always the same.

Under normal circumstances, the ability of flows to remain laminar depends on the degree to which viscous damping, deriving from the molecular diffusion of momentum, can erase the influences of individual perturbations. Since the Reynolds number, $R_e = UL/\nu$, where ν is the kinematic viscosity, U is a typical velocity of the flow, and L a typical length scale, characterizes the relative strength of viscous and inertial forces, it is the key parameter used in deciding whether or not a flow is likely to be turbulent. For low values of R_e viscous forces are dominant and the flow tends to be laminar. For larger values, a point is reached where a transition occurs in which disturbances are no longer damped out but rather, are amplified. Beyond this point a fully turbulent state results. Figure 1.1 illustrates the transition process in a turbulent jet. The flow is seen to undergo a dramatic change from a laminar state, through one containing organized oscillatory motions, finally, to fully turbulent conditions. In a bounded flow such as in a pipe for which $R_e = U_m d/\nu$, where U_m is the average mass flow velocity and d is the diameter, transition begins as low as $R_e \approx 200$ in the form of decaying turbulent slugs passing through the pipe [3], and generally a fully turbulent state occurs for $R_e > 2300$. It is not possible to be completely precise about these critical values since they are affected by upstream flow conditions and the smoothness of the boundary. With sufficiently small input disturbances and smooth walls, laminar flow in a pipe can be maintained to at least $R_e = 10^5$ [2].

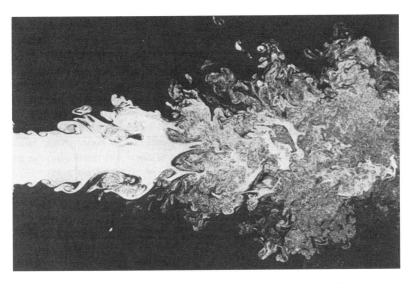

Fig. 1.1 *Transition to turbulence in a jet. (Courtesy of J.-L. Balint and L. Ong.)*

It is a fact of our experience that the Reynolds number encountered in many practical problems of concern to engineers and physicists is large, in fact well beyond the transitional range. For example, in the case of a 100 m-long submarine in seawater ($v = 1.044 \times 10^{-6}$ m^2/s), traveling at 15 knots, at the stern $R_e \approx 7.4 \times 10^8$, while at the rear of a 6 m-long car in air ($v = 1.525 \times 10^{-6}$ m^2/s) at 60 mph, $R_e \approx 1.1 \times 10^8$. The inevitable turbulence occurring in these and other flows must be accommodated in the design process. In the case of a submarine, turbulent forces and acoustic fields are of significant concern as well as the exact position of turbulent separated flow regions, which might lead to undesirable flow patterns affecting the propeller. Turbulent drag and side forces affect the economy and stability of road vehicles, and turbulent buffeting of the driver side window by vortices shed off the rear view mirror generates noise (see Fig. 1.2). Turbulent flow in the engine compartment affects cooling, and turbulence in the engine cylinder is necessary for effective combustion and reduced pollutant emissions. Atmospheric turbulence formed from thermal currents and gravity waves shed off topological features in stratified flow lead to the turbulence encountered by airplanes flying at low altitudes. Countless more examples exist: it is the ubiquity of turbulence in high-Reynolds-number flow that creates the need for the study of turbulence and its prediction.

A common characteristic of the flows depicted in Figs. 1.1 and 1.2 is the presence of rotational motion in the form of vortices. Large-scale vortical structures are visible in the outer edge of the jet. In the wake of the automobile the scales of the vortices and other turbulent motions are so small that the smoke marker in this region appears as a diffused blur. The dynamics of the energetic large-scale vortices have a significant influence on the physics of turbulent flow. Moreover, the importance and presence of vortices extend through all scales of turbulent motion, including the smallest

Fig. 1.2   *Visualization using smoke helps engineers see the flow patterns around vehicles, including the turbulence in their wakes. (Used with permission of the Volvo Car Corp., Aerodynamics Div.)*

scales. For example, the three-dimensional plot of vorticity vectors in a numerical simulation of turbulence shown in Fig. 1.3 suggests that embedded tubelike vortices form the essential fabric of turbulence at its smallest scale. It will often be seen in this book that the dynamics of the vorticity field represents the most convenient and succinct means for describing and understanding the behavior of turbulent fluid flow. Thus subsequent questions about the nature of turbulent transport phenomena and the physics of the energy field will find explanation in terms of vorticity dynamics.

The turbulent flow examples that have been mentioned thus far suggest that the data needed in analyzing turbulent fields can involve both mean statistics, as in the average forces on automobiles and submarines, or instantaneous quantities, such as the peak pressures responsible for sound generation on solid surfaces and the position of separated flow structures. In all cases, if the velocity and pressure fields in a turbulent flow can be obtained by one means or another, then besides knowing the instantaneous flow properties, it would be possible to compute average properties as well. As will be seen, however, it is usually a very difficult matter to predict turbulent fields, so much so that it is often the case that only mean statistics can be computed, and usually these are approximations produced by a partial analysis of the flow physics. It will thus become clear in the course of this book that knowledge of turbulence and how to predict it is incomplete. The boundary between the known and the unknown changes steadily, and the future can only bring improved insights and better predictive schemes.

Characterization of flows by their Reynolds number (e.g., between high and low Reynolds numbers) is an important factor to take into account when deriving solution strategies. There are a number of other fundamental categories into which flows may

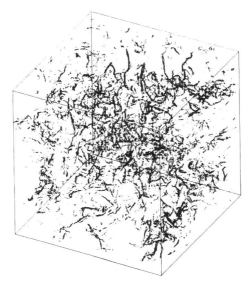

Fig. 1.3 *Vorticity vectors in a computer simulation of turbulent flow in a periodic cubic box. (From [5]. Reprinted with the permission of Cambridge University Press.)*

be divided in order to better suggest schemes for their measurement or prediction. One demarcation is between interior and exterior flows, as in the case of a pipe or engine cylinder versus the flow past an airplane. In the former, all parts of the flow generally are strongly affected by the presence of solid boundaries. For exterior flows, turbulence is often created at the boundary but then evolves downstream, free of its direct influence, while spreading into increasingly greater portions of the flow domain. Near boundaries, turbulent flow is often accompanied by high shear (i.e., large values of the mean velocity gradients). Such regions are a focal point of the physical processes governing the overall production and distribution of turbulence, and thus are worthy of extra attention in the discussions that follow. The opposite extreme from high-shear regions is *homogeneous turbulence*, where mean properties of the turbulence including mean velocity do not vary with position (i.e., are independent of translations of the coordinate axes). Little production of new turbulence takes place in these circumstances, and the flow is dominated by dissipation. In some instances, homogeneous turbulence is also *isotropic,* wherein, in addition to the independence of the mean turbulence properties to position, the turbulence displays no intrinsic directional preference.

The concept of isotropy implies that the turbulent flow must be homogeneous, since inhomogeneous flows have directional preferences, thus contradicting the assumption of isotropy. On the other hand, the condition of homogeneity does not imply that the flow must be strictly isotropic. In fact, another possibility is the case of *homogeneous shear flow,* in which there is a uniform mean velocity gradient everywhere, so that even though there is a directional preference, it is the same everywhere. In such flows, the statistical properties of the turbulence do not vary with position, with the sole exception of the mean velocity field. Homogeneous shear flows play a useful role in developing turbulence prediction schemes, since they allow attention to be focused on just specific parts of the governing equations. In addition to imposed shear, homogeneous flows with uniform rotations, plane strains, and combinations of these effects are also studied.

Finally, it should be noted that engineering flows are often *complex,* or *nonequilibrium,* in the sense that they contain unusual geometrical features or unsteadiness which causes the turbulent motion not to be explained easily in terms of relatively simple balances of physical effects, as would be the case, for example, in an attached boundary layer on a smooth wall. The latter type of flow is one of several that are often called *canonical,* since they are relatively simple and fairly well understood experimentally and numerically. Aspects of all these types of flows are discussed in subsequent chapters.

1.2 AVERAGING AND TURBULENT FIELDS

The average of a random field, such as the velocity in turbulent flow, is most naturally defined as an *ensemble average* over independent realizations of the flow field. Each realization may be viewed as one occurrence of an experiment, and an average over N of them is given by

$$\overline{\mathbf{U}}(\mathbf{x}, t) = \frac{1}{N} \sum_{j=1}^{N} \mathbf{U}^{j}(\mathbf{x}, t), \tag{1.1}$$

where, for example, $\mathbf{U} = (U_1, U_2, U_3)$ is the velocity field and $\mathbf{U}^j$ is the velocity measured in the jth experiment. $\mathbf{x} = (x_1, x_2, x_3)$ denotes position in three-dimensional space and t is the time. Throughout this book, vectors are denoted either in boldface, as in the case of $\mathbf{x}$ and $\mathbf{U}$ in (1.1), or equivalently, using index notation, as in U_i, $i = 1, 2, 3$. As the occasion arises, the notation (x, y, z) will also be used to represent (x_1, x_2, x_3) and (U, V, W) to denote (U_1, U_2, U_3).

Ensemble averaging conveniently commutes with time and space derivatives:

$$\overline{\frac{\partial \mathbf{U}}{\partial t}}(\mathbf{x}, t) = \frac{\partial \overline{\mathbf{U}}}{\partial t}(\mathbf{x}, t), \qquad \overline{\frac{\partial \mathbf{U}}{\partial x_i}}(\mathbf{x}, t) = \frac{\partial \overline{\mathbf{U}}}{\partial x_i}(\mathbf{x}, t), \tag{1.2}$$

which is of great benefit in theoretical analyses. In some instances, as in flow in an internal combustion engine cylinder, each four-stroke cycle of the engine may be considered to be one realization of the flow, in which case the average in (1.1) can readily be evaluated by experimental techniques.

It is more often the case, at least in experimental investigations, that (1.1) is impractical to implement because it would involve the laborious task of repeating the experiment many times. It thus is useful to consider alternatives. A particularly useful average is one over time, as in

$$\overline{\mathbf{U}}(\mathbf{x}, t) = \frac{1}{T} \int_{t-T/2}^{t+T/2} \mathbf{U}(\mathbf{x}, s) \, ds, \tag{1.3}$$

where T is a sufficiently long time interval and $\mathbf{U}(\mathbf{x}, s)$ is obtained from a continuous sampling of the flow velocity at $\mathbf{x}$ over the interval $(t - T/2, t + T/2)$. Equation (1.3) is conveniently evaluated in many standard experiments, and is most useful when the random field is *stationary* (i.e., the time average is independent of the interval T over which it is taken). In this case, the integral in (1.3) is independent of t. Time averaging loses legitimacy in nonstationary flows where the underlying mean signal changes over time. The engine flow is a good example of this. In such circumstances, (1.3) is a strong function of t and T and the first of the formulas in (1.2) is not formally valid since the time derivative does not commute with the integral in (1.3).

Another possibility is spatial averaging:

$$\overline{\mathbf{U}}(\mathbf{x}, t) = \frac{1}{V} \int_{\mathcal{V}} \mathbf{U}(\mathbf{x}', t) \, d\mathbf{x}', \tag{1.4}$$

where V is generally the volume of a region $\mathcal{V}$ surrounding the point $\mathbf{x}$, although in some cases, particularly flows with geometrical symmetries, $\mathcal{V}$ may be taken to be particular lines or surfaces in the flow field. Often, (1.4) is used in the analysis of numerical simulations such as of a channel flow, where symmetries allow for averaging over planes. This reduces the number of time steps or realizations of the flow needed to get converged statistics. Similar to the case of time averaging, depending on the particular volume $\mathcal{V}$ and the spatial variability of the mean field, commutation with spatial derivatives in the second relation in (1.2) will sometimes be legitimate and sometimes not.

Finally, note that it is plausible that ensemble, time, and spatial averages are equivalent when circumstances permit, although this conclusion lacks formal mathematical proof. Throughout the book each of these methods of averaging will be employed depending on the circumstances, and which type is in effect will be made clear. For most theoretical discussions it will be assumed that ensemble averaging is employed. In some contexts it will also be seen that it is useful to define partial averages, such as conditional averaging in which averages are taken only over flow events or phases of processes satisfying some predetermined criteria. In a similar vein, the large eddy simulation method discussed in Chapter 10 requires specialized averaging in the form of a filtering process.

For either ensemble, time, or space averaging, a turbulent velocity field may be decomposed according to

$$\mathbf{U} = \overline{\mathbf{U}} + \mathbf{u}, \tag{1.5}$$

where $\mathbf{u} \equiv \mathbf{U} - \overline{\mathbf{U}}$ is referred to as the *velocity fluctuation vector*. Note that where no confusion can result, such as in this instance, we refrain from indicating the dependence of $\mathbf{U}$ and the other fields on $\mathbf{x}$ and t. The average of the fluctuation is always zero (i.e., $\overline{\mathbf{u}} \equiv 0$), although this may not be exactly true in the case of spatial or time averaging unless the flow is appropriately homogeneous or steady. In laminar flow, by definition, $\mathbf{U} = \overline{\mathbf{U}}$, so that $\mathbf{u} = 0$.

Besides the mean, higher-order moments of the fluctuating velocity field may also be of interest: for example, $\overline{u_1^2}, \overline{u_2^2}, \overline{u_3^2}$, which are the variances of the components of $\mathbf{u}$. The sum of these yields the turbulent kinetic energy per unit volume, $K \equiv \rho\overline{u_i^2}/2$, where ρ is the density, and here and henceforth it is understood that repeated indices are summed from 1 to 3. More generally, the tensor

$$R_{ij} \equiv \overline{u_i u_j} \tag{1.6}$$

is the covariance of the random vector field $\mathbf{U}$. The components of $\overline{u_i u_j}$ have a clear physical interpretation as momentum fluxes, as shown in Section 2.1.

In many circumstances where averages are computed directly from the random field, our knowledge of the turbulent physics can be deepened by examining the associated probability density function (pdf). For example, a velocity component such as U_1 will occur over a range of values, with each one having a different likelihood of occurring. The cumulative effect is the average, $\overline{U}_1$, but the pdf tells how often values of U_1 in a particular range are likely to occur. Thus, if $p(U_1)$ is the pdf of U_1, then by definition, $p(U_1)\,dU_1$ is the probability that U_1 takes on a value U such that $U_1 \leq U \leq U_1 + dU_1$. Clearly,

$$\int p(U)\,dU = 1, \tag{1.7}$$

since U_1 must take on some value for each experiment. It also follows that

$$\overline{U}_1 = \int U p(U)\,dU, \tag{1.8}$$

$$\overline{u_1^2} = \int (U - \overline{U}_1)^2 p(U)\,dU, \tag{1.9}$$

and so on, for higher moments.

Probability density functions can also be written for multiple variables, in which case they are referred to as *joint probability density functions*. For example, for velocity components U_1 and U_2, $p(U_1, U_2)\,dU_1\,dU_2$ is the probability that both $U_1 \leq U \leq U_1 + dU_1$ and $U_2 \leq V \leq U_2 + dU_2$ are satisfied by a realization (U, V) of the velocity field. A notable use of the joint pdf is in analyzing correlations such as

$$\overline{u_1 u_2} = \iint (U - \overline{U}_1)(V - \overline{U}_2) p(U, V)\,dU\,dV, \tag{1.10}$$

which are important to the analysis of turbulent momentum transport near boundaries.

An important example of a probability density function is

$$p(U) = \frac{1}{(2\pi\sigma^2)^{1/2}} e^{-(U-\overline{U})^2/2\sigma^2}, \tag{1.11}$$

where σ is the standard deviation, which characterizes Gaussian random variables. Such variables make an appearance at several junctures in turbulent flow analyses; for example, it is often a good approximation to the fluctuating velocity field at a single point in turbulent flow, the main difference being that large fluctuation amplitudes are less likely to occur than in a true Gaussian. Equation (1.11) generalizes to accommodate processes that are jointly Gaussian and to Gaussian random fields (see Section 12.2) where the joint probability density function for variables at multiple locations in the flow field obeys a Gaussian form. It is one of the more profound attributes of turbulent flow fields that even though the velocity field may be approximately Gaussian at individual points in the flow, it does not constitute a Gaussian random field.

Besides second-order moments, other higher-order statistics of importance in turbulent flow analysis are the skewness and flatness, defined, respectively, via

$$S = \frac{\overline{u^3}}{\overline{u^2}^{3/2}} \tag{1.12}$$

and

$$F = \frac{\overline{u^4}}{\overline{u^2}^2}. \tag{1.13}$$

These quantities are particularly useful in helping to judge how far from Gaussianity a particular random variable might be, since for strict Gaussianity they satisfy $S = 0$ and $F = 3$.

In some circumstances it is useful to extend the averaging process to include the product of velocities at separate points, as in the two-point double and triple velocity correlation tensors, respectively, defined by

$$\mathcal{R}_{ij}(\mathbf{x}, \mathbf{y}, t) \equiv \overline{u_i(\mathbf{x}, t)u_j(\mathbf{y}, t)} \tag{1.14}$$

and

$$S_{ij,k}(\mathbf{x}, \mathbf{y}, t) \equiv \overline{u_i(\mathbf{x}, t)u_j(\mathbf{x}, t)u_k(\mathbf{y}, t)}. \tag{1.15}$$

Such correlations provide an opportunity to capture structural features of the turbulence which would be lost in one-point statistics. When $\mathbf{x} = \mathbf{y}$, $\mathcal{R}_{ij}(\mathbf{x}, \mathbf{x}, t) = \overline{u_i(\mathbf{x}, t)u_j(\mathbf{x}, t)} = R_{ij}(\mathbf{x}, t)$ [i.e., the single-point velocity covariance defined in (1.6)].

Letting $\mathbf{r} = \mathbf{y} - \mathbf{x}$ denote the position vector connecting $\mathbf{x}$ with $\mathbf{y}$, then as $\mathbf{r}$ varies, $\mathcal{R}_{ij}(\mathbf{x}, \mathbf{x} + \mathbf{r}, t)$ gives an indication of how the velocity field around $\mathbf{x}$ at a particular time t is "correlated." For small $\mathbf{r}$ the velocity field at $\mathbf{x}$ and $\mathbf{x} + \mathbf{r}$ will be quite similar and hence highly correlated, while for large $\mathbf{r}$, $\mathcal{R}_{ij}(\mathbf{x}, \mathbf{x} + \mathbf{r}, t)$ should be at or close to zero since there would not generally be a mechanism to create correlation over large distances. For $\mathbf{x}$ fixed it is convenient to simplify notation by defining $\mathcal{R}_{ij}(\mathbf{r}, t) \equiv \mathcal{R}_{ij}(\mathbf{x}, \mathbf{x}+\mathbf{r}, t)$. In homogeneous turbulence, correlations cannot depend on position, and in this case there is truly no $\mathbf{x}$ dependence in $\mathcal{R}_{ij}(\mathbf{r}, t)$.

$\mathcal{R}_{ij}(\mathbf{r}, t)$ is generally a bounded function of compact support, so its three-dimensional Fourier transform exists:

$$E_{ij}(\mathbf{k}, t) \equiv (2\pi)^{-3} \int_{\Re^3} e^{\imath \mathbf{r} \cdot \mathbf{k}} \mathcal{R}_{ij}(\mathbf{r}, t) \, d\mathbf{r}, \qquad (1.16)$$

where $\imath \equiv \sqrt{-1}$, $\mathbf{k}$ is the wavenumber vector, and $E_{ij}(\mathbf{k}, t)$ is referred to as the *energy spectrum tensor*. Corresponding to (1.16) is the inverse transform

$$\mathcal{R}_{ij}(\mathbf{r}, t) = \int_{\Re^3} e^{-\imath \mathbf{r} \cdot \mathbf{k}} E_{ij}(\mathbf{k}, t) \, d\mathbf{k}, \qquad (1.17)$$

where $d\mathbf{k} = dk_1 \, dk_2 \, dk_3$ is a differential volume in wavenumber space. Equations (1.16) and (1.17) represent one way to enter into a spectral analysis of turbulence in which the local fluid motion is decomposed according to the different scales which are acting, as represented by Fourier components $e^{\imath \mathbf{r} \cdot \mathbf{k}}$.

Setting $i = j$ in (1.17), summing over indices $i = 1, 2, 3$, dividing by 2, and setting $\mathbf{r} = 0$ gives the spectral decomposition of the turbulent kinetic energy

$$K(t) = \frac{1}{2} \int_{\Re^3} E_{ii}(\mathbf{k}, t) \, d\mathbf{k}, \qquad (1.18)$$

where $E_{ii}(\mathbf{k}, t)$ is the density of energy in wavenumber space, and it is understood that all quantities here have an implied dependence on $\mathbf{x}$. It is useful to collect the energy, which is typically scattered throughout $\mathbf{k}$ space, onto shells of fixed distance $k = |\mathbf{k}|$ from the origin. This is done by first rewriting (1.18) using spherical coordinates in wavenumber space:

$$K(t) = \int_0^\infty dk \left[\frac{1}{2} \int_{|\mathbf{k}|=k} d\Omega \, E_{ii}(\mathbf{k}, t) \right], \qquad (1.19)$$

where $d\Omega$ is an element of solid angle in $\mathbf{k}$ space and $d\mathbf{k} = d\Omega \, dk$. Next, the term in brackets is defined as the *energy density function* or *energy spectrum*:

$$E(k, t) \equiv \frac{1}{2} \int_{|\mathbf{k}|=k} E_{ii}(\mathbf{k}, t) \, d\Omega, \qquad (1.20)$$

yielding

$$K(t) = \int_0^\infty E(k, t)\, dk. \tag{1.21}$$

In this formula, $E(k, t)$ shows how the kinetic energy is distributed among the different scales of the flow. The nature of $E(k, t)$ in turbulent flows has long played a major role in the theoretical analysis of turbulence and will be considered at length in later chapters.

Two useful quantities that can be formed from $\mathcal{R}_{ij}(\mathbf{r}, t)$ are the longitudinal and transverse velocity correlation functions defined by

$$\overline{u_1^2} f(r) \equiv \mathcal{R}_{11}(r\mathbf{e}_1, t) \tag{1.22}$$

and

$$\overline{u_2^2} g(r) \equiv \mathcal{R}_{22}(r\mathbf{e}_1, t), \tag{1.23}$$

respectively (see Fig. 1.4). Here $\mathbf{e}_i$ represents a unit vector in the x_i direction for $i = 1, 2, 3$. For simplicity of notation we refrain from indicating the time dependence of f and g. Note that our use of the x_1 and x_2 directions is meant as an example; similar functions can be defined for any direction.

From their definitions it is clear that $f(0) = g(0) = 1$. Moreover, they must decay to zero as $r \to \infty$. Locally, near $r = 0$ they can be expanded in a Taylor series as in

$$f(r) = 1 + r\frac{df}{dr}(0) + \frac{r^2}{2!}\frac{d^2 f}{dr^2}(0) + \cdots. \tag{1.24}$$

The first three terms on the right-hand side form a parabolic approximation to f whose intersection with the r axis, as shown in Fig. 1.5, is used to define a length scale, λ, termed the *microscale*. Thus, by definition,

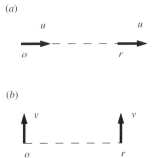

Fig. 1.4   *(a) Longitudinal and (b) transverse velocity correlations used in definitions of $f(r)$ and $g(r)$, respectively.*

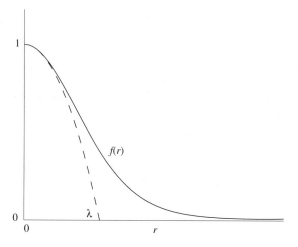

Fig. 1.5 *Microscale definition.*

$$0 = 1 + \lambda \frac{df}{dr}(0) + \frac{\lambda^2}{2!} \frac{d^2 f}{dr^2}(0), \qquad (1.25)$$

from which λ can be computed. The microscale is a measure of the scales of motion at which turbulent energy dissipation takes place, as shown in Section 7.3.1.

While λ is representative of the smaller scales in turbulent flow, a measure of the largest scales is given by the integral scale, defined by

$$\Lambda \equiv \int_0^\infty f(r) \, dr. \qquad (1.26)$$

The integral in (1.26) exists because $f(r)$ has bounded support. Λ is a measure of the size of the largest turbulent "eddies," meaning regions of the flow whose motion is in some sense interconnected. Ordinarily, events at the scale of Λ are very energetic since they are likely to be the direct result of the stirring mechanism that is responsible for turbulence generation.

In many circumstances it is most likely that time sequences of turbulent velocities at a given spatial point are known (e.g., from direct measurement using a fixed probe) rather than ensemble or volume information. In this case, it is not feasible to evaluate the energy density function per se. However, the data are sufficient to determine what is referred to as the one-dimensional energy spectrum, $E_{11}(\omega)$, where the angular frequency, ω, has units of hertz (i.e., cycles per second). $E_{11}(\omega)$ is defined via the Fourier transform of the time autocorrelation coefficient

$$\mathcal{R}_E(\tau) = \frac{\overline{u(t)u(t+\tau)}}{\overline{u^2(t)}}, \qquad (1.27)$$

where $u(t)$ is a measured time sequence of velocities at a fixed point.

In practice, the right-hand side of (1.27) would be evaluated using a time average, although the following discussion is simplified somewhat if it is viewed as an ensemble average. $\mathcal{R}_E(\tau)$ is related to its Fourier transform, $\widehat{\mathcal{R}}_E(\omega')$, where $\omega' = 2\pi\omega$ is the frequency, via the transform pair

$$\widehat{\mathcal{R}}_E(\omega') = \int_{-\infty}^{\infty} e^{-\imath\tau\omega'} \mathcal{R}_E(\tau)\, d\tau \tag{1.28}$$

and

$$\mathcal{R}_E(\tau) = \frac{1}{2\pi} \int_{-\infty}^{\infty} e^{\imath\tau\omega'} \widehat{\mathcal{R}}_E(\omega')\, d\omega'. \tag{1.29}$$

Evaluating (1.29) at $\tau = 0$ and defining

$$E_{11}(\omega) = 2\overline{u^2}\widehat{\mathcal{R}}_E(2\pi\omega) \tag{1.30}$$

gives, after a change of variables,

$$\overline{u^2} = \frac{1}{2} \int_{-\infty}^{\infty} E_{11}(\omega)\, d\omega. \tag{1.31}$$

If $\mathcal{R}_E(\tau)$ is symmetric [i.e., $\mathcal{R}_E(-\tau) = \mathcal{R}_E(\tau)$], as it would be if the random process is stationary so that (1.27) is independent of t, then (1.28) implies that

$$\widehat{\mathcal{R}}_E(\omega') = 2 \int_{0}^{\infty} \cos\,\tau\omega' \mathcal{R}_E(\tau)\, d\tau \tag{1.32}$$

and then that

$$\widehat{\mathcal{R}}_E(-\omega') = \widehat{\mathcal{R}}_E(\omega'), \tag{1.33}$$

so that (1.29) gives

$$\mathcal{R}_E(\tau) = \frac{1}{\pi} \int_{0}^{\infty} \cos\,\tau\omega' \widehat{\mathcal{R}}_E(\omega')\, d\omega'. \tag{1.34}$$

Thus $\mathcal{R}_E(\tau)$ and $\widehat{\mathcal{R}}_E(\omega')$ form a Fourier cosine transform pair. It also follows from (1.30), (1.31), and (1.33) that

$$\overline{u^2} = \int_{0}^{\infty} E_{11}(\omega)\, d\omega. \tag{1.35}$$

The fact that $\mathcal{R}_E(\tau)$ is easily evaluated from data from a single probe means that it is a relatively simple matter to determine the one-dimensional spectrum $E_{11}(\omega)$. In contrast, it is nearly impossible to directly measure $E(k, t)$. In Section 3.2.3.3

it is shown that in at least some circumstances it is possible to link the temporal and spatial variations in u. In this case, as shown in Section 7.6.3, a relationship between $f(r, t)$ and $\mathcal{R}_E(\tau)$ can be established and from this a means of estimating $E(k, t)$ from $E_{11}(\omega)$. Thus the one-dimensional spectrum $E_{11}(\omega)$ is a particularly useful quantity in characterizing turbulent flow.

1.3 TURBULENT FLOW ANALYSES

Traditionally, direct physical measurement of turbulence by experimental techniques has provided a large part of our current knowledge of turbulent flows. Physical experiments provide both descriptions of specific flow fields as well as data needed in verifying theoretical ideas. In some cases, careful scrutiny of measured data suggests the existence of general laws requiring theoretical justification. It will become clear below that it is more than a simple matter to perform fully satisfying measurements in turbulent flow. The flow sensors available for measuring properties of turbulence are beset with intrinsic difficulties of resolution, reliability, precision, and cost. Moreover, the flow variables that can be measured are often limited, and the kinds of flows that pose relatively few problems for experimental study are often not the ones that are encountered in engineering investigations or are sought to verify theoretical predictions.

Despite these obstacles, experimentalists over the last 100 years have accumulated a large body of knowledge about the nature of many individual turbulent flows and facets of the general physical laws governing them. Limited only by the characteristics of the sensors, physical experiments can achieve extremes of Reynolds number, geometrical complexity, and unsteadiness which so far have not been attainable by other methods. Moreover, advances across a wide technical front in electronics, microfabrication, optics, and computer control have led to significant new advances in devising experimental techniques. The extensive use of powerful laboratory workstations for data acquisition and analysis has made possible simultaneous multisensor measurements at several flow locations. Impressive recent advances augur well for very important new accomplishments in the future. Some indication of these is given in Chapter 3.

It is only in the modern era beginning in the 1980s that advances in computational science permitted numerical solutions to the Navier–Stokes equation to be obtained for turbulent conditions. These, referred to as *direct numerical simulations (DNS)*, have become an important and growing source of new information and insights about turbulence [1]. Within their range of applicability, which depends on having sufficient computer storage and speed to accommodate the full range of length and time scales active in a turbulent flow, they are generally unsurpassed for accuracy and wealth of information obtainable. They allow for complete and satisfying analyses of the relative importance of the terms in the averaged equations of motion. They permit examination of the time evolution of the three-dimensional pressure field or the life history of turbulent eddies or the calculation of multipoint correlations. As DNS continue to evolve, they are applied to an ever-widening range of important flows, such as jets, wakes, and mixing and boundary layers, as well as flows with increasingly complex geometries, such as concave walls, riblet surfaces, and free surfaces. By including additional equations for density, temperature, and so on, DNS have been extended to include physical phenomena such as heat and mass transfer, stratification, chemical reactions, two-phase flow, droplets, and combustion.

As will be seen later, the considerable computer resources needed for DNS limits the kinds of flows that can be reasonably well treated this way. Generally, high-Reynolds-number flows of the sort found in engineering design are excluded from DNS because they require too much memory and are not well suited to the use of fast numerical algorithms which take advantage of geometrical symmetries. The end result is that DNS are of greatest use in exploring the physics of low-Reynolds-number turbulence and from such studies provide aid in developing prediction strategies for more general flows.

The restrictions of DNS to relatively low Reynolds numbers has to do with the presence of a very wide range of dynamically significant scales of motion under high-Reynolds-number conditions. For reasons that will become evident later, this is a fundamental aspect of turbulence reflecting the large disconnect between the small scales of motion at which viscous smoothing processes are efficient, so that energy dissipation occurs, and the much larger scales, affected by the geometry of the flow, where turbulent energy resides. The bottom line in practical terms is that there is little hope in the foreseeable future of being able to solve the Navier–Stokes equation numerically for most turbulent flows of engineering interest, since the attainable resolution and speed of even the most advanced computers is inadequate once the Reynolds number is large.

A real economic incentive exists for developing computational and/or analytical techniques for simplifying and speeding up the prediction of important facets of turbulent flow needed in design work. In instances where for one reason or another DNS are not feasible or experimental methods are not practical, one has no choice but to attempt to calculate or compute approximate solutions to the dynamical equations. Since the goal is obtaining average quantities such as $\overline{\mathbf{U}}$, and, perhaps, K or $\overline{u_i u_j}$, without first having to calculate $\mathbf{U}$, it is natural to wonder if one can first average the equations of motion and then solve them directly for $\overline{\mathbf{U}}$ rather than first solving for $\mathbf{U}$ and then averaging. As we will see below, this approach is fraught with difficulties originating in the nonlinearity of the Navier–Stokes equation. The averaging process introduces unknown correlations in the averaged equations which themselves need additional equations to predict. For example, $\overline{u_i u_j}$ appears in the equation for $\overline{U}_i$, so that an additional relation for determining it would be needed to close the system of equations. However, the governing differential equations for quantities such as $\overline{u_i u_j}$ introduce even more unknowns (e.g., triple moments of the velocity field). The process is unending and is termed the *closure problem*. In essence, it reflects the fact that knowledge of a random field such as $\mathbf{U}$ is equivalent to knowing all its moments: $\overline{u_i u_j}$, $\overline{u_i u_j u_k}$, and so on. Thus an infinite set of coupled equations are needed to encompass all the physics going into the evolution of $\mathbf{U}$.

At its many different levels, turbulence theory attempts to look beyond the intricacies of individually evolving solutions to the Navier–Stokes equation to find general trends that can form the basis for physical laws governing turbulent flow. These can be either in a statistical sense, as in relations between averaged quantities, or in a description of dynamical events or processes that are common to all turbulent flows. Included in the latter are hypotheses concerning the structural makeup of turbulence and its evolution. Out of such considerations one hopes to resolve the closure problem: the hope is that at the price of introducing some approximation, the need for tracking an infinite number of moments can be short-circuited to a more manageable system of equations for just a few low-order moments, such as $\overline{U}_i$ and, perhaps, $\overline{u_i u_j}$. If successful, the endeavor yields accurate predictions; if not, the predictions are unacceptable. As will be seen, the record of such turbulent modeling is checkered, with some problems having been predicted accurately and others not. The general observation may be made that the more complex a flow is, the less likely it is to be well predicted by closure schemes. For example, there has been more success with attached vs. separated boundary layer flows. Despite the inaccuracies, it is often the case in engineering work that marginally accurate predictions have greater value than no predictions, so approaches capable of supplying the most accurate forecasts of flow conditions in a relative sense are still highly sought after.

An additional possibility in the realm of turbulent flow prediction has arisen in recent years as a compromise between computationally intensive DNS and approximate turbulence closure modeling. This involves large eddy simulations (LES), which are similar to DNS in that they produce a random approximation to the velocity field, and to closure models in that they incorporate theoretical models to represent fine-scale features of the flow which are not resolved in the numerical computation. Thus LES is a hybrid approach which attempts to resolve scales from the largest down to the smallest point that is practical given available computer memory and speed and then to model the influence of the *subgrid scales* on the resolved scales. The better the subgrid model is at accounting for the influence of small-scale motions on larger scales, the better the LES can be expected to be. It goes without saying that a LES becomes a DNS when all scales are resolved, so the importance of subgrid modeling to the success of the effort declines as resolution increases. As will be seen, the focus of much current research is on developing better and more physically accurate subgrid models so that better calculations can be done with coarser numerical representations.

Although traditional LES has been a grid-based approach, similar kinds of ideas can also be implemented by unconventional means, as in the class of methods referred to as *vortex methods*. In this grid-free approach, vortex elements are used to represent the turbulent field, thus offering potentially significant advantages in both the physical representation of turbulent dynamics and in numerical efficiency. Such methods offer the opportunity of utilizing alternative subgrid models suggested by theoretical analyses of vortex dynamics. An introduction to these methods is included in Chapter 10 within the context of a discussion of LES.

Finally, note that there is a long history of bringing the methods of statistical physics to bear on the turbulence problem. Such theories consider the fundamental properties of random solutions to the Navier–Stokes equation with the goal of finding out what statistical laws turbulent flows have in common. Generally, the nonlinearity of the Navier–Stokes equation makes the mathematics of such analyses formidable, and much of what has been accomplished is highly speculative. In Chapter 12 an introduction to some of the best known of such theories is given.

REFERENCES

1. Moin, P. and Mahesh, K. (1998) "Direct numerical simulations: a tool in turbulence," *Annu. Rev. Fluid Mech.* **30**, 539–578.

2. Pfenninger, W. (1961) "Boundary layer suction experiments with laminar flow at high Reynolds numbers in the inlet length of tubes by various suction methods," in *Boundary Layer and Flow Control* (G. V. Lachmann, Ed.), Pergamon Press, Oxford, pp. 961–980.

3. Tritton, D. J. (1988) *Physical Fluid Dynamics*, 2nd ed., Oxford University Press, Oxford, p. 287.

4. Tsuge, S. (1974) "Approach to origin of turbulence on basis of 2-point kinetic theory," *Phys. Fluids* **17**, 22–33.

5. Vincent, A. and Meneguzzi, M. (1991) "The spatial structure and statistical properties of homogeneous turbulence," *J. Fluid Mech.* **225**, 1–20.

2

Overview of Turbulent Flow Physics and Equations

It will come to be seen in the course of this book that a wide range of phenomena are associated with turbulent flow. Explanations for these ultimately lie in the physics of the momentum, energy, and vorticity fields. It is also often helpful in this regard to consider the dynamics of such quantities as the energy dissipation, multiple-point correlations, and the energy spectrum. Since numerous technological problems containing turbulent flow also involve the transport of mass and heat, consideration of scalar transport is also fundamental to the subject.

In this chapter we take a first look at the basic equations of turbulent flow corresponding to each of the aforementioned flow properties. The intent is to set the stage for subsequent discussions by developing a basic language with which the many separate issues may be addressed. In particular, the fundamental equations make clear the kinds of physical processes that need to be investigated, understood, or predicted when solving turbulent flow problems. Many of the themes enunciated here for the first time are returned to in more depth in subsequent chapters.

2.1 MOMENTUM EQUATION

2.1.1 Reynolds-Averaged Navier–Stokes Equation

The motion of an incompressible fluid is governed by Newton's second law of motion in the form of the Navier–Stokes equation [1]

$$\rho \left(\frac{\partial U_i}{\partial t} + U_j \frac{\partial U_i}{\partial x_j} \right) = \frac{\partial \sigma_{ij}}{\partial x_j} + \rho g_i, \tag{2.1}$$

where g_i is the gravitational acceleration vector and σ_{ij} is the stress tensor. For an incompressible Newtonian fluid, σ_{ij} obeys the constitutive law

$$\sigma_{ij} = -P\delta_{ij} + d_{ij}, \tag{2.2}$$

where P is the pressure; the Kronecker delta, δ_{ij}, is unity for $i = j$ and zero otherwise;

$$d_{ij} = 2\mu e_{ij} \tag{2.3}$$

is the deviatoric part of the stress tensor; μ is the viscosity; and

$$e_{ij} = \frac{1}{2}\left(\frac{\partial U_i}{\partial x_j} + \frac{\partial U_j}{\partial x_i}\right) \tag{2.4}$$

is the rate-of-strain tensor. With these definitions, the standard incompressible form of the Navier–Stokes equation is

$$\rho\left(\frac{\partial U_i}{\partial t} + U_j\frac{\partial U_i}{\partial x_j}\right) = -\frac{\partial P}{\partial x_i} + \mu\nabla^2 U_i + \rho g_i, \tag{2.5}$$

where the Laplacian $\nabla^2 \equiv \partial^2/\partial x_i^2$. A complete system of equations for U_i and P consists of (2.5) together with the conservation of mass or continuity equation for incompressible flow:

$$\frac{\partial U_i}{\partial x_i} = 0. \tag{2.6}$$

The impossibility of solving (2.5) and (2.6) under turbulent flow conditions, except numerically at low Reynolds numbers, has led to the practice of averaging these equations first and then devising means for solving the resulting system of equations for mean quantities $\overline{U}_i$ and $\overline{P}$. Proceeding along these lines, an approach first pursued systematically by Reynolds [9] over 100 years ago, (2.1) yields what is often referred to as the *Reynolds-averaged Navier–Stokes* (RANS) *equation*:

$$\rho\left(\frac{\partial \overline{U}_i}{\partial t} + \overline{U}_j\frac{\partial \overline{U}_i}{\partial x_j}\right) = \frac{\partial}{\partial x_j}\left(\overline{\sigma}_{ij} - \rho\overline{u_i u_j}\right), \tag{2.7}$$

where

$$\overline{\sigma}_{ij} = -\overline{P}\delta_{ij} + 2\mu\overline{e}_{ij} \tag{2.8}$$

is the mean stress tensor, $P = \overline{P} + p$, with p denoting the pressure fluctuation, and

$$\overline{e}_{ij} = \frac{1}{2}\left(\frac{\partial \overline{U}_i}{\partial x_j} + \frac{\partial \overline{U}_j}{\partial x_i}\right) \tag{2.9}$$

is the mean rate-of-strain tensor. The expression $\overline{u_i u_j}$ on the right-hand side of (2.7) is a by-product of averaging the nonlinear convection term in (2.1): namely, $U_j \partial U_i / \partial x_j$. Here and in the following, the gravitational body force term is omitted since it will not play a role in the subsequent discussion of incompressible turbulent motion. After substituting for (2.8), the RANS equation (2.7) becomes

$$\rho \left(\frac{\partial \overline{U}_i}{\partial t} + \overline{U}_j \frac{\partial \overline{U}_i}{\partial x_j} \right) = -\frac{\partial \overline{P}}{\partial x_i} + \mu \nabla^2 \overline{U}_i - \frac{\partial (\rho \overline{u_i u_j})}{\partial x_j}, \qquad (2.10)$$

which, apart from the extra term containing $\rho \overline{u_i u_j}$, is of the same form as the Navier–Stokes equation. Equation (2.10), together with the average of (2.6), namely,

$$\frac{\partial \overline{U}_i}{\partial x_i} = 0, \qquad (2.11)$$

forms a system of equations that one may hope to solve for $\overline{U}_i$ and $\overline{P}$. Note, incidentally, that subtracting (2.11) from (2.6) gives

$$\frac{\partial u_i}{\partial x_i} = 0; \qquad (2.12)$$

that is, the fluctuating velocity field satisfies the continuity equation in its own right.

The particular arrangement of terms in (2.7), which has $-\rho \overline{u_i u_j}$ grouped with the mean stress tensor, is purposeful. In fact, this makes clear that the physical effect represented by $-\rho \overline{u_i u_j}$ behaves in the equation of motion as if it were an additional stress acting on the mean field besides $\overline{\sigma}_{ij}$. It is referred to as the *Reynolds stress tensor* and henceforth is denoted as

$$\sigma_{ij}^T \equiv -\rho \overline{u_i u_j} = -\rho R_{ij}, \qquad (2.13)$$

where R_{ij}, defined in (1.6), is the one-point velocity covariance tensor. Note that R_{ij} is also often loosely referred to as the *Reynolds stress tensor*, and we will, on occasion, continue to follow this practice.

Equations (2.10) and (2.11) may be interpreted as governing the dynamics of a fictitious, incompressible fluid traveling at the mean velocity. By "fictitious" in this context is meant that there is no actual physical fluid that moves with the speed $\overline{U}_i$. If one were to write down a momentum balance for the fluid having velocity $\overline{U}_i$, then according to (2.7), one must recognize the existence of two internal surface forces: one coming from the average shear stress tensor $\overline{\sigma}_{ij}$ and the other coming from the Reynolds stress tensor. The end result of this force balance is to yield (2.7), in contrast to the standard force balance in a real fluid, which results in the Navier–Stokes equation (2.1). Unlike the case of a real fluid, the system of equations for $\overline{U}_i$ is not obviously closed since there is no direct means for relating σ_{ij}^T to $\overline{U}_i$ and $\overline{P}$. Finding such a "closure" to (2.10) and (2.11) via a constitutive law for σ_{ij}^T has been, and continues to be, a chief aim of turbulence modeling.

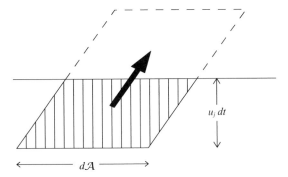

Fig. 2.1 *Shaded region constitutes the volume crossing a surface in time interval dt.*

Much physical understanding of the Reynolds stress comes from the fact that it can be interpreted as a turbulent momentum flux. To see this, consider a small surface of size $d\mathcal{A}$ traveling with the local mean velocity $\overline{U}_i$ and oriented in the x_j direction. The volume of fluid traveling across the (moving) surface $d\mathcal{A}$ in time dt is $u_j \, d\mathcal{A} \, dt$, as illustrated in Fig. 2.1. Multiplying this by the local momentum/volume ρU_i and averaging gives the mean flux of momentum crossing the surface (relative to the mean motion) as $\rho \overline{U_i u_j} \, d\mathcal{A} \, dt$. The momentum per area per second is then $\rho \overline{U_i u_j} = \rho \overline{u_i u_j}$.

This physical interpretation of the Reynolds stress fits in with our previous remark concerning the conservation of momentum for a fictitious fluid traveling with the mean velocity. In particular, consider a material volume[1] of the fictitious fluid, say $\mathcal{V}(t)$, with surface $\mathcal{A}(t)$. The rate of change of momentum in $\mathcal{V}(t)$ is equal to the sum of the mean shear stress acting on its surface, plus the net flux of momentum into the material volume. Mathematically, one has

$$\frac{\partial}{\partial t} \int_{\mathcal{V}(t)} \rho \overline{U}_i \, d\mathcal{V} = \int_{\mathcal{A}(t)} \overline{\sigma}_{ij} n_j \, d\mathcal{A} - \int_{\mathcal{A}(t)} \rho \overline{U_i u_j} n_j \, d\mathcal{A}, \qquad (2.14)$$

where n_j denotes the outward normal at any point on the surface $\mathcal{A}(t)$. Since $u_j n_j > 0$ at locations of outward flow on the bounding surface and $u_j n_j < 0$ for inward flow, the last integral in (2.14) captures the net flux of momentum *out* of the volume. It appears with a minus sign since a net outward flux of momentum is felt as a reduction in the total amount of momentum. Conversely, for a net gain of momentum, the integral will be negative. Applying the divergence theorem to the area integrals to convert them to volume integrals, and using the Reynolds transport theorem [5] to commute the time derivative with the time-dependent integration domain on the left-hand side of the equation gives

$$\int_{\mathcal{V}(t)} \left[\rho \left(\frac{\partial}{\partial t} \overline{U}_i + \overline{U}_j \frac{\partial \overline{U}_i}{\partial x_j} \right) - \frac{\partial}{\partial x_j} \left(\overline{\sigma}_{ij} - \rho \overline{u_i u_j} \right) \right] d\mathcal{V} = 0. \qquad (2.15)$$

---

[1]A material volume in this case is a purely abstract region of space containing a marked quantity of the fictitious fluid which can exchange momentum but not mass with the surrounding fluid.

Because this is true for arbitrary $\mathcal{V}(t)$, the integrand must be zero everywhere and (2.7) is recovered.

The physical interpretation that has been given to the Reynolds stress as a momentum flux is similar to that given to the stress tensor in the case of nondense gases. The principal difference between the two is that the former results from turbulent velocity fluctuations and the latter from velocity fluctuations at the molecular level. It proves helpful in what follows to review some aspects of this analogy and, in particular, how (2.2) is derived within the context of the kinetic theory of nondense gases [2]. This will later shed light on the validity of similar ideas commonly used in deriving models of σ_{ij}^T.

2.1.2   Molecular Momentum Transport

For the purposes of this discussion it is assumed that the hydrodynamic velocity field, $U_i(\mathbf{x}, t)$, of a nondense gas can be interpreted as being the result of averaging the velocities of atoms contained in a small sensing volume centered at $\mathbf{x}$ at time t. Assuming that $\mathbf{C}$ denotes the molecular velocities and $< \cdots >$ denotes averaging over molecules in a sensing volume; then $\mathbf{U} = < \mathbf{C} >$. Analogous to (1.5), it is then possible to make the decomposition

$$\mathbf{C} = \mathbf{U} + \mathbf{c}, \tag{2.16}$$

where $\mathbf{c}$ is the molecular velocity fluctuation vector. By definition, $< \mathbf{c} > = 0$.

Now consider a material surface of the fluid with area $d\mathcal{A}$ oriented in the x_j direction traveling with the continuum velocity $\mathbf{U}$. By the same argument as in Fig. 2.1, $c_j \, d\mathcal{A} \, dt$ is the volume of space adjacent to the moving surface that molecules with fluctuation velocities near c_j must lie within if they are to cross $d\mathcal{A}$ in time dt. The direction they cross is determined by c_j. If n is the number density of molecules (i.e., molecules/volume), then $nc_j \, dt \, d\mathcal{A}$ molecules cross $d\mathcal{A}$ in time dt with velocities near c_j. These molecules carry momentum mC_i, where m is the mass of a molecule. The result is that $nmC_i c_j \, dt \, d\mathcal{A}$ is the total momentum carried across the surface in time dt by molecules having fluctuation velocity c_j. Taking an average over all velocities and dividing by $d\mathcal{A} \, dt$ gives the momentum per area per second as $\rho < C_i c_j > = \rho < c_i c_j >$, where the density $\rho = mn$. The similarity of this expression to the Reynolds stress in (2.13) is evident.

Carrying these ideas further, now consider the momentum balance for a material volume of fluid, $\mathcal{V}(t)$. One has

$$\frac{\partial}{\partial t} \int_{\mathcal{V}(t)} \rho C_i \, d\mathcal{V} = - \int_{\mathcal{A}(t)} \rho < c_i c_j > n_j \, d\mathcal{A}, \tag{2.17}$$

in which the rate of change of momentum of the system is equal to the net flux of momentum into it, hence the minus sign on the right-hand side. Applying the divergence theorem to the right-hand side and the Reynolds transport theorem to the left-hand side yields

$$\int_{\mathcal{V}(t)} \rho \left[\left(\frac{\partial < C_i >}{\partial t} + < C_j > \frac{\partial < C_i >}{\partial x_j} \right) + \frac{\partial < c_i c_j >}{\partial x_j} \right] d\mathcal{V} = 0. \quad (2.18)$$

Since $\mathcal{V}(t)$ is arbitrary, the integrand must be zero, and one has

$$\rho \left(\frac{\partial < C_i >}{\partial t} + < C_j > \frac{\partial < C_i >}{\partial x_j} \right) = -\frac{\partial \rho < c_i c_j >}{\partial x_i}. \quad (2.19)$$

This is the Navier–Stokes equation since the momentum flux $-\rho < c_i c_j >$ may be identified with the stress tensor in a continuum model of the fluid:

$$\sigma_{ij} = -\rho < c_i c_j > . \quad (2.20)$$

In other words, for nondense gases, the interior surface forces as represented by σ_{ij} may be viewed as the average molecular momentum flux tensor.

A fundamental tenet of continuum mechanics states that the stress and rate-of-strain law (2.2) must hold for all fluids, both liquids and gases, as long as they are within the aegis of the continuum hypothesis. In particular, assuming only that the gas is not too rarefied, (2.2) must apply to $-\rho < c_i c_j >$, so that

$$- \rho < c_i c_j > = -P \delta_{ij} + \mu \left(\frac{\partial U_i}{\partial x_j} + \frac{\partial U_j}{\partial x_i} \right). \quad (2.21)$$

The conclusion may be reached that the Reynolds stress and its appearance in the equation for the mean velocity field (2.7) is philosophically very much consistent with the stress tensor—regarded as a momentum flux—and its appearance in the Navier–Stokes equation (2.1). In the one case the fluctuations are of the scale of turbulence, and in the other they are at the scale of molecular motions. For the stress tensor, the validity of the constitutive relation given in (2.2) has the effect of ensuring that the Navier–Stokes system (2.5) and (2.6) is a closed system (i.e., it represents four equations in the four unknowns U_i, $i = 1, 2, 3$ and P). Thus σ_{ij} is not an additional unknown. The pivotal question concerning the turbulent case is whether or not there is also such a complete and satisfying resolution of the closure problem; that is, is there a constitutive relation for the Reynolds stress tensor $\sigma_{ij}^T \equiv -\rho \overline{u_i u_j}$ analogous to (2.2) that renders a closure to the system of equations (2.10) and (2.11) for $\overline{U}_i$, $i = 1, 2, 3$ and $\overline{P}$?

The similarity in physical interpretation behind σ_{ij} and σ_{ij}^T as contained in (2.13) and (2.20) has played a big part in the development of theories for predicting the Reynolds stress. A principal line of reasoning is to expropriate (2.2) in a suitably generalized form so that it applies to σ_{ij}^T. It is sufficient for now to observe that the arguments leading to (2.2) do not simply generalize to the turbulent case. The story is much more complicated than this and is examined at length in Chapter 6.

It is clear from this discussion that predicting the Reynolds stress tensor is perhaps the most fundamental step to take in solving turbulent flow problems. Two principal routes toward its prediction have developed over the years. The first is the effort to understand its physics directly so that models can be proposed for its representation. This is an ongoing effort which has yet to reach a definitive conclusion, although there is some progress. An important advance has been in understanding the connection between Reynolds stress and the presence of structure in turbulent flows. Some aspects of this are described in Chapters 4 and 5.

The second approach is to regard the appearance of the Reynolds stress as the end result of the processes contained in its own governing equation. By modeling these and not the Reynolds stress itself, a closure can be deduced. In this effort, as well as the direct modeling approach, it is often necessary to account for the basic processes in turbulent flow that cause the production and dissipation of turbulent energy. Thus there is much interest in learning how to accurately model the turbulent kinetic energy and dissipation rate equations. We now introduce some of the important concepts that are encountered in the dynamical balances affecting these quantities.

2.2   TURBULENT KINETIC ENERGY AND ITS DISSIPATION RATE

2.2.1   Turbulent Kinetic Energy Equation

The energy in a unit volume of fluid resides in both the kinetic energy associated with the average motion or *mean drift* of molecules, $\frac{1}{2}\rho U_i^2$, and in the internal energy, ρe, associated with subatomic forces and molecular motion relative to the mean drift. e is the internal energy per mass, which in many circumstances, such as for liquids or perfect gases, can be expressed simply in terms of the temperature T and the specific heat [1]. In studying turbulent motion, our attention is most often focused on the average kinetic energy per volume, $\frac{1}{2}\rho \overline{U_i^2}$, and average internal energy per volume, $\rho \overline{e}$. For the incompressible case considered here, $\rho \overline{e}$ ordinarily can be determined independent of and subsequent to the solution of the RANS equations for $\overline{U}_i$. According to (1.5),

$$\tfrac{1}{2}\overline{U_i^2} = \overline{K} + K, \tag{2.22}$$

where $\overline{K} \equiv \frac{1}{2}\overline{U}_i^2$ is the kinetic energy per mass of the mean field (i.e., the kinetic energy of the fictitious flow whose velocity is $\overline{U}_i$) and $K = \frac{1}{2}\overline{u_i^2}$ is the kinetic energy per mass associated with the fluctuating velocity field.

The law governing kinetic energy of a fluid follows by taking a dot product of the Navier–Stokes equation (2.1) with the velocity

$$\rho \left(\frac{\partial U_i^2/2}{\partial t} + U_j \frac{\partial U_i^2/2}{\partial x_j} \right) = U_j \frac{\partial \sigma_{ij}}{\partial x_j}. \tag{2.23}$$

The term on the right-hand side is the total work per volume done by surface forces in changing kinetic energy [1]. It reflects differences in the stress field from one side of a material fluid element to the other that cause its acceleration and hence changes in its kinetic energy. This work is but one part of the total work done by surface forces. The other part is the deformation work, $\sigma_{ij}\partial U_i/\partial x_j$, which acts to change internal energy and appears in the first law of thermodynamics [1]:

$$\rho\left(\frac{\partial e}{\partial t} + U_j\frac{\partial e}{\partial x_j}\right) = \sigma_{ij}\frac{\partial U_i}{\partial x_j} + \frac{\partial}{\partial x_j}\left(k_e\frac{\partial T}{\partial x_j}\right), \tag{2.24}$$

which is to be solved for e. The last term in (2.24) reflects changes in internal energy resulting from molecular heat conduction, with k_e being the thermal conductivity. The deformation work term for incompressible flow is always nonnegative since

$$\sigma_{ij}\frac{\partial U_i}{\partial x_j} = 2\mu e_{ij}e_{ij} \geq 0, \tag{2.25}$$

a result that is obtained from (2.2), (2.4), and (2.6). It accounts for the generation or production of internal energy by the friction associated with layers of fluid sliding over each other. The sum of the work terms in (2.23) and (2.24) give the total work per volume by surface forces:

$$\frac{\partial(U_i\sigma_{ij})}{\partial x_j} = \sigma_{ij}\frac{\partial U_i}{\partial x_j} + U_i\frac{\partial\sigma_{ij}}{\partial x_j}, \tag{2.26}$$

an expression that appears in the equation governing the sum of kinetic and internal energy.

Keeping these results in mind, now average (2.23) and so obtain the equation for the total kinetic energy in turbulent flow as

$$\rho\left(\frac{\partial(\overline{K}+K)}{\partial t} + \overline{U}_j\frac{\partial(\overline{K}+K)}{\partial x_j}\right)$$
$$= \overline{U}_i\frac{\partial\overline{\sigma}_{ij}}{\partial x_j} + \overline{u_i\frac{\partial\sigma'_{ij}}{\partial x_j}} + \frac{\partial\overline{U}_i\sigma^T_{ij}}{\partial x_j} - \rho\frac{\partial\overline{u_i(u_j^2/2)}}{\partial x_i}, \tag{2.27}$$

where $\sigma'_{ij} \equiv \sigma_{ij} - \overline{\sigma}_{ij}$ is the fluctuating stress tensor. The first two terms on the right-hand side of (2.27) come from the work term on the right-hand side of (2.23), while the last two terms appear only after some manipulation of the nonlinear convection term. The first of these is evidently the total work done by the Reynolds stress on the mean field $\overline{U}_i$ [making an analogy to (2.26)], while the last term may be interpreted as the gradient of the turbulent flux of turbulent kinetic energy.

More sense can be made of (2.27) by considering the equations for $\overline{K}$ and K separately. The former, which follows from a dot product of (2.7) with $\overline{U}_i$, is

$$\rho\left(\frac{\partial\overline{K}}{\partial t} + \overline{U}_j\frac{\partial\overline{K}}{\partial x_j}\right) = \overline{U}_i\frac{\partial\overline{\sigma}_{ij}}{\partial x_j} + \overline{U}_i\frac{\partial\sigma^T_{ij}}{\partial x_j}. \tag{2.28}$$

Subtracting this from (2.27) gives the K equation:

$$\rho\left(\frac{\partial K}{\partial t} + \overline{U}_j \frac{\partial K}{\partial x_j}\right) = \overline{u_i \frac{\partial \sigma'_{ij}}{\partial x_j}} + \frac{\partial \overline{U}_i}{\partial x_j} \sigma^T_{ij} - \rho \frac{\overline{\partial u_i (u_j^2/2)}}{\partial x_i}. \tag{2.29}$$

It is seen that the first two work terms on the right-hand side of (2.27) respectively affect $\overline{K}$ and K. Furthermore, analogous to the way in which the work terms in (2.23) and (2.24) separately affected the kinetic and internal energies, it is seen that the total work done by the Reynolds stress, given by the third term on the right-hand side of (2.27), similarly divides into two parts, with one affecting $\overline{K}$ and the other K. The former is work causing accelerations of the mean field, so naturally, $\overline{K}$ is influenced. In the second part, K is changed by deformation of the mean velocity field by the Reynolds stress. If an analogy can be made between $\overline{K}$ and K on the one hand and the kinetic and internal energies on the other, one would expect that just as (2.25) shows that deformation work is always positive, one should also expect that the deformation work by the Reynolds stress in (2.29) leads to production of turbulent kinetic energy from the mean field. In fact, it will later be seen that in at least some important situations this is indeed the case.

As yet, nothing has been said about the fluctuating work term $\overline{u_i (\partial \sigma'_{ij}/\partial x_j)}$ in (2.29), but it turns out that it is of profound significance in turbulent flow. This is made evident by reworking it into a more useful form. Thus, define

$$d'_{ij} = 2\mu e'_{ij}, \tag{2.30}$$

where

$$e'_{ij} = \frac{1}{2}\left(\frac{\partial u_i}{\partial x_j} + \frac{\partial u_j}{\partial x_i}\right) \tag{2.31}$$

and note that

$$\sigma'_{ij} = -p\delta_{ij} + d'_{ij}. \tag{2.32}$$

In this case

$$\overline{u_i \frac{\partial \sigma'_{ij}}{\partial x_j}} = -\frac{\partial \overline{pu_i}}{\partial x_i} + \frac{\partial \overline{u_i d'_{ij}}}{\partial x_j} - \rho\epsilon_T, \tag{2.33}$$

where

$$\epsilon_T \equiv \frac{1}{\rho}\overline{\frac{\partial u_i}{\partial x_j}d'_{ij}} = 2\nu\overline{e'_{ij}e'_{ij}} \tag{2.34}$$

is the average dissipation rate (per unit mass) by internal friction of the kinetic energy associated with the fluctuating field. A simple manipulation then gives

$$\epsilon_T = \epsilon + \nu \overline{\frac{\partial u_i}{\partial x_j} \frac{\partial u_j}{\partial x_i}}, \tag{2.35}$$

where

$$\epsilon \equiv \nu \overline{\left(\frac{\partial u_i}{\partial x_j}\right)^2}. \tag{2.36}$$

In fact, by virtue of the identity

$$\overline{\frac{\partial u_i}{\partial x_j} \frac{\partial u_j}{\partial x_i}} = \frac{\partial}{\partial x_j} \left(\overline{u_i \frac{\partial u_j}{\partial x_i}} \right), \tag{2.37}$$

the second term on the right-hand side of (2.35) is zero in homogeneous turbulence. Consequently, ϵ may be thought of as the viscous dissipation rate in homogeneous isotropic turbulence (i.e., what is commonly referred to as the *isotropic dissipation rate*). Because of the further identity

$$\frac{1}{\rho} \frac{\overline{\partial u_i d'_{ij}}}{\partial x_j} = \nu \, \nabla^2 K + \nu \overline{\frac{\partial u_i}{\partial x_j} \frac{\partial u_j}{\partial x_i}}, \tag{2.38}$$

where $\partial u_j / \partial x_j = 0$ has been invoked, (2.33) becomes, after using (2.35) and (2.38),

$$\frac{1}{\rho} \overline{u_i \frac{\partial \sigma'_{ij}}{\partial x_j}} = -\frac{1}{\rho} \frac{\partial \overline{p u_i}}{\partial x_i} + \nu \, \nabla^2 K - \epsilon. \tag{2.39}$$

Since, according to (2.36), $\epsilon > 0$, it is clear that this part of the fluctuating work term acts as a sink of turbulent kinetic energy caused by the action of frictional viscous forces.

Incorporating (2.39) into (2.29) yields the standard form of the K equation, namely,

$$\frac{\partial K}{\partial t} + \overline{U}_j \frac{\partial K}{\partial x_j} = -\frac{\partial \overline{U}_i}{\partial x_j} R_{ij} - \epsilon - \frac{1}{\rho} \frac{\partial \overline{p u_i}}{\partial x_i} + \nu \, \nabla^2 K - \frac{\partial \overline{u_j (u_i^2/2)}}{\partial x_j}. \tag{2.40}$$

If this relation is integrated over a control volume (i.e., a fixed volume in space), it suggests the following physical picture: according to the left-hand side of the equation, the rate of change of turbulent kinetic energy in the control volume, plus its net rate of gain or loss by convection through the boundary, are balanced on the right-hand side by, in order, the rate of turbulent kinetic energy production, dissipation, pressure work, and viscous and turbulent diffusion through the control volume surface. Among these, the pressure work and viscous diffusion terms are normally significant only very near boundaries. Perhaps the most difficult aspect of solving (2.40) is in computing ϵ accurately.

2.2.2 ϵ Equation

The isotropic dissipation rate, ϵ, in turbulent flows depends intimately on how viscosity affects the local small-scale structure of turbulence. The particular distribution taken by ϵ in a turbulent flow has a large influence on the ultimate behavior of K. This will be evident in several flows considered subsequently, such as the problem of turbulent decay in the absence of production discussed in Section 7.3. In this case, ϵ is entirely responsible for the history of K. Virtually all contemporary closure schemes require prediction of ϵ, which they usually get through solution of its own modeled equation.

A relation governing ϵ is derivable from the Navier–Stokes equation (2.1) by straightforward (although tedious) manipulation. First, (2.1) is differentiated by $\partial/\partial x_j$, giving an equation for $\partial U_i/\partial x_j$. This is then multiplied by $2\nu\,\partial u_i/\partial x_j$ and the result averaged. In the course of this process, the time-differentiated term in (2.1) yields

$$2\nu\overline{\frac{\partial u_i}{\partial x_j}\frac{\partial^2 U_i}{\partial x_j \partial t}} = 2\nu\overline{\frac{\partial u_i}{\partial x_j}\frac{\partial^2 u_i}{\partial x_j \partial t}} = \frac{\partial \epsilon}{\partial t}. \tag{2.41}$$

Similar calculations may be done with the remaining terms. The result is traditionally written in the form

$$\frac{D\epsilon}{Dt} = P_\epsilon^1 + P_\epsilon^2 + P_\epsilon^3 + P_\epsilon^4 + \Pi_\epsilon + T_\epsilon + D_\epsilon - \Upsilon_\epsilon, \tag{2.42}$$

where

$$P_\epsilon^1 = -2\nu\overline{\frac{\partial u_j}{\partial x_i}\frac{\partial u_j}{\partial x_k}}\frac{\partial \overline{U}_i}{\partial x_k}, \tag{2.43}$$

$$P_\epsilon^2 = -\epsilon_{ij}\frac{\partial \overline{U}_i}{\partial x_j}, \tag{2.44}$$

$$P_\epsilon^3 = -2\nu\overline{u_k\frac{\partial u_i}{\partial x_j}}\frac{\partial^2 \overline{U}_i}{\partial x_k \partial x_j}, \tag{2.45}$$

$$P_\epsilon^4 = -2\nu\overline{\frac{\partial u_i}{\partial x_k}\frac{\partial u_i}{\partial x_j}\frac{\partial u_k}{\partial x_j}}, \tag{2.46}$$

$$\Pi_\epsilon = -2\nu\frac{\partial}{\partial x_i}\left(\overline{\frac{\partial p}{\partial x_j}\frac{\partial u_i}{\partial x_j}}\right), \tag{2.47}$$

$$T_\epsilon = -\nu\frac{\partial}{\partial x_k}\left(\overline{u_k\frac{\partial u_i}{\partial x_j}\frac{\partial u_i}{\partial x_j}}\right), \tag{2.48}$$

$$D_\epsilon = \nu\,\nabla^2\epsilon \tag{2.49}$$

$$\Upsilon_\epsilon = 2\nu^2 \overline{\left(\frac{\partial^2 u_i}{\partial x_j \, \partial x_k}\right)^2}, \tag{2.50}$$

and in (2.44),

$$\epsilon_{ij} = 2\nu \overline{\frac{\partial u_i}{\partial x_k} \frac{\partial u_j}{\partial x_k}} \tag{2.51}$$

is referred to as the *dissipation rate tensor*. Note that $\epsilon_{ii} = 2\epsilon$. For want of a more precise description of the underlying physics, the terms P_ϵ^i, $i = 1, \dots, 4$ are referred to as "production" terms. Of course, this is technically true only as long as they are positive, which is not always the case. It is somewhat easier to classify T_ϵ as a transport term and Υ_ϵ as the dissipation of dissipation term. D_ϵ accounts for viscous diffusion of ϵ, while Π_ϵ involves pressure and may or may not signify production of ϵ, depending on its sign. Π_ϵ is often referred to as a *pressure diffusion term*, although it is a misnomer to suggest that it represents an actual transport process.

Models for (2.42) typically include expressions accommodating each of the production, turbulent transport, pressure diffusion, and dissipation terms. This is fraught with difficulty since, to a much greater degree than was the case for the $\overline{U}_i$ and K equations, the terms in the ϵ equation are hard to comprehend physically and thus to model accurately. Significant help in this regard can be had by considering the ϵ equation in simpler circumstances, such as homogeneous turbulence, where many of its terms are identically zero. In Chapter 7 it will be seen that some important insights into P_ϵ^4 and Υ_ϵ can be had by analyzing the decay of isotropic turbulence and the development of K and ϵ in homogeneous shear flow. Much understanding of the ϵ equation also has come from a detailed appraisal of its entire budget computed from a channel flow DNS, discussed in Section 4.2.4.

2.3 REYNOLDS STRESS EQUATION

Section 2.1 made it clear that a model for the Reynolds stress R_{ij} is necessary to effect closure to the averaged Navier–Stokes equation. (For this discussion we prefer to use R_{ij} instead of σ_{ij}^T.) Although some theories attempt to model R_{ij} directly, others aim to develop closure to its own governing transport equation. For this reason, among others, it is of interest to consider some aspects of this relation.

An equation for R_{ij} is derived by averaging and adding together (2.5) written for U_i and multiplied by u_j, and (2.5) written for U_j and multiplied by u_i. The result is

$$\frac{\partial R_{ij}}{\partial t} + \overline{U}_k \frac{\partial R_{ij}}{\partial x_k} = -R_{ik} \frac{\partial \overline{U}_j}{\partial x_k} - R_{jk} \frac{\partial \overline{U}_i}{\partial x_k} - \epsilon_{ij} - \frac{\partial \beta_{ijk}}{\partial x_k} + \Pi_{ij} + \nu \, \nabla^2 R_{ij}, \tag{2.52}$$

where

$$\beta_{ijk} \equiv \overline{u_i u_j u_k} + \frac{1}{\rho} \overline{p u_i} \, \delta_{jk} + \frac{1}{\rho} \overline{p u_j} \, \delta_{ik} \tag{2.53}$$

and

$$\Pi_{ij} \equiv \frac{1}{\rho} \overline{p \left(\frac{\partial u_i}{\partial x_j} + \frac{\partial u_j}{\partial x_i} \right)}. \tag{2.54}$$

Note that a contraction of the indices in (2.52) and multiplication by 1/2 gives exactly (2.40). The first two terms on the right-hand side of (2.52), which have a form similar to the production term in (2.40), are given the same interpretation here. The third term, ϵ_{ij}, is the general tensorial dissipation term defined in (2.51), followed by the turbulent transport term containing the flux β_{ijk}, the pressure–strain term Π_{ij}, and the viscous diffusion term.

A potentially attractive feature of (2.52) is that the left-hand side of the equation, as well as the production terms, require no modeling. Thus, if (2.52) is solved in conjunction with the RANS equation, only the three quantities ϵ_{ij}, Π_{ij}, and β_{ijk} need to be modeled. In reality, experience shows that accurate solutions to (2.52) cannot generally be obtained without reasonably good models of the unclosed correlations. Finding such relations is far from a simple task. The dissipation and transport terms pose difficulties in the present context very similar to that of their cousins appearing in the K equation. On the other hand, the pressure–strain term has no equivalent in (2.40); in fact, the incompressibility condition guarantees that $\Pi_{11} + \Pi_{22} + \Pi_{33} = 0$, so Π_{ii} does not appear in the K equation. Π_{ij} acts to redistribute energy between components, $\overline{u_1^2}$, $\overline{u_2^2}$, and $\overline{u_3^2}$ without a change in the total energy. If the individual terms of Π_{ii} are nonzero, at least one must be positive and one negative. Away from boundaries, the tendency of the pressure–strain term is to bring $\overline{u_1^2}$, $\overline{u_2^2}$, and $\overline{u_3^2}$ closer to equality, such as occurs in isotropic turbulence. In fact, it can be shown that each of the individual terms of Π_{ij} is identically zero in isotropic turbulence. Modeling Π_{ij} in general anisotropic conditions is a great challenge, but one of particular importance if the individual Reynolds stress components are to be determined with any accuracy.

Additional insight into the pressure–strain term, and the route through which it is usually modeled, is gained by replacing the pressure fluctuation with its exact integral representation in terms of the velocity field and its gradients. This is accomplished by first taking the divergence of the Navier–Stokes equation (2.5) and subtracting from it the divergence of the RANS equation (2.10). This yields, after some simplification, the following Poisson equation for the pressure fluctuation field:

$$\frac{1}{\rho} \nabla^2 p = -2 \frac{\partial u_j}{\partial x_i} \frac{\partial \overline{U}_i}{\partial x_j} - \frac{\partial u_i}{\partial x_j} \frac{\partial u_j}{\partial x_i}. \tag{2.55}$$

For an infinite domain without boundaries, the fundamental solution of (2.55) is

$$p(\mathbf{x}, t) = \frac{\rho}{4\pi} \int_{\Re^3} \frac{1}{|\mathbf{x} - \mathbf{x}'|} \left(-2 \frac{\partial u_j}{\partial x_i} \frac{\partial \overline{U}_i}{\partial x_j} - \frac{\partial u_i}{\partial x_j} \frac{\partial u_j}{\partial x_i} \right) d\mathcal{V}, \tag{2.56}$$

where the term in parentheses is evaluated at $(\mathbf{x}', t)$. Additional surface integrals appear if boundaries are present. However, regardless of the circumstances, it is usual to consider just (2.56) when analyzing the pressure–strain term with a view toward its modeling.

Using (2.56), the pressure may be eliminated from Π_{ij}, giving

$$\Pi_{ij} = A_{ij} + M_{ijkl}\frac{\partial \overline{U}_k}{\partial x_l}, \tag{2.57}$$

where

$$A_{ij} = \frac{1}{4\pi} \int_{\Re^3} \frac{1}{|\mathbf{x} - \mathbf{x}'|} \overline{\left(\frac{\partial u_i}{\partial x_j} + \frac{\partial u_j}{\partial x_i}\right) \frac{\partial u_k}{\partial x_l} \frac{\partial u_l}{\partial x_k}} d\mathcal{V} \tag{2.58}$$

and

$$M_{ijkl} = \frac{1}{2\pi} \int_{\Re^3} \frac{1}{|\mathbf{x} - \mathbf{x}'|} \overline{\left(\frac{\partial u_i}{\partial x_j} + \frac{\partial u_j}{\partial x_i}\right) \frac{\partial u_l}{\partial x_k}} d\mathcal{V}. \tag{2.59}$$

These results are, strictly speaking, valid only in homogeneous shear flow, since it may be noted that the mean velocity gradient term in (2.56) has been taken outside the integral. The first of the terms in (2.57) is referred to as the *slow term* and the second as the *fast term*. This nomenclature refers to the fact that if one were suddenly to change the mean velocity field, it is the second of the terms in (2.57) which would immediately respond to the change, while the first, since it does not depend explicitly on the mean velocity field, would adjust more slowly as the turbulent velocity field evolves as a whole. In Chapter 8 we consider in depth some of the strategies that have been developed to turn (2.57) into a useful expression for modeling turbulent flow.

2.4 TWO-POINT CORRELATION TENSORS

Thus far the equations governing one-point moments of the velocity field such as $\overline{U}_i$, K, ϵ, and $\overline{u_i u_j}$ have been considered. It is clear that the physical laws governing these quantities are nontrivial. Moreover, it cannot be taken for granted that the subset of turbulence physics represented in these relations is all that is needed to be taken into account in formulating prediction schemes. In fact, one of the principal difficulties is that turbulent flow has a rich and varied interior structure—a large part of this having to do with interacting vortical motions—which is difficult to account for using a single-point statistical description. Thus it is reasonable to imagine that there may be something to be gained in our ability to understand and predict turbulence if one approaches the problem through multipoint statistics of the turbulent velocity field. Among the latter are $\mathcal{R}_{ij}(\mathbf{x}, \mathbf{y}, t)$ and $S_{ij,k}(\mathbf{x}, \mathbf{y}, t)$, introduced in (1.14) and (1.15), respectively. Unfortunately, in many situations the potential value of modeling turbulence through two-point correlations is offset by the added complexity of their governing equations. However, in simplified settings, such as isotropic turbulence, the multipoint equations turn out to be somewhat more tractable and are useful in studying some of the fundamental physical processes of turbulent flow. This has benefit in understanding turbulence in more general settings.

An equation governing $\mathcal{R}_{ij}(\mathbf{x}, \mathbf{y}, t)$ is derived by taking the average of $u_i(\mathbf{x}, t)$ times the jth component of the Navier–Stokes equation at $\mathbf{y}$ and adding to this the same quantity with i and j and $\mathbf{x}$ and $\mathbf{y}$ reversed:

$$\overline{u_i(\mathbf{x}, t) N S_j(\mathbf{y}, t)} + \overline{u_j(\mathbf{y}, t) N S_i(\mathbf{x}, t)}, \tag{2.60}$$

where $N S_i(\mathbf{x}, t)$ denotes the Navier–Stokes equation (2.5) at $\mathbf{x}, t$ written for U_i. The result is

$$\overline{\rho u_i(\mathbf{x}, t) \frac{\partial U_j}{\partial t}(\mathbf{y}, t)} + \overline{\rho u_j(\mathbf{y}, t) \frac{\partial U_i}{\partial t}(\mathbf{x}, t)}$$

$$+ \ \overline{\rho u_i(\mathbf{x}, t) U_k(\mathbf{y}, t) \frac{\partial U_j}{\partial y_k}(\mathbf{y}, t)} + \overline{\rho u_j(\mathbf{y}, t) U_k(\mathbf{x}, t) \frac{\partial U_i}{\partial x_k}(\mathbf{x}, t)}$$

$$= -\overline{u_i(\mathbf{x}, t) \frac{\partial p}{\partial y_j}(\mathbf{y}, t)} - \overline{u_j(\mathbf{y}, t) \frac{\partial p}{\partial x_i}(\mathbf{x}, t)} \tag{2.61}$$

$$+ \ \overline{\mu u_i(\mathbf{x}, t) \, \nabla^2 u_j(\mathbf{y}, t)} + \overline{\mu u_j(\mathbf{y}, t) \, \nabla^2 u_i(\mathbf{x}, t)}.$$

Using the definition of $\mathcal{R}_{ij}$ in (1.14), it follows that the first two terms on the left-hand side of (2.61) may be written as

$$\overline{\rho u_i(\mathbf{x}, t) \frac{\partial U_j}{\partial t}(\mathbf{y}, t)} + \overline{\rho u_j(\mathbf{y}, t) \frac{\partial U_i}{\partial t}(\mathbf{x}, t)} = \rho \frac{\partial \mathcal{R}_{ij}}{\partial t}(\mathbf{x}, \mathbf{y}, t) \tag{2.62}$$

since terms such as $\overline{u_i(\mathbf{x}, t) \, \partial \overline{U}_j(\mathbf{y}, t)/\partial t} \equiv 0$. The next two terms, coming from the advection term, give

$$\overline{\rho u_i(\mathbf{x}, t) U_k(\mathbf{y}, t) \frac{\partial U_j}{\partial y_k}(\mathbf{y}, t)} + \overline{\rho u_j(\mathbf{y}, t) U_k(\mathbf{x}, t) \frac{\partial U_i}{\partial x_k}(\mathbf{x}, t)}$$

$$= \overline{\rho u_i(\mathbf{x}, t) u_k(\mathbf{y}, t) \frac{\partial \overline{U}_j}{\partial y_k}(\mathbf{y}, t)} + \overline{\rho u_j(\mathbf{y}, t) u_k(\mathbf{x}, t) \frac{\partial \overline{U}_i}{\partial x_k}(\mathbf{x}, t)}$$

$$+ \ \overline{\rho \overline{U}_k(\mathbf{y}, t) u_i(\mathbf{x}, t) \frac{\partial u_j}{\partial y_k}(\mathbf{y}, t)} + \overline{\rho \overline{U}_k(\mathbf{x}, t) u_j(\mathbf{y}, t) \frac{\partial u_i}{\partial x_k}(\mathbf{x}, t)} \tag{2.63}$$

$$+ \ \overline{\rho u_i(\mathbf{x}, t) u_k(\mathbf{y}, t) \frac{\partial u_j}{\partial y_k}(\mathbf{y}, t)} + \overline{\rho u_j(\mathbf{y}, t) u_k(\mathbf{x}, t) \frac{\partial u_i}{\partial x_k}(\mathbf{x}, t)}.$$

The first two terms on the right-hand side of (2.63) are equal to

$$\rho \mathcal{R}_{ik}(\mathbf{x}, \mathbf{y}, t) \frac{\partial \overline{U}_j}{\partial y_k}(\mathbf{y}, t) + \rho \mathcal{R}_{jk}(\mathbf{y}, \mathbf{x}, t) \frac{\partial \overline{U}_i}{\partial x_k}(\mathbf{x}, t).$$

On the other hand, differentiation of (1.14) gives

$$\frac{\partial \mathcal{R}_{ij}}{\partial y_k}(\mathbf{x}, \mathbf{y}, t) = \overline{u_i(\mathbf{x}, t) \frac{\partial u_j}{\partial y_k}(\mathbf{y}, t)}$$

and similarly for x_k derivatives, so that the third and fourth terms on the right-hand side of (2.63) take the form of convection terms:

$$\rho \overline{U}_k(\mathbf{y}, t) \frac{\partial \mathcal{R}_{ij}}{\partial y_k}(\mathbf{x}, \mathbf{y}, t) + \rho \overline{U}_k(\mathbf{x}, t) \frac{\partial \mathcal{R}_{ij}}{\partial x_k}(\mathbf{x}, \mathbf{y}, t).$$

As far as the last two terms on the right-hand side of (2.63) are concerned, they may be written using (1.15) as

$$\rho \overline{u_i(\mathbf{x}, t) u_k(\mathbf{y}, t) \frac{\partial u_j}{\partial y_k}(\mathbf{y}, t)} = \rho \frac{\partial S_{jk,i}}{\partial y_k}(\mathbf{y}, \mathbf{x}, t) \tag{2.64}$$

and

$$\rho \overline{u_j(\mathbf{y}, t) u_k(\mathbf{x}, t) \frac{\partial u_i}{\partial x_k}(\mathbf{x}, t)} = \rho \frac{\partial S_{ik,j}}{\partial x_k}(\mathbf{x}, \mathbf{y}, t), \tag{2.65}$$

where the fact that

$$\overline{u_i(\mathbf{x}, t) \frac{\partial u_j}{\partial x_j}(\mathbf{x}, t) u_k(\mathbf{y}, t)} = 0$$

has been used as implied by incompressibility.

To treat the contribution to (2.61) from the terms containing pressure, introduce the two-point pressure–velocity correlation vector

$$\mathcal{K}_i(\mathbf{x}, \mathbf{y}, t) = \overline{u_i(\mathbf{x}, t) p(\mathbf{y}, t)} \tag{2.66}$$

and see that

$$\overline{u_i(\mathbf{x}, t) \frac{\partial p}{\partial y_j}(\mathbf{y}, t)} + \overline{u_j(\mathbf{y}, t) \frac{\partial p}{\partial x_i}(\mathbf{x}, t)} = \frac{\partial \mathcal{K}_i}{\partial y_j}(\mathbf{x}, \mathbf{y}, t) + \frac{\partial \mathcal{K}_j}{\partial x_i}(\mathbf{y}, \mathbf{x}, t).$$

Finally, it is easy to show that the viscous terms in (2.61) give

$$\overline{\mu u_i(\mathbf{x}, t) \nabla^2 u_j(\mathbf{y}, t)} + \overline{\mu u_j(\mathbf{y}, t) \nabla^2 u_i(\mathbf{x}, t)} = \mu \frac{\partial^2 \mathcal{R}_{ij}}{\partial y_k^2}(\mathbf{x}, \mathbf{y}, t) + \mu \frac{\partial^2 \mathcal{R}_{ij}}{\partial x_k^2}(\mathbf{x}, \mathbf{y}, t).$$

Putting together the results above, it is found that (2.61) becomes

$$
\frac{\partial \mathcal{R}_{ij}}{\partial t}(\mathbf{x}, \mathbf{y}, t) + \overline{U}_k(\mathbf{y}, t)\frac{\partial \mathcal{R}_{ij}}{\partial y_k}(\mathbf{x}, \mathbf{y}, t) + \overline{U}_k(\mathbf{x}, t)\frac{\partial \mathcal{R}_{ij}}{\partial x_k}(\mathbf{x}, \mathbf{y}, t)
$$

$$
= -\mathcal{R}_{ik}(\mathbf{x}, \mathbf{y}, t)\frac{\partial \overline{U}_j}{\partial y_k}(\mathbf{y}, t) - \mathcal{R}_{jk}(\mathbf{y}, \mathbf{x}, t)\frac{\partial \overline{U}_i}{\partial x_k}(\mathbf{x}, t)
$$

$$
- \frac{\partial S_{jk,i}}{\partial y_k}(\mathbf{y}, \mathbf{x}, t) - \frac{\partial S_{ik,j}}{\partial x_k}(\mathbf{x}, \mathbf{y}, t) - \frac{1}{\rho}\frac{\partial \mathcal{K}_i}{\partial y_j}(\mathbf{x}, \mathbf{y}, t)
$$

$$
- \frac{1}{\rho}\frac{\partial \mathcal{K}_j}{\partial x_i}(\mathbf{y}, \mathbf{x}, t) + \nu\frac{\partial^2 \mathcal{R}_{ij}}{\partial y_k^2}(\mathbf{x}, \mathbf{y}, t) + \nu\frac{\partial^2 \mathcal{R}_{ij}}{\partial x_k^2}(\mathbf{x}, \mathbf{y}, t).
$$

(2.67)

When $\mathbf{x} = \mathbf{y}$, $\mathcal{R}_{ij}(\mathbf{x}, \mathbf{x}, t) = R_{ij}(\mathbf{x}, t)$, and it is not difficult to show that (2.67) becomes identical with (2.52). This connection suggests that the first two terms on the right-hand side of (2.67) are production terms. The remaining terms correspond to various combinations of those in (2.52); there is no need to make a detailed correspondence.

Theories attempting to solve (2.67) are rare. Rather, of more importance is its simplified form, corresponding to idealized flow conditions such as homogeneity and isotropy. A large body of research into the nature of homogeneous turbulence takes an appropriate form of (2.67) as a starting point for analysis. This will also be done in our treatment of idealized flows in Chapter 7.

A fundamental consequence of homogeneity is that $\mathcal{R}_{ij}$ and $S_{ij,k}$ should depend only on the relative position of $\mathbf{x}$ and $\mathbf{y}$, not on their absolute positions. This implies that

$$
\mathcal{R}_{ij}(\mathbf{x}, \mathbf{y}, t) = \mathcal{R}_{ij}(\mathbf{y} - \mathbf{x}, t), \tag{2.68}
$$

$$
S_{ij,k}(\mathbf{x}, \mathbf{y}, t) = S_{ij,k}(\mathbf{y} - \mathbf{x}, t), \tag{2.69}
$$

and

$$
\mathcal{K}_i(\mathbf{x}, \mathbf{y}, t) = \mathcal{K}_i(\mathbf{y} - \mathbf{x}, t), \tag{2.70}
$$

where for convenience the same symbols $\mathcal{R}_{ij}$, $S_{ij,k}$, and $\mathcal{K}_i$ on the right-hand side are adopted; their applicability to homogeneous turbulence is implied by the appearance of one less argument. In each of (2.68), (2.69), and (2.70) it is most helpful to regard the correlation on their right-hand sides as being a function of $\mathbf{r}$ and t which is evaluated at $\mathbf{r} = \mathbf{y} - \mathbf{x}$, as in

$$
\mathcal{R}_{ij}(\mathbf{y} - \mathbf{x}, t) \equiv \mathcal{R}_{ij}(\mathbf{r}, t)|_{\mathbf{r}=\mathbf{y}-\mathbf{x}}. \tag{2.71}
$$

Now, say that one wishes to compute

$$\frac{\partial \mathcal{R}_{ij}}{\partial x_k}(\mathbf{x}, \mathbf{y}, t) \qquad (2.72)$$

in the case of homogeneous turbulence. Taking advantage of (2.68) and (2.71) and applying the chain rule gives

$$\frac{\partial}{\partial x_k}\left[\mathcal{R}_{ij}(\mathbf{y} - \mathbf{x}, t)\right] = \frac{\partial \mathcal{R}_{ij}}{\partial r_k}\Big|_{\mathbf{r}=\mathbf{y}-\mathbf{x}} \frac{\partial}{\partial x_k}(\mathbf{y} - \mathbf{x}) = -\frac{\partial \mathcal{R}_{ij}}{\partial r_k}(\mathbf{y} - \mathbf{x}, t). \qquad (2.73)$$

From such calculations it can be determined that in homogeneous turbulence

$$\frac{\partial \mathcal{R}_{ij}}{\partial y_k}(\mathbf{x}, \mathbf{y}, t) = \frac{\partial \mathcal{R}_{ij}}{\partial r_k}(\mathbf{y} - \mathbf{x}, t) = -\frac{\partial \mathcal{R}_{ij}}{\partial x_k}(\mathbf{x}, \mathbf{y}, t), \qquad (2.74)$$

$$\frac{\partial^2 \mathcal{R}_{ij}}{\partial y_k^2}(\mathbf{x}, \mathbf{y}, t) = \frac{\partial^2 \mathcal{R}_{ij}}{\partial r_k^2}(\mathbf{y} - \mathbf{x}, t) = \frac{\partial^2 \mathcal{R}_{ij}}{\partial x_k^2}(\mathbf{x}, \mathbf{y}, t), \qquad (2.75)$$

$$\frac{\partial S_{jk,i}}{\partial y_k}(\mathbf{y}, \mathbf{x}, t) = -\frac{\partial S_{jk,i}}{\partial r_k}(\mathbf{x} - \mathbf{y}, t), \qquad (2.76)$$

and

$$\frac{\partial S_{ik,j}}{\partial x_k}(\mathbf{x}, \mathbf{y}, t) = -\frac{\partial S_{ik,j}}{\partial r_k}(\mathbf{y} - \mathbf{x}, t). \qquad (2.77)$$

Equation (2.74) implies that if $\overline{U}_k(\mathbf{y}, t) = \overline{U}_k(\mathbf{x}, t)$, which is true in homogeneous turbulence, the two convection terms on the left-hand side of (2.67) sum to zero. Uniformity of $\overline{U}_i$ also implies that the two production terms on the right-hand side of (2.67) will be zero.

For homogeneous turbulence it also follows that

$$\frac{\partial \mathcal{K}_i}{\partial y_j}(\mathbf{x}, \mathbf{y}, t) = \frac{\partial \mathcal{K}_i}{\partial r_j}(\mathbf{y} - \mathbf{x}, t) \qquad (2.78)$$

and

$$\frac{\partial \mathcal{K}_j}{\partial x_i}(\mathbf{y}, \mathbf{x}, t) = \frac{\partial \mathcal{K}_j}{\partial r_i}(\mathbf{x} - \mathbf{y}, t). \qquad (2.79)$$

Putting together the various results, it is found that the two-point velocity correlation tensor in homogeneous turbulence is governed by the equation

$$\frac{\partial \mathcal{R}_{ij}}{\partial t}(\mathbf{r}, t) = \frac{\partial S_{jk,i}}{\partial r_k}(-\mathbf{r}, t) + \frac{\partial S_{ik,j}}{\partial r_k}(\mathbf{r}, t)$$
$$- \frac{\partial \mathcal{K}_i}{\partial r_j}(\mathbf{r}, t) - \frac{\partial \mathcal{K}_j}{\partial r_i}(-\mathbf{r}, t) + 2\nu \frac{\partial^2 \mathcal{R}_{ij}}{\partial r_k^2}(\mathbf{r}, t). \qquad (2.80)$$

Contracting the indices in (2.78), noting the definition of $\mathcal{K}_i$ in (2.70), and using the incompressibility condition gives

$$\frac{\partial \mathcal{K}_i}{\partial r_i}(\mathbf{r}, t) = 0. \tag{2.81}$$

Thus, after setting $j = i$, (2.80) becomes

$$\frac{\partial \mathcal{R}_{ii}}{\partial t}(\mathbf{r}, t) = \frac{\partial S_{ik,i}}{\partial r_k}(-\mathbf{r}, t) + \frac{\partial S_{ik,i}}{\partial r_k}(\mathbf{r}, t) + 2\nu \frac{\partial^2 \mathcal{R}_{ii}}{\partial r_k^2}(\mathbf{r}, t). \tag{2.82}$$

This relation is a useful starting point for the analysis of homogeneous isotropic turbulence, as will be seen in Chapter 7. It shows that the time rate of change of the trace of the two-point correlation depends on a balance between viscous diffusion, given in the last term, and the two terms depending on the two-point triple velocity correlation tensor. The later represent the critical process by which vortex stretching (discussed in Section 2.7.1) brings energy to small dissipative scales. This is a distinctive aspect of turbulent physics and (2.82) allows for its study in isolation from other flow phenomena.

2.5 SPECTRAL ANALYSIS

Equations (2.80) and (2.82) pave the way for a spectral analysis of homogeneous turbulence, in which the dynamical processes governing the evolution of the energy density $E_{ij}(\mathbf{k}, t)$ are examined. This gives information about how energy is produced, dissipated, and exchanged between different scales within turbulent flow. Here we lay out some of the general aspects of the spectral point of view and leave it to later chapters to give detailed results on its use in simplified flows such as isotropic turbulence.

Applying a Fourier transform, as in (1.16), to the terms in (2.80) yields the following dynamical equation for the evolution of the energy spectrum tensor $E_{ij}(\mathbf{k}, t)$ in homogeneous turbulence:

$$\frac{\partial E_{ij}}{\partial t}(\mathbf{k}, t) = T_{ij}(\mathbf{k}, t) + P_{ij}(\mathbf{k}, t) - 2k^2 \nu E_{ij}(\mathbf{k}, t), \tag{2.83}$$

where

$$T_{ij}(\mathbf{k}, t) = (2\pi)^{-3} \int_{\mathfrak{R}^3} e^{i\mathbf{r}\cdot\mathbf{k}} S_{ij}(\mathbf{r}, t) \, d\mathbf{r} \tag{2.84}$$

is the Fourier transform of the transfer term $S_{ij}(\mathbf{r}, t)$, defined as

$$S_{ij}(\mathbf{r}, t) = \frac{\partial S_{jk,i}}{\partial r_k}(-\mathbf{r}, t) + \frac{\partial S_{ik,j}}{\partial r_k}(\mathbf{r}, t), \tag{2.85}$$

and

$$P_{ij}(\mathbf{k}, t) = -(2\pi)^{-3} \int_{\Re^3} e^{i\mathbf{r}\cdot\mathbf{k}} \left[\frac{\partial \mathcal{K}_i}{\partial r_j}(\mathbf{r}, t) + \frac{\partial \mathcal{K}_j}{\partial r_i}(-\mathbf{r}, t) \right] d\mathbf{r} \qquad (2.86)$$

is the transformed pressure–velocity term. T_{ij} describes the rate at which a given wavenumber (or scale of motion) gains or loses energy by transfer of energy from or to the other scales. The physical process driving the energy exchange has to do with vorticity stretching, and T_{ij} is the form that this process takes in Fourier space. P_{ij}, which will later be seen is zero in isotropic turbulence, represents the influence of the pressure field in bringing initially anisotropic turbulence back toward isotropy. Finally, the viscous dissipation rate term in (2.83) is just k^2 times the energy spectrum tensor. This is an important result because it suggests that dissipation takes place at higher wavenumbers (i.e., smaller scales) than where most of the energy resides in turbulent flow. In other words, the multiplicative factor k^2 causes the peak in the dissipation rate term to be shifted to higher values of $\mathbf{k}$ relative to the peak of E_{ij}.

The last result can be made clearer through the following argument. Taking x_k and y_k derivatives of (2.68), similar to what was done in (2.74), it can be derived that

$$\overline{\frac{\partial u_i}{\partial x_k}(\mathbf{x}, t) \frac{\partial u_j}{\partial y_k}(\mathbf{y}, t)} = -\frac{\partial^2 R_{ij}}{\partial r_k^2}(\mathbf{y} - \mathbf{x}). \qquad (2.87)$$

Letting $\mathbf{y} \to \mathbf{x}$, setting $j = i$, and noting the definition of the isotropic dissipation rate ϵ gives

$$\epsilon = \nu \overline{\left(\frac{\partial u_i}{\partial x_k} \right)^2} = -\frac{\partial^2 R_{ii}}{\partial r_k^2}(0). \qquad (2.88)$$

Furthermore, contracting indices in (1.17) and differentiating twice by r_k gives

$$\frac{\partial^2 R_{ii}}{\partial r_k^2}(\mathbf{r}, t) = -\int_{\Re^3} k^2 E_{ii}(\mathbf{k}, t) e^{-i\mathbf{r}\cdot\mathbf{k}} \, d\mathbf{k}, \qquad (2.89)$$

and thus from (2.88)

$$\epsilon(t) = \nu \int_{\Re^3} k^2 E_{ii}(\mathbf{k}, t) \, d\mathbf{k}. \qquad (2.90)$$

As in the case of (1.21) the volume integration in (2.90) can be organized according to $d\mathbf{k} = d\Omega \, dk$, in which case (2.90) gives

$$\epsilon(t) = \nu \int_0^\infty k^2 E(k, t) \, dk, \qquad (2.91)$$

a relation that should be contrasted with (1.21).

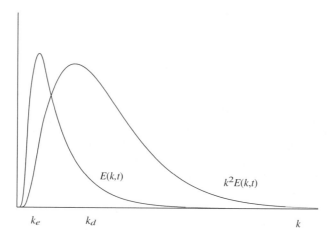

Fig. 2.2   *Spectral ranges of $E(k, t)$ and $k^2 E(k, t)$.*

Figure 2.2 shows how different spectral ranges underlie K and ϵ. Note that each value of k corresponds to Fourier modes $e^{-i\mathbf{k}\cdot\mathbf{r}}$ with $|\mathbf{k}| = k$, which may be viewed as essentially a sine or cosine variation in the fluctuating velocity field over a distance $\sim 1/k$. Thus small k is associated with variations in the velocity over large distances, while large k corresponds to variations over small distances. This allows Fig. 2.2 to be interpreted as showing that the turbulent energy associated with the fluid velocity field tends to be concentrated at larger scales than those at which energy dissipation takes place. Often, the scale of motion corresponding to the maximum of $E(k, t)$ is related to that of the stirring mechanism (i.e., the mechanism causing turbulence production). Regardless of how turbulence is generated, one may assume that the greatest concentration of energy accumulates around a certain wavenumber, say k_e. A distance $l_e = 1/k_e$ may be associated with motions at this wavenumber. In physical space it is helpful to think of eddies or vortical motions with size l_e.

At a larger wavenumber, say k_d, the dissipation process is at a peak. Here the dissipative motions can be associated with the movement of fine-scale eddies or vortices. Clearly, there must be a mechanism by which energy is brought to the small scales to be dissipated. This is represented in physical space by the nonlinear term in (2.80) and by the correlation T_{ij} in Fourier space.

As the study of energy transfer first evolved, it was thought that the process by which energy travels from large to small scales occurred in an *energy cascade*, a one-way street in which the counterflow of energy from small to large scales was insignificant in comparison to the movement of energy in the opposite direction. It is now better understood that the energy flow in both directions is significant, with just a net excess of energy moving to small scales to be dissipated. From the perspective of vortices in the flow, energy moves to small scales through vortex stretching and folding and moves to larger scales when vortices compress, recombine, or reconnect into larger vortices.

The two viewpoints on energy transfer reflect opposite perspectives on the equilibrium nature of turbulent motion. The cascade theory is one of disequilibrium, with a focus being on the changes of individual vortices as energy moves to small scales. If energy transfer is significant in both directions, an equilibrium system of vortices is to be expected where both vortex merger and stretching takes place.

For large Reynolds numbers it is expected that the distance between k_e and k_d will increase, because k_d will be larger for k_e fixed. For sufficiently high Reynolds number, a separation of scales will develop (i.e., a gap will form between the E and $k^2 E$ curves in Figure 2.2). In the gap there is neither a lot of energy residing nor significant dissipation. It may be imagined that energy is merely in transit between states: either from large to small scales in the traditional viewpoint or in both directions in the modern conception. When T_{ij} is later examined in more depth, it will become clear that the spectral approach offers a means for examining how individual groups of wavenumbers interact to cause transfer. A number of the theories covered in Chapter 12 make hypotheses about the statistical nature of this process.

2.6 TURBULENCE SCALES

It is helpful in our analysis and understanding of turbulence to develop a quantitative picture of the relative and absolute size of the scales taking part in the energy decay process. Among other reasons, this is beneficial to experimentalists and numerical analysts who want to ensure that probe resolution or numerical grid resolution is appropriate to the flow they are studying.

The famous Russian mathematician Kolmogorov [6,7] pioneered a scale analysis that has had a profound influence on the way in which the energy spectrum in turbulence is understood. In this, he first presupposed that at high Reynolds numbers— and away from the immediate influence of boundaries—the small, dissipation scales acquire the character of a *universal equilibrium* whose statistical and structural properties are assumed to be common to all turbulent flows. In other words, while the stirring force that creates turbulence will surely vary from flow to flow and will affect the turbulence characteristics, the small-scale/high-wavenumber motions at which dissipation takes place develop a common form for all flows. If this is true, it can be argued that the equilibrium state should be scaled by the viscosity ν and dissipation rate ϵ. In this case, the length scale

$$\eta \equiv \frac{\nu^{3/4}}{\epsilon^{1/4}} \tag{2.92}$$

and time scale

$$t_d = \left(\frac{\nu}{\epsilon}\right)^{1/2}, \tag{2.93}$$

known as the Kolmogorov length and time scales, respectively, should be good yard-sticks of dissipative phenomena. Additionally, a velocity scale $v_d = (\nu\epsilon)^{1/4}$ can be formed. It is evident from our previous discussion of k_d, as in Fig. 2.2, that

$$\eta \sim \frac{1}{k_d} \tag{2.94}$$

even if they are not exactly equal. Some experimental work [4,12] has shown that $k_d \approx \alpha/\eta$ for $\alpha = 0.1 \rightarrow 0.15$, and, in fact, that most of the dissipation takes place for $k < 0.5/\eta$. Thus η, in a real sense, represents the smallest excited scales in turbulent flow.

Micro and macro scales, λ and Λ, were defined in Section 1.2. It is of considerable interest to see how these are related to η. Similarly, one can enquire as to how the velocity scale, v_d, is related to the root-mean-square (rms) velocity components as typified by $u_{rms} = \sqrt{\overline{u^2}}$. Relations between these quantities stem from considering ϵ from two different points of view and then bringing them together.

From the definition of ϵ in (2.36) and the microscale λ in (1.25), it will be derived in Chapter 7 that for isotropic turbulence

$$\epsilon \sim \nu \frac{u_{rms}^2}{\lambda^2}. \tag{2.95}$$

This relation looks at ϵ from the perspective of the small-scale motions at which dissipation is actualized in turbulent flows. A second viewpoint comes from considering ϵ as the rate at which energy is extracted from the energy-containing scales. The time necessary for a measurable amount of energy to be lost from such scales depends on both how much turbulent energy there is, as characterized by u_{rms}, and the physical extent of such eddies as represented by l_e (or equivalently Λ). The reasoning is that it would take at least this length for a turbulent eddy to show appreciable change. The scales l_e and u_{rms} combine to form a time scale $\sim l_e/u_{rms}$. Moreover, since ϵ has units of energy per second, one is then led to the scaling

$$\epsilon \sim \frac{u_{rms}^2}{l_e/u_{rms}} = \frac{u_{rms}^3}{l_e}. \tag{2.96}$$

The two expressions (2.95) and (2.96) thus suggest that

$$\nu \frac{u_{rms}^2}{\lambda^2} \sim \frac{u_{rms}^3}{l_e}, \tag{2.97}$$

and consequently,

$$\frac{l_e}{\lambda} \sim R_\lambda, \tag{2.98}$$

where $R_\lambda = u_{rms}\lambda/\nu$ is a turbulence Reynolds number. For many laboratory flows, R_λ can be as large as 100, showing that in these cases there is a difference in size of up to two orders of magnitude between l_e and λ. Multiplying the numerator and denominator on the left-hand side of (2.98) by u_{rms}/ν gives

$$R_e \sim R_\lambda^2, \tag{2.99}$$

or

$$\sqrt{R_e} \sim R_\lambda, \tag{2.100}$$

where $R_e = u_{\text{rms}} l_e / \nu$ is a turbulence Reynolds number based on the physical size of the flow domain.

The ratio between λ and η can be found from (2.92) and (2.95) and is

$$\frac{\eta}{\lambda} \sim \frac{1}{\sqrt{R_\lambda}} \sim \frac{1}{R_e^{1/4}}, \tag{2.101}$$

showing that η is generally smaller than λ but not so much so. In fact it can be seen that λ is a reasonable measure of the scales where most of the dissipation takes place. From (2.98) and (2.101) it follows that

$$\frac{l_e}{\eta} \sim R_\lambda^{3/2} \sim R_e^{3/4}, \tag{2.102}$$

which is the ratio of the largest to smallest scales in the flow. A similar calculation gives

$$\frac{u_{\text{rms}}}{v_d} \sim R_\lambda^{1/2} \sim R_e^{1/4}. \tag{2.103}$$

Equation (2.102) is the basis for an important estimate that can be made concerning the numerical cost of computing turbulent flow. Since η is the smallest scale in the flow, a full numerical simulation on a mesh would generally require a mesh spacing $\sim \eta$ to resolve the flow details. On the other hand, the spatial extent of the flow domain is $\sim l_e$, so in any one direction, approximately l_e/η mesh points are required. A three-dimensional mesh would then have to be $\sim (l_e/\eta)^3$ in size. In view of (2.102) it follows that the number of mesh points in a fully resolved turbulent flow simulation has a $R_e^{9/4}$ dependence on Reynolds number.

In a numerical calculation the elapsed time for a single iteration can be taken to be $\sim \eta/u_{\text{rms}}$, since this reflects the shortest time period needed to be resolved. At the same time, l_e/u_{rms} reflects time variations of the large-scale motions. The ratio of these two scales, $\sim l_e/\eta$, reflects roughly how many time steps must be computed in a typical turbulent flow simulation in order to determine average quantities. According to (2.102), this means that $O(R_e^{3/4})$ time steps are required in the simulation, and since there are $O(R_e^{9/4})$ grid points, the total computational effort involves $O(R_e^3)$ operations. The conclusion is reached that the cost of doing a turbulence simulation increases very rapidly with Reynolds number. Thus, even though simulations incorporating 512^3 or more mesh points are currently feasible and have successfully represented flows with $R_e \sim 10^4$, to get a factor of 10 higher in R_e requires at least 1000 times more computer power. For still larger Reynolds numbers the outlook is even more bleak. This explains why there is little expectation that DNS will become a practical tool for simulating high Reynolds number flows anytime soon.

2.6.1   Inertial Subrange

A second far-reaching idea of Kolmogorov was that of an *inertial subrange* consisting of a section of wavenumber space between k_e and k_d where energy cascades toward small scales without significant dissipation or production. Such a cascade in this range of wavenumbers would depend on just ϵ and not ν. Kolmogorov argued that this has an important consequence for the form of the energy spectrum function $E(k, t)$. Because $E(k, t)$ has units of (length)^3/sec^2, the only form of $E(k, t)$ dimensionally consistent with a scaling in terms of k and ϵ is given by

$$E(k, t) \sim k^{-5/3} \epsilon^{2/3}, \tag{2.104}$$

or with a Kolmogorov constant C_K,

$$E(k, t) = C_K k^{-5/3} \epsilon^{2/3}. \tag{2.105}$$

This prediction of a $-5/3$ spectrum is amenable to experimental verification, and, in fact, has been observed to occur in a wide range of turbulent flows at high Reynolds number [10,11] with the typical value $C_K = 1.4$. Results accumulated from many different experiments in different types of turbulent flows and covering a very wide range of wavenumbers are shown in Fig. 2.3. Plotted here is the scaled one-dimensional energy spectrum function $E_{11}(k_1)$ (see Section 1.2).[2] That $E_{11}(k_1)$ should also satisfy a $-5/3$ law will become apparent once a relation between E_{11} and E is developed in Chapter 7. Figure 2.3 clearly shows the $-5/3$ inertial subrange in k_1 spectra with the wavenumber extent of the inertial subrange increasing with R_λ. That these quite different flows collapse to a common spectral form at higher wavenumbers provides support for Kolmogorov's hypothesis regarding the universality of the small scales of turbulence.

While the prediction of the $-5/3$ energy spectrum is confirmed experimentally, it is ironic to note that the original Kolmogorov argument in support of its existence is now generally considered to be invalid [3,8]. The criticisms center on the cascade assumption, which has energy tumbling in a disequilibrium to small scales and on the observed intermittency of the regions of significant dissipation. In other words, most of the dissipation in turbulent flows appears not to take place relatively uniformly over physical space but rather, it is concentrated in small regions where the dissipation level is large. Such physics is incompatible with a simple scaling argument. Trying

---

[2]Frequency has been converted to wavenumber via Taylor's hypothesis (Section 3.2.3.3).

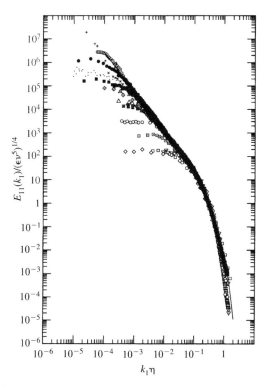

Fig. 2.3 *Experimental demonstration of the $-5/3$ law. (From [10]. Reprinted with permission of Cambridge University Press.)*

to find a new explanation for the $-5/3$ law, one that is more compatible with current insights into the decay process, is an issue of much concern and interest. This will be touched upon to some extent in Chapter 12.

2.7 VORTICITY EQUATION

It will come to be seen that many aspects of turbulent flow are best understood in terms of the dynamics of the vorticity field, $\boldsymbol{\Omega} \equiv \nabla \times \mathbf{u}$. Consequently, it is natural to wonder if there might be some advantages to analyzing and predicting turbulent flows by casting the problem directly in terms of the averaged vorticity instead of momentum. In fact, such a perspective underlies the early work of G. I. Taylor, who, in his vorticity transport theory [13,14], analyzed turbulence through the vorticity flux instead of the Reynolds stress. It is interesting to note that much of his work actually predates later Reynolds stress modeling.

The approach taken by Taylor was to consider the average Navier–Stokes equation as a starting point for analysis, but to use it in its rotational form, in which the Reynolds stress is replaced by the vorticity flux. This depends on applying the identity

$$\frac{\partial \overline{u_i u_j}}{\partial x_j} = \frac{\partial K}{\partial x_i} - \epsilon_{ijk}\overline{u_j \omega_k} \tag{2.106}$$

to (2.10), yielding

$$\frac{\partial \overline{U}_i}{\partial t} + \overline{U}_j \frac{\partial \overline{U}_i}{\partial x_j} = -\frac{\partial(\overline{P}/\rho + K)}{\partial x_i} + \nu \nabla^2 \overline{U}_i + \epsilon_{ijk}\overline{u_j \omega_k}. \tag{2.107}$$

Here $\omega_i = \Omega_i - \overline{\Omega}_i$ and ϵ_{ijk} is the alternating tensor (i.e., $\epsilon_{ijk} = \pm 1$ if i, j, k are in cyclic or anticyclic order, and zero otherwise). Equation (2.107) contains the vorticity flux $\overline{u_i \omega_j}$, representing the rate at which the jth component of vorticity is transported in the ith direction by the fluctuating velocity field. It should be noted that the appearance of K with the pressure in (2.107) normally poses no hardship: one can solve for $\overline{P}/\rho + K$ as an unknown in (2.107) in the same way that $\overline{P}$ is determined in (2.10). Since $K = 0$ on solid surfaces, having computed $\overline{P}/\rho + K$ actually gives $\overline{P}$ on boundaries so that forces can readily be obtained.

To perhaps more fully capitalize on the potential advantages offered by the vorticity field perspective, it is necessary to consider its own equation derived by taking the curl of (2.5), namely,

$$\frac{\partial \Omega_i}{\partial t} + U_j \frac{\partial \Omega_i}{\partial x_j} = \Omega_j \frac{\partial U_i}{\partial x_j} + \nu \nabla^2 \Omega_i. \tag{2.108}$$

In contrast to (2.5), (2.108) has no pressure dependency, and moreover, it displays the physics of the vorticity field in its most natural setting. In particular, it is seen that the rate of change of vorticity of a fluid particle, as given on the left-hand side is determined by vortex stretching and reorientation given by the first term on the right-hand side and viscous diffusion by the last term. The stretching of vorticity is such an integral part of the physics of turbulent motion that we now digress for a brief look at why this physical interpretation is given to the term $\Omega_j(\partial U_i/\partial x_j)$ appearing in (2.108).

2.7.1 Origin of the Stretching Term

Consider the motion of two nearby material points $\mathbf{a}(t)$ and $\mathbf{b}(t)$ in a flow $\mathbf{U}(\mathbf{x}, t)$, and let

$$\mathbf{r}(t) = \mathbf{b}(t) - \mathbf{a}(t) \tag{2.109}$$

denote the vector pointing out their relative positions. Let $l(t) = |\mathbf{r}(t)|$ denote the distance between the points, which is assumed to be very small, and let $\mathbf{s}(t) = \mathbf{r}(t)/|\mathbf{r}(t)|$ be a unit vector pointing in the direction of $\mathbf{r}$. Thus

$$\mathbf{r}(t) = l(t)\mathbf{s}(t). \tag{2.110}$$

Since **a** travels with the local fluid velocity, it satisfies

$$\frac{d\mathbf{a}}{dt} = \mathbf{U}(\mathbf{a}(t), t), \tag{2.111}$$

in which case, after integration from an initial time $t = 0$, where the particle is at $\mathbf{a}(0)$, one has

$$\mathbf{a}(t) = \mathbf{a}(0) + \int_0^t \mathbf{U}(\mathbf{a}(t'), t') \, dt'. \tag{2.112}$$

Considering just the slight movement of **a** occurring after a small time interval Δt, (2.112) suggests that **a** is well approximated by

$$\mathbf{a}(\Delta t) = \mathbf{a}(0) + \mathbf{U}(\mathbf{a}(0), 0) \, \Delta t. \tag{2.113}$$

Similarly, for **b** the approximation

$$\mathbf{b}(\Delta t) = \mathbf{b}(0) + \mathbf{U}(\mathbf{b}(0), 0) \, \Delta t \tag{2.114}$$

holds. According to (2.109), $\mathbf{b}(0) = \mathbf{a}(0) + \mathbf{r}(0)$, and, since by assumption $l(0)$ is small, it is legitimate to write the Taylor series expansion:

$$\mathbf{U}(\mathbf{b}(0), 0) = \mathbf{U}(\mathbf{a}(0), 0) + (\nabla \mathbf{U})\mathbf{r}(0). \tag{2.115}$$

Substituting this into (2.114) and subtracting (2.113) gives

$$\mathbf{r}(\Delta t) = \mathbf{r}(0) + (\nabla \mathbf{U})\mathbf{r}(0) \, \Delta t. \tag{2.116}$$

Substituting for **r** using (2.110), dividing by $l \, \Delta t$, taking the limit as $\Delta t \to 0$ gives

$$\frac{d\mathbf{s}}{dt} + \frac{1}{l}\frac{dl}{dt}\mathbf{s} = (\nabla \mathbf{U})\mathbf{s}. \tag{2.117}$$

Because **s** is a unit vector,

$$\mathbf{s} \cdot \frac{d\mathbf{s}}{dt} = \frac{1}{2}\frac{d(\mathbf{s} \cdot \mathbf{s})}{dt} = 0. \tag{2.118}$$

Thus $d\mathbf{s}/dt$, which is the rate of change of the orientation of **r**, is perpendicular to **r** itself.

To use these results in understanding the stretching term in (2.108), assume that **s** lies in the direction of the vorticity vector, so that $\mathbf{s} = \mathbf{\Omega}/|\mathbf{\Omega}|$. In this case, after multiplying through by $|\mathbf{\Omega}|$ and changing to index notation, (2.117) becomes

$$\frac{1}{l}\frac{dl}{dt}\Omega_i + |\mathbf{\Omega}|\frac{d(\Omega_i/|\mathbf{\Omega}|)}{dt} = \Omega_j \frac{\partial U_i}{\partial x_j}. \tag{2.119}$$

This shows that the stretching term is equivalent to the sum of the two effects on the left-hand side of this equation. The first is in the direction of the local vorticity vector and is proportional to the fractional rate at which a local fluid line element is stretching (or contracting). The effect on the vorticity in this case is to make it either larger or smaller, depending on whether l is increasing or decreasing. According to (2.118), the second term on the right-hand side of (2.119) indicates a change in vorticity by reorientation of the local vorticity in a direction normal to its current direction. No change in the overall strength of the vorticity is implied by this term.

To see the implications of (2.119) in a more concrete setting, imagine that the vorticity vector at a point lies in the x direction, so that $\Omega_2 = \Omega_3 = 0$. Then the Ω_1 equation (2.108) has one stretching term,

$$\Omega_1 \frac{\partial U}{\partial x}, \tag{2.120}$$

and this must contribute toward a gain or loss of Ω_1 by pure stretching (or compression)—as given by the first term on the right-hand side of (2.119)—since the second term is orthogonal to the x_1 direction. On the other hand, the Ω_2 equation has the stretching term

$$\Omega_1 \frac{\partial V}{\partial x}, \tag{2.121}$$

while the Ω_3 equation has

$$\Omega_1 \frac{\partial W}{\partial x}. \tag{2.122}$$

These terms clearly originate in the second term in (2.119) and represent reorientation of the Ω_1 vorticity into the y and z directions, respectively, by shearing motions caused by variations of V and W in the x direction.

In the general case, arbitrarily oriented vorticity filaments are simultaneously stretched or compressed *and* reoriented by the shearing motions. That is, the two processes represented in (2.119) act simultaneously to change vorticity. This can be readily observed for a short vortex segment by tracking the relative motion of its two ends. This property is used to advantage in vortex methods, described in Chapter 10, which attempt to capitalize on the fact that tubelike vortical structures play a major role in the dynamics of turbulent flow.

2.7.2 Averaged Vorticity Equation

The mean vorticity may be determined as the solution to the equation resulting from an average of (2.108):

$$\frac{\partial \overline{\Omega}_i}{\partial t} + \overline{U}_j \frac{\partial \overline{\Omega}_i}{\partial x_j} = \overline{\Omega}_j \frac{\partial \overline{U}_i}{\partial x_j} + \overline{\omega_j \frac{\partial u_i}{\partial x_j}} + \frac{\partial}{\partial x_j} \left(\nu \frac{\partial \overline{\Omega}_i}{\partial x_j} - \overline{u_j \omega_i} \right). \tag{2.123}$$

In the last term on the right-hand side it is seen that viscous vorticity diffusion is augmented by turbulent vorticity diffusion given by $\overline{u_j \omega_i}$. The two terms before this account for vortex stretching and reorientation by the mean and fluctuating fields, respectively.

Closure to (2.123) nominally requires a theory to enable calculation of the transport and stretching correlations, $\overline{u_i \omega_j}$ and $\overline{\omega_j (\partial u_i / \partial x_j)}$. However, since the vorticity is solenoidal (i.e., it is divergence free), it follows that

$$\overline{\omega_j \frac{\partial u_i}{\partial x_j}} = \frac{\partial \overline{u_i \omega_j}}{\partial x_j}, \tag{2.124}$$

so that, technically speaking, only a closure is needed for the flux correlation, as is the case for (2.107). Indeed, (2.123) could be derived by taking a curl of (2.107). In Section 6.2, after some tools are developed with which to analyze the stretching correlation, it will be shown that it is not necessarily advantageous to pursue the approach suggested by (2.124). In fact, there are some potential advantages to direct modeling of the stretching terms as they appear in (2.123).

In two-dimensional turbulence the stretching term is identically zero, and vorticity transport is the only correlation in need of modeling. Taylor's initial effort [13] at devising a vorticity transport closure was devoted to this case. In later work [14], he attempted to extend the approach to three-dimensional turbulence, although this did not turn out as satisfactorily as comparable models of the Reynolds stress correlation. However, modern treatments of vorticity transport, discussed in Chapter 6, have exposed and corrected the difficulties underlying Taylor's analysis, so that, at least in principle, vorticity transport models are also a feasible choice in modeling turbulent flow. There is some evidence that the advantages of vorticity transport modeling are real, although they have yet to achieve widespread application.

2.7.3 Enstrophy Equation

The enstrophy, $\zeta \equiv \overline{\omega_i^2}$, has a relationship to $\overline{\Omega}$ similar to that of K to $\overline{U}$. Moreover, the identity

$$\frac{\epsilon}{\nu} = \zeta + \overline{\frac{\partial u_i}{\partial x_j} \frac{\partial u_j}{\partial x_i}} \tag{2.125}$$

suggests that ϵ and ζ have much in common. In fact, the last term in (2.125) is identically zero in homogeneous turbulence, implying that in this special circumstance ζ exactly equals ϵ/ν. Even in the presence of sizable mean shear, evidence from DNS suggests that the approximation $\zeta \approx \epsilon/\nu$ is extremely good, as will be seen in Fig. 4.17. As a consequence, it is generally acceptable to use $\nu\zeta$ interchangeably with ϵ when developing turbulence models. The advantage of using enstrophy in place of ϵ as an unknown comes from the opportunity it provides to consider the closure of the ζ equation instead of the ϵ equation. The former, although similar to the latter in many ways, has some subtle differences that make its modeling potentially easier.

One obvious difference is that the ζ equation does not have pressure terms, as will now be seen.

The exact ζ equation is derivable from (2.108) by multiplying it as written for Ω_i by ω_i and averaging. The result, after some significant manipulation, is

$$\frac{D\zeta}{Dt} = P_\zeta^1 + P_\zeta^2 + P_\zeta^3 + P_\zeta^4 + T_\zeta + D_\zeta - \Upsilon_\zeta, \qquad (2.126)$$

where

$$P_\zeta^1 = 2\overline{\omega_i \omega_k} \frac{\partial \overline{U}_i}{\partial x_k},$$

$$P_\zeta^2 = 2\overline{\omega_i \frac{\partial u_i}{\partial x_k}} \Omega_k,$$

$$P_\zeta^3 = -2\overline{u_k \omega_i} \frac{\partial \overline{\Omega}_i}{\partial x_k},$$

$$P_\zeta^4 = 2\overline{\omega_i \omega_k \frac{\partial u_i}{\partial x_k}},$$

$$T_\zeta = -\frac{\partial}{\partial x_k} \overline{(u_k \omega_i \omega_i)},$$

$$D_\zeta = \nu \nabla^2 \zeta,$$

$$\Upsilon_\zeta = 2\nu \overline{\frac{\partial \omega_i}{\partial x_k} \frac{\partial \omega_i}{\partial x_k}}.$$

As in the case of the ϵ equation, the terms in (2.126) can be categorized as production, transport, diffusion, and dissipation terms. A nice property of (2.126) is that P_ζ^2 and P_ζ^3 are built on the transport and stretching correlations, which would previously have been modeled for use in (2.123). The difficulty of analyzing the remaining terms is comparable to their counterparts in the ϵ equation. More will be said about the physics and modeling of this equation in later sections.

2.8   SCALAR TRANSPORT

The ability of turbulent flow to effect the transport and mixing of contaminant species is one of its most significant properties. By definition, passive scalars do not affect the underlying turbulent flow causing their transport. This is the case with dilute concentrations of contaminants or the transport of low levels of heat in incompressible fluids. In this book we do not consider the more general case in which two-way coupling occurs (e.g., if buoyancy is present or the contaminant consists of heavier particles or chemically reacting species). In such and similar cases the evolution of the velocity field is affected by the presence of contaminant.

For the purposes of this discussion we denote an arbitrary scalar contaminant field as $C(\mathbf{x}, t)$, having units of contaminant per volume. This may be mass per volume in the case of a marked species or internal energy per volume in the case of the diffusion of heat. If $d\mathcal{V}$ represents a small volume of fluid around a point $\mathbf{x}$ at time t, the amount of contaminant contained in $d\mathcal{V}$ is by definition $C(\mathbf{x}, t)\, d\mathcal{V}$.

Assuming that the scalar diffuses according to a Fickian diffusion law, then in the case of a contaminant species, the diffusion rate in units of scalar per area per second is

$$-\mathcal{D}\frac{\partial C}{\partial x_i}, \tag{2.127}$$

where $\mathcal{D}$ is the coefficient of mass diffusion. (Alternatively, if C denotes internal energy per volume, $\mathcal{D}$ represents the thermal diffusivity, $\alpha \equiv k/\rho c_p$, where k is the thermal conductivity and c_p is the specific heat at constant pressure.) Conservation of scalar contaminant implies that C satisfies the convective diffusion equation

$$\frac{\partial C}{\partial t} + U_j \frac{\partial C}{\partial x_j} = \mathcal{D}\,\nabla^2 C + q, \tag{2.128}$$

where q is a source density in units of scalar per volume per second.

In a turbulent flow one generally wishes to predict the average scalar field $\overline{C}(\mathbf{x}, t)$, whose evolution is governed by the average form of (2.128), namely,

$$\frac{\partial \overline{C}}{\partial t} + \overline{U}_j \frac{\partial \overline{C}}{\partial x_j} = \frac{\partial}{\partial x_j}\left(\mathcal{D}\frac{\partial \overline{C}}{\partial x_j} - \overline{u_i c}\right) + q, \tag{2.129}$$

where q is considered to be nonrandom. As in our previous examples of such equations, closure to (2.129) rests on predicting the scalar transport flux rate $\overline{u_i c}$. In Chapter 11 we consider in some detail the analysis and modeling of this quantity.

Equations (2.128) and (2.129) reflect an Eulerian view of the diffusion process in that they represent a balance of field quantities. Alternatively, the diffusion problem can be cast in a Lagrangian framework involving the movement of particles of scalar contaminant. In the molecular diffusion case, as pertains to (2.128), a statistical theory of the random migrations of fluid particles is available in the theory of Brownian motion. This provides an alternative, and physically suggestive, means for solving diffusion problems. Analogously, for turbulent flow, where (2.129) is applicable, the contaminant also moves on random paths through the flow, although the statistical process is not rigorously given by that of Brownian motion. Nevertheless, there is still much to be gained by making a connection between the Lagrangian and Eulerian viewpoints, as will be seen in Chapter 11. In the next section we describe the relationship between the Eulerian and Lagrangian points of view for the case of molecular diffusion and provide motivation for similar results in the turbulent case in the following section. These results are exploited in Chapter 11 to gain a solution to some diffusion problems in turbulent flow.

2.8.1 Molecular Transport

Consider the diffusion of a fixed amount, Q, of marked fluid released at $t = 0$ from a point source at $\mathbf{x} = 0$ (i.e., what is commonly referred to as a *puff*). This problem is formally equivalent to solving (2.128) with the condition

$$q = Q\delta(t)\delta(\mathbf{x}), \qquad (2.130)$$

where δ denotes the Dirac delta function. $\delta(t)$ has units of inverse time and satisfies $\int \delta(t - s) f(s) \, ds = f(t)$ for all functions $f(t)$. Similarly, $\delta(\mathbf{x})$ has units of inverse volume and satisfies $\int \delta(\mathbf{x} - \mathbf{y}) f(\mathbf{y}) \, d\mathbf{y} = f(\mathbf{x})$, for all functions $f(\mathbf{x})$. The forcing given by (2.130) can be usefully approximated by having $q = Q/(dt \times d\mathcal{V})$ for a small time interval dt around $t = 0$ and volume $d\mathcal{V}$ around $\mathbf{x} = 0$, while $q = 0$ for $t > dt$ and/or $\mathbf{x}$ not in $d\mathcal{V}$. After release, the marked fluid convects downstream in the laminar flow while diffusing outward at a rate determined by $\mathcal{D}$. This phenomenon is formally equivalent to Brownian motion, in which large molecules (e.g., dust) suspended in a carrier fluid follow chaotic trajectories due to their collisions with molecules of the host fluid. In other words, one expects that the diffusion of the marked fluid after its release into the flow can be viewed as equivalent to an ensemble of random Brownian paths, one path associated with the motion of a single molecule released at $t = 0, \mathbf{x} = 0$ in each realization of the molecular field. The paths are characterized according to the probability density function, $\mathcal{P}(\mathbf{x}, t)$, in units of inverse volume associated with the movement of molecules from the origin at time zero to $\mathbf{x}$ at time t. $\mathcal{P}(\mathbf{x}, t) \, d\mathcal{V}$ is then the probability that a molecule originally at $\mathbf{x} = 0$ at $t = 0$ will lie within the volume $d\mathcal{V}$ of $\mathbf{x}$ at time t. The link between the Lagrangian and Eulerian viewpoints is then through the following equivalent descriptions of a puff of contaminant:

$$C(\mathbf{x}, t) \, d\mathcal{V} = Q\mathcal{P}(\mathbf{x}, t) \, d\mathcal{V}, \qquad (2.131)$$

where both sides represent the amount of contaminant found within the volume $d\mathcal{V}$ of $\mathbf{x}$ at time t. This interpretation of the right-hand side follows from the fact that it represents the fraction of Q that has made it to $d\mathcal{V}$ and resides there at time t. Since $d\mathcal{V}$ is arbitrary, (2.131) implies that

$$C(\mathbf{x}, t) = Q\mathcal{P}(\mathbf{x}, t). \qquad (2.132)$$

For any time $t > 0$, integration of (2.132) over all space recovers the total mass of contaminant injected into the fluid [i.e., $Q = \int_{\Re^3} C(\mathbf{x}, t) \, d\mathcal{V}$, since $\int_{\Re^3} \mathcal{P}(\mathbf{x}, t) \, d\mathcal{V} = 1$]. To the extent that one can determine $\mathcal{P}$, one will have a window into how C must be distributed. As will be shown in Chapter 11, in some circumstances there are sound statistical arguments giving $\mathcal{P}$; in these cases one gains knowledge of C as well.

2.8.2 Turbulent Scalar Transport

The discussion of Section 2.8.1 is valuable because it also suggests a means for establishing a connection between the Lagrangian and Eulerian viewpoints in the case of turbulent scalar transport. Thus consider a puff of contaminant released into a turbulent flow. Similar to our consideration of molecular transport, the average behavior of many such turbulent puffs (i.e., the composite puff formed as an average over an ensemble of puffs) can be described via a collection of random paths leaving from the puff source: one path for each realization of the turbulent field. In this instance, the paths are associated with collections of marked fluid particles moving in the turbulent flow, not the trajectories of molecules in the previous case. In analogy to (2.131), one has

$$\overline{C}(\mathbf{x}, t)\, d\mathcal{V} = Q\mathcal{P}(\mathbf{x}, t)\, d\mathcal{V}, \tag{2.133}$$

wherein both sides of the equation represent the amount of scalar transferred, on average, to within $d\mathcal{V}$ of $\mathbf{x}$ at time t. On the right-hand side, $\mathcal{P}(\mathbf{x}, t)$ is the pdf of the displacement of fluid particles in the turbulent fluid starting from $\mathbf{x} = 0$ at $t = 0$. $\mathcal{P}\, d\mathcal{V}$ is the probability that fluid paths end up in $d\mathcal{V}$, so $Q\mathcal{P}\, d\mathcal{V}$ is the part of Q that manages to arrive at and reside in $d\mathcal{V}$ at time t. As in the laminar case, q appearing in (2.129) corresponds to the rate of release of Q amount of contaminant per unit volume into the fluid over the small time interval dt.

Since $d\mathcal{V}$ is arbitrary, (2.133) becomes

$$\overline{C}(\mathbf{x}, t) = Q\mathcal{P}(\mathbf{x}, t), \tag{2.134}$$

showing that the problem of determining concentration of the puff at $\mathbf{x}$, t is equivalent to the problem of finding $\mathcal{P}(\mathbf{x}, t)$. In situations where $\mathcal{P}(\mathbf{x}, t)$ can be reasonably well determined, the approach typified by (2.134) succeeds in avoiding the need to solve (2.129), including inventing a model for $\overline{u_i c}$. An interesting sidelight of this discussion is that, in some cases, knowledge about $\overline{C}$ originally coming from P via (2.134) can be used in (2.129) to gain knowledge about the scalar flux rate. This has potential benefit in improving our understanding of transport phenomena. Further discussion on these points is deferred until Chapter 11.

REFERENCES

1. Batchelor, G. K. (1967) *Introduction to Fluid Dynamics*, Cambridge University Press, London.

2. Chapman, S. and Cowling, T. G. (1952) *The Mathematical Theory of Non-uniform Gases*, 2nd ed., Cambridge University Press, London.

3. Chorin, A. J. (1994) *Vorticity and Turbulence*, Springer-Verlag, New York.

4. Grant, H. L., Stewart, R. W. and Moilliet, A. (1962) "Turbulence spectra from a tidal channel," *J. Fluid Mech.* **12**, 241–263.

5. Gurtin, M. E. (1981) *An Introduction to Continuum Mechanics*, Academic Press, New York.

6. Kolmogorov, A. N. (1941) "Dissipation of energy in locally isotropic turbulence," *C. R. Acad. Sci. URSS* **32**, 19–21.

7. Kolmogorov, A. N. (1941) "The local structure of turbulence in an incompressible viscous fluid for very large Reynolds numbers," *C. R. Acad. Sci. URSS* **30**, 301–305.

8. Lesieur, M. (1997) *Turbulence in Fluids,* 3rd ed., Kluwer Academic, Dordrecht, The Netherlands.

9. Reynolds, O. (1895) "On the dynamical theory of incompressible viscous fluids and the determination of the criterion," *Philos. Trans. R. Soc. Ser. A* **186**, 123–164.

10. Saddoughi, S. G. and Veeravalli, S. V. (1994) "Local isotropy in turbulent boundary layers at high Reynolds number," *J. Fluid Mech.* **268**, 333–372.

11. Sreenivasan, K. R. (1995) "On the universality of the Kolmogorov constant," *Phys. Fluids* **7**, 2778–2784.

12. Stewart, R. W. and Townsend, A. A. (1951) "Similarity and self-preservation in isotropic turbulence," *Philos. Trans. R. Soc. London Ser. A* **243**, 359–386.

13. Taylor, G. I. (1915) "Eddy motion in the atmosphere," *Philos. Trans. R. Soc. London* **215**, 1–26.

14. Taylor, G. I. (1932) "The transport of vorticity and heat through fluids in turbulent motion," *Proc. R. Soc. Ser. A* **135**, 685–705.

3

Experimental and Numerical Methods

3.1 INTRODUCTION

A large part of what is known about turbulent flow has been gleaned from physical experiments carried out over the last half-dozen decades. Experimental facilities continue to proliferate today to meet the demand for knowledge about how particular flows behave. Despite the inroads of numerical prediction schemes in a number of specialized flow categories, direct physical measurement of turbulent flow is and will continue to be the most reliable means of analyzing many engineering flows into the foreseeable future. The purpose of this chapter is to give an introduction to the kinds of measurement techniques that are the most widely used in experimental investigations.

The kinds of techniques that are used for measuring flows vary depending on whether the flow medium is a gas or liquid. For the former, hot-wire anemometry (HWA) still is the most commonly used laboratory method for measuring the turbulent velocity field. To be useful, the mean velocities and turbulence intensities of the gas must not be too high. The gas should be approximately isothermal and free of contaminating particles. With these conditions met, HWA yields statistical properties of the velocity field with good accuracy and usually good resolution. Indeed, using small arrays of hot-wire sensors in various configurations, experimentalists are able to measure the vorticity vector components, the elements of the strain-rate and dissipation-rate tensors [21,45,49,50] and other complex properties of interest in turbulence flow analysis. With larger hot-wire arrays, some of the properties of the structures found in turbulent shear flows have been measured [10,22].

Hot wires are usually not robust enough to withstand the forces that would be exerted on them in liquid flows. Hot-film anemometry, with sensors that are thin metal layers deposited over quartz cylinders, is the alternative sometimes used in liquid flows. Also useful for measuring velocity components in liquid and multiphase flows is laser-Doppler velocimetry (LDV), in which light is reflected off seed particles that follow the movements of the turbulent fluid. This approach is particularly effective for liquid flows, since the necessary seed particles can often be found as natural impurities that are known to follow the turbulent motions more faithfully than in air or gas flows under most circumstances.

Various optical methods that allow measurement of the velocity components in full planar fields have been developed in recent years. The most widely used of these for both gas and liquid flows is particle image velocimetry (PIV). Variations of this method, such as particle tracking velocimetry and speckle velocimetry, are also used and will be considered here as well. It should be noted that while PIV has been extended to three dimensions using holography, rapid planar scanning, and, with limited spatial resolution, three-dimensional particle tracking, such methods are very expensive and difficult to implement.

In addition to velocity components, it is often necessary to measure variations in the temperature or scalar contaminant fields in turbulent flows. Small "cold"-wire sensors, discussed in Section 3.2.4, are the most common means of measuring local temperature fluctuations. Scalar concentration is measured locally by specially designed probes, or by using optical imaging techniques on planar fields, as discussed briefly in Section 3.5.

Finally, in Section 3.6 we introduce the typical numerical schemes used to solve the Navier–Stokes equations under turbulent flow conditions (i.e., the numerical basis of DNS). The inclusion of this material in a chapter on measurement techniques is meant to reflect the present role that DNS has in turbulence research as a "numerical wind tunnel." Thus DNS is used to learn things about turbulent flow behavior, in much the same way as physical experiments.

3.2 HOT-WIRE AND HOT-FILM ANEMOMETRY

The hot-wire method is based on the simple physical principle that the amount of cooling experienced by a heated wire can be related to the local flow velocity. A small-diameter metal wire sensor, usually tungsten, platinum, or a platinum alloy, is heated above the ambient temperature of the flow by an electrical current. Wire diameters typically vary in the range 0.5 to 5 μm and the lengths from 0.15 to 1.5 mm, depending on the application and spatial resolution required. Heat from the wire is transferred to the flow, the rate of which depends nonlinearly on the flow velocity. The same principle applies for hot-film sensors, where the heat is transferred to the flow from the heated thin metal film deposited over a quartz supporting core substrate. Photographs of some commercial hot-wire and hot-film probes are shown in Fig. 3.1.

If the temperature change is not too great, to a good approximation the resistance of a hot-wire or hot-film sensor depends linearly on its temperature and is given by

$$R_s = R_f[1 + \alpha(T_s - T_f)], \tag{3.1}$$

where T_s is the heated sensor temperature, T_f is fluid temperature, R_s and R_f are the resistances of the sensor at the corresponding temperatures, and α is the temperature

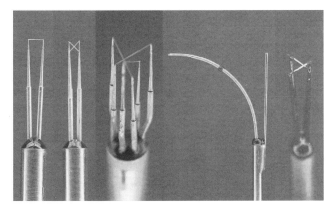

Fig. 3.1   *Hot-wire and hot-film probes. (Photographs provided by and used with permission of TSI, Inc.)*

resistance coefficient with units of reciprocal degrees (approximately 0.004 K^{-1} for tungsten and platinum). Thus the instantaneous variation of the resistance of the sensor wire can be related to, among other factors, the instantaneous flow velocity cooling the wire, at least to the extent that the resistance variation accurately follows the high-frequency velocity variations.

The operation of a hot-wire or hot-film sensor is governed by the differential equation

$$\frac{dQ}{dt} = P - F, \tag{3.2}$$

which describes the thermal energy balance. Here Q is the internal energy of the sensor, $P \equiv IE \equiv I^2 R_s$ is the electrical input power to the sensor, with I the current through the sensor, E the voltage drop across the sensor, and R_s the sensor resistance. In this, electrical energy is converted into thermal energy through joule heating of the wire at a rate P which is a function of the sensor current I and resistance R_s (or temperature T_s). F represents the total rate of heat transferred from the sensor and is given by

$$F = q_c + q_p + q_s + q_r, \tag{3.3}$$

where q_c is the heat transfer rate due to convection from the sensor by the flow, q_p is the heat transfer rate due to conduction to the prongs supporting the ends of the sensors, q_s is the conductive heat transfer rate to the quartz substrate (for hot-films only), and q_r is the heat transfer rate due to radiation from the sensor.

For a given sensor and a constant fluid temperature T_f, the rate of thermal energy transferred to the fluid is governed by the sensor temperature T_s and the fluid velocity U, while q_r and q_p also depend primarily on the sensor temperature. Thus, taking $Q \equiv cmT_s$, where c is the specific heat and m is the mass of the sensor, (3.2) can be rewritten as

$$cm\frac{dT_s}{dt} = P(I, T_s) - F(U, T_s), \tag{3.4}$$

where U is the component of the velocity vector normal to the sensor. In Section 3.2.2 we discuss how to account for the fact that the velocity vector is directed at random angles to the sensor axis in turbulent flow and that its component tangential to the sensor has a secondary cooling effect.

Equation (3.4) indicates that the variation of the sensor temperature, and thus its resistance, depends on (1) the sensor temperature itself, (2) the current heating the sensor, and (3) the local flow velocity cooling the sensor. The sensor temperature and current, which are the independent variables in $P(I, T_s)$, can also, of course, be related to the voltage drop across the sensor through $E \equiv IR_s$ and (3.1). For (3.4) to be useful, it is necessary to obtain an explicit expression for $F(U, T_s)$, and the system must be arranged so that *either* the sensor current I *or* the sensor temperature T_s is constrained not to vary. This is necessary so that the velocity variation can be related to only one of these two variables. These two modes of anemometry operation, constant current and constant temperature, are discussed in more detail in Sections 3.2.4 and 3.2.5. Furthermore, as discussed in Section 3.2.4, if the system can be arranged so that the thermal energy balance of the sensor described by (3.4) is much more affected by temperature changes of the flow ΔT_f than by changes in the velocity ΔU, the probe becomes a thermometer that can be used to measure local, instantaneous temperature fluctuations.

Limiting the discussion now to hot wires, q_s in (3.3) is zero. Additionally, if the aspect ratio of the wire $\ell/d \geq 200$, where ℓ is the wire length and d is its diameter, studies show that the conduction to the prong supports q_p is quite small. The small residual heat transfer between the prongs and the fluid can be accounted for in calibration when the prong diameters are large compared to those of the sensors. The aspect ratio requirement must be balanced, of course, with the competing requirements that the sensors be able to resolve the small scales of the turbulence and do not disturb the flow too much. Calculation of the radiative heat transfer rate for most hot-wire applications shows that $q_r/q_c << 0.001$, so q_r also can usually be neglected.

Thus, for hot-wire sensors, the heat transfer for common applications is due almost entirely to convection. Convective heat transfer can be further subdivided into free and forced convection. The former is due to fluid motion induced by buoyancy forces that arise from the heating of the fluid by the hot wire and its consequent change of density. To maximize sensitivity to the flow velocity, it is desirable that free convection be negligibly small. Moreover, in turbulent flow, the respective contributions of free and forced convection will vary unpredictably, making it difficult to establish a connection between cooling and velocity. Generally, free convection is small [12] if the Reynolds number (based on the velocity cooling the sensor and the sensor diameter) satisfies

$$R_{e_d} > 2G_r^{1/3}, \tag{3.5}$$

where the Grashof number, $G_r \equiv gd^3\beta(T_w - T_f)/\nu^2$, characterizes buoyancy effects, β is the coefficient of volumetric thermal expansion of the gas, T_w is the sensor wire temperature, and v is the kinematic viscosity of the fluid. After rearranging (3.5), the criterion for neglecting free convection can be stated as

$$\frac{U}{2}\left[\frac{1}{gv\beta(T_w - T_f)}\right]^{1/3} > 1. \tag{3.6}$$

Thus, for low flow speeds cooling the sensor, one must have small temperature differences $T_w - T_f$ in order to keep free convection small. For typical hot-wire operating conditions in air, free convection can be neglected for local flow speeds above about 30 cm/s.

3.2.1 Forced Convection Hot-Wire Cooling

If all the restrictions on the application of (3.4) to particular experimental conditions mentioned previously are met, the cooling of the sensor wire is by forced convection heat transfer, with a rate given by

$$q_c = hS(T_w - T_f), \tag{3.7}$$

where h is the convective heat transfer coefficient and $S = \pi d\ell$ is the surface area of the sensor wire of length ℓ and diameter d. Unfortunately, h is *not* a constant; it varies with many flow and sensor properties. Usually, hot-wire heat transfer is given in nondimensional form as a Nusselt number, which expresses the ratio of the convective heat transfer rate to the fluid from the wire to that by conduction alone. This is expressed as

$$N_u = \frac{hd}{k_f}, \tag{3.8}$$

where k_f is the thermal conductivity of the fluid. N_u indicates the difference in the heat transfer rate to the fluid when the fluid is moving compared to what it would be if the fluid were stationary relative to the sensor.

The object now is to relate N_u to the velocity of the fluid. In the most general case the functional form [14] expressing the dependency of N_u on the flow and sensor properties is

$$N_u = N_u\left(R_{e_d}, P_r, M_a, G_r, K_n, \frac{\ell}{d}, r_T, \gamma, \theta\right), \tag{3.9}$$

where, in addition to the parameters already defined, P_r is the Prandtl number, M_a is the Mach number, K_n is the Knudsen number, $r_T = (T_w - T_f)/T_f$ is the temperature overheat ratio, γ is the ratio of specific heats of the fluid, and θ is the angle between the normal to the axis of the sensor and the velocity vector at the midpoint of the sensor. Fortunately, the effects of M_a, G_r, K_n, and ℓ/d are negligible for appropriate choices of operating conditions and sensor dimensions. The Knudsen number does not play a role if the molecular mean free path is less than about 1.5% of the wire diameter, in which case the flow can be considered to be a continuum. Although typical hot wires are near the limit of this assumption, residual Knudsen number effects are accounted for in calibration. Mach number effects (compressibility) are negligible for $M_a \lesssim 0.3$. If the sensor diameter is large compared to the Kolmogorov length scale of the flow, the sensor roughness length scale also needs to be included in (3.9), but this is usually not the case. Finally, for the sake of this discussion, it is assumed that the flow is normal to the sensor (angle of attack $\theta = 90°$), and we take up the general case in Section 3.2.2.

With these restrictions, (3.9) reduces to

$$N_u = N_u(R_{e_d}, P_r, r_T) \tag{3.10}$$

for incompressible flows. For gas flows with nearly constant Prandtl number in the range 0.7 to 1.0, the Nusselt number can be expressed as

$$N_u = \left[M(r_T) + N(r_T) R_{e_d}^n \right] \left(1 + \frac{r_T}{2} \right)^m, \tag{3.11}$$

which is called *King's law of convective cooling* [28], formulated at the beginning of the twentieth century for convective heat transfer due to flow past an infinitely long cylinder. M and N are constants for a fixed overheat ratio r_T in a given gas. However, their values have been found [29] to depend on the Prandtl number as $M = 0.42 P_r^{0.26}$ and $N = 0.57 P_r^{0.33}$, where the fluid properties that make up the Prandtl number are determined at an average "film" temperature, $(T_w + T_f)/2$. Thus $M(r_T)$ and $N(r_T)$ have different values for fluids with different Prandtl numbers. In practice, they should be determined by calibration.

If the wire temperature is held constant and the fluid temperature changes, M and N change. M is particularly sensitive to changes in r_T. This has two practical consequences. First, if the temperature of the fluid changes with time, the convective heat transfer rate q_c, expressed by N_u, will not depend only on the velocity U (embedded in R_{e_d}). Additionally, even if the fluid temperature is virtually constant during an experiment, unless it is the same as that occurring during calibration, the constants M and N deviate from the calibrated values.

In addition to M and N, the exponent n also depends on the Prandtl number of the fluid. It was taken to be 0.5 in air by King, but subsequent investigations have placed it in the range $0.33 \leq n \leq 0.52$, depending on the speed of the flow. The exponent m of the term in parentheses on the far right of (3.11), called the *temperature loading factor*, is quite small and is often taken to be zero, consistent with the assumptions of King.

If it is assumed that $m = 0$ and (3.7) and (3.8) are substituted into (3.11), there results

$$q_c = \left(M' + N' U^n \right) (T_w - T_f), \tag{3.12}$$

where M' and N' are new constants that incorporate M and N and their products with other constants, and U is the component of the velocity vector normal to the sensor. From (3.12) it is apparent that the forced convective heat transfer rate of the hot wire depends explicitly on both the local flow velocity cooling the wire and the temperature difference between the wire and the fluid.

If the *current* of the sensor wire is constrained to be constant, its temperature (and resistance) changes will vary with the velocity fluctuations that cause the convective heat transfer. However, since $dT_s/dt \neq 0$, for this case, (3.4) indicates that a phase lag and an amplitude attenuation will occur because of the changing amount of internal energy stored in the sensor. This is the major drawback of the constant-current operating mode.

On the other hand, if the *temperature* of the wire is constrained to be constant and the fluid temperature changes are negligible, the heat transfer rate will depend on the local flow velocity alone. Moreover, since $dT_s/dt \approx 0$ in this case, (3.4) shows that the forced convection heat transfer rate (3.12) can be equated to the power supplied to the sensor wire, namely $P = E^2/R_w$, and the temperature difference can be related to the wire resistance difference using (3.1) to obtain

$$\frac{E^2}{R_w} = \left(M' + N'U^n\right)\frac{R_w - R_f}{\alpha R_f} \tag{3.13}$$

or

$$E^2 = A + BU^n, \tag{3.14}$$

where the new constants A and B are the products of M' and N' with $(R_w - R_f)/\alpha R_f$. This is the form in which King's law is usually given.

To account for the sensor wire cooling from the velocity component tangential to the sensor as well as to that from the normal component, an "effective" velocity U_e is usually assumed, so that (3.14) becomes

$$E^2 = A + BU_e^n. \tag{3.15}$$

More general functional forms relating E and U_e have also been used. In particular, the effective velocity U_e is frequently expressed as a function of a qth-order polynomial of the voltage E, as in

$$U_{e_j} = \sum_{m=1}^{q} A_{mj} E_j^{m-1}, \tag{3.16}$$

where A_{mj} are the polynomial coefficients for the jth sensor in a multisensor probe. In [33] a fourth-order polynomial in E_j is equated to $U_{e_j}^2$ to obtain

$$U_{e_j}^2 = A_{1j} + A_{2j}E_j + A_{3j}E_j^2 + A_{4j}E_j^3 + A_{5j}E_j^4, \tag{3.17}$$

where the coefficients A_{mj} are determined by calibration. Other functions, such as splines, have also been used for this purpose. It is even possible simply to create a table of ordered values of E, U_1, U_2, and U_3 to establish the relationship between the measured voltage and the velocity components, although this does not necessarily have to be single valued.

3.2.2 Effective Cooling Velocity

Until now the angle of attack of the velocity vector to the sensor and the directional sensitivity of the sensor have been ignored by assuming the velocity vector always to be normal to the sensor. For turbulent flows this is, of course, not true, because the local velocity vector is oriented randomly relative to the sensor, and this orientation changes rapidly with time. In addition to the velocity component normal to the sensor axis that cools it primarily, it is necessary to take into account the heat transfer due to the component tangential to the sensor axis, which is a secondary effect but still significant. In fact, it is customary to decompose the velocity vector into three components: (1) U_n normal to the sensor axis and in the plane of the prongs supporting it; (2) U_b normal to both the sensor axis and the plane of the prongs, called the binormal component; and (3) U_t tangential to the sensor axis. Thus the velocity vector $\mathbf{V}$ cooling the sensor can be written as

$$\mathbf{V} = U_n\mathbf{n} + U_b\mathbf{b} + U_t\mathbf{t}, \tag{3.18}$$

where $\mathbf{n}$, $\mathbf{b}$, and $\mathbf{t}$ are unit vectors in the normal, binormal, and tangential directions with respect to the sensor as shown in Fig. 3.2.

A widely used relation, known as *Jorgensen's law* [24], expresses the square of the "effective" velocity magnitude cooling the sensor as a weighted sum of the squares of the components in this decomposition, namely,

$$U_e^2 = U_n^2 + h^2U_b^2 + k^2U_t^2, \tag{3.19}$$

where the respective weighting factors for the three components are unity, h^2, and k^2. The value of h depends on, among other influences, the aerodynamic blockage of the flow by the prongs and is usually close to unity. The value of k depends on the aspect ratio of the sensor, ℓ/d; for $\ell/d \approx 200$, $k \approx 0.2$.

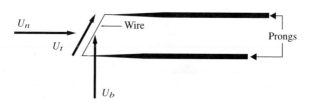

Fig. 3.2 *Orientations of normal (U_n), tangential (U_t), and binormal (U_b) velocity components with respect to a hot-wire or hot-film sensor.*

It is often more convenient to decompose $\mathbf{V}$ into components in a laboratory coordinate system:

$$\mathbf{V} = U_1\mathbf{i} + U_2\mathbf{j} + U_3\mathbf{k}, \tag{3.20}$$

with unit vectors $\mathbf{i}$, $\mathbf{j}$, and $\mathbf{k}$ in the streamwise, cross-stream, and spanwise (x_1, x_2, and x_3) directions, respectively. For an arbitrary orientation of $\mathbf{V}$, the two decompositions (3.18) and (3.20) can be related by

$$U_n = n_1 U_1 + n_2 U_2 + n_3 U_3, \tag{3.21}$$

$$U_b = b_1 U_1 + b_2 U_2 + b_3 U_3, \tag{3.22}$$

and

$$U_t = t_1 U_1 + t_2 U_2 + t_3 U_3, \tag{3.23}$$

where the coefficients n_i, b_i, and t_i ($i = 1, 2, 3$) can be written in terms of sines and cosines of the angles of inclination of the sensors to the laboratory coordinate system axes. Substituting U_n, U_b, and U_t from (3.21)–(3.23) into (3.19) yields, for the jth sensor of a multisensor probe array, a general expression for the effective cooling velocity in terms of the velocity components in the laboratory coordinate system:

$$U_{e_j}^2 = a_{1_j} U_{1_j}^2 + a_{2_j} U_{2_j}^2 + a_{3_j} U_{3_j}^2 + a_{4_j} U_{1_j} U_{2_j} + a_{5_j} U_{1_j} U_{3_j} + a_{6_j} U_{2_j} U_{3_j}, \tag{3.24}$$

where the coefficients a_{n_j} ($n = 1, \ldots, 6$) are products of the geometry coefficients, n_i, b_i, and t_i, in (3.21) through (3.23) with the weighting factors, h and k, in (3.19). In practice, the a_{n_j} coefficients in (3.24) are determined by calibration, as discussed in Section 3.2.6.

3.2.3 Multisensor Probes

3.2.3.1 *Velocity Component Measurements* To obtain more than one component of velocity simultaneously using hot wires or hot films requires more than one sensor. For example, a two-sensor array is frequently used to obtain the two components of velocity in a plane parallel to the planes of the two pairs of supporting prongs. The two sensors are usually arranged in an X- or V-array configuration, with the sensors typically inclined at 45° to the streamwise mean flow, as shown in Fig. 3.3. The orientation of two-sensor probes is usually so that the sensors are in the x_1–x_2 or x_1–x_3 planes in the laboratory coordinate system. In this case either the (U_1, U_2) or (U_1, U_3) velocity component pairs are measured simultaneously. For the sake of this discussion, assume that the sensors of such a probe lie in the x_1–x_2 plane. Because there are only two sensors, the cooling effect of the binormal component of velocity, U_b, which is identical to U_3 in this orientation, must be neglected in order to have a determinate set of equations with which to solve for U_1 and U_2. Neglecting U_b and assuming that (3.17) applies, then (3.24) reduces to the pair of equations

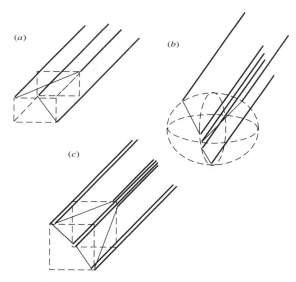

Fig. 3.3 *Sketches of (a) two-, (c) three-, and (c) four-sensor hot-wire probes.*

$$a_{1_j} U_{1_j}^2 + a_{2_j} U_{2_j}^2 + a_{4_j} U_{1_j} U_{2_j} = A_{1_j} + A_{2_j} E_j + A_{3_j} E_j^2 + A_{4_j} E_j^3 + A_{5_j} E_j^4 \qquad (3.25)$$

($j = 1, 2$) in the two unknown velocity components U_1 and U_2.

In addition to the necessity of neglecting binormal cooling, it is also clear from (3.25) that it must be assumed that the values of the velocity components, U_1 and U_2, which are considered to be cooling the sensors, are the same at each sensor. That is, the spatial variation of the instantaneous velocity field must be neglected over the distance between the sensors. The errors in the measured cross-stream velocity components and the Reynolds shear stress incurred when using X-array probes that result from both neglecting binormal cooling and the spatial variation of the velocity field can be substantial [37].

To include the effects of binormal cooling, at least a third sensor must be incorporated into the sensor array of the probe. Probes with three sensors in circular orientation, 120° apart, and with four sensors in orthogonal X- and V-array pairs, as sketched in Fig. 3.3, are typically used. With such probes, all three velocity components, U_1, U_2, and U_3, can, in principle, be measured. However, even for these probes, the effects of the spatial variation of the velocity field over the probe measuring region must be neglected. The errors resulting from this assumption for a four-sensor array have been examined [37].

3.2.3.2 Simultaneous Velocity Vector and Velocity Gradient Tensor Measurements
To account for the spatial variation of the velocity field in the hot-wire or hot-film cooling equations, at least to first order, a minimum of nine sensors must be used. Such a probe was built, tested, and used in a turbulent boundary layer

[5,50]. Later, an improved design [21] was used to make extensive boundary layer measurements. Subsequently, probes with three symmetric arrays of four sensors each ($j = 1, \ldots, 12$) have been developed [45,49,52]. A photograph of a 12-sensor probe is shown in Fig. 3.4, where the 24 prongs supporting the 12 sensors are contained within a small measuring region of about 2.5 mm diameter in the cross-stream plane.

Because of the spatial variation of the velocity field, for any multisensor probe the velocity components U_{1_j}, U_{2_j}, and U_{3_j} that cool each of the j sensors are different. To first order in the dimension of the probe, the differences in velocity depend on the local gradients. Assuming that the latter are constant within the probe measuring region at a given instant, the velocity components occurring at the midpoint of each sensor can be estimated from a first-order Taylor series expansion. Taking this to be about the probe centroid lying in the cross-stream plane, which is perpendicular to the probe axis and passes through the centers of each sensor, gives

$$U_{1_j} = U_{1_o} + C_j \frac{\partial U_1}{\partial y} + D_j \frac{\partial U_1}{\partial z}, \tag{3.26}$$

$$U_{2_j} = U_{2_o} + C_j \frac{\partial U_2}{\partial y} + D_j \frac{\partial U_2}{\partial z}, \tag{3.27}$$

and

$$U_{3_j} = U_{3_o} + C_j \frac{\partial U_3}{\partial y} + D_j \frac{\partial U_3}{\partial z}, \tag{3.28}$$

where U_{1_o}, U_{2_o}, and U_{3_o} are the velocity components at the probe centroid. The coefficients C_j and D_j ($j = 1, \ldots, 12$) represent the x_2 and x_3 displacements from the centroid, which must be measured accurately. Substituting (3.26)–(3.28) into (3.25), neglecting gradient product terms, and rearranging, yields 12 nonlinear algebraic equations for nine unknowns: the three velocity components U_{1_o}, U_{2_o}, and U_{3_o} and the

Fig. 3.4   *Twelve-sensor hot-wire probe to measure velocity and velocity gradient components.*

six velocity gradients in the cross-stream plane, $\partial U_1/\partial y$, $\partial U_2/\partial y$, $\partial U_3/\partial y$, $\partial U_1/\partial z$, $\partial U_2/\partial z$, and $\partial U_3/\partial z$. The 12 equations can be expressed as

$$
\begin{aligned}
f_j \equiv & -P_j + U_{1_o}^2 + 2C_j U_{1_o} \frac{\partial U_1}{\partial y} + 2D_j U_{1_o} \frac{\partial U_1}{\partial z} \\
& - k_{2j} \left[U_{2_o}^2 + 2C_j U_{2_o} \frac{\partial U_2}{\partial y} + 2D_j U_{2_o} \frac{\partial U_2}{\partial z} \right] \\
& - k_{3j} \left[U_{3_o}^2 + 2C_j U_{3_o} \frac{\partial U_3}{\partial y} + 2D_j U_{3_o} \frac{\partial U_3}{\partial z} \right] \\
& - k_{4j} \left[U_{1_o} U_{2_o} + C_j \left(U_{1_o} \frac{\partial U_2}{\partial y} + U_{2_o} \frac{\partial U_1}{\partial y} \right) + D_j \left(U_{1_o} \frac{\partial U_2}{\partial z} + U_{2_o} \frac{\partial U_1}{\partial z} \right) \right] \\
& - k_{5j} \left[U_{1_o} U_{3_o} + C_j \left(U_{1_o} \frac{\partial U_3}{\partial y} + U_{3_o} \frac{\partial U_1}{\partial y} \right) + D_j \left(U_{1_o} \frac{\partial U_3}{\partial z} + U_{3_o} \frac{\partial U_1}{\partial z} \right) \right] \\
& - k_{6j} \left[U_{2_o} U_{3_o} + C_j \left(U_{2_o} \frac{\partial U_3}{\partial y} + U_{3_o} \frac{\partial U_2}{\partial y} \right) + D_j \left(U_{2_o} \frac{\partial U_3}{\partial z} + U_{3_o} \frac{\partial U_2}{\partial z} \right) \right] \\
= & \ 0,
\end{aligned}
\tag{3.29}
$$

where $k_{nj} = a_{nj}/a_{1j}$, $n = 2, \ldots, 6$, $j = 1, \ldots, 12$, and P_j is the right-hand side of (3.25). The system of equations (3.29) can be solved at each time step by minimizing the error function given by $\sum f_j^2$ using Newton's method.

The streamwise gradients that do not appear in (3.29) must be determined by transforming time derivatives into derivatives in the x direction using Taylor's hypothesis, which is discussed in the next section. The conclusion is reached that with a 12-sensor probe not only are the velocity components determined, but this is done such that the effects of flow nonuniformity to first order over the sensing area of the probe are taken into account. Such probes also give a reasonably good estimate of the terms in the velocity gradient tensor. From this, important characteristics of turbulent flows, such as vorticity, strain rate, and turbulent kinetic energy dissipation rate, can be determined. Note that other means exist for solving the coupled system of equations appearing for 9- and 12-array probes [21,45,48,49].

3.2.3.3 *Taylor's Frozen Turbulence Hypothesis*   To provide a basis for the measurement of streamwise derivatives of flow variables, Taylor [44] proposed that for short time intervals, turbulence can be assumed to be "frozen" as it convects past a probe at a fixed point in space. With this assumption, changes in the measured velocity components as a function of time can be viewed as proportional to their respective spatial changes. Mathematically, this can be expressed as

$$
\frac{dU_i}{dx} = -\frac{1}{U_c} \frac{dU_i}{dt},
\tag{3.30}
$$

where U_c is the convection speed of the frozen turbulence, usually taken to be the local mean velocity, although the presence of solid boundaries has a significant effect on the best choice for this parameter [25].

Examination of (3.30) makes it clear that Taylor's hypothesis is similar to equating the acceleration of a fluid particle to zero at the instant it reaches the probe sensor, that is,

$$\frac{\partial U_i}{\partial t} + U_1 \frac{\partial U_i}{\partial x} + U_2 \frac{\partial U_i}{\partial y} + U_3 \frac{\partial U_i}{\partial z} = 0. \tag{3.31}$$

Rearranging (3.31) yields

$$\frac{\partial U_i}{\partial x} = -\frac{1}{U_1} \left(\frac{\partial U_i}{\partial t} + U_2 \frac{\partial U_i}{\partial y} + U_3 \frac{\partial U_i}{\partial z} \right). \tag{3.32}$$

Setting $U_c = U_1$ and neglecting the last two terms on the right-hand side, (3.32) reduces to (3.30), so it can be considered to be a refined form of Taylor's hypothesis. This suggests that the underlying assumption of Taylor's hypothesis is that the particles of fluid passing the probe are not accelerating (i.e., the pressure and viscous forces offset each other). This is a radical assumption that is unlikely to hold true for much of the time. Nevertheless, sufficiently far away from solid boundaries so that the viscous forces and the turbulence intensities are not too high, Taylor's hypothesis does lead, on average, to good approximations of the streamwise derivatives of the velocity components. For example, good magnitude and phase correspondence between the dU_i/dx gradients and $-1/\overline{U}_1(dU_i/dt)$ in a turbulent channel flow occurs in and above the buffer layer [40]. Somewhat surprisingly, including the additional terms in (3.32) compared to (3.30) does not significantly improve the accuracy of the dU_i/dx determination [31]. Figure 3.5 shows a time-series measurement taken with a 12-sensor probe in a turbulent mixing layer. The signals shown in the figure correspond to dU_1/dx from (3.30), from (3.32), and from the continuity equation, $dU_1/dx = -(dU_2/dy + dU_3/dz)$. Clearly, the signal from the simple Taylor's hypothesis and from its refined form are very similar. They are also similar to the signal determined from the continuity equation much of the time, with large deviations occurring only over short intervals.

Besides enabling the estimation of streamwise derivatives, Taylor's hypothesis may be used to transform one-dimensional frequency spectra to streamwise wavenumber spectra. Such an approximation of the streamwise wavenumber is given by

$$k_x \approx \frac{2\pi f}{U_1}, \tag{3.33}$$

where f is the frequency of the fluctuations. Applications of this idea appear below.

3.2.4 Constant-Current Operation

Before the availability of high-gain feedback amplifiers, anemometers were operated in the constant-current mode, which typically incorporates a circuit such as that

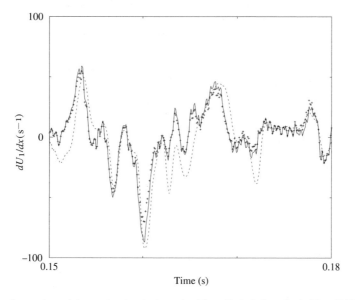

Fig. 3.5 *Comparison of time-series signals determined from Taylor's hypothesis [Eqs. (3.30) —— and (3.32) +++] and from the continuity equation (· · ·) using mixing-layer data from a 12-sensor probe. (From [31].)*

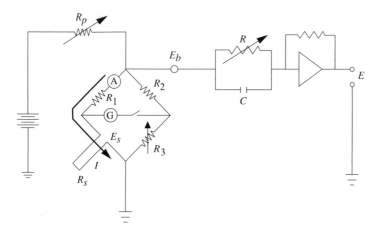

Fig. 3.6 *Constant-current hot-wire anemometer circuit. (Adapted from [38].)*

shown in Fig. 3.6. In this, current is supplied to the Wheatstone bridge by the power source, while the current amplitude is set and held to be virtually constant by the variable-current control potentiometer R_p, whose resistance is large compared to that of the sensor. Because the current is constant, by adjusting the potentiometer R_3 on one side of the Wheatstone bridge, more or less current is forced through the other side, where the sensor R_s is one of the resistors. This is done by adjusting the potentiometer R_3 until the bridge ratio relationship

$$\frac{R_s}{R_1} = \frac{R_3}{R_2} \qquad (3.34)$$

is satisfied, as indicated by a galvanometer, G. Here R_s denotes the resistance of the probe, prongs, and lead wires. At this point the nominal operating temperature of the sensor is set as is its overheat ratio, which is the ratio of the nominal operating temperature (or resistance) to its temperature (or resistance) at ambient conditions. When exposed to the turbulence, the sensor is cooled by varying degrees as the velocity fluctuates thus changing its temperature and resistance. The current through the sensor can be measured by an ammeter, A, in that leg of the bridge.

The changes in resistance of the sensor also causes the voltage E_b from the top of the bridge to ground to change with velocity. After being amplified, E_b becomes the output signal of the circuit. Amplification is necessary because the mean voltage signal from the bridge is about 50 mV, while its fluctuations are an order of magnitude less than that for the usual laboratory operating conditions.

As a result of the thermal inertia of the sensor, changes in sensor temperature (and resistance) in response to the fluid velocity are attenuated and have a phase lag, both of which increase with increasing frequency of the velocity fluctuations. The extent of the attenuation and phase lag can be estimated by determining the linearized response of the sensor to a small sinusoidal oscillation of the Nusselt number with an angular frequency ω and constant amplitude $\mathcal{A}$ (see [6] for a more complete development). The change in the resistance of the sensor to this periodic perturbation is

$$\Delta R_s = \frac{\mathcal{A}}{1 + \omega^2 \mathcal{M}^2} (\cos \omega t + M\omega \sin \omega t), \qquad (3.35)$$

where M is a time constant of the sensor, depending on its geometry, material properties, and operating conditions. Equation (3.35) shows that the amplitude is attenuated by a factor of $1/(1 + \omega^2 M^2)$, and the phase is shifted by $M\omega \sin \omega t$. For a typical tungsten wire sensor of 5μm diameter and 1.25 mm length under normal operating conditions, $M \approx 0.6$ ms. For $\omega << 1/M$ the attenuation and phase lag become very small. The response attenuation and phase lag at higher frequencies can be compensated for by placing a simple circuit before the amplifier, such as that shown in Fig. 3.6, whose nonlinear response characteristics can be set to be the inverse of that of the sensor. Unfortunately, the response characteristics of the sensor are not constant but change with the flow speed and overheat ratio because the time constant M varies. Thus the capacitor in the compensation circuit must be changed accordingly. This is a major drawback of the constant-current mode of operation.

The constant-current mode is seldom used today except for measurements of temperature fluctuations. For this purpose the sensor is operated at nearly zero temperature overheat and thus is known as a *cold sensor*. When measuring temperature the current-control potentiometer is set to provide very low currents through the bridge. The sensor is heated very little and is quite insensitive to the changing flow velocity. However, it still is relatively sensitive to temperature fluctuations in the fluid.

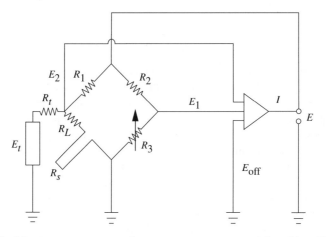

Fig. 3.7   *Constant-temperature hot-wire anemometer circuit. (Adapted from [8].)*

3.2.5   Constant-Temperature Operation

The first constant-temperature (or resistance) hot-wire circuits were tried in the 1940s [8]. These incorporate a circuit of the generic form indicated in Fig. 3.7. In this, the sensor is also one of the resistances of a Wheatstone bridge. The idea here is to have the differential amplifier sense any imbalance of the bridge voltage, $E_1 - E_2 \neq 0$, due to the resistance change of the sensor caused by the flow. The small voltage difference associated with the imbalance is amplified and a transistor feeds back the current through the bridge necessary to bring the circuit into balance. If this is done rapidly enough so that the bridge remains always nearly in balance, the temperature and resistance of the sensor remain virtually constant. Amplifiers used in these circuits have high gains (i.e., they respond to very small voltage differences) and high-frequency responses. Typical gains are about 1000 and frequency responses are in the range 10 to 15 kHz, in contrast to only a few hundred hertz for constant-current anemometers.

The constant-temperature circuit is operated by first finding the resistance of the sensor at the temperature of the fluid. This is done by running the circuit in an open-loop configuration and forcing a small externally supplied current through the bridge. This heats the sensor, bringing it out of balance and causing the amplifier to produce a voltage. This voltage, with no additional offset, is proportional to the unbalanced bridge voltage difference. By adjusting the potentiometer R_3, the bridge is brought back into balance and the sensor resistance R_s is then calculated easily from (3.34).

Now the loop is closed and an offset voltage is added to the amplifier output voltage, which is zero at this point in the operation. The offset voltage, aided by a transistor, supplies a mean current, typically 5 to 50 mA, to the bridge and raises the sensor temperature, which can then be adjusted to any desired overheat simply by varying the potentiometer. To obtain the current levels required, the resistances on the right side of the bridge, R_2 and R_3, are usually much larger than those on the left side. The bridge ratio is specified as R_2/R_1 and varies between 1 and 50, with a usual value of 10. The higher the overheat, the greater the sensitivity of the sensor to velocity changes. However, if the sensor wire is heated too much, it can oxidize or even burn out. Furthermore, free convection caused by the heated sensor near walls and thermal crosstalk between sensors of a multisensor probe are greater at higher overheat values. Therefore, the overheat ratio must be chosen judiciously based on the specific application. Overheat values tend to range from a low of 1.1 to a high of 2.0.

When the sensor is heated to its desired operating temperature, the differential amplifier again balances the bridge. After placing the probe in a turbulent flow, the amplifier adjusts the voltage at the top of the bridge E automatically to bring the bridge back into balance as the velocity changes. The ability of the circuit to follow velocity changes depends on the characteristics of the amplifier and the sensor.

As discussed in Section 3.2.1, the dependence of the voltage drop across the sensor on the effective cooling velocity can be expressed by King's law (3.15) or some other curve fit, such as a polynomial [e.g., (3.16) and (3.17)]. The voltage at the top of the bridge easily is seen to be proportional to the voltage drop across the sensor, so the former can replace the latter in the response functions. This is the fluctuating voltage that is measured in an experiment as it responds to the turbulent velocity fluctuations.

3.2.6 Calibration

The coefficients in (3.15) and (3.29) are functions of the fluid temperature, the properties of the electronic circuits, and the material properties of the sensors. Since these aspects may change with time, it is standard practice to calibrate hot-wire and hot-film sensors just before and just after data acquisition in a turbulent flow experiment. This provides current values of the coefficients and ensures that they have not changed significantly during the course of the experiment. Static calibration of a single sensor probe is straightforward: the sensor is exposed to a uniform flow with fixed U and at the same ambient temperature as that of the turbulent flow to be investigated. The speed of the calibration flow is varied over the full range of values expected in the experiment, and the bridge voltage is measured at each speed. If King's law is being used, for example, then E^2 can be plotted against U^n and a least-squares error fit used to determine the values of A and B with an iteration of n. If a polynomial is used instead of King's law, the polynomial coefficients can also be found.

For multisensor probes with sensors inclined to the mean flow, in addition to varying the speed of the calibration flow, the angle of attack of the probe with respect to the flow must be varied through the full range expected in the experiment. This can be achieved by systematically varying the orientation of the probe to the flow, or vice versa.

Static calibration does not provide information about the frequency response of the probe to a rapidly varying turbulent flow. Determining this directly by varying the flow speed or moving the probe at high frequencies is usually not possible, so an alternative is necessary. One way to do this is by simulating a velocity field perturbation artificially and then observing the response of the anemometer circuit. Often, a square-wave test signal with voltage E_t is added to one side of the bridge to bring it suddenly out of balance, as illustrated in Fig. 3.7. The voltage response E at the top of the bridge is then monitored on an oscilloscope, and the response optimized until the maximum flat frequency response is obtained. A flat frequency response is desired because it assures that the relationship between the measured output and input amplitudes is independent of frequency.

3.3 LASER-DOPPLER VELOCIMETRY

Laser-Doppler velocimetry (LDV) is an optical method for flow measurement utilizing the Doppler principle that coherent (laser) light reflected from a moving particle exhibits a frequency relative to a fixed observer that depends, not only on the known laser light source wavelength, but also on the velocity of the moving particle. Thus, if one can measure in a turbulent flow the frequency of the light scattered from tiny particles chosen to have sufficiently small diameter and appropriate density so that they follow the fluid motion faithfully, the local fluid velocity can be determined. Unlike hot-wire or hot-film anemometry, LDV is a nonintrusive method except for the presence of the seed particles. Although LDV is a relatively recent measurement technique [54], it has rapidly become a widely used alternative to hot-wire and hot-film anemometry for obtaining quantitative turbulent velocity data, particularly, but not exclusively, in liquid flows.

3.3.1 Reference Beam Method

To understand the basic principles of operation of LDV, consider Fig. 3.8a. This shows a light-reflecting particle, with velocity, $\mathbf{V}$, passing through a beam of coherent light of frequency, f_s, with some of the reflected light being received by a stationary photodetector. The angle between the axis of the laser beam and the direction of the particle's motion is α, while β is that between the line of sight from the photodetector to the particle and the direction the particle is moving. $\mathbf{i_s}$ and $\mathbf{i_r}$ are, respectively, unit vectors along the laser beam (source) axis and the photodetector (receiver) line of sight to the particle. With this configuration, the frequency of the light scattered from the moving particle is

$$f_p = f_s \left(1 - \frac{\mathbf{i_s} \cdot \mathbf{V}}{c} \right), \tag{3.36}$$

where c is the speed of light, $V_s \equiv \mathbf{i_s} \cdot \mathbf{V} = V \cos \alpha$ is the component of the particle velocity in the direction of the laser beam axis, and $V = |\mathbf{V}|$. f_p is seen to be made up of two parts: the original illuminating laser beam frequency, f_s, and the Doppler shift frequency due to the component of the particle motion along the laser beam

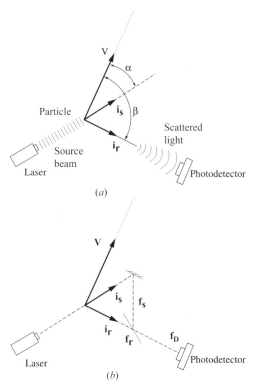

Fig. 3.8 *(a) Light received by a photodetector scattered from a particle passing through a laser beam, and (b) LDV reference beam method optical arrangement. (Adapted from [16]. Reprinted with permission of Teubner.)*

axis, $f_s(\mathbf{i_s} \cdot \mathbf{V})/c$. Consequently, if the particle is moving directly along the axis of the laser, the frequency of the light scattered from the particle is $f_p = f_s(1 \pm V/c)$, depending on whether the particle is moving toward $(+)$ or away from $(-)$ the laser. Far enough away from a particle, the reflected light can be considered to move as a spherical wave with frequency f_p. Therefore, the frequency observed at the stationary photodetector is

$$f_r = f_p \left(1 + \frac{\mathbf{i_r} \cdot \mathbf{V}}{c} \right). \tag{3.37}$$

Thus, if the particle is moving directly along the line of sight from the photodetector to the particle, the reflected light frequency observed at the photodetector is $f_r = f_p(1 \pm V/c)$, depending on whether the particle is moving toward $(+)$ or away from $(-)$ it. On the other hand, if the particle is moving perpendicular to the line of sight, $f_r = f_p$.

When (3.36) is substituted into (3.37), it is found that

$$f_r = f_s \left[1 + \frac{\mathbf{V} \cdot (\mathbf{i_r} - \mathbf{i_s})}{c} \right]$$

$$= f_s + \frac{\mathbf{V} \cdot (\mathbf{i_r} - \mathbf{i_s})}{\lambda_s},$$

(3.38)

where λ_s is the wavelength of the laser beam. According to (3.38), the frequency observed at the photodetector is made up of three parts: the original illuminating laser beam frequency, f_s, the Doppler shift frequency due to the component of the particle motion along the laser beam axis, $(\mathbf{V} \cdot \mathbf{i_s})/\lambda_s$, and the frequency due to the component of particle velocity along the line of sight from the photodetector to the particle, $(\mathbf{V} \cdot \mathbf{i_r})/\lambda_s$. The Doppler shifted frequency, f_r, has two desirable features: its magnitude depends *linearly* on the magnitude of the component of the particle velocity in the $\mathbf{i_r} - \mathbf{i_s}$ direction, and it also depends on the *sign* of the component in this direction. If f_r can be measured, any velocity component desired can be measured simply by specifying the direction of $\mathbf{i_r} - \mathbf{i_s}$ through the choice of optical axes. Furthermore, the sign of the velocity can be found by determining whether f_r is greater or less than f_s.

Unfortunately, photodetectors cannot directly measure the Doppler-shifted frequency they receive because f_r is normally outside the range of their frequency response. This problem can be circumvented, however, by a technique called *optical mixing* (also called *optical heterodyne detection*), whereby f_s is subtracted from f_r to obtain the total Doppler shift frequency alone:

$$f_D \equiv f_r - f_s = \frac{\mathbf{V} \cdot (\mathbf{i_r} - \mathbf{i_s})}{\lambda_s}.$$

(3.39)

This much smaller frequency, f_D, can be measured by a photodetector. As illustrated in Fig. 3.8b, with the help of a beamsplitter and filter, the subtraction can be accomplished by bringing the unshifted original laser light and the Doppler-shifted reflected light onto the surface of the photodetector, which then adds them and squares their sum. All product terms resulting from squaring have frequencies on the order of f_s (i.e., typically about 10^{14} Hz), which are higher than those to which the photodetector can respond except the term at the Doppler frequency, f_D, which is on the order of 10^6 Hz (for example, when a He–Ne laser $f_s = 4.74 \times 10^{14}$ Hz is used to illuminate particles with speeds of about 3 m/s). The output signal from the photodetector oscillates with this frequency, which can be measured [2].

As mentioned, for preset orientations of the laser beam axis and the line of sight of the photodetector and a chosen laser wavelength λ_s, the velocity component of the particle in the $\mathbf{i_r} - \mathbf{i_s}$ direction can be determined. The major drawback in this reference beam LDV method is that it requires a very small photodetector aperture in order to avoid mixing light received from different directions and thus with different Doppler frequency shifts. Such mixing will occur if a lens is used to focus the light from different directions on a larger photodetector surface. On the other hand, for a small aperture or, equivalently, if a mask is placed in front of the lens to exclude reflected light that is not nearly parallel to the line of sight of the photodetector, too little light is received and, therefore, the output signal from the photodetector is weak.

A variation, called the *dual scatter method*, is much like the reference beam method in its optical arrangement. The difference is that in this approach the light arriving on the photodetector surface from scattering in two directions is combined rather than having the scattered light combine with the reference beam. This method also requires a small aperture, and thus it and the reference beam method are rarely used in practice. It is clear, however, from the foregoing that unlike HWA, LDV has the very substantial advantage of not requiring calibration of the measurement system against a known flow.

3.3.2 Dual-Beam Method

In this configuration, the beam from a laser with wavelength λ_s is split into two separate beams, oriented along different axes, that are subsequently brought together to cross at the measurement point, as shown in Fig. 3.9a. The unit vectors in the beam directions are $\mathbf{i}_{s_1}$ and $\mathbf{i}_{s_2}$, and the unit vector along the line of sight of the photodetector is, as before, denoted by $\mathbf{i_r}$. The Doppler-shifted frequencies of the light scattered off a particle moving through the measurement point from each of the beams are

$$f_{p_1} = f_s \left(1 - \frac{\mathbf{i}_{s_1} \cdot \mathbf{V}}{c} \right) \tag{3.40}$$

and

$$f_{p_2} = f_s \left(1 - \frac{\mathbf{i}_{s_2} \cdot \mathbf{V}}{c} \right). \tag{3.41}$$

The two frequencies observed at the photodetector from the Doppler-shifted light scattered from the particle are

$$f_{r_1} = f_{p_1} \left(1 + \frac{\mathbf{i_r} \cdot \mathbf{V}}{c} \right) \tag{3.42}$$

and

$$f_{r_2} = f_{p_2} \left(1 + \frac{\mathbf{i_r} \cdot \mathbf{V}}{c} \right). \tag{3.43}$$

Since the two laser beams are oriented in different directions relative to the particle motion, the scattered light frequencies f_{p_1} and f_{p_2} are different, as are the frequencies f_{r_1} and f_{r_2} in (3.42) and (3.43). As with the reference beam and dual scatter methods, when the two beams are brought to the surface of the photomultiplier, they are added and squared. The terms resulting from the squaring have frequencies

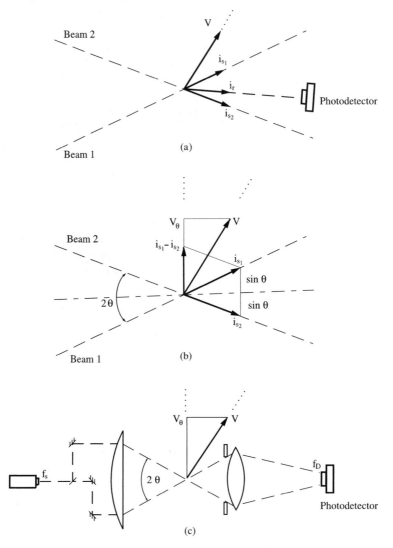

Fig. 3.9 *(a) Light received by a photodetector scattered from a particle passing through a measurement point where two laser beams cross, (b) geometry of this optical arrangement, and (c) LDV dual-beam-method optical arrangement. (Adapted from [16]. Reprinted with permission of Teubner.)*

above the response capabilities of the photodetector except for the Doppler shift frequency:

$$f_D \equiv f_{r_1} - f_{r_2} = f_s \left(1 + \frac{\mathbf{i_r} \cdot \mathbf{V}}{c}\right)\left(1 - \frac{\mathbf{i_{s_1}} \cdot \mathbf{V}}{c} - 1 + \frac{\mathbf{i_{s_2}} \cdot \mathbf{V}}{c}\right). \qquad (3.44)$$

If (3.44) is expanded and the quadratic terms are neglected, it reduces to

$$f_D = \frac{f_s}{c} \mathbf{V} \cdot (\mathbf{i}_{s_2} - \mathbf{i}_{s_1})$$

$$= \frac{\mathbf{V} \cdot (\mathbf{i}_{s_2} - \mathbf{i}_{s_1})}{\lambda_s} \tag{3.45}$$

or

$$f_D = \frac{2V_\theta \sin\theta}{\lambda_s}, \tag{3.46}$$

where 2θ is the angle between the axes of the two laser beams, as seen in Fig. 3.9b, and V_θ is the component of the particle velocity in the direction $i_{s_2} - i_{s_1}$ (i.e., perpendicular to the bisector of the angle 2θ). From (3.45) it is clear that the Doppler frequency measured by the dual beam method does not depend on the orientation of the line of sight of the photodetector. Rather, it depends only on the orientation axes of the laser beams, the velocity of the light-scattering particle, and the wavelength of the laser. From the sketch of the dual-beam optical configuration in Fig. 3.9c, its advantage becomes clear compared to the reference beam method illustrated in Fig. 3.8b. In particular, all the light rays scattered by a particle that can be focused by a lens on the photodetector surface contribute to the intensity of the signal. Thus a larger photodetector aperture can be used, and the output signal intensity is much greater than that of the reference beam and dual scatter optical configurations. For this reason, the dual-beam method is almost always preferred over the other methods. A dual-beam system is classified as *forward scattering* if the light collection lens and photodetector is forward of the source beams and particle, as in Fig. 3.9c. If the lens collects light scattered backward from the particle and the photodetector is off the optical axis, it is a *backscattering* system. The latter arrangement can often reduce the background light seen by the photodetector.

Whether or not a Doppler burst can be detected above a particular threshold level depends on the power of the laser, the size and scattering properties of the particle, where it passes through the nominal measurement volume, the light collection optics, and the type and quality of the photodetector.

3.3.2.1 *Sign of Velocity Component*   The two beams used in the dual-beam method are generally obtained by splitting the laser light into two parts with frequencies $f_{s_1} = f_{s_2} = f_s$. A sign change in the particle velocity component in the $\mathbf{i}_{s_2} - \mathbf{i}_{s_1}$ direction produces a $180°$ phase change in the photodetector output signal, but this is difficult to discriminate. To overcome this problem in those flows for which the velocity components of interest may change sign, it is the usual practice to shift the frequency of one of the beams, say f_{s_2}, by a constant frequency f_o. This is done so that fluctuations about the zero level of the measured velocity component are transduced into frequency fluctuations about the base frequency f_o, as illustrated in Fig. 3.10. In this case the output of the photodetector is a sine wave with frequency

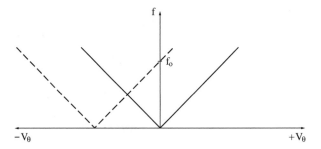

Fig. 3.10 *Frequency shift of one of the two beams in the LDV dual-beam method to enable determination of the sign of the velocity component.*

$$f_{pd} = f_o + \frac{\mathbf{V} \cdot (\mathbf{i}_{s_2} - \mathbf{i}_{s_1})}{\lambda_s}. \tag{3.47}$$

Whether f_{pd} is greater or less than f_o determines the sign of the velocity component. The frequency shift f_o is achieved using rotating diffraction gratings or with Pockels, Kerr, or as is most often the case, Bragg cells.

3.3.2.2 Spatial Resolution of Measurement Volume

The twin laser beams of the dual-beam method are focused on the point in the flow where the velocity component is to be measured. The two beams are nearly cylindrical at the measurement

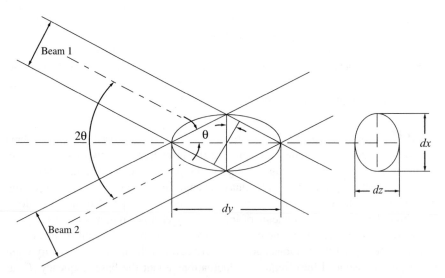

Fig. 3.11 *Measurement volume of two intersecting laser beams in a LDV dual beam system. (Adapted from [16]. Reprinted with permission of Teubner.)*

location, so their intersection forms an ellipsoid, as illustrated in Fig. 3.11. The dimensions of the ellipsoid in a typical experiment (i.e., one that has not been specially designed for high spatial resolution) are $dx = dz \approx 0.1$ mm and $dy \approx 0.8$ mm. With effort, these dimensions can be reduced significantly. For example, a measuring volume ellipsoid with a 35 μm diameter and 66 μm length has been achieved [13].

The ellipsoid formed by the intersection of the laser beams defines the nominal measurement volume. For certain types of photodetector output signal processors, such as spectrum analyzers and correlators, or when the signal is due to the presence of many particles in the measurement volume simultaneously, the size of the nominal volume is a good measure of the spatial resolution. However, if the signal processor is a counter or frequency tracker and the Doppler signal is from a single particle, the effective measurement volume may vary, depending on several factors [2]. Additionally, if the measurement volume ellipsoid is too large, most dual-beam systems employ a pinhole placed before the photodetector aperture to exclude light from sources other than the measurement volume and to restrict the effective field of view from which the scattered light is collected. This results in a smaller effective measurement volume.

3.3.3 Multicomponent Velocity Measurements

In principle, more than one component of velocity can be measured with LDV simply by using additional optical systems oriented in directions that yield the desired components. The main problem with this is being able to distinguish the signals from the separate systems. If the reference beam method is used, which velocity component is measured is determined by the orientation of the line of sight of the photodetector relative to the laser beam axis, so recognition of different components is rather easily done in this case. With the dual-beam method it is more difficult, but it can be accomplished using one of three means of distinguishing the laser beams: different colors, different frequencies, and different polarizations. If, for example, two components are to be measured, the blue and green light from an argon-ion laser can be separated, split into pairs of dual beams in orthogonal planes, and focused on the measurement point. A pair of separate receiving optics and photodetectors that filter out the scattered light from the undesired color will each produce Doppler burst signals. These correspond, respectively, to the velocity component measured by each system in the plane of their beams and perpendicular to the bisector of their included angle and to each other.

Three component measurements require even more complicated optical arrangements. One possibility is to use the dual-beam method to measure two components, say U and W, in the x–z plane, and use the reference beam method to measure V in the y direction orthogonal to this plane.

3.3.4 LDV Particles

The choice of particle size, density, and concentration for use in LDV is important. Particles that are too large do not respond to high-frequency fluctuations of the local instantaneous continuum flow field, while particles that are too small are affected by Brownian motion within the fluid. The frequency response time constant of particles has been estimated [17] to first order to be

$$\tau = \frac{d_p^2 \rho_p}{18 \mu_f}, \tag{3.48}$$

where the subscripts p and f denote particle and fluid, respectively. As a practical matter, the particle time constant should be such that $f_{3\text{-dB}} = 1/2(\pi \tau)$ is greater than the greatest frequency of the turbulent velocity fluctuations [2]. $f_{3\text{-dB}}$ is the 3-dB frequency at which the particle follows a sinusoidal velocity variation with an amplitude that is 0.707 of the fluid velocity amplitude. Another consideration is that the particles should be neutrally buoyant (i.e., have a density matching that of the carrier fluid so that they do not rise or sink). Moreover, if the particle concentration is too great, the Doppler signal becomes a complex amalgam of superposed responses to many scattering sites, which reduces the mixing and diminishes the signal. Large particle concentrations can even change the material properties of the fluid. On the other hand, if the concentration is too small, it becomes difficult to construct a true measure of the time-varying velocity field because too few particles flow through the measuring volume of the system.

Naturally occurring aerosols in airflows and hydrosols in water flows can sometimes be used as light-scattering particles for LDV applications. In air, aerosols often perform well except at very high speeds. Sometimes water or oil droplets in air are used. For water flows, if forward scatter is employed, hydrosols may be an appropriate particle. In the case of backscatter operation, which is often preferable for reasons cited previously, larger particles, about 5 μm or greater in diameter, have to be used with higher-power lasers in order to obtain sufficient scattered light flux energy. Commercially available particles at these sizes have the added advantage of having a narrow size distribution.

3.3.5 Photodetectors

Two types of photodetectors are commonly used: photomultipliers and photodiodes. Photodiodes perform well if the light flux levels are high, but they have little or no gain, making them unsuitable for low light levels. In contrast, photomultipliers, because of their high gain, perform well with low light flux levels but are at risk of burning out if the levels are too high. The quantum efficiency of photodiodes is about four times that of photomultipliers for visible light, but photodiodes are considerably smaller and require smaller power supplies. In the final analysis, the higher sensitivity of photomultipliers is the primary reason they are generally used for LDV applications [2].

The total flux of light energy received by a photodetector has several parts. It is the sum of the Doppler light flux, the *pedestal* light flux, and in the case of the reference beam method, the light flux due to the reference beam. Added to these fluxes is any background light flux from the room lights, reflections, laser flare, and radiation from

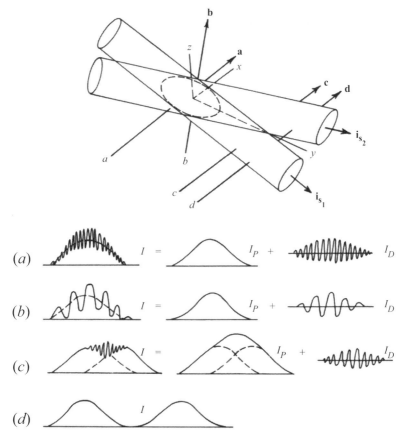

Fig. 3.12   *Types of LDV signals resulting from various particle trajectories composed of pedestals and Doppler bursts. (Adapted from [2]. Copyright 1996. Reproduced by permission of Routledge, Inc., part of The Taylor & Francis Group.)*

the fluid. The pedestal is the pulse generated by the photodetector corresponding to the sum of the pulses that would occur from each of the individual beams. The Doppler pulse is superimposed on the pedestal, as illustrated in Fig. 3.12, and can be separated from it by suitable high-pass filtering. The background light flux, on the other hand, is detrimental to the quality of the measurements.

In addition to these considerations, photodetectors themselves produce two types of inherent noise: electronic noise and shot noise. Electronic noise does not depend on the light flux received by the photodetector, but shot noise does. Shot noise arises from random photoelectron emissions in response to the light flux at the photodetector surface.

It may be concluded that after the pedestal signal has been removed, the output from the photodetector is made up of a part corresponding to the desired Doppler signal plus a part due to the electronic and shot noise, the latter being increased by the background light striking the photodetector surface. To improve performance, it is necessary to enhance the signal and reduce the noise. The background light can be reduced by a careful setup of very clean optical components and by blocking unwanted light. Some sources of electronic noise can be guarded against, but the best way to maintain a high signal-to-noise ratio is to choose particles that produce a strong signal.

3.3.6 Photodetector Signal Processing

An ideal Doppler frequency signal is sketched in Fig. 3.13a. Contiguous Doppler bursts are displayed corresponding to the continuous passage of individual particles through the measurement volume, one after the other. Their frequencies depend on the velocities of the particles, and their amplitudes depend on both the size and light reflecting efficiency of the particles and on the part of the measurement volume they

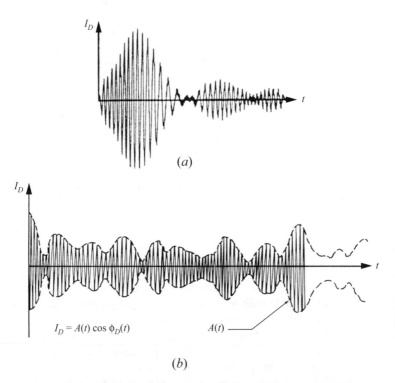

Fig. 3.13 *(a) Ideal LDV Doppler burst signal. (Adapted from [16]. Reprinted with permission of Teubner.) (b) Typical Doppler burst signal for high concentration of light-scattering particles. A(t) is the random amplitude and $\theta_D(t)$ is the random phase of the signal. (Adapted from [2]. Copyright 1996. Reproduced by permission of Routledge, Inc., part of The Taylor & Francis Group.)*

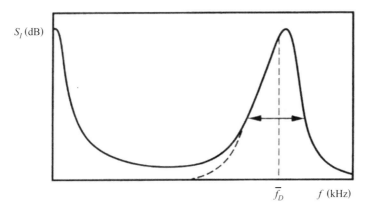

Fig. 3.14 *Power spectrum of typical LDV photodetector signal. (Adapted from [16]. Reprinted with permission of Teubner.)*

traverse. The foregoing suggests, however, that this ideal signal is never achieved in practice: There will always be noise to contend with and the presence of the low-frequency pedestal. Additionally, if the particle concentration is low, the signal will contain time intervals during which no Doppler bursts occur. On the other hand, both the amplitude and phase of the signal become random if the particle concentration is too high, because the signal is a superposition of many overlapping bursts occurring randomly in time. This situation is illustrated in Fig. 3.13*b*.

A sketch of the power spectrum corresponding to a typical photodetector signal is shown in Fig. 3.14. The centroid of the peak on the right corresponds to the mean Doppler frequency $\overline{f_D}$, which is a measure of the mean flow velocity at the measurement volume. The width of this part of the spectrum gives some indication of the magnitude of the fluctuations in f_D. It is not an exact relation since noise effects can widen the spectrum, as can the variability of the residence time of the particles in the measurement volume (even in laminar flow, this region is not a discrete spike). Apart from these complicating effects, this part of the spectrum can be viewed as corresponding to the probability density distribution of the velocity component measured by the LDV system. The peak on the left at zero frequency corresponds to the pedestal signal, which can clearly be low-pass filtered. The spectrum does not drop to zero power between the two peaks because of the electronic and shot noise embedded in the photodetector signal.

To obtain the instantaneous velocity from an individual Doppler burst, the signal must be processed further. This is done through a variety of methods, which are based on burst spectrum analyzers, counter processors, and frequency trackers. These three methods are now briefly described, although the latter two are now used infrequently.

3.3.6.1 *Burst Spectrum Analyzers*

Burst spectrum analyzers digitize the high-pass-filtered photodetector signal and then obtain the Doppler frequency by applying the fast Fourier transform (FFT) algorithm to segments of the discrete signal.

Commercial systems that can process input frequencies as high as 160 MHz with bandwidths as large as 120 MHz are available which are suitable for analyzing high-speed turbulent flows. From the burst Doppler frequency the instantaneous velocity component is determined, and then stored, along with the time of its occurrence, thereby creating a discrete time-series signal of the desired velocity component. The information in the time series is inherently irregularly spaced since it depends on the passages of the particles through the sensing volume at irregularly spaced intervals.

3.3.6.2 *Counter Processors*

Counter processors have been used primarily with low-burst-density signals (i.e., data with gaps), but they may also be used with high-burst-density signals if the signal-to-noise ratio is sufficiently high. In this method, the photodetector signal is first bandpass filtered to remove the pedestals and as much of the high-frequency noise as possible. Then a Schmidt trigger converts the signal to a square wave of the same frequency. By setting a high- and a low-amplitude threshold, the beginning and end of the burst signal can be specified and no square waves are generated by the zero crossing of noise during the gaps between bursts. The thresholds also effectively define the measurement volume and the rate with which data are obtained. Figure 3.15 illustrates the beginning and end times of the bursts as determined by the thresholds.

One means of finding the frequencies of the Doppler bursts is to determine the time interval T_1 corresponding to a fixed number of cycles N_1 so that

$$f_{D_1} = N_1/T_1. \tag{3.49}$$

Simultaneously, a longer time interval, T_2, corresponding to a larger number of cycles of the burst is determined, and $f_{D_2} = N_2/T_2$ is compared to f_{D_1} for verification. This is called the *N-cycle mode* and can result in more than one measurement per

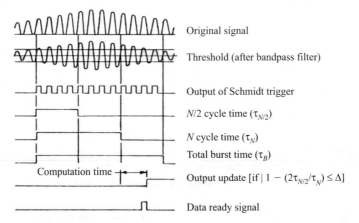

Fig. 3.15   *LDV Doppler burst signal processing using a counter. (Adapted from [2]. Copyright 1996. Reproduced by permission of Routeldge, Inc., part of The Taylor & Francis Group.)*

Doppler burst. The total burst mode measures the time interval of the entire burst, T_B, corresponding to the number of cycles in the burst, N_B. In this case, $f_D = N_B/T_B$. This method is preferable for low burst densities.

3.3.6.3 Frequency Trackers

Frequency trackers are of two types: phase-locked loop and frequency-locked loop. Since the latter have some advantages over the former, just the case of frequency-locked-loop trackers in determining the Doppler frequency is described here. As with the counter processors, the photodetector signal is first bandpass-filtered to remove the pedestals and high-frequency noise. It is then fed into a mixer that multiplies the signal with a sine wave, of frequency f_M, originating in a voltage-controlled oscillator. The product consists of two sine waves with frequencies $f_M - f_D$ and $f_M + f_D$. The product signal is then further bandpass-filtered at a center frequency f_C. The filter and f_M are chosen so that $f_C \approx f_M - f_D$ and $f_M + f_D$ is removed, leaving only the difference frequency. A frequency discriminator generates a feedback voltage proportional to $f_C - (f_M - f_D)$, and this voltage, in turn, controls the oscillator frequency f_M, increasing or decreasing it so that $f_C - (f_M - f_D)$ remains constant. The analog voltage of the frequency discriminator that assures the constant-frequency difference is thus proportional to the Doppler frequency of each Doppler burst and can be digitized and stored as a measure of the instantaneous velocity in the measurement volume. Frequency trackers work best when the burst density is high. When used with low-burst-density signals, the voltage-controlled oscillator keeps f_M constant until a new burst is encountered.

3.3.7 Velocity Bias Correction

The fact that LDV velocity component data are randomly spaced in time gives rise to a bias in computed statistical properties such as the mean and rms values [34]. Thus if it is assumed that particles are homogeneously distributed throughout the flow field, more high-speed seed particles than low-speed particles will pass through the measurement volume per unit time. In this case, the sample of velocity components obtained from LDV measurements will be biased toward high speeds. This problem can be avoided only if the particle seeding density is so high that the data can be sampled at a constant rate consistent with the Nyquist frequency (i.e., at least twice the highest frequency present in the turbulent fluctuations). However, it has been shown [19] that the bias can easily be corrected by weighting the velocity component values by the time they reside in the measurement volume. Figure 3.16 is a schematic of a discrete time series of velocities with resident time durations τ_B^i, represented by the widths of the crosshatched zones. For a time series with s samples, the corrected mean and nth-order fluctuation moments are given, respectively, by

$$\overline{U}_k = \sum_{i=1}^{s} U_k(t_i)\tau_B^i \bigg/ \sum_{i=1}^{s} \tau_B^i \qquad (3.50)$$

and

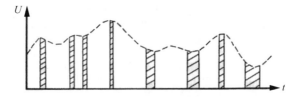

Fig. 3.16 *Velocity component time series with crosshatched zone widths indicating the resident time of the scattering particles in the LDV measurement volume. (From [19].)*

$$\overline{u_k^n} = \sum_{i=1}^{s} [U_k(t_i) - \overline{U}_k]^n \tau_B^i \Big/ \sum_{i=1}^{s} \tau_B^i. \tag{3.51}$$

3.4 PLANAR FLOW FIELD VELOCIMETRY

Hot-wire and laser-Doppler velocimetry are limited in practice to simultaneous measurements at only a relatively small number of locations in a flow field. Of more use in many applications are simultaneous planar measurements of the velocity components (i.e., measurement of the velocity on a fine grid of points covering a plane surface). The technique of planar flow field velocimetry has arisen in recent years and is now developed to a stage where commercial systems are available. Extensions of planar velocimetry to full three-dimensional flow field measurements have also been made in a few laboratories using holography, rapid planar scanning, and with limited spatial resolution, three-dimensional particle tracking. At the present time, such three-dimensional methods are difficult to use, quite expensive, and, generally speaking, are still in the development phase.

Planar field velocimetry usually estimates velocities by measuring the displacement of seeded particles over a short time interval. Letting $\mathbf{X}(t)$ denote the path of such a particle, its velocity, $\mathbf{V}(t)$, is estimated from the simple first-order differencing

$$\mathbf{V}(t) \approx \frac{\Delta \mathbf{X}(t)}{\Delta t} = \frac{\mathbf{X}(t + \Delta t) - \mathbf{X}(t)}{\Delta t}. \tag{3.52}$$

$\Delta \mathbf{X}$ is found by *pulsed light velocimetry*, in which the positions of marker particles in a plane are noted at times t and $t + \Delta t$ by illuminating them through successive pulses of laser light and capturing the images either on film or via a charge-coupled device (CCD). Equation (3.52) is evaluated for the two components of velocity in the plane of the laser light sheet at the location of the particle. This is a postprocessing step in which the particle images at the two times are matched with each other. Figure 3.17 shows a typical experimental arrangement in which a thin particle-laden volume of fluid, of thickness Δz_o, is illuminated by a sheet of light. Particle images are recorded by a camera having its optical axis perpendicular to the light sheet and its focal plane

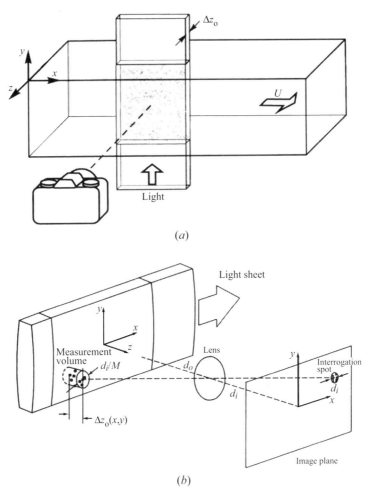

Fig. 3.17 *Typical planar pulsed light velocimetry optical arrangement, showing camera and laser light sheet illuminated plane containing scattering particles. [(a) Adapted from [16]. Reprinted with permission of Teubner. (b) Adapted from [2]. Copyright 1996. Reproduced by permission of Routledge, Inc., part of The Taylor & Francis Group.]*

located at the midpoint of the thickness of the light sheet. As discussed in more detail in Section 3.4.1.3, Δz_o needs to be large enough so that a sufficient number of particles remain within it between laser pulses, but not so large that the error due to projecting three-dimensional motion onto a two-dimensional plane is unacceptably large. Of course, the light sheet thickness must also comport with the depth of focus of the recording camera.

Pulsed light velocimetry can be classified into two types [1], as illustrated in Fig. 3.18. These depend on whether the markers are particles or are molecules mixed with the fluid media that can be made visible by photochromic or fluorescent means. Here

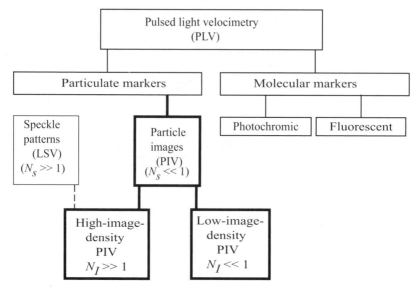

Fig. 3.18 *Classification of pulsed light velocimetry. (From [1] with permission from the Annual Review of Fluid Mechanics, Vol. 23, © 1991 by Annual Reviews www.AnnualReviews.org.)*

the discussion considers particles only, since they are now the most common markers for quantitative planar velocimetry. Particle methods may be classified further according to what is called their source density, the average number of particles residing in the "resolution cell" at a given time

$$N_s = C \, \Delta z_o \frac{\pi d_i^2}{4M^2}, \tag{3.53}$$

where C is the number of particles per unit volume, M is the magnification of the lens, and d_i is the particle image diameter. The latter depends not only on the particle diameter and the lens magnification, but also on the point response function of the lens. The resolution cell is thus the small cylinder formed by the intersection of the circle of diameter d_i/M formed by the projection of the particle image back into the fluid through the light sheet. The diameter of this cylinder is usually larger than the true particle diameter because of the diffraction-limiting properties of most lenses. The more N_s exceeds unity, the greater the probability that more than one particle resides in a resolution cell. When this occurs, the images of multiple particles overlap in the image plane. If the light is coherent and the source density is large (i.e., $N_s \gg 1$), the random phases of the overlapping images create patterns everywhere in the image plane called *laser speckle*. Speckle patterns move with the local group of overlapping particles that create them, and thus the velocity components of the patterns can be determined. This method is know as *laser speckle velocimetry* (LSV). If, on the other hand, the source density is small enough so that images of individual particles can be discerned everywhere in the image plane (i.e., $N_s \ll 1$), measuring their displacement over a short time in order to determine the velocity components in the image plane is known as *particle image velocimetry* (PIV). PIV is now much more commonly used than LSV [1], because in many experiments it is difficult or undesirable to maintain the very high particle concentrations required for LSV.

PIV itself may be further subclassified as high image density (HID) or low image density (LID). For HID/PIV there are many particles in an "interrogation cell" at any time, but not enough to cause speckle. An interrogation cell is defined as a subregion of the recorded field of view over which the postprocessing analysis determines the velocity vector. Rather than attempting to track the motion of all the many individual particles in the interrogation cell, for the HID method the displacement of the entire group is determined via a correlation technique. By contrast, for LID/PIV the particle number density is low enough so that only a few particles are located in the interrogation region at any one time, and they are tracked individually to determine the velocity components. LID/PIV is also often called *particle tracking velocimetry* (PTV).

A measure used to determine which of the PIV modes is present for a given experiment [1] is the *image density*,

$$N_I = \frac{C \, \Delta z_o \, \pi d_I^2}{4M^2}, \tag{3.54}$$

where d_I is the diameter of an interrogation spot on the image such as that shown in Fig. 3.19. Projected back through the fluid, it forms a cylindrical cell of diameter d_I/M in the light sheet (the shape of the cell may be different from a cylinder if so desired). $N_I \ll 1$ corresponds to the low-image-density limit in which the probability of finding more than one particle in the interrogation cell is much less than finding only one particle. At the other extreme, when $N_I \gg 1$, the probability is high that many particles occupy the interrogation cell at a given time. Figure 3.19 illustrates low- and high-density images suitable for PIV as well as a speckle image used in LSV.

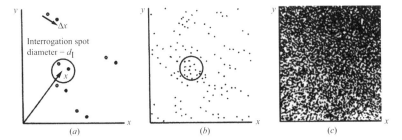

Fig. 3.19   (a) Low- and (b) high-particle-density images used in particle image velocimetry (PIV), as well as (c) a speckle image used for laser speckle velocimetry (LSV). (From [2]. Copyright 1996. Reproduced by permission of Routledge, Inc., part of The Taylor & Francis Group.)

3.4.1 Particle Image Velocimetry

Nd:YAG lasers are commonly used for PIV in airflows in order to have sufficient light, although ruby or copper vapor lasers sometimes suffice. Typically, 100 mJ of energy per pulse or more is needed, depending on the flow speed and optics. The accuracy with which the magnitude of the particle displacements in the X–Y image plane,

$$\Delta X = U \, \Delta t \, M$$

$$\Delta Y = V \, \Delta t \, M,$$

$$(3.55)$$

can be measured increases with the ratio $\Delta D / d_i$, where U and V are the components of the particle speed in the light sheet focal plane, and $\Delta D = \sqrt{\Delta X^2 + \Delta Y^2}$. Thus, in addition to following the flow better, small particles give greater measurement accuracy.

The accuracy with which U and V are determined is greatly affected by the choice of the time interval between pulses Δt, the parameter that is easiest to control in experiments. It should be small enough so that ΔX and ΔY are small compared to the smallest scales to be resolved in the flow, but still large enough to be determined accurately. In practice, Δt is chosen so that $\Delta D \approx 20$ to 30 times larger than d_i.

In water, because its refractive index is higher than that of air, particles of a given size scatter light 10 or more times less effectively than in air. Thus more light and/or larger particles are necessary. For experiments with 1- to 20-μm particles in air and water, aid in choosing particle sizes depending on camera f-numbers, light sheet dimensions, laser pulse energies, and film or CCD camera sensitivities is available [3].

Laser pulse durations are typically about 20 ns for airflows (i.e., short enough to obtain sharp images even at relatively high speeds). In water flow, where speeds are usually considerably lower, the pulse durations are typically about 100 μs.

3.4.1.1 Spatial Resolution
A quantity that is useful for estimating the dependence of the spatial resolution of a PIV experiment on particle concentration is given [2] by

$$\frac{\bar{r}}{\lambda_T} = \frac{0.55}{N_\lambda^{1/3}},$$

$$(3.56)$$

which is the ratio of the mean spacing between the sampled particles to the length scale of the small structures of the flow. Here the *data density* $N_\lambda = C\lambda_T^3$ and λ_T is the Taylor microscale of the turbulent flow. Equation (3.56) is useful for judging whether or not the data will be finely enough spaced so that it may be differentiated numerically to determine the vorticity and strain-rate components. In particular, with some qualifications, it is reasonable to assume that $N_\lambda = (4M^2\lambda_T^3/\pi \, \Delta z_o \, d_I^2)N_I$, so that if Δz_o and d_I are made to be of order λ_T or smaller, the HID/PIV limit $N_I >> 1$ implies that the spatial resolution is good.

It has also been pointed out [1] that the choice of recording media greatly affects the spatial resolution and thus also the accuracy of pulsed light anemometry. Standard video arrays have 500×500 pixels, although high-resolution video cameras can provide as many as 2048×2048 pixels. Digital cameras are also now readily available with 1024×1024 pixel resolution. This is still relatively low compared to standard 35-mm, 300-line/mm technical film, which gives $10{,}500 \times 7500$ pixels. In fact, the additional resolution of film compared to video results in approximately 10 times better measurement resolution and accuracy. Video and digital camera PIV is increasingly popular, however, because the images can be directly digitized, thus eliminating the laborious and time-consuming step of interrogating images stored on film. Furthermore, the high-resolution capability of film can be effectively utilized only with optics that take full advantage of it.

3.4.1.2 *Image Shifting* For experiments in which regions of flow reversal occur, there may be a degree of ambiguity in deciding which is the first and which is the second illuminated particle image. To circumvent this problem, a technique called *image shifting*, analogous to frequency shifting in LDV, can be used. Thus, between the two pulses, the image can be shifted a prescribed distance that causes all the displacements to appear to be in one direction only. A sketch of such a shift in the streamwise direction, denoted as ΔX_o, is shown in Fig. 3.20. For this example, after the images have been analyzed and the velocity components determined, the artificial velocity associated with the shift, $U_S = \Delta X_o / (M\ \Delta t)$, is subtracted from the streamwise component.

3.4.1.3 *Out-of-Plane Velocity Component* The effects of the out-of-plane velocity component is a potentially serious problem for planar velocimetry. Particularly for thin light sheets, it can occur that a substantial number of the particles are traveling fast enough with a component in the direction perpendicular to the plane of the light sheet to be illuminated only by the first or second pulse, not by both pulses. In these cases, the in-plane velocity components cannot be determined, and their neglect biases the measured statistical properties as determined from the particles that are illuminated by both pulses. The number of problematical particles can be made vanishingly small by reducing Δt, but this can be done only with some sacrifice

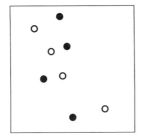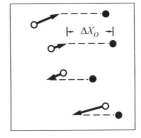

Fig. 3.20   *Image shifting. (From [16]. Reprinted with the permission of Teubner.)*

in the accuracy of the measured ΔX and ΔY image displacements, as mentioned above.

The out-of-plane motion leads to another error [42], due to the fact that the X–Y position of the particle image on the recorded image plane depends on its position in the out-of-plane direction and not just on its x–y position in the illuminating light sheet. Since the z-direction position is not known for two-dimensional PIV, an error is introduced.

3.4.1.4 Determination of Particle Image Displacement   With *high-image-density particle image velocimetry* (HID/PIV), many particles are found in each interrogation spot at any given time. Correlation techniques are used to determine the average displacement of the entire group of particles over the interval between pulses, Δt. If the images are recorded on video or digital camera, single-exposure images are suitable. On the other hand, although single-exposed photographic film frames can be used, because of registration errors that can occur between successive frames, it is better to double-expose a single film frame with the two pulses. In fact, multiple exposures can also be used, although for simplicity the following discussion is limited to double-exposure film frames.

Figure 3.21*a* contains a video frame image of smoke released from a slot on the wall of a turbulent boundary layer flow containing a wall-mounted obstacle. Figure 3.21*b* shows a pair of successive images separated by time $\Delta t = 0.225$ ms from a PIV subfield of the larger image. To determine the average particle displacements over small regions of the subfield pair of images, they have been subdivided into 49 interrogation spots over which the particle speeds and directions are approximately constant.

Let **D** denote the position vector to the center of each pixel in the image plane with components X and Y. Then the *cross-correlation* of the pixel light intensity $I(\mathbf{D})$ within each interrogation spot is found by shifting the interrogation spot of the first image over the second image by increments $\Delta\mathbf{D} = \Delta\mathbf{X} + \Delta\mathbf{Y}$. The product of the light intensities of the superimposed pairs of pixels from the interrogation spot of the first image and the corresponding region in the second image is then found and integrated over the spot area, giving

$$R(\Delta\mathbf{D}) = \int_{\text{spot}} I(\mathbf{D})I(\mathbf{D} + \Delta\mathbf{D})\, d\mathbf{D}. \tag{3.57}$$

It is clear that if all the particle images within the interrogation spot were displaced with the same magnitude and direction to their locations in the second image, an image shift $\Delta\mathbf{D}$ could be found where $R(\Delta\mathbf{D})$ would show perfect correlation apart from the unpaired images caused by particles entering or leaving the interrogation spot through in-plane or out-of-plane motion during the interval between pulses. On the other hand, if significant velocity gradients occur within the interrogation spot, the particle displacements within the spot will vary, leading to lower correlation values in $R(\Delta\mathbf{D})$ and diminished accuracy. In fact, when the cross-correlation does work, it shows a distinct peak rising significantly above the noise level of the correlation map.

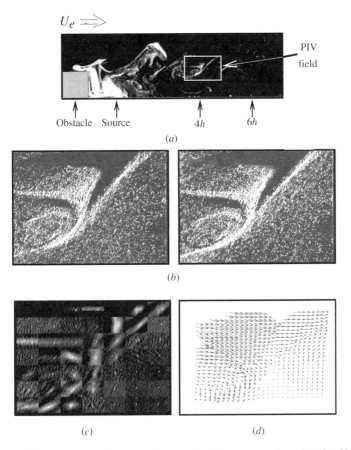

Fig. 3.21 *(a) Video image of smoke released from wall slot downstream of an obstacle of height h in a turbulent boundary layer; (b) successive PIV images of subzone indicated in video image; (c) map of cross-correlation of the successive PIV images; (d) resulting planar velocity vector field. (From [47]. Reprinted with permission of Cambridge University Press.)*

In Fig. 3.21c the white dots visible in most of the interrogation spots are such correlation peaks. They indicate the X and Y locations in the interrogation spots where maximum correlation occurs between the original particle images and the displaced particle images. The component distances ΔX and ΔY from the center of the interrogation spot to this location are the measured displacements. These can be determined with subpixel accuracy [32,42] using various methods. One particular approach that works well for narrow correlation peaks is to fit a function (often, a Gaussian is chosen) to three adjoining correlation values in each direction centered on the maximum correlation value in the image plane. The location of the maximum value of the function can then be determined to within approximately 0.1 pixel without too much computational effort.

Figure 3.21*d* shows the planar vector field resulting from this processing procedure. More vectors have been found than the 49 corresponding to the original grid of interrogation spots. This is done by choosing spots that are displaced fractional increments of the original grid. The overlap of the spots that resulted in the vectors in Fig. 3.21*d* was 50%.

If a double-exposed image is used rather than two successive single-exposure images, the process is one of *autocorrelation*. In this case (3.57) describes displacements, over itself, of the interrogation spot consisting of the group of double-exposed particle images. A large center peak, corresponding to $\Delta \mathbf{D} = 0$ will be seen, as in the example in Fig. 3.22, because each particle image correlates with itself. The other two peaks that are displaced symmetrically with respect to the self-correlation peak result from particle images from the second exposure being correlated with those from the first, and vice versa. The average particle displacement within the interrogation spot is then found as the distance between the centroid of the center, self-correlation peak, and one or the other secondary correlation peaks.

For either the cross-correlation or autocorrelation methods, poor-quality images, large in-plane and/or out-of-plane image loss, and/or large spatial velocity gradients within the interrogation region may result in spurious peaks that yield *false* velocity vectors. Techniques have been developed to detect and remove these errors. One such technique is to require that for a correlation peak to be considered valid, it must rise above a threshold that is chosen to have a value well above the correlation

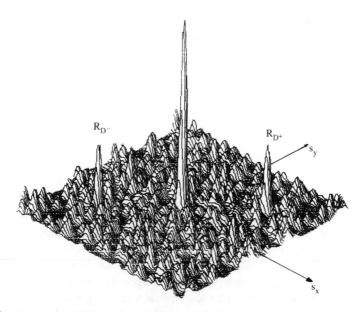

Fig. 3.22   *Autocorrelation function for double-exposed images. (From [2]. Copyright 1996. Reproduced by permission of Routledge, Inc., part of The Taylor & Francis Group.)*

noise. Another technique is to require that each vector not deviate from its immediate neighbors more than a prescribed amount.

To obtain optimum performance from double-pulse autocorrelation HID/PIV, it is generally recommended [1] that several conditions be satisfied. First, the average number of particles in the interrogation volume should be

$$N_I \geq 10 - 20. \tag{3.58}$$

Second,

$$\sqrt{U^2 + V^2} \, \Delta t \leq \tfrac{1}{4} d_I / M \tag{3.59}$$

(i.e., the in-plane particle displacement) should be less than one-fourth of the projected interrogation volume diameter. Third, the out-of-plane displacement should be less than one-fourth of the light-sheet thickness:

$$W \, \Delta t \leq \tfrac{1}{4} \Delta z_o. \tag{3.60}$$

The fourth criterion requires that the maximum velocity difference over the interrogation volume be no more than 20% of the average velocity of the particles in the interrogation volume:

$$|\Delta \mathbf{V}| / |\mathbf{V}| \leq 0.2. \tag{3.61}$$

Adhering to these constraints assures that 90 to 95% of the interrogation spots yield valid velocity vectors with small bias errors.

For video or digital camera HID/PIV, registration errors are not a consideration. In this case, single-exposure images can be used in which the second interrogation spot is chosen to be larger than the first to remove the in-plane loss of pairs. This can also be accomplished by shifting the second interrogation spot within the plane using knowledge of the average velocity of the particles in the first spot to predict where the second spot should be located. If there is a mean motion in the out-of-plane cross-stream direction, the second spot may be shifted in the direction perpendicular to the light sheet to reduce the out-of-plane loss of particle pair images.

With *low-image-density particle image velocimetry* (LID/PIV), where $N_I < 1$, it is probable that two images located in the same interrogation spot are from the same rather than from different particles. Thus, determination of the velocity components in such an interrogation spot simply reduces to determining the displacements of the individual particles. This is often called *particle tracking velocimetry* (PTV). Obviously, if the image density becomes too large, the probability of falsely pairing images becomes unacceptably large. A value of $N_I = 0.2$ is a good compromise between the need to eliminate false velocity vectors and the need to determine a sufficient number of vectors to define the flow field [2]. Postprocessing procedures can be used to eliminate many false vectors. Figure 3.23 shows a velocity vector field determined with PTV for open-channel flow.

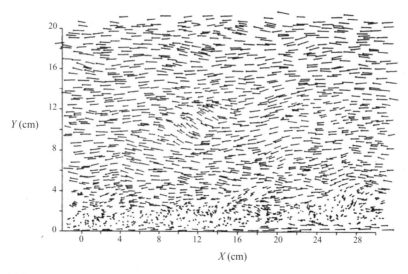

Fig. 3.23   *Planar velocity vector field from open-channel flow obtained with particle tracking velocimetry. (From [46]. Copyright 1984. Springer-Verlag.)*

3.5   CONCENTRATION MEASUREMENT METHODS

As suggested previously, one of the defining characteristics of turbulence is its ability to effect the transport of heat and mass within a flow field. Often, measurements of temperature and mass concentration fluctuations are useful for understanding turbulent dispersion and mixing processes. The ideal, which is especially challenging to accomplish, is to measure the velocity simultaneously with the scalar field at the same physical location so that the flux can be evaluated. In this section we consider techniques for measuring scalar fields with and without the velocity.

The most common method of measuring turbulent temperature fluctuations is with cold-wire anemometry, described briefly in Section 3.2.4. Also possible for local point concentration measurements [7,53] are hot-wire and hot-film anemometry, although principally optical methods are used today. In their early stages of development, optical methods were limited to local points [18,30,36], but they have now been extended to line and planar concentration field measurements.

3.5.1   Line Measurements

The most common type of line measurements, particularly for studies of reacting gases with combustion applications, is *Rayleigh light scattering* (RLS) [20,41], which is based on resonant scattering (i.e., no change in wavelength) of the incident light by molecules. For an isothermal constant-pressure gas consisting of two species, the RLS intensity from a unit volume of the gas mixture can be expressed as

$$I = N I_0 (\sigma_1 X_1 + \sigma_2 X_2), \tag{3.62}$$

where N is the total number of molecules in the unit volume, I_0 is the intensity of the source light in the measurement volume, σ_1 and σ_2 are the RLS cross sections of species molecules 1 and 2, and X_1 and X_2 are the mole fractions of species 1 and 2 in the measurement volume. Conservation of mass requires that $X_1 + X_2 = 1$. If I_0 is constant and σ_1 and σ_2 are known, the mole fractions X_1 and X_2 can be found by measurement of I. For calibration purposes, relative values of intensity usually are obtained by imaging the RLS light onto a suitable detector [15] and recording responses for two known concentrations. Most often, samples of the unmixed gases 1 and 2 are used for this purpose. The relative values of RLS intensity for the two gases are denoted as I_1 and I_2. Following these calibrations, the mole fraction of gas 1 in an arbitrary mixture of gases 1 and 2 is determined by recording the detector response for RLS from the measurement volume (i.e., I) and substituting this in

$$X_1 = \frac{I - I_1}{I_2 - I_1}, \tag{3.63}$$

which is easily derived from (3.62). High-frequency response and spatial resolution are possible for this type of concentration measurement technique [15].

3.5.2 Planar Measurements

A useful technique for obtaining planar measurements of the scalar concentration field in water is based on the phenomenon of laser-induced fluorescence (LIF). In this case a fluorescent dye is mixed with and carried as a passive scalar in the flow. When illuminated by a laser, the dye absorbs incident light at one wavelength and reemits it at a different wavelength. The reemitted light intensity is proportional to the dye concentration at the point of measurement according to the Beer–Lambert law. Under certain conditions the relationship is linear. The application of LIF to quantitative scalar concentration measurements has been developed in numerous investigations [43].

The principal difficulty with LIF is that light gets absorbed all along the affected path: between the recording camera and the point of measurement, L_f, and between this point and the source of the incident light ray, L_i. A nearly linear proportionality between the recorded light intensity and the concentration can be obtained by using a weak dye concentration and by minimizing the total optical path. The light intensity I arriving at the camera that results from an instantaneous scalar concentration $C(x, y, z, t)$, at the measurement point can be expressed as

$$I = A \times C(x, y, z, t) \times F(\text{absorption}). \tag{3.64}$$

A is a constant absorption coefficient and F is a space- and time-dependent function of L_i, L_f, ε_i, and ε_f, where the latter two parameters are, respectively, the absorption coefficients for the two parts of the optical path.

In the case of airflow experiments, although fluorescent particles have been used, a more common approach to measuring scalar contaminant dispersal is by measuring the particle number density of markers like the dry incense smoke shown in Fig. 3.21. In this experiment, which required two separate optical systems, the smoke particles are used for both velocity field measurements via PIV and concentration field measurements based on the principle of Mie light scattering. The latter technique works by relating light intensity levels to particle concentration number density. For the case of monodispersed particles (i.e., the particles all have the same diameter), the relationship is linear. In practice, it is often the case that polydispersed particles are used (i.e., they cover a range of diameters). In this case [4], the intensity of the scattered light is proportional to $\sum d_i^2 N_i$, where N_i is the number density of particles of diameters d_i. Such proportionality holds provided that the absorption of light along the optical path, geometrical distortions, and background illumination effects have been taken into account. It remains to relate the scattered light to the concentration in this case. One way to do this is by first establishing that the properties of polydispersed particles are equivalent to that of monodispersed particles of an ideal, equivalent diameter. This connection has been investigated for the dispersion of polydispersed incense smoke from a wall line source in a smooth wall boundary layer flow [51], from which the linear relationship

$$\frac{\overline{C}_{eq}}{\overline{C}_{eq(max)}} \equiv \frac{d_{eq}^2 \overline{N}_{eq}}{d_{eq}^2 \overline{N}_{eq(max)}} \equiv \frac{\sum d_i^2 \overline{N}_i}{\sum d_i^2 \overline{N}_{i(max)}} \approx \frac{\overline{I}}{\overline{I}_{max}} \qquad (3.65)$$

has been derived between the relative mean concentration for the monodispersion of "equivalent" particles of diameter d_{eq} and the relative mean light intensity from the polydispersed particles. Here, "max" denotes the maximum mean values of the light intensity and of the concentration number density in the illuminated field of view. Furthermore, it should be noted that this relationship is found to hold instantaneously as well as on average.

According to (3.65), the distribution of the relative intensity of the scattered light, $\overline{I}/\overline{I}_{max}$, from polydispersed particles is the same as that from monodispersed particles of size d_{eq} with a relative particle concentration number density $\overline{N}_{eq}/\overline{N}_{eq(max)}$. A second ideal characterization of the particles that is useful for analysis is to imagine that the size distribution of the polydispersed particles remains constant throughout the plume. In this case, the instantaneous relative measured intensity $\overline{I}/\overline{I}_{max}$ is also linearly related to the relative particle concentration number density $\overline{C}_{eq}/\overline{C}_{eq(max)}$ [4]. In summary, the measured relative instantaneous and mean light intensities give relative concentration levels in the plume for two equivalent experiments: one with monodispersed particles and the other with polydispersed particles that have a constant size distribution throughout the plume. In other words, a mapping exists that can transform a polydispersed particle experiment into one or the other of these two ideal experiments.

3.6   NUMERICAL METHODS

In this section we describe some of the essential aspects entailed in numerical schemes for simulating turbulent flow. First, some general ideas about one particularly success-ful numerical scheme for solving the Navier–Stokes equation, a projection method, is presented. Then some details on how this approach can be used in simulating periodic flow in a cubic box and channel are described. In these instances the focus is primar-ily on spectral methods which have been the main route to developing DNS codes. It should be noted, however, that there has been much recent interest in performing DNS calculations using finite difference and other schemes that are better suited than spectral methods to irregular geometries.

3.6.1   General Considerations

According to the Helmholtz decomposition theorem [23], an arbitrary three-dimen-sional vector field $\mathbf{V}(\mathbf{x})$ in a domain $\mathcal{D}$ with boundary $\partial\mathcal{D}$ can be written uniquely as

$$\mathbf{V}(\mathbf{x}) = \mathbf{U}(\mathbf{x}) + \nabla\phi(\mathbf{x}), \tag{3.66}$$

where $\mathbf{U}(\mathbf{x})$ is divergence-free, that is,

$$\nabla \cdot \mathbf{U} = 0, \tag{3.67}$$

and satisfies

$$\mathbf{U} \cdot \mathbf{n} = 0 \tag{3.68}$$

on $\partial\mathcal{D}$, where $\mathbf{n}$ is the outward normal and ϕ is a scalar field that can be determined to within an arbitrary constant. Taking the divergence of (3.66), it follows that ϕ satisfies the Poisson equation

$$\nabla^2\phi = \nabla \cdot \mathbf{V}, \tag{3.69}$$

while (3.66) and (3.68) show that ϕ obeys the Neumann condition

$$\frac{\partial\phi}{\partial n} = \mathbf{V} \cdot \mathbf{n} \tag{3.70}$$

on boundaries. Equations (3.69) and (3.70) are enough to determine ϕ to within a constant value.

These observations are useful in developing schemes for solving the Navier–Stokes equation numerically. In particular, consider the following nondimensional form of (2.5) with the body force omitted:

$$\frac{\partial U_i}{\partial t} + \frac{\partial P}{\partial x_i} = -U_j\frac{\partial U_i}{\partial x_j} + \frac{1}{R_e}\nabla^2 U_i. \tag{3.71}$$

The left-hand side may be viewed as the unique Helmholtz decomposition of the vector on the right-hand side. In particular, the term $\partial U_i / \partial t$ satisfies (3.67) and (3.68), while P takes on the role of ϕ in (3.66).

For any vector $\mathbf{V}$, it is possible to define a projection operator $\mathcal{P}$ which maps $\mathbf{V}$ to the divergence-free part: namely, $\mathbf{U}$ (i.e., $\mathcal{P}\mathbf{V} = \mathbf{U}$) of its unique Helmholtz decomposition. In this case, (3.71) can be rewritten in projection form as

$$\frac{\partial U_i}{\partial t} = \mathcal{P}\left(-U_j \frac{\partial U_i}{\partial x_j} + \frac{1}{R_e}\nabla^2 U_i\right). \tag{3.72}$$

This relation forms the basis for an effective means for sequencing the determination of the unknowns U_i and P in numerical solutions to (3.71) [11]. First, (3.72) is solved without the projection to get a preliminary field, say $\partial V_i / \partial t$. Then the divergence-free part of this is extracted and set equal to $\partial U_i / \partial t$. Formally, one proceeds as follows: consider the following second-order accurate time discretization of (3.72) without the projection:

$$\frac{V_i - U_i^n}{\Delta t} = -\left(\frac{3}{2}U_j^n \frac{\partial U_i^n}{\partial x_j} - \frac{1}{2}U_j^{n-1}\frac{\partial U_i^{n-1}}{\partial x_j}\right) + \frac{1}{2R_e}\nabla^2(V_i + U_i^n), \tag{3.73}$$

where U_i^n refers to the nth time step. The first grouping of terms on the right-hand side of (3.73) is an Adams–Bashforth discretization of the convection term [26,35,39], and the last group is a standard Crank–Nicholson [39] approximation to the diffusion term. In setting up (3.73), it is assumed that at the start of each time step, $V_i = U_i^n$. Next, the Helmholtz decomposition is approximated numerically by taking

$$\frac{V_i - U_i^n}{\Delta t} = \frac{U_i^{n+1} - U_i^n}{\Delta t} + \frac{\partial \phi}{\partial x_i} \tag{3.74}$$

or, after simplification,

$$U_i^{n+1} = V_i - \Delta t \frac{\partial \phi}{\partial x_i}, \tag{3.75}$$

where U_i^{n+1} must satisfy a discrete form of (3.67) together with the boundary condition (3.68).

The scheme combining (3.73) and (3.75) must be consistent with the Navier–Stokes equation (i.e., in the limit as $\Delta t \to 0$, these equations should reduce to the Navier–Stokes equation). In fact, adding (3.73) and (3.74) together gives

$$\frac{U_i^{n+1} - U_i^n}{\Delta t} + \frac{\partial P}{\partial x_i}$$

$$= -\left(\frac{3}{2}U_j^n \frac{\partial U_i^n}{\partial x_j} - \frac{1}{2}U_j^{n-1}\frac{\partial U_i^{n-1}}{\partial x_j}\right) + \frac{1}{2R_e}\nabla^2(U_i^{n+1} + U_i^n), \tag{3.76}$$

where the pressure satisfies

$$\frac{\partial P}{\partial x_i} = \frac{\partial \phi}{\partial x_i} + \frac{1}{2R_e} \nabla^2 (U_i^{n+1} - V_i).$$ (3.77)

Equation (3.76) is a fully consistent second-order accurate approximation to the Navier–Stokes equation.

In this numerical scheme, (3.73) is first solved for V_i and then (3.68) and (3.75) are used to solve for U_i^{n+1} and ϕ. ϕ is found first by solving the Poisson equation

$$\nabla^2 \phi = \frac{1}{\Delta t} \frac{\partial V_i}{\partial x_i}$$ (3.78)

resulting from taking the divergence of (3.75), and then U_i^{n+1} is determined from (3.75).

Substituting (3.75) into (3.77) gives

$$\frac{\partial P}{\partial x_i} = \frac{\partial \phi}{\partial x_i} - \frac{1}{2R_e} \nabla^2 \left(\Delta t \frac{\partial \phi}{\partial x_i} \right)$$ (3.79)

or

$$\frac{\partial}{\partial x_i} \left(P - \phi + \frac{\Delta t}{2R_e} \nabla^2 \phi \right) = 0,$$ (3.80)

from which it is found that

$$P = \phi - \frac{\Delta t}{2R_e} \nabla^2 \phi.$$ (3.81)

Thus P can be readily found from (3.81), once ϕ is computed.

To apply the projection algorithm described here to the solution of specific problems, decisions have to be made concerning how the spatial derivatives in the equations are to be approximated. A wide range of alternatives exist, including spectral, finite difference, finite volume, and finite element methods. Once a particular approach is selected, specific forms of the boundary conditions for V_i and ϕ can be derived. In the following sections we consider how this algorithm can be specialized to computing turbulent flow in a box with periodic boundary conditions and then to turbulent channel flow.

3.6.2 Periodic Flow in a Cubical Domain

Consider turbulent flow in a cubical domain of dimension L with periodic conditions in all directions. The problems suitable for study with this geometry generally involve the response of homogeneous isotropic turbulence to a given random forcing field or its decay from an initial state. Consequently, in this discussion it is assumed that $\overline{U}_i = \overline{P} = 0$, so that, for example, $U_i = u_i$ and $P = p$.

The assumption of periodicity suggests that it is advantageous to seek the solution of (3.73), (3.75), and (3.78) in the context of a spectral method (i.e., one in which Fourier series are used to represent the unknown quantities and the discrete equations become relations governing the evolution of Fourier coefficients). The foundation for this approach is the fact that the trigonometric functions

$$e^{\imath \mathbf{k} \cdot \mathbf{x}}, \tag{3.82}$$

where

$$\mathbf{k} = 2\pi \mathbf{n}/L, \tag{3.83}$$

$\imath = \sqrt{-1}$, and $\mathbf{n}$ denotes the set of integer triples (n_1, n_2, n_3) running from $-\infty \rightarrow +\infty$, form a complete set of basis functions. That is, any periodic function on the cube can be expressed as a Fourier series:

$$u_i(\mathbf{x}, t) = \sum_{\mathbf{k}} \widehat{u}_i(\mathbf{k}, t) e^{\imath \mathbf{k} \cdot \mathbf{x}}, \tag{3.84}$$

where the summation is over all values of $\mathbf{k}$ and the countable set of vectors $\widehat{u}_i(\mathbf{k}, t)$ are the Fourier coefficients. The Fourier coefficients give information about the degree to which specific wavenumbers in the decomposition (3.84) are excited. The larger the magnitude of $\mathbf{k}$, the finer the scale of motion that is represented by (3.82).

The Fourier coefficients are obtained from $u_i(\mathbf{x}, t)$ via the inverse transform

$$\widehat{u}_i(\mathbf{k}, t) = \frac{1}{L^3} \int_{\mathcal{V}_L} u_i(\mathbf{x}, t) e^{-\imath \mathbf{k} \cdot \mathbf{x}} \, d\mathbf{x}, \tag{3.85}$$

so that knowledge of the complete set of Fourier coefficients is equivalent to knowledge of $u_i(\mathbf{x}, t)$, and vice versa. Here the integration domain $\mathcal{V}_L$ is the cube of side L.

Since $u_i(\mathbf{x}, t)$ is real, the series in (3.84) must equal its complex conjugate, $u_i^*(\mathbf{x}, t)$, that is,

$$u_i(\mathbf{x}, t) = u_i^*(\mathbf{x}, t) = \sum_{\mathbf{k}} \widehat{u}_i^*(\mathbf{k}, t) e^{-\imath \mathbf{k} \cdot \mathbf{x}} = \sum_{\mathbf{k}} \widehat{u}_i^*(-\mathbf{k}, t) e^{\imath \mathbf{k} \cdot \mathbf{x}}, \tag{3.86}$$

where the last equality comes about by changing $\mathbf{k}$ to $-\mathbf{k}$. Subtracting the two series (3.84) and (3.86) and grouping terms according to $\mathbf{k}$ gives

$$\sum_{\mathbf{k}} [\widehat{u}_i^*(\mathbf{k}, t) - \widehat{u}_i(-\mathbf{k}, t)] e^{-\imath \mathbf{k} \cdot \mathbf{x}} = 0. \tag{3.87}$$

Since Fourier modes are linearly independent, it follows that the term in brackets must be zero; hence

$$\widehat{u}_i^*(\mathbf{k}, t) = \widehat{u}_i(-\mathbf{k}, t), \tag{3.88}$$

as a consequence of the fact that the velocity is real-valued.

It has previously been suggested that the process of energy dissipation is generally most effective at the smallest scales in turbulent flow. Consequently, beyond a dissipation range of scales (or equivalently, wave number, $\mathbf{k}$), it can be expected that there will be no turbulent energy. This means that in any given turbulent flow, the values of $\mathbf{k}$ for which the Fourier coefficients are nonzero are finite, and a largest excited value of $\mathbf{k}$ can be found. In view of this, it may be legitimate to seek a numerical solution for the velocity field in the form of a truncated Fourier representation,

$$u_i(\mathbf{x}, t) = \sum_{-N/2}^{N/2} \widehat{u}_i(\mathbf{k}, t) e^{\iota \mathbf{k} \cdot \mathbf{x}}, \tag{3.89}$$

where by this notation it is meant that only values of $\mathbf{k}$ for which

$$|n_i| \leq N/2, \qquad i = 1, 2, 3 \tag{3.90}$$

are kept in the sum. In practice, it can be expected that truncation of the Fourier expansion will be legitimate under the relaxed requirement that modes $|n_i| > N/2$ contain just a small amount of energy, not that they have exactly zero energy.

In obtaining a spectral solution to (3.73), (3.75), and (3.78), truncated Fourier series such as (3.89) are assumed for all variables, including V_i and ϕ. The equations to be solved for the Fourier coefficients $\widehat{u}_i$, $\widehat{V}_i$, and $\widehat{\phi}_i$ can be developed in several ways. One convenient approach is a Fourier–Galerkin method in which the orthogonality property of the basis functions (3.82) is used to get the desired relations. This may be implemented by taking the Fourier transform of the discrete equations [i.e., by applying the operation in (3.85) to each of the terms in the equations]. This entails the use of identities such as

$$\frac{\widehat{\partial V_i}}{\partial x_j} = \iota k_j \widehat{V}_i, \tag{3.91}$$

which is readily derived from (3.85) by integrating by parts and using the periodicity condition. Moreover, applying (3.91) twice gives the result

$$\widehat{\nabla^2 V_i}(\mathbf{k}, t) = -k^2 \widehat{V}_i(\mathbf{k}, t). \tag{3.92}$$

Applying a Fourier transform to the advection terms in (3.73) gives

$$\widehat{u_j \frac{\partial u_i}{\partial x_j}}(\mathbf{k}, t) = \frac{1}{L^3} \int_{V_L} u_j(\mathbf{x}, t) \frac{\partial u_i}{\partial x_j}(\mathbf{x}, t) e^{-\iota \mathbf{k} \cdot \mathbf{x}} \, d\mathbf{x}. \tag{3.93}$$

Noting from (3.84) the identity

$$\frac{\partial u_i}{\partial x_j}(\mathbf{x}, t) = \iota \sum_{\mathbf{k}} k_j \widehat{u}_i(\mathbf{k}, t) e^{\iota \mathbf{k} \cdot \mathbf{x}} \tag{3.94}$$

and substituting (3.84) and (3.94) into the right-hand side of (3.93) yields

$$\widehat{u_j \frac{\partial u_i}{\partial x_j}}(\mathbf{k}, t) = \iota \sum_{\mathbf{l}} \sum_{\mathbf{m}} l_j \widehat{u}_i(\mathbf{l}, t) \widehat{u}_j(\mathbf{m}, t) \frac{1}{L^3} \int_{V_L} d\mathbf{x} \, e^{-\iota(-\mathbf{l}+\mathbf{k}-\mathbf{m}) \cdot \mathbf{x}}$$

$$= \iota \sum_{\mathbf{l}} l_j \widehat{u}_i(\mathbf{l}, t) \widehat{u}_j(\mathbf{k} - \mathbf{l}, t), \tag{3.95}$$

where the important identity

$$\frac{1}{L^3} \int_{V_L} e^{-\iota \mathbf{k} \cdot \mathbf{x}} \, d\mathbf{x} = \begin{cases} 1 & \mathbf{k} = 0 \\ 0 & \mathbf{k} \neq 0, \end{cases} \tag{3.96}$$

which expresses the orthogonality property of the functions (3.82), has been used. Equation (3.95) has the interesting property that the contributions to the advection term at $\mathbf{k}$ come only from triads of wavenumbers (i.e., $\mathbf{l}$ and $\mathbf{k} - \mathbf{l}$, which together with $\mathbf{k}$ form the vertices of a triangle, as shown in Fig. 3.24). More is said about this in Sections 7.7 and 12.5.

Collecting together the results above, the transformed version of (3.73) is

$$\frac{\widehat{v}_i(\mathbf{k}) - \widehat{u}_i{}^n(\mathbf{k}, t)}{\Delta t} = -\iota \left[\frac{3}{2} \sum_{\mathbf{l}} l_j \widehat{u}_i^n(\mathbf{l}, t) \widehat{u}_j^n(\mathbf{k} - \mathbf{l}, t) \right.$$

$$\left. - \frac{1}{2} \sum_{\mathbf{l}} l_j \widehat{u}_i^{n-1}(\mathbf{l}, t) \widehat{u}_j^{n-1}(\mathbf{k} - \mathbf{l}, t) \right] \tag{3.97}$$

$$- \frac{1}{2R_e} k^2 [\widehat{v}_i(\mathbf{k}, t) + \widehat{u}_i^n(\mathbf{k}, t)],$$

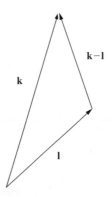

Fig. 3.24   *Triads of wavenumbers.*

where the sum over **l** is limited to just those values for which **l** and **k** − **l** both satisfy (3.90).

In the same vein, (3.78) becomes

$$-k^2 \widehat{\phi}(\mathbf{k}, t) = \frac{1}{\Delta t} \imath k_j \widehat{v}_j(\mathbf{k}, t), \tag{3.98}$$

and (3.75) and (3.81), respectively, give

$$\widehat{u}_i^{n+1}(\mathbf{k}, t) = \widehat{v}_i(\mathbf{k}, t) - \imath \, \Delta t \, k_i \widehat{\phi}(\mathbf{k}, t) \tag{3.99}$$

and

$$\widehat{p}(\mathbf{k}, t) = \widehat{\phi}(\mathbf{k}, t) + \frac{\Delta t}{2R_e} k^2 \widehat{\phi}(\mathbf{k}, t). \tag{3.100}$$

The numerical scheme is implemented by carrying out each of these operations in sequence: First, (3.97) is solved for $\widehat{v}_i(\mathbf{k}, t)$, then (3.98) for $\widehat{\phi}(\mathbf{k}, t)$, (3.99) for $\widehat{u}_i^{n+1}(\mathbf{k}, t)$, and finally, (3.100) for $\widehat{p}(\mathbf{k}, t)$.

The most time-consuming part of this scheme is the evaluation of the sum in the nonlinear terms in (3.97). In fact, direct evaluation of these terms is prohibitively expensive if N is large, since for a three-dimensional problem the amount of computational labor is at best proportional to N^4 [9]. It may be noted that in the original physical space, evaluation of the nonlinear term would not be particularly expensive, since it requires only a simple multiplication. Of course, this assumes that the velocity and its derivatives are evaluated at physical space locations. To access this advantage, a fast means must be found for transforming between the Fourier coefficients $\widehat{u}_i(\mathbf{k}, t)$ and the velocity in physical space, $u_i(\mathbf{x}, t)$. In fact, for this purpose the fast Fourier transform (FFT) [9] may be used to map between $\widehat{u}_i(\mathbf{k}, t)$ and the quantity $u_i^*(\mathbf{k}, t)$, denoting a numerical approximation to $u_i(\mathbf{x_k}, t)$ at the physical points $\mathbf{x_k}$, which constitute a uniformly spaced mesh covering the cube domain. Thus, in this pseudospectral scheme, the velocities are transformed from Fourier space to physical space, the nonlinear term is evaluated, and then the result is transformed back to Fourier space. The algorithm is quite efficient and makes the overall scheme practical.

A careful analysis of the pseudospectral approach shows that it is subject to *aliasing* errors associated with the use of the discrete transform. In essence, spurious contributions to modes within the truncated range are created by the nonlinear term as it attempts to send energy to nonexistent high-wavenumber modes. A common means for preventing aliasing errors is referred to as the *two-thirds rule*. In this, the truncated Fourier expansion, which covers modes from $-N/2$ to $N/2$, is embedded in a larger vector of modes extending from $-M/2$ to $M/2$, where $M \geq (3/2)N$. The extra modes are set to zero, so, in effect, two-thirds of the new vector contains the Fourier modes of interest. When the transfer term is evaluated using the larger vector, it can be shown that the aliasing errors are identically zero [i.e., the end result is identical to the nonlinear terms appearing in (3.97)]. Of course, dealiasing increases the numerical cost somewhat, but its cost is not prohibitive and it is an essential feature in maintaining physical accuracy. The effectiveness of the pseudospectral method is evident in the fact that calculations with $N = 256$ or 512 are routine, and some studies with 1028 have been performed.

3.6.3 Application to Channel Flow

The algorithm described in Section 3.6.1 can be used for numerically simulating turbulent channel flow after special provision is made for the channel geometry. The computational domain in the streamwise, wall-normal, and spanwise directions, denoted as (x, y, z), respectively, is taken to be $L_1 \times 2 \times L_3$, it being assumed that lengths are scaled by the half-channel height. Periodic boundary conditions are imposed in the x and z directions, and nonpenetration and nonslip conditions are at $y = \pm 1$.

As in the case of a periodic box, the symmetries of the channel suggest the use of a pseudospectral method that can deliver enhanced accuracy for the same computational effort as finite difference or other approaches. In view of the periodicity in the streamwise and spanwise directions, it is natural to incorporate a planar Fourier representation on the surfaces parallel to the boundaries. To accommodate the finite extent of the wall-normal direction, an expansion in Chebyshev polynomials, among other possibilities, is appropriate. Thus, in place of (3.89), the velocity field is expressed through the truncated Fourier–Chebyshev expansion:

$$u_i(\mathbf{x}, t) = \sum_{n_1=-N_1/2}^{N_1/2} \sum_{n_2=0}^{N_2} \sum_{n_3=-N_3/2}^{N_3/2} \widehat{u}_i(\mathbf{k}, t) e^{i\mathbf{k}^* \cdot \mathbf{x}^*} T_{n_2}(y), \qquad (3.101)$$

where $\mathbf{k}^* = (k_1, k_3)$ and $\mathbf{x}^* = (x, z)$ are two-dimensional vectors, and

$$T_n(y) = \cos(n \cos^{-1} y) \qquad (3.102)$$

is the Chebyshev polynomial of order n. Similar expansions may be defined for the other variables in the numerical system (i.e., V_i, ϕ, and P). As before, it should be remembered that $\mathbf{k}^*$ is shorthand for the wavenumbers $(2\pi n_1^*/L1, 2\pi n_3^*/L3)$, where $\mathbf{n}^* = (n_1, n_3)$ are integer pairs, and it is to be understood that $k_2 = n_2$.

There are several ways in which (3.101) can be used in the context of the scheme incorporating (3.73), (3.75), and (3.78). The most common strategy, due to its relative simplicity, is to use a spectral collocation method [9]. Thus, imagine that at any time step, the Fourier–Chebyshev coefficients $\widehat{u}_i(\mathbf{k}, t)$ are known for all variables. Then expansions such as (3.101) are substituted into the right-hand sides of the equations of the numerical scheme and evaluated at the collocation points (i.e., a set of points where the equations are forced to hold exactly). The end result is that the quantities, u_i, V_i, and ϕ have been determined at the collocation points according to (3.73), (3.75), and (3.78). Collocation points are chosen in such a way that one can transform back and forth between the Fourier–Chebyshev coefficients and the physical variables at these points. To complete the iteration cycle, the inverse mapping is performed to get the updated Fourier–Chebyshev coefficients and the next cycle can begin.

It is evident that an essential aspect of this scheme is having the capability of rapidly mapping back and forth between the Fourier–Chebyshev coefficients and their physical space counterparts on the collocation mesh. This means that the collocation points need to be selected in such a way that a FFT can be used for the mapping as it was in Section 3.6.2. One choice that is acceptable in this regard is to use a uniform mesh (e.g., $x_j = j \, \Delta x$, $j = 1, \ldots, N_1$ and $z_j = j \, \Delta z$, $j = 1, \ldots, N_3$, where $\Delta x = L_1/N_1$ and $\Delta z = L_3/N_3$) on the planes parallel to the wall, with the collocation points

$$y_j = \cos \frac{\pi j}{N_2}, \qquad j = 0, \ldots, N_2 \tag{3.103}$$

in the wall-normal direction.

The rationale for choosing (3.103) can be made clear through considering a one-dimensional example of the use of Chebyshev polynomials as basis functions. Thus consider a function

$$u(y, t) = \sum_{n=0}^{N_2} \widehat{u}(n, t) T_n(y). \tag{3.104}$$

At the collocation points (3.103) this gives

$$u(y_i, t) = \sum_{n=0}^{N_2} \widehat{u}(n, t) T_n(y_i) = \sum_{n=0}^{N_2} \widehat{u}(n, t) \, \cos \frac{\pi i n}{N_2}, \tag{3.105}$$

where the second sum follows after using the definition (3.102). Equation (3.105) is a discrete Fourier cosine series that may be evaluated using an FFT to give the physical space quantities at the collocation points. Associated with it is the inverse transform

$$\widehat{u}(n, t) = \frac{2}{C_n N_2} \sum_{i=0}^{N_2} \frac{1}{C_i} u(y_i, t) \, \cos \frac{\pi i n}{N_2}, \tag{3.106}$$

where $C_0 = C_N = 2$ and $C_n = 1$, $n = 1, \ldots, N_2 - 1$. These results generalize to the full three-dimensional transform, in which case a FFT can be used to evaluate the discrete Fourier transforms in the x and z directions together with the cosine transform in the y direction.

Two other aspects of this method are notable. In the first place, the nonlinear terms in (3.73) are evaluated by the same pseudospectral means as in Section 3.6.2, except that this time the FFT appropriate to the Fourier–Chebyshev expansion is used. Thus the nonlinear term is evaluated by transforming quantities into physical space through the Fourier–Chebyshev fast transform, multiplying them together, and then transforming back into spectral space using the inverse transform. In this, dealiasing is usually done only in the Fourier directions.

A second important consideration concerns the evaluation of terms in (3.73), (3.75), and (3.78) containing first or second derivatives in the y direction. Such terms, after substitution of series such as (3.101), will contain derivatives of the Chebyshev polynomials. To maintain efficiency of the algorithm, such quantities must be evaluated rapidly. It turns out, however, that through use of the recursion relations for Chebyshev polynomials, it is possible to replace the evaluation of the derivatives with evaluation of the polynomials themselves. The end result is that the expressions containing derivatives revert to summations over cosines as in (3.105) but with specially computed coefficients. The details of this calculation are discussed in [9] and [39].

The distribution of points in the wall-normal direction given by (3.103) gives greater resolution near the boundaries than in the vicinity of the channel centerline. This is consistent with the desire to provide extra resolution in the near-wall region, where relatively small-scale phenomena exert a significant influence on the overall dynamics of the flow field. For example, many small-scale vortical motions occur next to the boundary which drive transport processes and turbulence production. The success of a DNS calculation often depends on how well resolved such features are and the use of Chebyshev mapping aids in this process.

A method similar to the one outlined here was employed in the pioneering DNS of channel flow by Kim et al. [27], which established the value of numerical simulation methods in understanding the physics of turbulent flow. This calculation was carried out on a domain with $L_1 = 4\pi$ and $L_3 = 2\pi$. The number of grid points in the streamwise, wall-normal, and spanwise directions were $192 \times 129 \times 160$, so that the grid spacing was $\Delta x^+ \approx 12$, $\Delta z^+ \approx 7$, and Δy^+ varied from 0.05 near the wall to 4.4 at the centerline. The decision on how wide and long the computational domain should be was based on several considerations. In particular, for the streamwise direction, one would like the streamwise correlation functions, such as $\mathcal{R}_{11}(x)$, to be zero when the separation is approximately half the streamwise domain length. The motivation is that if the streamwise length is too short, then through the periodic conditions, the downstream flow will be interacting with its own upstream and downstream images. A similar consideration applies in the spanwise direction. Moreover, it is interesting to note that in this case, if the domain is made too narrow, the computed solutions have been found to collapse to a laminar state. As will be seen below, this is a particularly striking example of the existence of structure near the boundary whose evolution is essential for the maintenance of turbulence.

3.7 SUMMARY

The unifying theme of this chapter has been its consideration of methodologies for observing the properties of turbulent flows. In this spirit, a section on numerical methods for DNS is included here since, as has become increasingly the case in recent years for those relatively restricted instances when DNS is feasible (i.e., affordable), it is the method of choice for learning about the properties of turbulence. Indeed, a seductive benefit of the numerical wind tunnel is its capacity to provide exhaustive information about every facet of a turbulent flow, including such exotic quantities as simultaneous pressure and velocity moments and Lagrangian information.

As suggested here and in the scaling argument in Section 2.6, the encroachment of DNS into the territory formerly the exclusive realm of physical experiments will be very slow going, and for most engineering flows physical experiments remain the only way that one can hope to obtain relatively definite information about turbulent flows. In fact, as demonstrated above, experimental techniques continue to improve as technologies for data acquisition and analysis continue to advance. Improvements continue to be felt in probe resolution so that measured data become more accurate and high Reynolds numbers can be studied. In addition, a range of methodologies have been devised, and are continuing to be developed, for measuring the velocity and velocity gradient fields on planes and even within volumes. In this sense there is a convergence of capabilities between physical experiments and DNS.

REFERENCES

1. Adrian, R. J. (1991) "Particle imaging techniques for experimental fluid mechanics," *Annu. Rev. Fluid Mech.* **23**, 261–304.

2. Adrian, R. J. (1996) "Laser velocimetry," in *Fluid Mechanics Measurement*, 2nd ed. (R. J. Goldstein, Ed.), Taylor & Francis, Washington, D.C., pp. 175–290.

3. Adrian, R. J. and Yao, C. S. (1987) "Pulsed laser technique application to liquid and gaseous flows and the scattering power of seed materials," *Appl. Opt.* **24**, 44–52.

4. Ayrault, M. and Simoëns, S. (1995) "Turbulent concentration determination in gas flow using multiple CCD cameras," *J. Flow Visualization Image Process.* **2**, 195–208.

5. Balint, J.-L., Wallace, J. M. and Vukoslavčević, P. (1991) "The velocity and vorticity vector fields of a turbulent boundary layer. 2. Statistical properties," *J. Fluid Mech.* **228**, 53–86.

6. Blackwelder, R. F. (1981) "Hot-wire and hot film anemometry," in *Methods of Experimental Physics: Fluid Dynamics* (R. J. Emerich, Ed.), Academic Press, New York.

7. Brown, G. L. and Rebollo, M. R. (1972) "Small, fast response probe to measured composition of a binary gas mixture," *AIAA J.* **10**, 649–652.

8. Bruun, H. H. (1995) *Hot-Wire Anemometry*, Oxford University Press, New York.

9. Canuto, C., Hussaini, M. Y., Quarteroni, A., and Zang, T. A. (1988) *Spectral Methods in Fluid Dynamics,* Springer-Verlag, Heidelberg.

10. Chen, C.-H. P. and Blackwelder, R. P. (1978) "Large scale motion in a turbulent boundary layer: study using temperature contamination," *J. Fluid Mech.* **89**, 1–31.

11. Chorin, A. J. (1968) "Numerical solution of the Navier–Stokes equations," *Math. Comput.* **22**, 745–762.

12. Collis, D. C. and Williams, M. J. (1959) "Two-dimensional convection from heated wires at low Reynolds numbers," *J. Fluid Mech.* **6**, 357–384.

13. Compton, D. A. and Eaton, J. K. (1996) "A high-resolution laser Doppler anemometer for three-dimensional turbulent boundary layers," *Exp. Fluids* **22**, 111–117.

14. Corrsin, S. (1963) "Experimental methods," *Handbuch der Physik* **8.2**, Springer-Verlag, Berlin, 524–590.

15. Dowling, D. R., Lang, D. B. and Dimotakis, P. E. (1989) "An improved laser Rayleigh scattering photodetection system," *Exp. Fluids* **7**, 435–440.

16. Eckelmann, H. (1997) *Einführung in die Strömungsmeßtechnik*, B.G. Teubner, Stuttgart, Germany.

17. Feller, W. V. and Meyers, J. F. (1975) "Development of a controllable particle generator for LDV seeding in hypersonic wind tunnels," in *Proc. Minnesota Symposium on Laser Anemometry*, (E. R. G. Eckert, Ed.), University of Minnesota, Minneapolis, Minn., pp. 342–357.

18. Gad-el-Hak, M. and Morton, J. B. (1979) "Experiments on the diffusion of smoke in isotropic turbulent flow," *AIAA J.* **177**, 558–562.

19. George, W. K. (1977) "Limitations to measuring accuracy inherent in the laser Doppler signal," in *The Accuracy of Flow Measurements by Laser Doppler Methods: Proc. LDA-Symposium, Copenhagen* (P. Buchave et al., Eds.), Hemisphere Publishing, Washington, D.C., pp. 20–63.

20. Graham, S. C., Grant, A. J. and Jones, J. M. (1974) "Transient molecular concentration measurements in turbulent flows using Rayleigh light scattering," *AIAA J.* **12**, 1140–1142.

21. Honkan, A. and Andreopoulos, Y. (1997) "Vorticity, strain-rate and dissipation characteristics in the near-wall region of turbulent boundary layers," *J. Fluid Mech.* **350**, 29–96.

22. Hussain, A. K. M. F. and Hayakawa, M. (1987) "Eduction of large scale organized structures in a turbulent plane wake," *J. Fluid Mech.* **189**. 193–229.

23. Johnson, R. W. (Ed.) (1998) *Handbook of Fluid Dynamics*, CRC Press, Boca Raton, Fla., p. A-64.

24. Jorgensen, F. (1971) "Directional sensitivity of wire and fiber film probes," *DISA Inf.* **11**, 31–37.

25. Kim, J. and Hussain, F. (1993) "Propagation velocity of perturbations in turbulent channel flow," *Phys. Fluids A* **5**, 695–706.

26. Kim, J. and Moin, P. (1985) "Application of a fractional-step method to incompressible Navier–Stokes equations," *J. Comput. Phys.* **59**, 308–323.

27. Kim, J., Moin, P. and Moser, R. (1987) "Turbulence statistics in fully developed channel flow at low Reynolds number," *J. Fluid Mech.* **177**, 133–166.

28. King, L. V. (1914) "On the convection of heat from small cylinders in a stream of fluid," *Philos. Trans. R. Soc. Ser. A* **214**, 373–432.

29. Kramers, H. (1946) "Heat transfer from spheres to flowing media," *Physica* **12**, 61–80.

30. Lee, J. and Brodkey, R. S. (1963) "Light probe for measurement of turbulent concentration fluctuations," *Rev. Sci. Instrum.* **34**, 1086–1090.

31. Loucks, R. (1998) "An experimental examination of the velocity and vorticity fields in a plane mixing layer," Ph.D. dissertation, University of Maryland.

32. Lourenco, L. and Krothapalli, A. (1995) "On the accuracy of velocity and vorticity measurements with PIV," *Exp. Fluids* **18**, 421–428.

33. Marasli, B., Nguyen, P. and Wallace, J. (1993) "A calibration technique for multiple sensor hot-wire probes and its application to vorticity measurements in the wake of a circular cylinder," *Exp. Fluids* **15**, 209–218.

34. McLaughlin, D. K. and Tiederman, W. G. (1973) "Biasing correction for individual

realization of laser anemometer measurements in turbulent flows," *Phys. Fluids* **16**, 2082–2088.

35. Moin, P. and Kim, J. (1980) "On the numerical solution of time-dependent viscous incompressible fluid flows involving solid boundaries," *J. Comput. Phys.* **35**, 381–392.

36. Nye, J. O. and Brodkey, R. S. (1967) "Light probe for measurement of turbulent concentration fluctuations," *Rev. Sci. Instrum.* **38**, 26–28.

37. Park, S.-R. and Wallace, J. M. (1992) "The influence of instantaneous velocity gradients on turbulence properties measured with multi-sensor hot-wire probes," *Exp. Fluids* **16**, 17–26.

38. Perry, A. E. (1982) *Hot-Wire Anemometry*, Clarendon Press, Oxford.

39. Peyret, R. and Taylor, T. D. (1985) *Computational Methods for Fluid Flow*, Springer-Verlag, New York.

40. Piomelli, U., Balint, J.-L. and Wallace, J. M. (1989) "On the validity of Taylor's hypothesis for wall-bounded turbulent flows," *Phys. Fluids A* **1**, 609–611.

41. Pitts, W. M. and Kashiwagi, T. (1984) "The application of laser-induced Rayleigh light scattering to the study of turbulent mixing," *J. Fluid Mech.* **141**, 391–429.

42. Raffel, R., Willert, C. and Kompenhans, J. (1998) *Particle Image Velocimetry*, Springer-Verlag, New York.

43. Simoéns, S. and Ayrault, M. (1994) "Concentration flux measurements of a scalar quantity in turbulent flows," *Exp. Fluids* **16**, 273–281.

44. Taylor, G. I. (1938) "The spectrum of turbulence," *Proc. R. Soc. London Ser. A* **164**, 476–490.

45. Tsinober, A., Kit, E. and Dracos, T. (1992) "Experimental investigation of the field of velocity gradients in turbulent flows," *J. Fluid Mech.* **242**, 169–192.

46. Utami, T. and Uedo, T. (1984) "Visualization and picture processing of turbulent flow," *Exp. Fluids* **2**, 25–32.

47. Vinçont, J.-Y., Simoëns, S., Ayrault, M. and Wallace, J. M. (2000) "Passive scalar dispersion in a turbulent boundary layer from a line source at the wall and downstream of an obstacle," *J. Fluid Mech.* **424**, 127–167.

48. Vukoslavčević, P. and Dragan, D. (2000) *Multiple Hot-Wire Probes: Measurements of Turbulent Velocity and Vorticity Vector Fields*, Montenegrin Academy of Science, Podgorica, Yugoslavia.

49. Vukoslavčević, P. and Wallace, J. M. (1996) "A 12-sensor hot-wire probe to measure the velocity and vorticity vectors in turbulent flow," *Meas. Sci. Technol.* **4**, 176–178.

50. Vukoslavčević, P., Wallace, J. M. and Balint, J.-L. (1991) "The velocity and vorticity vector fields of a turbulent boundary layer. 1. Simultaneous measurement by hot-wire anemometry," *J. Fluid Mech.* **228**, 25–52.

51. Wallace, J. M., Bernard, P. S., Chiang, K. S. and Ong, L. (1995) "Contaminant dispersal in bounded turbulent shear flow," *Proc. 13th Symposium on Energy Engineering Science*, Argonne National Laboratory Conference, pp. 177–185.

52. Wallace, J. M. and Foss, J. F. (1995) "The measurement of vorticity in turbulent flow," *Annu. Rev. Fluid Mech.* **27**, 469–514.

53. Way, J. and Libby, P. A. (1970) "Hot-wire probes for measuring velocity and concentration in helium–air mixtures," *AIAA J.* **8**, 976–978.

54. Yeh, Y. and Cummins, H. Z. (1964) "Localized fluid flow measurements with an He–Ne laser spectrometer," *Appl. Phys. Lett.* **4**, 150–153.

4

Properties of Bounded Turbulent Flows

The techniques presented in Chapter 3 have been used by many scientists and engineers to study a wide range of turbulent flows. Often, just the averaged velocity field or forces induced by turbulence have been measured or computed. In some cases, particularly for flows that are relatively easy to produce in the laboratory or simulate numerically, very extensive measurements of the flow properties have been made. These can include the Reynolds stresses and even the complete set of terms in the momentum, energy, and dissipation equations. Occasionally, more complex measurements are taken, such as two-point correlations and conditional averages. Numerical simulations have also been used to determine quantities that are difficult to obtain experimentally, such as the properties of moving fluid particles or the shape of iso-contours of pressure and vorticity.

Besides the body of experimental and numerical results that give an indication of what turbulence is about, some significant theoretical insights into turbulent flow have also emerged. These help make sense of the trends observed in flow measurements or simulation. Theoretical descriptions of flows are most likely when simplifying features of turbulence are present, such as a regular geometry. For example, in the absence of complications caused by surface features, it may be possible to develop similarity solutions for some flows or parts of flows.

It is convenient to divide our description of turbulent flows and their properties into those that evolve under the direct influence of solid boundaries, such as in a channel, pipe, and boundary layer, and those that do not, such as a wake, jet, and mixing layer. Thus this chapter delves into bounded flows, while free shear flows are considered in Chapter 5. In these two chapters some important and specific results about turbulence are presented which help to give insights into its nature. It is not possible to survey the full range of results obtained from the many flows that have been measured and analyzed over the years. Instead, this presentation is limited to a few of the most significant flows by virtue of their generality and their usefulness in demonstrating additional aspects of turbulence. Taken together, these chapters provide much useful insight into turbulent flow and how it may be studied. Later it will become clear that the knowledge presented here plays a significant role in efforts to predict turbulent flows.

4.1   INTRODUCTION

The presence of a solid boundary has a profound effect on turbulence, not the least of which is the fact that it often provides the place where "new" turbulence is generated. Flows near walls have a number of distinctive properties that are common to many otherwise dissimilar flows. Thus the three canonical bounded flows considered here, the channel, pipe, and zero-pressure-gradient flat plate boundary layer, have very similar characteristics close to the wall, whereas away from the wall they have notable differences. For example, the central region of a pipe or channel flow is turbulent, while the region outside turbulent boundary layers is often potential and lacks turbulent fluctuations.

The results described here suggest that flow near simple walls has the three distinct regions illustrated in Fig. 4.1: a relatively thin layer adjacent to the boundary where the flow dynamics are strongly influenced by viscosity; a relatively large *intermediate zone* whose velocity field is often the subject of similarity models; and an outer layer (e.g., near the center of a pipe or channel flow, or the free-stream edge of a boundary layer), which is more or less independent of the direct influence of the boundary. The layer closest to the wall is further divided into an exceedingly thin *viscous sublayer* flush with the bounding surface and a *buffer layer* bridging it with the similarity region. The buffer region is home to many of the most interesting dynamical processes of turbulent flow, such as turbulence creation.

Even though, as will be seen, much is known about the physics of wall-bounded flow, some aspects remain a source of significant controversy. Most notably, there continues to be debate as to the nature of the turbulent structures in the buffer layer and the proper similarity theory to describe the intermediate layer. Here we present some of the opposing views on these issues and leave it to future experiments and computations to one day provide more definitive answers to these questions.

We proceed in this development by considering channel flow first. This has the simplest mathematics and is the most accessible to numerical simulation. Consequently, it has been a focal point of efforts to discover how bounded flows are constructed, and we present a number of insightful results from these studies. Next, attention is turned to pipe flow, which has been at the center of a recent controversy concerning the similarity properties of the intermediate layer. Following this turbulent boundary layers are considered in light of the previous discussion of channel and pipe flow. Then, drawing primarily upon results from numerical studies of channel flow and experimental studies of boundary layers, important questions about the structural makeup of the wall region of turbulent flows are considered.

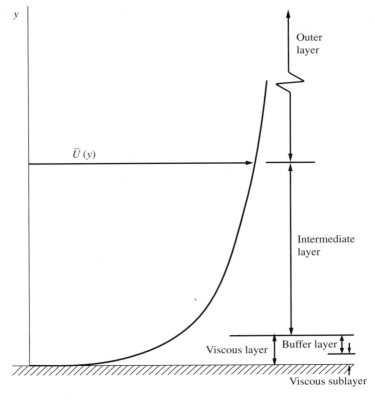

Fig. 4.1   *Conceptualization of boundary layer regions (for clarity, not drawn to proportionate scale).*

4.2   CHANNEL FLOW

4.2.1   Force Balance

The great simplicity of channel flow derives from its being fully developed (i.e., that the average velocity moments do not vary in the streamwise direction). As a result, for flow in the x direction in a channel occupying the region $0 \leq y \leq h$, as shown in Fig. 4.2, $\overline{\mathbf{U}}(\mathbf{x}, t) = (\overline{U}(y), 0, 0)$ and (2.10) gives

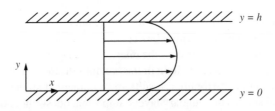

Fig. 4.2   *Geometry of channel flow.*

$$0 = -\frac{\partial \overline{P}}{\partial x} + \frac{d}{dy}\left(\mu\frac{d\overline{U}}{dy} - \rho\overline{uv}\right) \tag{4.1}$$

$$0 = -\frac{\partial \overline{P}}{\partial y} - \rho\frac{d\overline{v^2}}{dy} \tag{4.2}$$

for the x and y momentum equations, respectively. Taking an x derivative of each of these equations shows that $\partial\overline{P}/\partial x$ is independent of both x and y (i.e., that it is constant everywhere in fully developed channel flow).

Integration of (4.2) outward from the wall at a fixed x yields

$$\overline{P}(x, y) = \overline{P}(x, 0) - \rho\,\overline{v^2}(y) \tag{4.3}$$

since $\overline{v^2} = 0$ at a fixed solid boundary. Thus it is seen that at any cross section of the channel, $\overline{P}$ is a minimum where $\overline{v^2}(y)$ is a maximum. This contrasts with the laminar case, where pressure is constant across every cross section. Note as well that the pressure difference between two downstream locations is independent of y [i.e., $\overline{P}(x, y) - \overline{P}(x + L, y) = \overline{P}(x, 0) - \overline{P}(x + L, 0)$] since the streamwise pressure gradient is constant everywhere.

Integration of (4.1) over the area $0 \leq x \leq L, 0 \leq y \leq h$ yields the force balance

$$\Delta\overline{P}\,h - 2\tau_w L = 0, \tag{4.4}$$

where

$$\tau_w = \mu\frac{d\overline{U}}{dy}(0) \tag{4.5}$$

is the (positive) wall shear stress and

$$\Delta\overline{P} = -L\frac{\partial\overline{P}}{\partial x} = \overline{P}(x, 0) - \overline{P}(x + L, 0) \tag{4.6}$$

is the pressure drop between x locations. In deriving (4.4), the symmetry of channel flow is used to conclude that $d\overline{U}/dy(0) = -d\overline{U}/dy(h)$. According to (4.4), the net pressure force on a segment of the channel flow is balanced by the mean viscous shear at the wall surface, which acts to retard the fluid. Note that even though the Reynolds stress, $\overline{uv}$, does not appear explicitly in (4.4), it nonetheless exerts an influence on the force balance through its effect on τ_w.

Figure 4.3 shows the average velocity in channel flow normalized by the mean centerline velocity. For comparison sake, a similarly scaled parabolic laminar velocity profile (i.e., Poiseulle flow) is shown, which solves (4.1) without the Reynolds stress term. In fact, an integration of the laminar form of (4.1) gives

$$\mu dU/dy = \tau_w(1 - 2y/h). \tag{4.7}$$

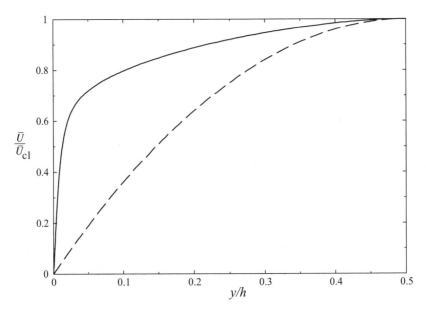

Fig. 4.3   *Average velocity in channel flow of width h scaled by mean centerline velocity U_{cl}. —, turbulent flow; − −, laminar flow.*

Thus the laminar shear stress increases in magnitude linearly from the centerline toward either wall. The peak is at the boundaries, where the greatest retardation force is exerted on the fluid. If the shear stress is viewed in terms of a molecular flux, as in Section 2.1, its positivity in the lower half of the channel means that according to (2.20), at any position there is a net molecular flux of streamwise momentum toward the lower wall. Since the flux increases as one gets closer to the wall, any finite thickness layer of fluid parallel to the boundary is experiencing a net loss of momentum by diffusion. This is balanced by the acceleration caused by the streamwise pressure gradient. In the laminar flow, equally sized layers of fluid experience a similar momentum loss by transport regardless of where they are located.

By its very different shape, the turbulent velocity profile suggests that the underlying physics of the turbulent momentum balance is quite different from that of the laminar case. In particular, there is a penetration of high-speed fluid toward the wall that does not occur in the laminar case, leading to a much higher mean shear at the bounding surface and thus, according to (4.4), the need for a greater pressure differential to drive fluid through the channel for the same flow rate. As will now be shown, it is the presence and action of the Reynolds stress that leads to this outcome.

According to (4.4) and (4.6),

$$-\frac{\partial \overline{P}}{\partial x} = \frac{2}{h}\tau_w. \tag{4.8}$$

Substituting into (4.1) and integrating from 0 to y gives

$$\mu \frac{d\overline{U}}{dy} - \rho\overline{uv} = \tau_w \left(1 - \frac{2y}{h}\right),$$ (4.9)

showing that the total shear stress as given on the left-hand side—which is the sum of the viscous and Reynolds stresses—varies linearly across the channel, just as the viscous stress alone does in the laminar case. Here, however, the viscous and Reynolds stresses do not, individually, vary linearly.

It is helpful in what follows to scale quantities in terms of parameters characterizing flow near boundaries. For this purpose it is traditional to define a *friction velocity*,

$$U_\tau \equiv \left(\frac{\tau_w}{\rho}\right)^{1/2},$$ (4.10)

and a corresponding length scale, ν/U_τ. The y coordinate scaled by ν/U_τ, denoted as

$$y^+ = \frac{U_\tau y}{\nu},$$ (4.11)

makes possible a common scaling of the near-wall flow in channels, pipes, and boundary layers. In fact, in terms of such "wall" units, the physics of these disparate flows looks remarkably similar. The ratio of channel half-width, $h/2$, to ν/U_τ forms the Reynolds number

$$R_\tau = \frac{U_\tau h}{2\nu},$$ (4.12)

which may be interpreted as the number of wall length scales at the channel centerline. Nondimensionalization of (4.9) with respect to U_τ and ν/U_τ gives

$$\frac{d\overline{U}^+}{dy^+} - \overline{uv}^+ = 1 - \frac{y^+}{R_\tau},$$ (4.13)

where $\overline{U}^+ = \overline{U}/U_\tau$ and $\overline{uv}^+ = \overline{uv}/U_\tau^2$. A plot of (4.13) across the channel is given in Fig. 4.4 for a computation with $R_\tau = 590$. The total stress varies linearly across the channel and is everywhere equal to the sum of the mean viscous and Reynolds stresses. The importance of the mean viscous stress is confined to a thin region near the boundary. Thus only layers of fluid very close to the wall feel the retardation effect of the solid boundary at the molecular level. Away from the wall there is no such effect. Instead, the molecular momentum flux is replaced by a turbulent momentum flux as represented by the Reynolds stress. This is antisymmetric across the channel with $\overline{uv}^+ < 0$ on the lower half and $\overline{uv}^+ > 0$ on the upper half. Thus, like the molecular flux, the turbulent momentum flux is always toward the wall in each half of the channel. The location of the peak turbulent momentum flux, which depends on Reynolds number, occurs in this case at $y^+ \approx 42$. $\overline{uv}^+$ varies almost exactly linearly throughout the central region of the channel where the molecular shear stress is negligible.

It may be inferred from (4.13) that as R_τ increases, so does the extent of the region (in y^+ units) where the right-hand side of the equation is nearly constant. In terms of (4.9) this means that in this region, while the viscous shear stress decreases rapidly and the Reynolds shear stress increases rapidly, their sum changes only slowly and is approximately equal to the wall shear stress. Beyond the point where the viscous stress is significant, a region may be discerned where the Reynolds stress is approximately constant and equal to the wall shear stress. The region in which this occurs, often called the *constant stress layer*, may be quite a large region if the Reynolds number is high.

The situation in Fig. 4.4 is reflected in the balance of forces in turbulent channel flow, which thus has a different character from that of the laminar case. In particular, viscous forces are significant only near the wall, and forces originating in turbulent momentum transport dominate the remainder of the channel. The end result is, however, the same: all layers of fluid experience a net loss of momentum which is compensated for by acceleration from the mean pressure force.

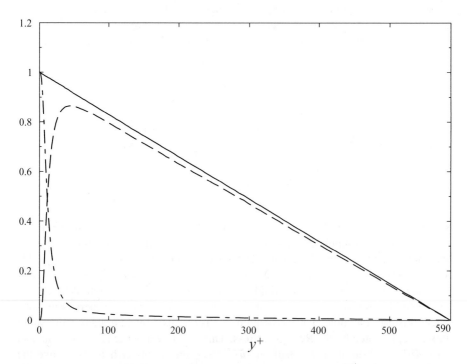

Fig. 4.4 *Decomposition of total stress in turbulent channel flow.* $— \cdot —$, $d\overline{U}^+/dy^+$; $— —$, $\overline{uv}^+$; $—$, $1 - y^+/R_\tau$. *(From [49].)*

These observations are reflected in the balance of terms in the mean momentum equation, (4.1), which in terms of wall units becomes

$$0 = \frac{1}{R_\tau} + \frac{d^2\overline{U}^+}{dy^{+2}} - \frac{d\overline{uv}^+}{dy^+}, \qquad (4.14)$$

with the help of (4.8). Note that $1/R_\tau$ is the pressure force term in this equation. The three terms in (4.14) are plotted in Fig. 4.5, where it is seen that by far the largest forces occur near the wall surface. Away from the wall the acceleration caused by the mean pressure balances the net transport of momentum toward the wall by turbulent fluctuations. Near the wall there is a very large gain in momentum by turbulent transport: specifically, between the wall surface and the point where $|\overline{uv}|$ is a maximum. This effect, together with the small pressure acceleration, are balanced by momentum loss caused by molecular diffusion to the wall surface. Thus the presence of very high wall shear in turbulent flow is clearly a direct consequence of the ability of turbulent eddying motions to force momentum close to the wall surface.

4.2.2 Flow Regimes

Evaluating (4.1) at $y = 0$ and using (4.8) gives

$$\frac{d^2\overline{U}}{dy^2}(0) = -\frac{2}{h}\frac{d\overline{U}}{dy}(0), \qquad (4.15)$$

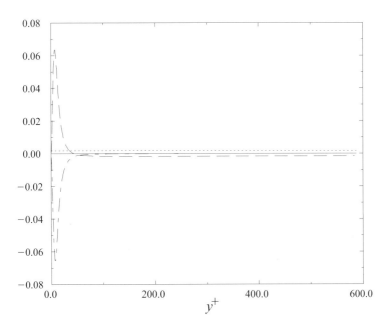

Fig. 4.5 *Decomposition of mean momentum equation (4.14) in turbulent channel flow.* — · —, *viscous force;* — —, *turbulent transport;* · · ·, *pressure force. (From [49].)*

since the identity

$$\frac{\partial \overline{uv}}{\partial y} = \overline{\frac{\partial u}{\partial y} v} + \overline{u \frac{\partial v}{\partial y}} \tag{4.16}$$

implies that $\partial \overline{uv}/\partial y = 0$ at the boundary. By differentiating (4.1) with respect to y and using the fact that $\partial^2 \overline{uv}/\partial y^2$ is also zero at the boundary, it follows that

$$\frac{d^3 \overline{U}}{dy^3}(0) = 0. \tag{4.17}$$

Substituting (4.15) and (4.17) into the Taylor series expansion

$$\overline{U}(y) = \sum_{n=0}^{\infty} \frac{y^n}{n!} \frac{d^n \overline{U}}{dy^n}(0) \tag{4.18}$$

gives

$$\overline{U}(y) = h \frac{d\overline{U}}{dy}(0) \left(\frac{y}{h} - \frac{y^2}{h^2} \right) + O\left[\left(\frac{y}{h} \right)^4 \right], \tag{4.19}$$

where the "big oh" notation, as in $f = g + O[x]$, means that $|f - g| \leq C_1 |x|$ for some constant C_1. After scaling, (4.19) becomes

$$\overline{U}^+(y^+) = y^+ - \frac{(y^+)^2}{2R_\tau} + \cdots, \tag{4.20}$$

where the next term in the series is proportional to $(y^+)^4$. Equation (4.20) makes clear that close to the wall the relation

$$\overline{U}^+(y^+) = y^+ \tag{4.21}$$

should hold to a high degree of accuracy. In fact, experiments (e.g. [2]) show that (4.21) is very accurate until $y^+ \approx 5$. According to (4.20), at $y^+ = 5$ (4.21) is satisfied to within 1.25% for $R_\tau = 200$ and to within 0.03% for $R_\tau = 8000$.

The region of linear mean velocity characterizes the viscous sublayer, where viscosity has a predominant influence on the dynamics of the flow. Figure 4.4 shows that $\overline{uv}$ makes only a very small contribution to the total stress in this region. In fact, experiment shows that the product uv is highly intermittent at any given point within the sublayer. Its large-amplitude fluctuations are due primarily to occasional penetrations of relatively fast moving parcels of fluid ($u > 0$) that succeed in moving close to the wall ($v < 0$), as discussed in Section 4.5.

Figure 4.6 shows the variation of $\overline{U}^+$ with y^+ in semilogarithmic form. The data are from an oil channel flow [66] with a thick viscous sublayer at $R_\tau = 187$. Also shown are DNS channel flow results [37,49] at $R_\tau = 180$ and 590 and experimental

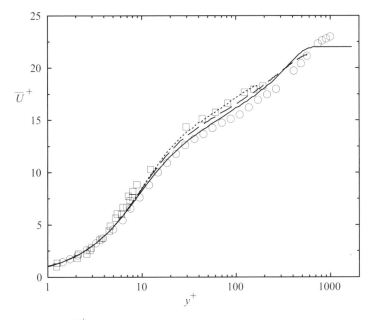

Fig. 4.6   *Mean velocity $\overline{U}^+$.* □, *channel flow measurements [66], $R_\tau = 187$;* ○, *boundary layer measurements [33], $R_\tau = 1050$;* —, *DNS of boundary layer [60], $R_\tau = 650$;* ⋯, *DNS of channel flow [37], $R_\tau = 180$;* − −*[49], $R_\tau = 590$.*

and DNS boundary layer results [33,60] at $R_\tau \approx 1050$ and $R_\tau \approx 650$. Both the experimental and simulation data confirm the essential linearity of the velocity in the viscous sublayer. Here and elsewhere in this discussion of channel flow, data from boundary layers are included to illustrate commonalities in the wall region. The average boundary layer thickness δ is used to determine R_τ for such data.

The mean velocity in Fig. 4.6 rises very rapidly from zero at the wall surface. For a 13.2-cm-thick boundary layer [33], it climbs to about 22% of the free-stream velocity at the upper edge of the sublayer, which is less than 1 mm thick.

Outside the viscous sublayer the Reynolds shear stress rises rapidly through the buffer layer. Here, both the viscous and Reynolds shear stresses are important. As shown in Fig. 4.4, the Reynolds shear stress is about 20% of the total shear stress at the lower end of the buffer layer ($y^+ \approx 5$) and increases to about 85 to 90%, depending on Reynolds number, at the upper end ($y^+ \approx 30$). At this position in the flow, which is still only about 4 mm above the surface in the boundary layer experiment [33], the mean velocity is approximately 55% of the free-stream speed. No theoretically based expression has been derived for the mean velocity distribution in the buffer layer, but several empirical fits to the data have been proposed (e.g., [61]).

The *intermediate layer*, beginning at $y^+ \approx 30$ and extending outward to an upper bound depending on the Reynolds number, is also referred to as the *fully turbulent layer*, *overlap layer*, or *logarithmic layer*. The first of these designations refers to the dominance of the Reynolds stress in this region. The overlap layer designation reflects the constraint that the intermediate layer must be consistent with scaling of both the near-wall layer and the outer region. The logarithmic layer designation, of course, supposes that $\overline{U}$ in this domain can be represented by a universal logarithmic law. The validity of the latter result is at the center of a recent controversy in which it has been asserted that a power law is a more appropriate description of this layer. In Section 4.3, which is devoted to pipe flow, we describe this controversy in some detail. Beyond the intermediate layer is found the core region of the channel. Here $\overline{uv}$ is relatively small and changes sign at the centerline, and $\overline{U}$ is almost constant.

4.2.3   Velocity Moments

An interesting facet of wall region flow is illustrated in Fig 4.7, showing the turbulent kinetic energy, K, together with the three normal components of the Reynolds stress tensor in channel flow at $R_\tau = 590$ from a DNS [49] and from LDV measurements in a boundary layer [33] at $R_\tau = 1050$.

According to the figure, K peaks relatively close to the boundary in the neighborhood of $y^+ = 15$ in the buffer layer. Clearly, this is an indication that turbulent activity is at a maximum in this region. The large K comes about primarily from $\overline{u^2}$; $\overline{v^2}$ and $\overline{w^2}$ are relatively much smaller. The component $\overline{v^2}$, which reflects the magnitude of velocity fluctuations normal to the surface, is heavily damped by its presence. In

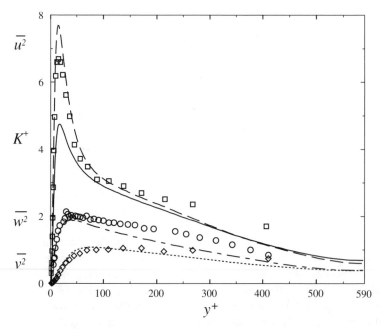

Fig. 4.7 *Kinetic energy and normal stresses. Channel flow DNS [49], boundary layer measurements [33]. $- -$ and □, $\overline{u^2}^+$; $\cdots$ and ◇, $\overline{v^2}^+$; $- \cdot -$and ○, $\overline{w^2}^+$; $—$, K^+.*

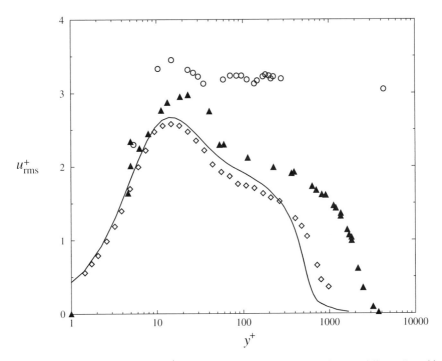

Fig. 4.8 *Measurement and DNS of u_{rms}^+ in channel and boundary layer flows at different Reynolds numbers.* — *[60], $R_\tau = 650$; ◇ [33], $R_\tau = 1050$; ▲ [38], $R_\tau = 2750$; ○, [21], $R_\tau = O(10^6)$.*

the core region there is a semblance of isotropy, suggesting that the special influence of the wall has diminished by this point. The particular anisotropic form taken by the normal stresses near the boundary is a signature of an underlying dynamics of the turbulence which is in need of explanation. Experiments in channel and boundary layer flow (see Fig. 4.8) show that, with increasing Reynolds number, the peak in u_{rms}^+ is less sharp and a plateau appears in the fully turbulent layer. However, the maximum value is reached in about the same y^+ region, even at the very high Reynolds numbers in the atmospheric surface layer, as determined in a field experiment in the Utah desert [21].

As noted previously, $\overline{U}$ is linear near the boundary out to approximately $y^+ = 5$. Since u_{rms}^+ has the Taylor series expansion about $y^+ = 0$,

$$u_{rms}^+ = a_1^+ y^+ + a_2^+ y^{+2} + a_3^+ y^{+3} + \cdots, \tag{4.22}$$

where $a_1^+ \equiv \partial u_{rms}^+ / \partial y^+(0)$, $a_2^+ \equiv (1/2) \, \partial^2 u_{rms}^+ / \partial y^{+2}(0)$, and so on, it is natural to wonder to what distance from the boundary can u_{rms}^+ also be reasonably well modeled as a linear function of y^+ [i.e., to what distance the leading term in (4.22) is dominant]. An equivalent way to look at this question is through the ratio $u_{rms}^+ / \overline{U}^+$, which may be shown from (4.20) and (4.22) to have the series expansion

$$\frac{u_{\text{rms}}^+}{\overline{U}^+} = a_1^+ + y^+ \left(\frac{a_1^+}{2R_\tau} + a_2^+ \right) + \cdots . \tag{4.23}$$

Thus the degree to which u_{rms}^+ behaves linearly near the wall is reflected in the constancy of $u_{\text{rms}}^+/\overline{U}^+$ for small y^+.

The first studies exploring these questions were physical experiments which appeared to show that $u_{\text{rms}}^+/\overline{U}^+$ was constant for only a very short distance (i.e., approximately $y^+ = 0.1$). Later DNS calculations [37], however, suggested that the distance was much longer, in fact a significant fraction of the viscous sublayer. The disagreement prompted a second look at the experiments [2] which revealed the presence of subtle errors in the original measurements caused by instrument heat transfer to the boundary. When compensated for, the physical experiments had excellent agreement with computations yielding the numerical values $a_1^+ \approx 0.4$, $a_2^+ \approx 0.07$, and $a_3^+ \approx 0.008$, so that near the wall,

$$\frac{u_{\text{rms}}^+}{\overline{U}^+} = 0.4 + y^+ \left(\frac{0.2}{R_\tau} + 0.07 \right) + \cdots . \tag{4.24}$$

In this case, linearity of u_{rms}^+ is maintained until $y^+ \approx 2 \to 3$.

Similar to (4.22), it may be shown that

$$w_{\text{rms}}^+ = c_1^+ y^+ + c_2^+ y^{+2} + c_3^+ y^{+3} + \cdots , \tag{4.25}$$

where $c_1^+ \equiv \partial w_{\text{rms}}^+/\partial y^+(0)$, and so on. However, because v_{rms}^+ and $\partial v_{\text{rms}}^+/\partial y^+$ are zero at the wall (the latter is a consequence of incompressibility), it follows that

$$v_{\text{rms}}^+ = b_2^+ y^{+2} + b_3^+ y^{+3} + \cdots . \tag{4.26}$$

where $b_2^+ \equiv (1/2) \, \partial^2 v_{\text{rms}}^+/\partial y^{+2}(0)$, and so on. The values of $c_1^+ \approx 0.25$ and $b_2^+ \approx 0.011$ have been found in computations [37]. In Fig. 4.9, experimental [33] and DNS [49,60] channel and boundary layer data for u_{rms}^+, v_{rms}^+, and w_{rms}^+ in the sublayer are plotted. It is clear from this that the leading-order terms in the Taylor series expansions are good approximations to these quantities very near the wall. The uncertainty in accurately measuring the very small values of y^+ and v_{rms}^+ in the viscous sublayer causes the experimental values of v_{rms}^+/y^{+2} to deviate from the DNS results in this region.

4.2.4 Turbulent Kinetic Energy and Dissipation Rate Budgets

Attention is now turned to the balance of physical processes leading to the turbulent kinetic energy distribution in Fig. 4.7. Application of the general K equation (2.40) to channel flow yields

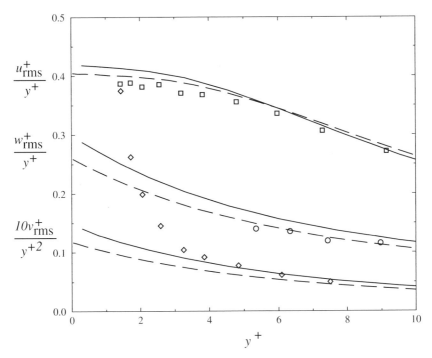

Fig. 4.9 *Near-wall trend in normal Reynolds stresses in a boundary layer measured at $R_\tau = 1050$ [33]. $\square$, u_{rms}^+/y^+; $\circ$, w_{rms}^+/y^+; $\diamond$, $10v_{rms}^+/y^{+2}$; —, DNS boundary layer at $R_\tau = 650$ [60]; – –, DNS channel flow at $R_\tau = 590$ [49].*

$$0 = -\overline{uv}\frac{d\overline{U}}{dy} - \epsilon - \frac{1}{\rho}\frac{d\overline{pv}}{dy} + v\frac{d^2 K}{dy^2} - \frac{d\overline{vu_j^2}/2}{dy}, \qquad (4.27)$$

where it may be recalled that the terms on the right-hand side account for kinetic energy production, dissipation, pressure work, and viscous and turbulent transport, respectively. The energy budget for (4.27) is plotted in Fig. 4.10 from a simulation at $R_\tau = 590$. In this figure, the terms are in wall units. For $y^+ > 30$ it is seen that the turbulent kinetic energy distribution is maintained almost exclusively as a balance between production and dissipation. The largest production rate occurs at $y^+ \approx 12$ (i.e., close to the peak in K itself). The peak in ϵ, on the other hand, is at the wall, and there is an interesting local plateau off the boundary.

Turbulent transport of kinetic energy is important mainly near the wall. It is negative in the range $8 < y^+ < 30$ and positive within $y^+ \approx 8$ of the wall, suggesting that much of the turbulent energy produced in the buffer layer is transfered toward the boundary. At the wall surface, the rate of viscous diffusion is balanced by that of viscous dissipation since the other terms in (4.27) are identically zero. In other

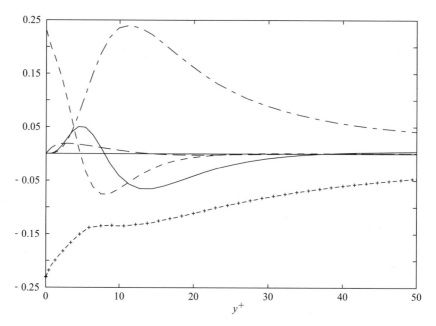

Fig. 4.10 *Turbulent kinetic energy budget in channel flow* $R_\tau = 590$ *scaled with* ν *and* u_τ. *— · —, production; − + −, dissipation; — —, pressure work; − −, viscous diffusion; —, turbulent transport. (From [49].)*

words, molecular diffusion brings energy toward the surface, where it is dissipated. In the wall region outside the viscous sublayer, the energy balance is more complex and involves transfer, production, dissipation, and pressure work.

The location of peak production occurs where the magnitude of the Reynolds shear stress is rapidly increasing with y, while that of the mean gradient in $\overline{U}$ is decreasing rapidly (see Fig. 4.4). A good estimate of this location can be determined by substituting for $\overline{uv}^+$ in the production term using (4.13) to give

$$-\overline{uv}^+ \frac{d\overline{U}^+}{dy^+} = \left(\tau^+ - \frac{d\overline{U}^+}{dy^+}\right) \frac{d\overline{U}^+}{dy^+}, \qquad (4.28)$$

where $\tau^+ = 1 - y^+/R_\tau$ is the total stress. Then the maximum of the production term occurs where the y^+ derivative of (4.28) vanishes. A calculation gives approximately

$$\left(1 - 2\frac{d\overline{U}^+}{dy^+}\right) \frac{d^2\overline{U}^+}{dy^{+2}} = 0 \qquad (4.29)$$

after terms of $O(R_\tau^{-1})$ are dropped. Since $d^2\overline{U}^+/dy^{+2} \neq 0$, as is clear from Figure 4.5, the peak production occurs approximately when

$$\frac{d\overline{U}^+}{dy^+} = \frac{1}{2} = -\overline{uv}^+, \tag{4.30}$$

where the last equality comes from (4.28), assuming that $\tau^+ \approx 1$. The point where (4.30) is satisfied is clearly visible in Fig. 4.4 at $y^+ \approx 12$. This location is not very much affected by the Reynolds number. Figure 4.11 shows experimental boundary layer values for the turbulent kinetic energy production rate [33] and compares them to boundary layer DNS and channel flow DNS results [49,60]. The peak occurs at $y^+ \approx 12$, in agreement with Eq. (4.30). Also shown are the distributions of the turbulent kinetic energy dissipation rate. Evidently, the behavior of both the production and dissipation rates in boundary layers are virtually the same as for channel flow near the wall.

Another interesting facet of the statistics comes from the balance of terms in the ϵ equation, which is shown in Fig. 4.12 for $R_\tau = 590$. Despite the simplifications inherent in channel flow, the individual terms in (2.42) retain the same degree of complexity that they have in the general case, and it is difficult to provide a physical interpretation of their trends. From the figure it is seen that the turbulent vortex

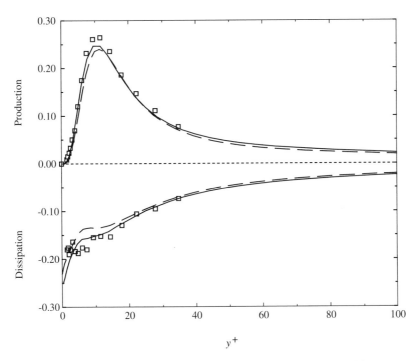

Fig. 4.11 *Comparison of turbulent boundary layer and channel flow production and dissipation rates scaled with v and u_τ. □, boundary layer measurements at $R_\tau = 1050$ [33]; —, DNS boundary layer at $R_\tau = 650$ [60]; – –, DNS channel flow $R_\tau = 590$ [49].*

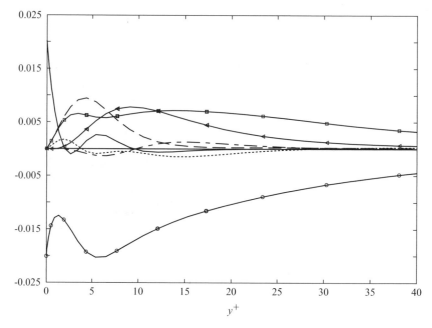

Fig. 4.12 ϵ *equation budget in channel flow at* $R_\tau = 590$ *scaled with* ν *and* u_τ. $-\,-$, P_ϵ^1; ◁, P_ϵ^2; $-\cdot-$,
P_ϵ^3; □, P_ϵ^4; ○, $-\Upsilon_\epsilon$; —, D_ϵ; $\cdots$, $\Pi_\epsilon + T_\epsilon$. *(From [49].)*

stretching term P_ϵ^4 balances the dissipation rate Υ_ϵ away from the channel walls. The
production terms P_ϵ^1 and P_ϵ^2 are significant within approximately 25 wall units of the
surface. Somewhat less important is P_ϵ^3 and the sum of the pressure and transport
terms, $\Pi_\epsilon + T_\epsilon$. Near the wall, $-\Upsilon_\epsilon$ has a local minimum off the surface. At the
boundary, $-\Upsilon_\epsilon$ and D_ϵ are in balance. The factors affecting ϵ near the boundary are
complicated and a very great challenge to modeling.

4.2.5 Reynolds Stress Budget

A numerical evaluation at $R_\tau = 590$ of (2.52) for the individual normal stresses
$\overline{u^2}$, $\overline{v^2}$, and $\overline{w^2}$ is shown in Fig. 4.13. The sum of these budgets, divided by 2, gives
the K equation budget in Fig. 4.10. The budget for $\overline{u^2}$ has much in common with
that of K, apart from the extra loss deriving from the pressure–strain term. For
example, viscous diffusion matches dissipation at the surface. The $\overline{v^2}$ and $\overline{w^2}$ budgets,
on the other hand, are noticeably different. In particular, losses due to dissipation
are balanced mostly by the pressure–strain term, which thus takes on the role of
primary production term. In fact, the production term in (2.52), $-R_{ik}(\partial \overline{U}_j/\partial x_k) -$
$R_{jk}(\partial \overline{U}_i/\partial x_k)$, is nonzero in a channel flow only when $i = j = 1$, and this is exactly
twice the production term in the K equation. Thus, there can be no direct production
of $\overline{v^2}$ or $\overline{w^2}$ from the mean flow. Rather, they are produced as a by-product of the
redistribution of energy from $\overline{u^2}$ to the other normal Reynolds stress components.

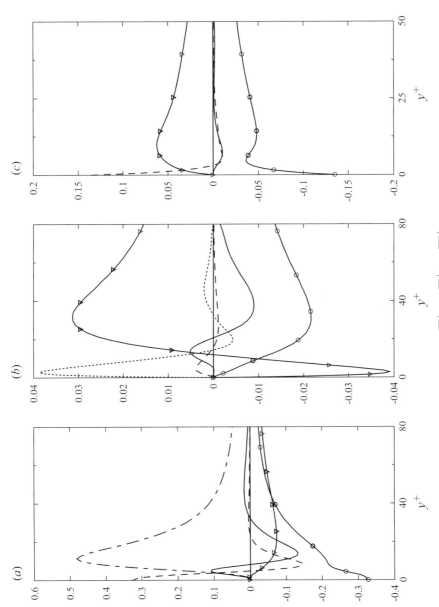

Fig. 4.13 *Normal stress budgets in channel flow for $R_\tau = 590$: (a), $\overline{u^2}^+$; (b), $\overline{v^2}^+$; (c), $\overline{w^2}^+$. — · —, production; ○, dissipation; — · —, pressure work; — —, viscous diffusion; ——, turbulent diffusion; ▽, pressure–strain. Note the difference in scales of the ordinates. (From [49].)*

The redistribution role of pressure–strain is highlighted in Fig. 4.14, showing a plot of the pressure–strain term in each of the normal Reynolds stress equations. The sum of these terms is identically zero (i.e., as was seen before, pressure–strain has no effect on the turbulent kinetic energy). It is interesting to note that moving toward the core region of the pipe, the lost energy in $\overline{u^2}$ is redirected in equal amounts to $\overline{v^2}$ and $\overline{w^2}$. Near the surface, energy redistribution is entirely into the spanwise direction, with much of it coming from the wall normal direction. From considering Fig. 4.13, it is evident that the source of $\overline{v^2}$ is the pressure–work term. However, the net effect of the two pressure terms in the $\overline{v^2}$ equation is relatively small, because they very nearly cancel in the vicinity of the surface. Finally, the $\overline{w^2}$ balance is similar to that of $\overline{u^2}$ next to the wall in that dissipation is equal and opposite to viscous diffusion.

The equation governing the balance of Reynolds shear stress in channel flow is deduced from (2.52) in the form

$$0 = -\overline{v^2}\frac{dU}{dy} - \epsilon_{12} - \frac{d\overline{uv^2}}{dy} - \frac{1}{\rho}\frac{d\overline{pu}}{dy} + \Pi_{12} + \nu\frac{d^2\overline{uv}}{dy^2}, \qquad (4.31)$$

where the first term on the right-hand side accounts for production, followed by dissipation,

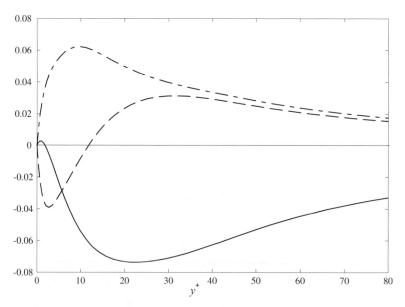

Fig. 4.14 *Pressure–strain term for $R_\tau = 590$. —, $\overline{u^2}^+$ equation; – –, $\overline{v^2}^+$ equation; — · —, $\overline{w^2}^+$ equation. (From [49]).*

$$\epsilon_{12} = 2\nu \overline{\frac{\partial u}{\partial x_k} \frac{\partial v}{\partial x_k}},$$

turbulent transport, pressure–work, pressure–strain,

$$\Pi_{12} \equiv \frac{1}{\rho} \overline{p \left(\frac{\partial u_1}{\partial x_2} + \frac{\partial u_2}{\partial x_1} \right)},$$

and viscous diffusion. The distribution across the channel of the terms in (4.31) is shown in Fig. 4.15. Here, since $\overline{uv} < 0$, the production term is negative. Interestingly enough, this is balanced chiefly by the pressure–strain term, since unlike the case of normal stresses, the dissipation rate is relatively insignificant. In fact, it has been pointed out [44] that the combination of ϵ_{12} and the viscous diffusion term in (4.31) is small throughout the channel, suggesting that the balance of effects leading to changes in $\overline{uv}$ is not strongly dependent on viscosity. These results also make clear the importance of correctly modeling the pressure–strain term if (4.31) is to be used as the basis for developing a predictive scheme. As in the case of the $\overline{v^2}$ balance, the sum of the pressure terms nearly cancels in the vicinity of the wall. This is an argument for

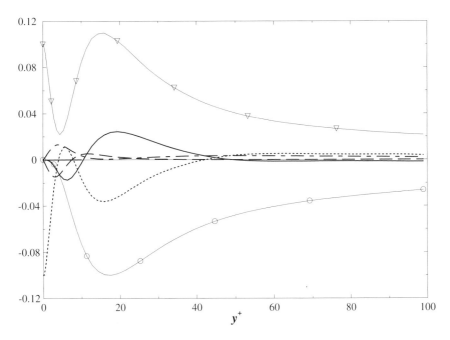

Fig. 4.15 *Reynolds shear stress budget in channel flow for $R_\tau = 590$ scaled with ν and u_τ. ○, production; — · —, dissipation; · · ·, pressure work; − −, viscous diffusion; —, turbulent diffusion; ▽, pressure–strain. (From [49].)*

using the combined pressure terms, instead of considering them separately. In some discussions this approach is adopted [44].

4.2.6   Enstrophy and Its Budget

The decomposition of the enstrophy $\zeta = \overline{\omega_1^2} + \overline{\omega_2^2} + \overline{\omega_3^2}$ into its components and scaled with ν and u_τ for channel flow is shown in Fig. 4.16. DNS channel flow data at $R_\tau = 590$ [49] and experimental boundary layer data at $R_\tau = 1135$ [52] are compared. The value of u_τ used to normalize the experimental data has been corrected following [52]. To an even greater extent than the normal Reynolds stresses shown in Fig. 4.7, the mean-squared vorticity components appear to be isotropic away from the wall. In fact, as close as $y^+ = 40$ there is little difference between them. Since, by definition, $\overline{\omega_1^2}(0) = \overline{(\partial w/\partial y)^2}(0)$, $\overline{\omega_2^2}(0) = 0$, and $\overline{\omega_3^2}(0) = \overline{(\partial u/\partial y)^2}(0)$, the dominance of $\overline{\omega_3^2}$ at and near the surface reflects the greater shearing in the streamwise direction than in the transverse direction. It is noteworthy that $\sqrt{\overline{\omega_3^2}}$ reaches 40% of the magnitude of $\overline{\Omega}_3 \equiv -\partial\overline{U}/\partial y$ at the wall surface.

Also, like the decomposition of K in Fig. 4.7, the anisotropic pattern of the enstrophy components, particularly outside the viscous sublayer, should in some sense be an indication of the presence of an underlying structure in the wall region flow. Since the components of ζ are squared vorticity components, it may be intuited that the

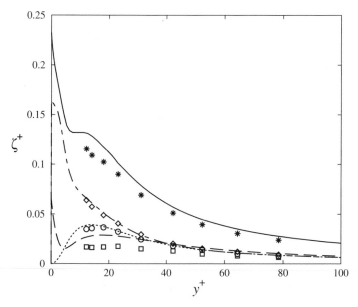

Fig. 4.16   *Comparison of enstrophy and its components in channel flow at $R_\tau = 590$ [49] and in a boundary layer at $R_\tau = 1135$ [52]. $--$ and □, $\overline{\omega_1^2}^+$; $\cdots$ and ○, $\overline{\omega_2^2}^+$; $-\cdot-$ and ◇, $\overline{\omega_3^2}^+$; $-$ and $*$, $\zeta^+ = \overline{\omega_1^2}^+ + \overline{\omega_3^2}^+ + \overline{\omega_3^2}^+$.*

trends in Fig. 4.16 are to some degree associated with the orientation of vortices in the boundary region. By this reckoning, it appears that there is a mechanism causing an increase in streamwise vorticity beyond its local minimum at the upper edge of the viscous sublayer. In the same location there is a noticeable rise in wall-normal vorticity. These observations are consistent with our later analysis of the boundary layer structure in Section 4.5.

As mentioned in Section 2.7.3, the enstrophy has a close relationship to ϵ, being exactly equal to ϵ/ν in homogeneous turbulence. In channel flow the exact relationship (2.125) simplifies to

$$\frac{\epsilon}{\nu} = \zeta + \frac{d^2\overline{v^2}}{dy^2}. \tag{4.32}$$

A plot of ϵ/ν and ζ, scaled in wall units, from a DNS [49] is shown in Fig. 4.17. Good agreement with these results are found from boundary layer experiments and DNS [33,52,60] (see Figs. 4.11 and 4.16). Thus, despite the presence of significant shearing in the channel flow and boundary layer, the two quantities are virtually identical. It thus appears that they can be used almost interchangeably in representing the dissipation rate.

The budget for the ζ equation in channel flow at $R_\tau = 145$ is shown in Fig. 4.18. Comparing Fig. 4.18 to Fig. 4.12 reveals much in common between these balances. The vortex stretching term P_ζ^4 balances enstrophy dissipation away from the

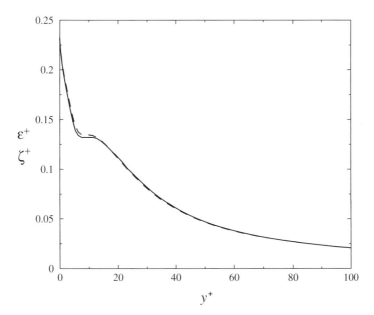

Fig. 4.17 *Comparison of enstrophy and kinetic energy dissipation rate in channel flow at* $R_\tau = 590$. $--$, ϵ^+; $—$, ζ^+. *(From [49].)*

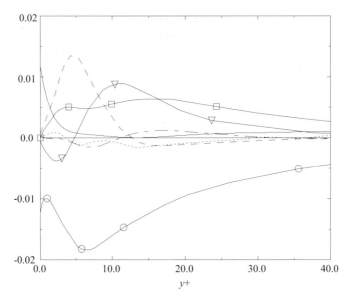

Fig. 4.18 ζ *equation budget in channel flow for* $R_\tau = 145$ *scaled with* v *and* u_τ. $- -$, P_ζ^1; $\triangledown$, P_ζ^2; $- \cdot -$, P_ζ^3; $\square$, P_ζ^4; $\cdots$, T_ζ; $—$, D_ζ; $\circ$, $-\Upsilon_\zeta$. (From [23].)

boundary. At the wall, enstrophy dissipation balances its diffusion. The production terms are qualitatively similar to their counterparts in the ϵ equation. They are concentrated for the most part in the buffer layer, with P_ζ^2 having the largest peak value, occurring just at the boundary between the viscous sublayer and the buffer layer. The turbulent transport and viscous diffusion terms are small compared to the production terms.

Many of the features of channel flow discussed in this section can be expected to carry over to other bounded flows, especially close to fixed walls. This has already been demonstrated with data comparisons in many of the figures in this section. To carry the discussion further, pipe flow is now considered where some important measurements at high Reynolds number are available with which to help reach conclusions about the properties of flow in the intermediate layer.

4.3   PIPE FLOW

In view of the widespread commercial importance of fluid transport in pipes, there has long been considerable interest in understanding the properties of such flows. Here the main concern is with the mean velocity field in fully developed conditions [i.e., when entrance effects have died out so that the mean velocity field consists of only the streamwise component $\overline{U}(r)$]. Under these conditions the averaged streamwise momentum equation in cylindrical coordinates simplifies to

$$0 = -\frac{\partial \overline{P}}{\partial x} + \frac{1}{r}\frac{d}{dr}\left(\mu r \frac{d\overline{U}}{dr} - \rho r \overline{u v_r}\right), \tag{4.33}$$

where $r : 0 \rightarrow R$ is the radial coordinate outward from the center of the pipe, R is the pipe radius, and the velocity fluctuation vector $\mathbf{u} = (u, v_r, v_\theta)$ in the axial, radial, and azimuthal (θ) directions, respectively. As in the case of channel flow, the mean axial pressure gradient is constant everywhere.

It is helpful in the following discussion to introduce a coordinate $y \equiv R - r$ which is normal to the pipe surface and points toward the pipe center (i.e., opposite to the direction of the r coordinate). In terms of y, the mean velocity is given by $\overline{U}^*(y) = \overline{U}(R - y)$. However, since it will always be clear from the context whether $\overline{U}$ or $\overline{U}^*$ is being referred to, henceforth, for simplicity, the symbol $\overline{U}$ is used in all cases.

In terms of y, the wall shear stress

$$\tau_w = \mu \frac{d\overline{U}}{dy}(0), \tag{4.34}$$

and integrating (4.33) over a pipe cross section yields, equivalent to (4.8),

$$-\frac{\partial \overline{P}}{\partial x}\pi R^2 = 2\pi R \tau_w. \tag{4.35}$$

Thus, if τ_w is known (e.g., from knowledge of $\overline{U}$), (4.35) gives the pressure drop. At the same time, the volumetric flux of fluid through the pipe

$$Q = 2\pi \int_0^R \overline{U}(r)r\,dr \tag{4.36}$$

can be determined from $\overline{U}$, and then also the average mass flow velocity, for constant density, $U_m \equiv Q/\pi R^2$. Thus, for any given distribution of $\overline{U}$, unique values of $R_e \equiv U_m R/\nu$ and $R_\tau = U_\tau R/\nu$ are predicted, and there is an implied functional relationship between them. In a similar vein, one can establish a functional relationship between Q and the pressure drop. Empirically determined relationships covering these quantities constitute the standard Moody diagram used in engineering design work [48,67].

4.3.1 Scaling of the Intermediate Layer

It will be tacitly assumed that the physics of pipe flow, for small y, is similar to that of the channel flow. Thus (4.21) should hold in the viscous sublayer, and flow in the buffer layer should have characteristics very similar to those in the channel. Equation (4.21) is an example of similarity behavior in that it holds for all Reynolds numbers. It is consistent with the hypothesis that the mean velocity near the boundary should have the general functional dependence

$$\overline{U} = f(y, \tau_w, \nu, \rho). \qquad (4.37)$$

Because the only two quantities in (4.37) depending on mass are τ_w and ρ, they must combine to form U_τ. Furthermore, it may be observed that only U_τ and ν depend on time, so they also must combine and form the length ν/U_τ. This dimensional reasoning leads to the conclusion that

$$\frac{\overline{U}}{U_\tau} = f(y^+), \qquad (4.38)$$

which is known as the *law of the wall*. Very near the wall, (4.38) reduces to (4.21). The relevancy of (4.38) to flow farther from the wall, in the intermediate layer, is a subject of current debate, which is considered next.

Before doing so, however, note in passing that in the core region of the pipe the relevant parameters that affect $\overline{U}$ are U_τ (or equivalently U_m), y, and R (i.e., ν should not be directly involved because velocity gradients are small). A similarity analysis combined with experimental results indicate that the core region can be described through the *velocity defect law*,

$$\frac{\overline{U}_{cl} - \overline{U}(y)}{U_\tau} = g(\xi), \qquad (4.39)$$

where $\overline{U}_{cl}$ is the mean centerline velocity and $\xi \equiv y/R$ is a similarity variable. The validity of (4.39) extends well beyond the core region, in fact, to encompass much of the intermediate region as well. This description of the outer region generally applies in channel, pipes, and boundary layers.

Thus far it has been shown that close to the wall a law of the wall governs the mean flow in which the linear relation (4.21) holds, while far from the boundary a velocity defect law (4.39) is well supported by experiment. Neither of these results are controversial. In the intermediate layer, however, including the region beginning at the outer edge of the buffer region (i.e., the constant-stress region) and extending to the core region, there is a lack of consensus as to the proper basis on which to develop a similarity solution. Since the original work of von Kármán [64] and Prandtl [54] in the 1930s, it has long been believed that the intermediate layer is best described by a logarithmic law associated with an assumption of complete similarity. In recent years an argument with corroborating experimental comparisons has been put forth with the view that a more appropriate description is that of a power law associated with an assumption of "incomplete similarity." Each of these alternatives is now considered in turn.

4.3.2 Log Law

In essence, the *log law* represents an extension of the law of the wall to the intermediate layer. A standard argument in this direction runs along the following lines: As was seen in the analysis of channel flow, the Reynolds shear stress at the upper end of the buffer layer accounts for almost all the total stress and is approximately equal to the shear stress at the wall. Now $\overline{uv}$, which represents a momentum flux toward the wall, is associated with the mean velocity gradient $d\overline{U}/dy$, since there will be no flux in a uniform flow (i.e., when $\overline{U}$ is constant). Later, in Chapter 6, this association is considered in great depth. In view of the dominance of $\overline{uv}$ in the constant-stress layer, it may be argued that viscosity does not have as significant an influence on the dynamics here as it does on the region closer to the wall.[1] At the same time, it appears reasonable to assume that turbulence in the intermediate layer, in view of its proximity to the boundary, is independent of the scale of the pipe, R. This reasoning leads to the conclusion that the only quantities upon which $d\overline{U}/dy$ should depend, assuming that it obeys a similarity law, are τ_w, the distance from the wall, y, and ρ. In other words,

$$\frac{d\overline{U}}{dy} = f(y, \tau_w, \rho). \tag{4.40}$$

Note that this argument is framed in terms of $d\overline{U}/dy$, instead of $\overline{U}$, to avoid constraints imposed by the nonslip condition at the boundary (i.e., $\overline{U}$ is strongly influenced by viscosity at the surface). As was noted in our derivation of the law of the wall, τ_w and ρ combine to give U_τ, and it may be concluded that the mean velocity derivative should satisfy

$$\frac{d\overline{U}}{dy} \sim \frac{U_\tau}{y}. \tag{4.41}$$

Introducing a dimensionless constant of proportionality κ, known as the *Kármán constant*, (4.41) becomes

$$\frac{d\overline{U}}{dy} = \frac{U_\tau}{\kappa y}. \tag{4.42}$$

Expressing (4.42) in wall units and integrating gives

$$\overline{U}(y^+) = \frac{1}{\kappa} \ln y^+ + B, \tag{4.43}$$

where B is a second "universal" constant. Equation (4.43) is the logarithmic form of the law of the wall given by (4.38).

A more formal means of deriving the same result is the overlap argument of Millikan [45]. This assumes that the defect law (4.39) remains valid in the constant-stress layer, where the law of the wall (4.38) should also hold. In other words, there is an overlap region where both laws should be valid. If this is so, it follows from (4.38) and (4.39) that

[1] However, viscosity *does* enter the argument via τ_w (see [3]).

$$f(y^+) = \frac{U_{cl}}{U_\tau} - g(\xi). \tag{4.44}$$

Differentiating this with respect to y gives

$$\frac{df}{dy^+}(y^+)\frac{U_\tau}{\nu} = -\frac{dg}{d\xi}(\xi)\frac{1}{R}. \tag{4.45}$$

However, $y^+/\xi = U_\tau R/\nu$, so (4.45) becomes

$$y^+\frac{df}{dy^+}(y^+) = -\xi\frac{dg}{d\xi}(\xi). \tag{4.46}$$

The only way that (4.46) can remain consistent with the similarity assumptions in the intermediate and core regions is if each side of the equation is constant: this prevents multivaluedness in the sense that, for a fixed value of ξ, the right-hand side of (4.46) cannot have a different value each time R_τ, and hence y^+, is changed. Setting the constant to $1/\kappa$, it follows that (4.46) gives

$$y^+\frac{df}{dy^+}(y^+) = \frac{1}{\kappa}, \tag{4.47}$$

which once again gives the log law (4.43). The right-hand side of (4.46) indicates that the defect law is logarithmic in the overlap region as well.

An important consideration in applying the log law is the question of the roughness of the boundary surface. The arguments leading to (4.43) assume perfectly smooth surfaces, conditions that are never realized in experiments. In fact, elaborations of the log law can be developed to accommodate roughness, by including another parameter, ϵ/R, called the *relative roughness*, in (4.38). Here ϵ is the average height of the roughness elements. Further consideration of roughness may be found in the literature [56].

Many studies have offered values of the constants κ and B in (4.43). Values of $\kappa = 0.41$ and $B = 5.5$ yield the best overall fit of many experiments. However, if one is willing to make some adjustments in these values, it is possible to get a better fit of any particular experiment or group of experiments. For example, Fig. 4.19 presents a comparison of the log law with channel, pipe, and boundary layer flow data at several Reynolds numbers in which $B = 5.1$. It is interesting to note that in all cases the good fit at lower y^+ breaks down at a point beyond the constant-stress layer, where $\overline{U}$ rises above the log-law prediction. Part of the argument in favor of the power law description of $\overline{U}$ is that it provides an explanation for this observed trend, as discussed in the next section.

4.3.3   Power Law

While the log law has by and large acquired the status of an accepted truth about bounded turbulent flows, it has done so without being reconciled with persistent observations that the power law

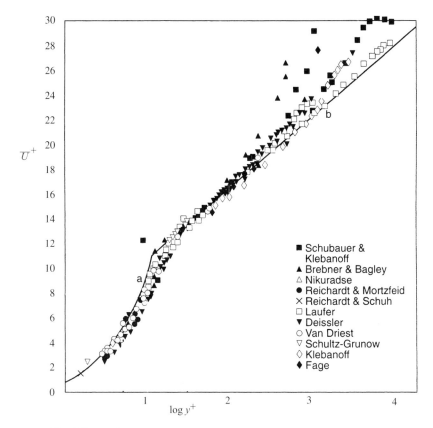

Fig. 4.19 *Log law versus experiment at several Reynolds numbers. (From [47].)*

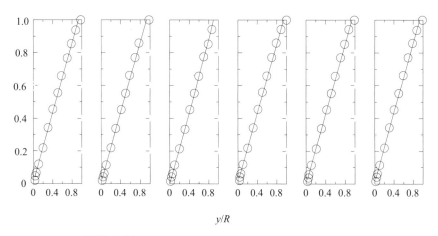

Fig. 4.20 *Plots of $(\overline{U}/\overline{U}_{\max})^{1/\alpha}$ in pipe flow for empirically fitted exponents, α. From left to right $1/\alpha = 6.0, 6.6, 7.0, 8.8, 10.0,$ and $10.0,$ and the Reynolds numbers are $4 \times 10^3, 2.3 \times 10^4, 1.1 \times 10^5, 1.1 \times 10^6, 2 \times 10^6,$ and $3.2 \times 10^6.$ (From [56], p. 563.)*

$$\overline{U}(y^+) = \beta y^{+^\alpha},\tag{4.48}$$

with appropriately selected parameters β and α, is effective in accounting for the mean velocity field across large regions of pipe flow, in fact, regions larger than that for which log law behavior is observed [31]. The only caveat is that the parameters β and α are found to vary with Reynolds number [i.e., (4.48) is not an example of the law of the wall, (4.38)]. Figure 4.20 shows plots of the normalized pipe mean velocity raised to the power $1/\alpha$, with α empirically determined. Linearity of the plotted function implies power law behavior, which is seen to apply over a significant range of Reynolds numbers and covers almost the entire radius of the pipe. In fact, such comparisons were standard in the era before introduction of the log law.

Historical preference for the log law stems from the Reynolds number independence of its coefficients and the relative simplicity of the similarity argument from which it follows. In contrast, by its need to be adjusted for Reynolds number and the absence of theoretical justification, the power law has had more of the character of an empirical relation. This point of view may need to be revised in the face of a recent advance in developing a theoretical framework with which to justify a power law, including the Reynolds number dependence of its coefficients. From this point of view, Reynolds number-dependent coefficients in (4.48) are required by the physics of the intermediate layer and are as such unavoidable. At its heart, the log law/power law question hinges on the validity of the law of the wall assumption (4.38) in the intermediate layer. Specifically, whether viscosity exerts an influence at large Reynolds numbers apart from its appearance in τ_w, is the decisive question. Note that it is already evident in (4.20) that the Reynolds number may potentially have some influence within the buffer layer. The question is whether this influence is likely to extend beyond this to the intermediate layer.

Since the scaling assumption behind the log law may, in fact, be false, it is legitimate to consider the consequences of alternative scaling theories such as the following generalization of (4.40):

$$\frac{d\overline{U}}{dy} = f(y, \tau_w, \nu, \rho, R),\tag{4.49}$$

where now dependence of the intermediate layer on ν and R is not excluded *a priori*. By dimensional arguments similar to those given earlier, this leads to the result

$$\frac{d\overline{U}}{dy} = \frac{U_\tau}{y} f(y^+, R_\tau)\tag{4.50}$$

with $R_\tau = RU_\tau/\nu$. Since, as shown previously, R_τ is functionally related to R_e, one is at liberty to replace (4.50) with

$$\frac{d\overline{U}}{dy} = \frac{U_\tau}{y} f(y^+, R_e).\tag{4.51}$$

The question of power law versus log law is now seen to turn on the behavior of $f(y^+, R_e)$. In effect, the log law presupposes that for $y^+ >> 1$ and $R_e >> 1$, $f(y^+, R_e)$ can be replaced by a constant $1/\kappa$. In the language of asymptotics, this is to say that as $y^+ \to \infty$, $R_e \to \infty$, $f(y^+, R_e) \to 1/\kappa$, or more succinctly,

$$f(\infty, \infty) = \frac{1}{\kappa}. \tag{4.52}$$

In this, all v and R dependencies vanish at high Reynolds number. This is an assumption of complete similarity.

In contrast to (4.52), a different behavior of $f(y^+, R_e)$ as $y^+ \to \infty$, $R_e \to \infty$ may be postulated. Specifically, assume that it has a power series of the form

$$f(y^+, R_e) = \beta(R_e) y^{+\alpha(R_e)} + \cdots, \tag{4.53}$$

where the exponents of subsequent terms in the series (not written) have decreasing and eventually, negative powers of y^+. For large y^+ and R_e, the second and subsequent terms in (4.53) are eclipsed by the first term. Since our interest is confined to the realm of large y^+ and R_e, where just the first term in (4.53) dominates, it follows from (4.53) and (4.51) that

$$\frac{d\overline{U}^+}{dy^+} = \frac{\beta(R_e)}{y^+} y^{+\alpha(R_e)}. \tag{4.54}$$

The assumption contained in (4.53) that leads to (4.54) has been described (see [5]) as incomplete similarity in y^+ and no similarity in R_e. It is, like (4.52), an assumption that requires verification.

Integrating (4.54) with respect to y^+ yields

$$\overline{U}^+(y^+) = \frac{\beta(R_e)}{\alpha(R_e)} y^{+\alpha(R_e)} + C_2, \tag{4.55}$$

and consistency with experimental observations implies that $C_2 = 0$. Therefore, (4.55) reduces to a power law with Reynolds number–dependent coefficient and exponent.

Comparison of (4.55) with measurements of $\overline{U}$ at different Reynolds numbers provides a means for deducing $\alpha(R_e)$ and $\beta(R_e)$. This task is simplified somewhat if it can be established a priori how α and β should depend on R_e as it increases to infinity. A first, reasonable hypothesis is that α and β are constants in the limit of infinite R_e. This motivates the assumption that

$$\alpha(R_e) = \alpha_0 + \alpha_1(R_e), \qquad \beta(R_e) = \beta_0 + \beta_1(R_e), \tag{4.56}$$

where $\alpha_1(R_e)$, $\beta_1(R_e) \to 0$ as $R_e \to \infty$. Note that a more formal justification for these hypotheses comes from examining the asymptotic state of turbulence in the limit of zero viscosity (see [5]). Substituting (4.56) into (4.55) yields

$$\overline{U}^+(y^+) = \frac{(\beta_0 + \beta_1(R_e))}{(\alpha_0 + \alpha_1(R_e))} y^{+\alpha_0 + \alpha_1(R_e)}$$

$$= \frac{(\beta_0 + \beta_1(R_e))}{(\alpha_0 + \alpha_1(R_e))} e^{(\alpha_0 + \alpha_1(R_e)) \ln y^+}.$$

(4.57)

An important clue to obtaining acceptable values of α_0 and α_1 comes from considering the behavior of (4.57) in the limit as $\nu \to 0$. In particular, consider pipe flow for a fixed $\partial \overline{P}/\partial x$. According to (4.35), τ_w remains constant as $\nu \to 0$, and so does U_τ. Since $\overline{U}$ is bounded, $\overline{U}^+$ is bounded, so the left-hand side of (4.57) is bounded as $\nu \to 0$. Consequently, the right-hand side must also be bounded as $y^+ \to \infty$ and $R_e \to \infty$. To keep the exponential term in (4.57) bounded, in this case, it must be that $\alpha_0 = 0$ and that $\alpha_1(R_e) \ln y^+$ converges to a bounded constant. The choice $\alpha_1(R_e) \sim 1/\ln R_e$ as $R_e \to \infty$ is sufficiently general to allow $\alpha_1(R_e) \ln y^+$ to approach a nonzero limit, as against a scaling which forced it to converge to zero.[2] This choice also appears to be most in agreement with experiments. Thus it is assumed that

$$\alpha_1(R_e) = \frac{\alpha_1}{\ln R_e},$$

(4.58)

where, for convenience, the symbol α_1 is also used to denote the proportionality constant on the right-hand side. It is also convenient to assume that $\beta(R_e)$ enjoys the same dependence on R_e as does α as $R_e \to \infty$, so it is also postulated that

$$\beta(R_e) = \beta_0 + \frac{\beta_1}{\ln R_e},$$

(4.59)

where β_0 and β_1 are also constants.

An additional rationale for accepting (4.58) and (4.59) is that expressions of this form, in the limit as $R_e \to \infty$, are independent of the scales used in defining the Reynolds number, so, in fact, α_1, β_0, and β_1 take on a universal character. Thus, if R_e were defined differently (e.g., using a different velocity or length scale), one would have to replace R_e in (4.58) and (4.59) by γR_e, where γ is a constant. However, $\ln \gamma R_e = \ln \gamma + \ln R_e$, which converges to $\ln R_e$ as $R_e \to \infty$, so the parameters α_1, β_0, and β_1 remain unchanged.

Using (4.57) and summarizing the arguments thus far, it has been deduced that

$$\overline{U}^+(y^+) = (\beta_0 \ln R_e + \beta_1) y^{+\alpha_1/\ln R_e},$$

(4.60)

where a factor α_1 is assumed absorbed into the constants β_0 and β_1. Equation (4.60) is a power law description of the intermediate layer of pipe flow. Just as the constants in (4.43) are determined by empirical fit, so is it necessary to find values for α_1, β_0, and

---

[2]As in the log-law case, it is assumed that the limiting condition at $R_e \to \infty$ is quickly reached for finite y^+ even when $R_e < \infty$.

β_1 by comparing (4.60) with experimental data. Such an effort [5] based on tabulated date of Nikuradze [51] over the Reynolds number range, $4 \times 10^3 \rightarrow 3.24 \times 10^6$, gives the approximate values

$$\alpha_1 = 1.5, \qquad \beta_0 = 0.578, \qquad \beta_1 = 2.5. \qquad (4.61)$$

Within the extent of the Nikuradze data, (4.60) provides a succinct way of describing the measured mean velocity at different Reynolds numbers. A particularly suggestive way of seeing how well the power law can account for these data is shown in Fig. 4.21, in which the number pairs ($\ln y^+$, ψ), where

$$\psi \equiv \frac{\ln R_e}{\alpha_1} \ln \left(\frac{\overline{U}^+}{\beta_0 \ln R_e + \beta_1} \right), \qquad (4.62)$$

are plotted from many experiments. In fact, the equation $\psi = \ln y^+$ is just another way of writing (4.60) so that to the extent that (4.60) is true, the data should appear on a single line of unit slope in the ($\ln y^+$, ψ) plane. Figure 4.21 shows that this condition appears to be met except for the region closest to the wall, where the power law is not expected to apply. There are also a small number of contrary data points whose validity has been called into question [5]. The collapse of data in Fig. 4.21 into a line is similar in quality to that achieved by the log law for the identical data set [5]. The conclusion may be drawn that, at least for the experiments performed by Nikuradze, (4.60) appears to be as legitimate a means for describing the data trends as the traditional log law.

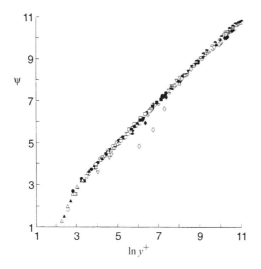

Fig. 4.21 ψ versus $\ln y^+$. *Data from 16 different Reynolds numbers from 4×10^3 to 3.24×10^6 measured in [51] (From [5]. Reprinted with permission of ASME International.).*

A further step in establishing the validity of the power law is to demonstrate that the departures from log behavior that occur at sufficiently large y^+ for any given experiment, as illustrated in Fig. 4.19, occur in a way that can be predicted from the power law. Two main points may be mentioned. First of all, for the fixed values of the constants in (4.61), Eq. (4.60) gives a one-parameter family of curves, one for each value of R_e. As shown by the solid line in Fig. 4.22, the individual power law curves have an envelope which is remarkably similar to the log law, with constants $\kappa = 0.41$ and $B = 5.5$. In fact, a slight adjustment of the constant B in (4.43) from 5.5 to 5.1—the value that was used to fit the pipe flow experiments in Fig. 4.19—brings the two curves to virtual coincidence. This helps account for some of the difficulties in interpreting the behavior of experimental results. Second, an expression has been derived for the angle at which it can be expected that the individual curves depart from the envelope or, equivalently, log-law line, as y^+ increases. This behavior is confirmed to some extent by the data in Fig. 4.19.

These two tendencies are also evident in a series of recent high-Reynolds-number range pipe flow experiments [70,71], as shown in Fig. 4.23. These data encompass both the Reynolds number range in the Nikuradze experiments as well as much higher values. Where the new data correspond to the Reynolds numbers of the older experiments, it is also found to well satisfy (4.60). For the higher Reynolds number range, some disagreement occurs, although by an adjustment of the constants, a result similar to that in Fig. 4.21 is obtained. It has been suggested that a similar shift in the log-law constants caused by pipe roughness may also explain the need to shift constants in the power law. In contrast to this point of view, it has been pointed out that great lengths have been taken to ensure the smoothness of the pipe used in the new experiments. It is not possible to arbitrate between these viewpoints until such time as new data are obtained for high-Reynolds-number pipe flow, although the appropriateness of the power law over the Reynolds number range covered by both sets of experiments seems to be a possibility which has to be seriously considered.

4.4 BOUNDARY LAYER FLOW

The examination of channel flow in Section 4.2 made clear that the region closest to the solid surface is the most critical for the generation and maintenance of turbulent fluid motion. Thus it is particularly interesting to see if more can be learned about what goes on in this part of the flow field. Due to the relative ease with which boundary layers can be generated in the laboratory, the standard zero-pressure-gradient boundary layer formed by a uniform flow over a flat surface has been a favorite of experimentalists intent on exploring the workings of turbulent flows. Such studies have revealed much about the transition of laminar flow to the turbulent state. Moreover, it was in boundary layer research that the first inkling of the presence of coherent structures in turbulent flow became evident. Turbulent boundary layers also appear in a wide variety of natural and practical engineering flows, which contributes to their popularity as a subject of study. In this section some of the general properties of

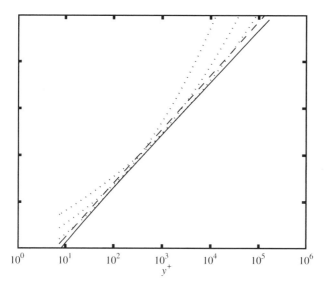

Fig. 4.22 *Comparison of log law with envelope of power law. —, envelope; — —, log law; · · ·, power law with* $R_e = 1.67 \times 10^4, 2.05 \times 10^5, 3.24 \times 10^6$.

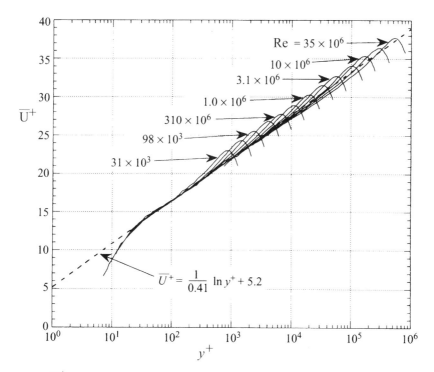

Fig. 4.23 $\overline{U}^+$ *for high-Reynolds-number experiments. (From [70]. Reprinted with the permission of Cambridge University Press.)*

turbulent boundary layers are considered, followed by a discussion of their structure in the next section.

A turbulent boundary layer consists of a thin region adjacent to solid surfaces where the fluid velocity varies from zero at the wall to the free-stream velocity U_∞, as shown in Fig. 4.24. Under normal circumstances, the boundary layer growing on a smooth flat plate will initially be laminar and thus well described by the Blasius similarity solution [56]. At the downstream x location where $\text{Re}_x \equiv xU_\infty/\nu \approx 3 \times 10^6$, transition to the turbulent state typically occurs. Transition to turbulence can be moved upstream in experiments through the use of roughness elements or trip wires on the wall.

To illustrate what a turbulent boundary layer looks like, Fig. 4.25 is a photograph of a smoke-marked turbulent flow over a flat plate in a laboratory wind tunnel. Smoke is introduced both upstream of the image at the interface of the free-stream irrotational flow with the boundary layer and at the wall surface. An obvious property of the flow revealed this way is the presence of apparently strong vortical motions.

The outer edge of the boundary layer shown in Fig. 4.25 is highly corrugated (i.e., there is a discernible, irregular boundary between fluid that appears to be turbulent and nonturbulent). At any y location near the interface, the passage of turbulent rotational fluid alternates with that of nonturbulent irrotational fluid. This is referred to as *intermittency*, and it is one of the significant differences between boundary layer and channel or pipe flows.

Because of intermittency, the thickness of the turbulent fluid layer at a given x position changes in time. Any notion of boundary layer thickness must therefore necessarily be statistical. One obvious definition, commonly denoted as $\delta(x)$, is the distance from the boundary where the mean velocity is within 1% of the free-steam velocity U_∞ [i.e., $\overline{U}(x, \delta) = 0.99U_\infty$]. Another possibility is the *displacement thickness*, $\delta^*(x)$, defined by

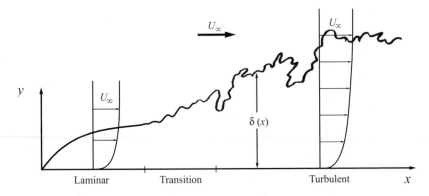

Fig. 4.24 *Turbulent boundary layer over a flat plate.*

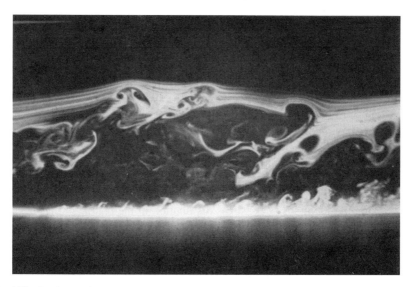

Fig. 4.25   *Smoke visualization of turbulent boundary layer at $R_\theta = 3000$. (From [65]. Copyright 1983. Reproduced by permission of the publisher, ASCE.)*

$$\delta^*(x) \equiv \int_0^\infty \left(1 - \frac{\overline{U}(x, y)}{U_\infty(x)} \right) dy \tag{4.63}$$

as the distance the fixed boundary would have to be displaced normal to itself (into the fluid) in order for a flow at constant velocity U_∞ to have the same local mass flux passing the surface as the actual flow. Finally, the *momentum thickness* θ is defined as

$$\theta(x) \equiv \int_0^\infty \frac{\overline{U}(x, y)}{U_\infty(x)} \left(1 - \frac{\overline{U}(x, y)}{U_\infty(x)} \right) dy. \tag{4.64}$$

This is the thickness a layer of fluid (traveling at speed U_∞) would have to have for it to have a momentum flux equal to that lost by the retarding effect of the boundary [56].

   Probe measurements reveal that intermittency occurs between $0.5 \lesssim y/\delta \lesssim 1.3$. The intermittency function, γ, defined as the ratio of the time when the flow at a given point is completely turbulent to the total sample time, is shown in Fig. 4.26a for measurements in a $R_\theta \approx 1130$ boundary layer. Here $R_\theta = U_\infty \theta/\nu$ is a commonly used Reynolds number for boundary layer flows. Shown in Fig. 4.26b is the distribution of the Reynolds shear stress, averaged in three ways: (1) for the total sample time; (2) only during periods of turbulent, rotational flow; and (3) only during periods of nonturbulent, irrotational flow. In the intermittent region, it is evident that nearly all the Reynolds shear stress occurs during turbulent flow intervals. In fact, when the conventionally defined Reynolds shear stress is divided by the intermittency factor, the results agree with the values averaged in the manner of (2).

As has been suggested, channel, pipe, and boundary layer flows have much common behavior near fixed boundaries. This is illustrated in Fig. 4.27, showing a comparison of the Reynolds shear stress in a zero-pressure-gradient turbulent boundary layer with that from a channel flow. The curves are at different values of R_τ, including a water flow experiment at $R_\tau = 1050$ [33] and DNS calculations at $R_\tau = 150,\ 325,$ and 650 [60]. The exact channel flow expression in (4.13) is shown for $R_\tau = 1050$ using the slope of the mean velocity profile from the experimental boundary layer data [33]. Here $R_\tau = U_\tau \delta / \nu$ for a boundary layer. The different curves are in very good agreement, confirming the assertion that confined and unconfined bounded flows share a common Reynolds shear stress distribution in the region sufficiently close to the wall.

Also shown in Fig. 4.27 is the variation of viscous shear stress, $d\overline{U}^+/dy^+$, and the total stress at the various Reynolds numbers. As predicted by (4.13), the layer of almost constant total shear stress extends to higher y^+ values as R_τ increases, as does the location of the maximum Reynolds shear stress. The boundary layer measurements shown in Fig. 4.8 establish that for higher Reynolds numbers, the rms streamwise velocity decreases from its distinct peak in the buffer layer to a quasi-plateau that begins at the lower edge of the fully turbulent layer ($y^+ \approx 30$,

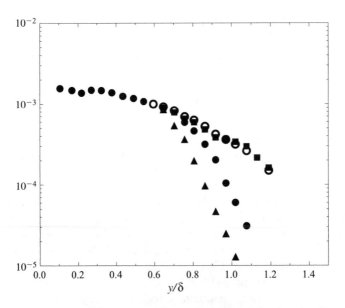

Fig. 4.26   (a) Intermittency factor in a turbulent boundary layer.

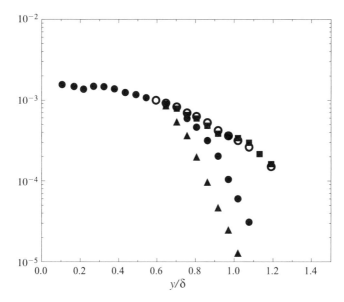

Fig. 4.26   *(b) ●, $\overline{uv}$; ■, $\overline{uv}$ for turbulent intervals; ▲, $\overline{uv}$ for irrotational intervals; ○, $\overline{uv}/\gamma$. All Reynolds shear stress data scaled by U_∞^2. (From [9].)*

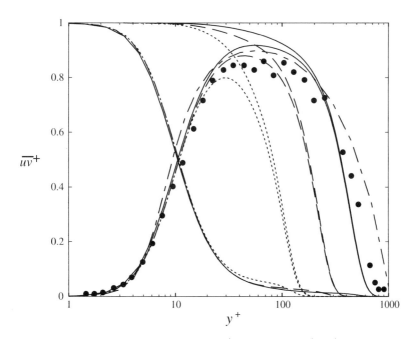

Fig. 4.27   *Distributions of the Reynolds shear stress, $\overline{uv}^+$, viscous stress $d\bar{u}^+/dy^+$, and total stress, τ^+, in turbulent boundary layers. ●, experiment at $R_\tau = 1050$ [33];DNS [60], $\cdots$, $R_\tau = 150$; $--$, $R_\tau = 325$; —, $R_\tau = 650$; $-\cdot-$, channel flow Eq. (4.13) for $R_\tau = 1050$.*

$y/\delta \approx 0.1$). Here the rms values diminish more slowly with increasing y^+. The plateau region becomes less distinct with decreasing Reynolds number until it is almost imperceptible at very low but still turbulent Reynolds numbers. It has been shown [44] that this trend is mirrored in the production term in the $\overline{u^2}$ equation. The distributions of $v_{rms} = \sqrt{\overline{v^2}}$ and $w_{rms} = \sqrt{\overline{w^2}}$ do not have very distinct peaks; rather, they have broad maxima in the lower part of the fully turbulent layer. On the whole, the trends are similar to the channel flow case in Fig. 4.7.

4.4.1 Outer Layers

The occurrence of intermittency distinguishes the outer region of boundary layers from that of channel and pipe flows. Moreover, unlike channel and pipe flows, boundary layers are not fully developed in the sense that $\overline{U}$ varies in the streamwise direction. Moreover, $\overline{V} > 0$ near walls since the incompressibility condition means that the retardation of streamwise momentum by the wall must be balanced by mean convection away from the surface. Despite this, the streamwise variation of mean velocity moments is generally much smaller than their lateral variation, that is,

$$\frac{\partial}{\partial x} << \frac{\partial}{\partial y}. \tag{4.65}$$

Because of this, streamwise changes in the outer layer boundary layer can be accommodated by scaling quantities in terms of $U_\infty(x)$ and $\delta(x)$.

Following along the lines of the analysis of the core region in pipe flow in Section 4.3.1, a velocity defect law of the form

$$U_\infty - \overline{U}(y) = F\left(y, \delta, \rho, U_\tau, \frac{dP_\infty}{dx}\right) \tag{4.66}$$

may be assumed in boundary layer flow. Here x dependence is not indicated explicitly. The phrase "velocity defect" is used since (4.66) describes the deficit of the velocity in the outer layer relative to the free-stream velocity. In this region, viscosity has no direct role, and the streamwise pressure gradient, dP_∞/dx, is included [15,16] to make the analysis more general. Using dimensional analysis, it follows that

$$\frac{U_\infty - \overline{U}(y)}{U_\tau} = F\left(\frac{y}{\delta}, \frac{\delta}{\rho U_\tau^2}\frac{dP_\infty}{dx}\right), \tag{4.67}$$

where the dimensionless group,

$$\frac{\delta}{\rho U_\tau^2}\frac{dP_\infty}{dx},$$

is essentially equivalent to the *Clauser equilibrium parameter*, which has the same form but with δ replaced by δ^*. Experiments suggest that the mean velocity of boundary layers with different pressure gradients collapse to a single profile for equivalent

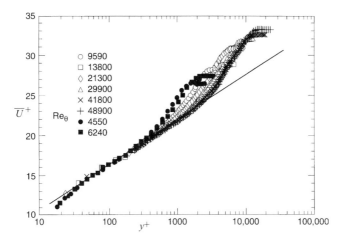

Fig. 4.28   *Comparison of $\overline{U}$ measured in a zero-pressure-gradient boundary layer with the log law* $\overline{U}^+ = (1/0.41)\,\ln y^+ + 5.0$ *(solid line). (From [50]. Copyright 1995 by the American Institute of Aeronautics and Astronautics. Reprinted with permission.)*

values of the equilibrium parameter. Such boundary layers are said to be in turbulent equilibrium with each other.

Figure 4.28 compares some experimental measurements of $\overline{U}$ in a zero-pressure-gradient boundary layer with the log law. As expected from the foregoing discussion, $\overline{U}$ deviates from the logarithmic profile in the outer flow region $0.1 \leq y/\delta \leq 1$. However, the figure also suggests that it may be useful in determining F to require (4.67) to match the logarithmic law of the wall in the fully turbulent region. This is the overlap argument seen in Section 4.3.2. In the present case this means that in the overlap region,

$$\frac{1}{\kappa}\ln y^+ + B = U_\infty^+ - F\left(\frac{y}{\delta}\right) \tag{4.68}$$

or

$$F\left(\frac{y}{\delta}\right) = \left(U_\infty^+ - \frac{1}{\kappa}\ln\delta^+ - B\right) - \frac{1}{\kappa}\ln\frac{y}{\delta}. \tag{4.69}$$

This may be viewed as giving the functional form of F for relatively small y/δ. It then proves helpful when y/δ is in the outer flow region to rewrite F in the form

$$F\left(\frac{y}{\delta}\right) = \left(U_\infty^+ - \frac{1}{\kappa}\ln\delta^+ - B\right) - \frac{1}{\kappa}\ln\frac{y}{\delta} - \frac{\Pi}{\kappa}W\left(\frac{y}{\delta}\right), \tag{4.70}$$

where the function $W(y/\delta)$ represents the perturbation of $F(y/\delta)$ away from the log function. Here Π is a parameter. In fact, Π can be determined from the condition that $F(1) = 0$, which is implied by (4.67), so that

$$\frac{\Pi}{\kappa} W(1) = U_\infty^+ - \frac{1}{\kappa} \ln \delta^+ - B. \tag{4.71}$$

It is usual to select the arbitrary constant $W(1) = 2$. Equation (4.67) then yields, using (4.70), *Coles' law of the wake* [17]:

$$\overline{U}^+ = \frac{1}{\kappa} \ln y^+ + B + \frac{\Pi}{\kappa} W\left(\frac{y}{\delta}\right). \tag{4.72}$$

From these relations it is not hard to show that $W(y/\delta)/W(1)$ can also be interpreted as the ratio

$$\frac{\overline{U}^+ - 1/\kappa \ln y^+ - B}{U_\infty^+ - 1/\kappa \ln \delta^+ - B}, \tag{4.73}$$

representing the velocity deficits with respect to the log law. Coles [17] proposed an empirical sine function fit to the wake function:

$$W\left(\frac{y}{\delta}\right) \approx 2 \sin\left(\frac{\pi}{2} \frac{y}{\delta}\right). \tag{4.74}$$

With this, (4.72) does a good job of representing the mean velocity in the outer region of turbulent boundary layers.

Equation (4.71) indicates that Π has an x dependence through U_∞^+ and δ^+. For the zero pressure gradient, smooth-wall boundary layer, measurements [56] show that $\delta U_\infty/\nu = 0.37(xU_\infty/\nu)^{4/5}$ and $U_\infty^+ = 5.89(xU_\infty/\nu)^{1/10}$, from which it can be ascertained that Π varies relatively slowly with x. For example, with $\kappa = 0.4$ and $B = 5.1$, while Re_x ranges from $5 \times 10^6 \to 10^7$, Π varies from 0.48 to 0.63. Typically, the value $\Pi = 0.55$ is used.

4.4.2 Power Law

In the preceding discussion of the outer region of the turbulent boundary layer, the traditional log law representation of the intermediate layer has been used where they join. However, in the same way that a power law has been offered as a potentially superior model of pipe flows, a power law has been proffered [4] as a more physically correct representation of the intermediate layer in boundary layer flows. Some credence for this is seen in Fig. 4.29, consisting of a log-log plot of the mean velocity in a zero-pressure-gradient boundary layer. Many similar plots for other Reynolds numbers have been made, and they show similar characteristics. In particular, it is seen that two separate straight-line plots represent the data well through much of the boundary layer. Each region appears to be separately representable by a power law. If it can be believed that the flow in the intermediate layer of boundary layers, pipes, and channels should have a common form, such as has long been the assumption with respect to the log law, it may be imagined that the power law region in Fig. 4.29 closest to the wall should in some sense be equivalent to (4.60).

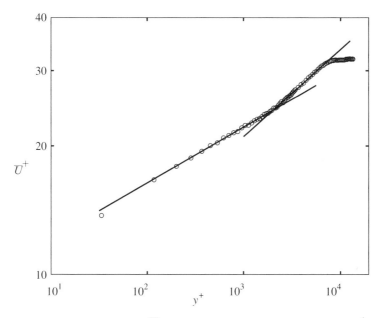

Fig. 4.29 *Representative log-log plot of $\overline{U}$ in boundary layer flow. (Data at $R_\theta = 3 \times 10^4$ from [4]. Copyright 1997 National Academy of Sciences, U.S.A.)*

To make a precise connection between these power laws, it is necessary to have a Reynolds number to incorporate in (4.60) that is appropriate to boundary layer flow. However, it is not self-evident what this should be, since the boundary layer lacks a clear-cut externally imposed characteristic length such as pipe diameter. To proceed in making the comparison, the problem can be turned around by using (4.60) to solve for an effective Reynolds number. To do this, a straight line is fitted to the data in the intermediate layer, such as seen in Fig. 4.29. This yields estimates of the constants α and β in a power law of the form

$$\frac{\overline{U}}{U_\tau} = \beta y^{+\alpha}. \tag{4.75}$$

From these, a Reynolds number, $\mathrm{Re_{BL}}$, may be computed separately from the equations $\beta = \beta_0 \ln \mathrm{Re_{BL}} + \beta_1$ and $\alpha = \alpha_1 / \ln \mathrm{Re_{BL}}$ implied by (4.60). As it turns out, in most cases the two predictions of $\ln \mathrm{Re_{BL}}$ agree closely, suggesting that there may be some validity to (4.60) in boundary layer flow. For example, for the data in Fig. 4.29, the two estimates of $\ln \mathrm{Re_{BL}}$ are 11.5 and 11.25.[3] It is natural to wonder what the relationship might be between $\mathrm{Re_{BL}}$ and such traditionally defined Reynolds numbers as R_θ. Unfortunately, the relationship is not easily discerned since the value

---

[3]The fact that this small discrepancy implies a much larger difference in $\mathrm{Re_{BL}}$ awaits further explanation.

of $\mathrm{Re_{BL}}$ found from two different experiments at a similar R_θ can be significantly different.

The second, or outer, intermediate layer is found from the experimental data to be approximately of the form

$$\frac{\overline{U}}{U_\tau} = C y^{+^m}, \tag{4.76}$$

where C is a constant and the exponent m is generally close to 1/5 and shows some slight Reynolds number dependence. Moreover, experimental results suggest that the existence of the second power law region depends on the level of free-stream turbulence in the outer layer. In particular, as the free-stream turbulence increases, the second power law appears to fade away. Apparently, when the background turbulence is too large, it ceases to exist.

If it can be assumed that $\mathrm{Re_{BL}}$ is reasonably well defined, there should be a meaningful length scale associated with it, even if there is no scale implied by the geometrical configuration. Judging by the behavior of $\overline{U}$ in figures such as Fig. 4.29, it is apparent that there is a naturally appearing length scale in the presence of the sharply defined distance, say L, where the inner power law ends and the outer power law begins. This distance may be interpreted physically as one at which the lingering effects of viscosity—which are very pronounced in the viscous sublayer and not totally lost in the intermediate layer where (4.75) is valid—fully end.

The question whether L may be associated with the length scale implied by the existence of $\mathrm{Re_{BL}}$ has been addressed [6] by comparing it systematically to the Reynolds number $U_\infty L / v$ computed from a large number of experimentally measured velocity profiles. The provisional conclusion has been drawn that there is some measure of a close relationship between these scales, and thus a Reynolds number based on L may have a role in the near-wall power law formula. It is to be expected that many new results concerning this and other facets of the boundary layer scaling will be forthcoming in the near future. For example, it is of interest to find out what relationship there might be between $\mathrm{Re_{BL}}$ and the standard boundary layer Reynolds numbers, such as R_θ.

4.5 STRUCTURE OF BOUNDED TURBULENT FLOWS

If the randomness of turbulent flow is viewed from the perspective of the velocity signal at a fixed point in the field, it is easy to form the impression that turbulence is primarily a local phenomenon, similar, perhaps, to the randomness associated with the movements of molecules in a gas. This viewpoint regarding turbulence dynamics was, in fact, widely held up until the middle of the twentieth century, and encouraged statistical analyses of turbulent motion such as those considered in Chapter 7, where ideal flows are discussed. A series of experimental discoveries beginning in the 1950s changed this view of turbulence to one in which it is now recognized that there are "structural" features embedded in the randomness of turbulent flow. Moreover, these apparently make major contributions to the dynamics of turbulence.

An indication that turbulence might have structure is evident in the peculiar distribution of the normal Reynolds stress near boundaries, such as was shown in Fig. 4.7. The significant anisotropy between the streamwise and spanwise fluctuation velocity components and their maxima near the boundary is not readily explained by the mere presence of a solid wall, as is, for example, the obvious damping effect that the wall has on the wall-normal velocity component. Thus it is not difficult to imagine the presence of unseen agents in the flow whose signature would be the observed Reynolds stress distributions.

Since the 1950s, a number of experiments of increasing sophistication have revealed a degree of connectedness in the motions occurring in bounded turbulent flows. For example, it was shown that velocities at different spatial and/or temporal positions can be highly correlated. Once again, it is easiest to imagine that such phenomena are the result of coherent events within the seemingly random turbulent field. Substantial effort has been expended in attempting to get a deeper understanding of the wall-layer physics that might be responsible for these observations. New experimental techniques have been devised and accompanied by numerical studies using channel and boundary layer simulations. In recent decades, the accumulated knowledge of wall region flow has coalesced around a few rather closely related ideas about the structural composition of the boundary layer. Central to these is the idea that bounded turbulent flows are composed of *coherent structures* (i.e., flow elements or eddies showing a considerable degree of organization and repetitiveness). These objects or events are the key to understanding how these flows function, so it is natural to ask the following questions:

1. What are the structures in bounded turbulent flows (i.e., their physical characteristics)?
2. By what mechanisms are they created and maintained?
3. How does their presence lead to such important dynamical properties as mass, heat, and momentum transport?

Although it is hard to give definitive answers to these questions, nonetheless, there is much that can be said to illuminate all three areas. In this section some major results concerning the first two of these questions are presented. Later, in Chapter 6, after discussing the nature of turbulent transport and its connection to the Reynolds stress, we will be in a position to consider answering the third question. Our first concern is now to give an account of some early studies which led to the idea that there is coherency in bounded turbulent flows.

4.5.1 Low-Speed Streaks, Bursts, and Shear Layers

One of the earliest clues that there must be something special about turbulent flow next to boundaries was the observation of *low-speed streaks* in the near-wall region of

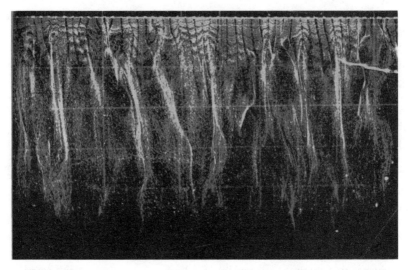

Fig. 4.30 *Visualization of low-speed streaks at $y^+ = 4.5$. (From [40]. Reprinted with the permission of Cambridge University Press.)*

turbulent flows. These consist of elongated, streamwise-oriented regions of low-speed fluid very close to the boundary [40]. They have been observed to extend more than 1000 viscous lengths in the streamwise direction. Figure 4.30 shows such low-speed streaks in a water flow that is marked with micro-air bubbles created by electrolysis from a thin wire stretched across the boundary layer in the spanwise direction at $y^+ = 5$ above the wall. Numerous investigations (e.g., [26,58]) have confirmed that, on average, the low-speed streaks are spaced about 100 viscous lengths apart, although with a large standard deviation. Recently, this nominal streak spacing has been observed visually at very high Reynolds numbers in the range $R_\theta = 10^6 \rightarrow 10^7$ in an experiment carried out in the atmospheric surface layer above the Utah desert [39].

It has been observed further [40] that low-speed streaks can migrate away from the wall (i.e., "lift-up,") and show signs of oscillating in the buffer layer $8 \le y^+ \le 12$, until suffering an abrupt "breakup" in the region $10 \le y^+ \le 30$. During lift-up, locally inflectional velocity profiles such as are prone to instability have been observed. The lift-up and oscillation followed by breakup is called a turbulent *burst*. These observations of the occurrence of a streaky structure and its bursting put a remarkably different face on the boundary layer dynamics than had been held previously. The importance of these phenomena is assured by the discovery that nearly all the production of turbulent kinetic energy from the wall out to $y^+ = 100$ occurs during the bursting periods of the flow.

A number of experiments in bounded flows revealed the presence of an additional structural feature referred to as *localized shear layers* [1,8,18,41,66], occurring mainly in the viscous and buffer regions and often over low-speed streaks. In these, two regions of fluid with significantly different velocity are situated next to or above each other, causing a region of high shear between them. These are likely to originate from the movement of low-speed fluid away from the wall, as happens in bursts or under other circumstances.

Later studies of bursts [36] observed that the oscillatory growth phase of the burst sequence is associated with concentrated streamwise vortical activity. The latter may be regarded as essentially streamwise or quasi-streamwise vortices. By *quasi-streamwise* is meant vortices not exactly aligned in the streamwise direction but with their main orientation in this direction. Subsequently, study of organized structures evolved around the two themes of cyclically occurring instabilities leading to the turbulence production observed in bursts and the dynamics of streamwise-oriented vortical structures. It is not hard to imagine, however, that these two phenomena might, in fact, just be different means of viewing the same thing. In such a view, low-speed streaks are but one fingerprint of streamwise vortices, which also cause bursting motions, and then, by extension, localized shear layers. This hints at the advantages of adopting a vortical view of bounded flows, because with it one is better able to unify the variously observed structural elements and thus better understand the boundary layer dynamics.

The presence of low-speed fluid ($u < 0$) departing, or "ejecting" from the boundary vicinity ($v > 0$), such as is found in bursts, is consistent with the observation of a negative Reynolds shear stress, $\overline{uv} < 0$, in the wall region. Since the converse, events in which $u > 0$ and $v < 0$ (i.e., high-speed fluid approaching the wall), also leads to negative shear stress, it is natural to consider that such motions may also be present near the boundary. Indeed, a visualization experiment [18] in which a translating movie camera filmed the motion of submicron particles seeded in a turbulent pipe flow at Reynolds number $R_\tau = 652$ observed coherent motions that were out of phase with the ejection motions near the wall and had a velocity nearly equal to or somewhat greater than the local mean. These "sweep" motions were parallel to or at an angle toward the wall, sometimes penetrating the viscous sublayer. Another early flow visualization [25] observed that not only were ejection motions large contributors to Reynolds shear stress, but so were apparent "inrushes" of high-speed fluid.

Further insight into the ejection and sweep motions comes from investigating their relative importance at different distances from the wall surface [66,69]. In this, the uv product signal is separated into its four possible sign combinations: quadrant Q1 ($u > 0, v > 0$); Q2 ($u < 0, v > 0$); Q3 ($u < 0, v < 0$); and Q4 ($u > 0, v < 0$). Quadrants Q2 and Q4 are associated with ejection and sweep motions, respectively.

Figure 4.31 shows the contribution of each quadrant to the total Reynolds shear stress measured in a physical turbulent channel flow experiment [66] and in DNS [37,49]. Although there is some disagreement between the experiment and the simulation values for $y^+ \lesssim 10$, they both show that in the buffer layer at $y^+ \approx 15$, Q2 and Q4 motions contribute equally. Closer to the wall, Q4 motions (i.e., sweeps), dominate, whereas Q2 motions (i.e., ejections), dominate farther from the wall. The Q4 motions account for most of the highly intermittent fluctuations in the viscous sublayer.

To summarize, it may be concluded that the Reynolds shear stress appears to be closely associated with ejection and sweep motions, which have a relationship to

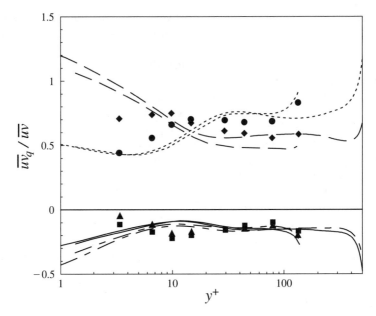

Fig. 4.31 *Quadrant decomposition of Reynolds shear stress in a channel flow. DNS at $R_\tau = 180$ [37] and 590 [49] and experiment [66] at $R_\tau = 187$:* ■ *and* — · —, *Q1;* ● *and* · · ·, *Q2;* ▲ *and* —, *Q3;* ♦ *and* − −, *Q4.*

low-speed streaks, bursts, and quasi-streamwise vortices. We now consider some of the first studies which attempted to identify and characterize the properties of bounded flow structure.

4.5.2 Two-Point Measurements

Some of the earliest bounded flow studies sought to discover how the turbulent field is interconnected (i.e., over what distances and times are its phenomena related or correlated). Such measurements [24] of the two-point correlation coefficient tensor,[4]

$$\mathcal{R}^*_{\alpha\alpha}(\mathbf{r}) \equiv \frac{\overline{u_\alpha(\mathbf{x}, t)u_\alpha(\mathbf{x} + \mathbf{r}, t)}}{\overline{u_\alpha{}^2(\mathbf{x}, t)}}, \tag{4.77}$$

are shown in Fig. 4.32. They reveal that both in the fully turbulent layer at $y/\delta \approx 0.05$ and in the outer layer at $y/\delta \approx 0.7$, there is substantial correlation in $\mathcal{R}^*_{11}(x, 0, 0)$ even for distances greater than the boundary layer thickness.[5] To have correlation over such distances suggests the presence of structures of this scale, since only then could one

---

[4]Here and in the following, Greek letters are used as indices to indicate that the summation convention is being suppressed.
[5]Note that for the purposes of Fig. 4.32, x, y, and z denote the displacement of $\mathbf{r}$ in the corresponding coordinate direction.

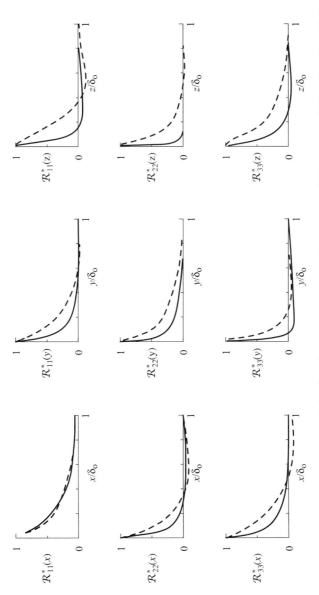

Fig. 4.32 *Two-point correlation measurements in a turbulent boundary layer:* − −, $y/\delta_o \approx 0.7$; —, $y/\delta_o \approx 0.05$. *Here* $\delta_o = 0.7\delta$. *(Adapted from [24]. Reprinted with the permission of Cambridge University Press.)*

explain how the intrinsic randomness between u at two such widely separated points can be held at bay.

For the other components of (4.77), the correlations are considerably larger in the outer layer than they are in the fully turbulent layer. This suggests that whatever structures may be present tend to scale with distance from the wall, even if at the same time they are of a relatively long physical extent in the streamwise direction.

Other early measurements of two-point space-time velocity correlations [19,20] mapped out iso-correlation contours over the x–z plane. High correlations occurred over regions with aspect ratio of 45 elongated in the streamwise direction. Moreover, significant correlations were observed between near-wall points [e.g., $y/\delta = 0.03$ (approximately $y^{+} = 40$)] and points in the outer layer at least as far as $y/\delta = 0.3$. It was also observed that perturbations of the streamwise velocity convected downstream at approximately the local mean velocity. These experiments show that turbulent activity near the wall has a connection to activity far from it and that near-wall events convect downstream for long distances. In both instances the presence of structures is suggested.

Two-point space-time correlation measurements have been used to reveal the evolution of events across the turbulent boundary layer. In one set of such experiments [9], a probe was fixed at $y_o/\delta = 0.03$ ($y^{+} \approx 24$), and a second probe was moved in the direction normal to the wall at the same streamwise location. Figure 4.33 shows the resulting correlation map, where increasing negative time can be viewed as downstream distance by using Taylor's hypothesis (Section 3.2.3.3). The locus of maximum correlation as a function of time delay is shown as the dashed line in the figure.

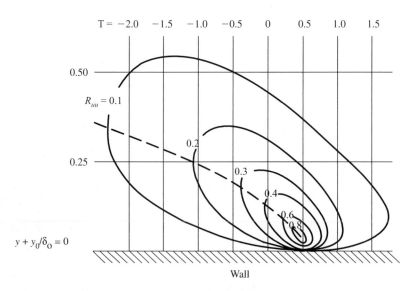

Fig. 4.33   *Space-time correlations of the streamwise velocity fluctuations. Fixed probe at $y = 0$ and wall at $y = -0.03\delta$. (From [9].)*

Just as had been found previously [19,20], the correlation remains significant far into the outer flow. The event revealed here appears to be associated with slow-moving fluid being ejected away from the boundary. It should be noted that a similar space-time correlation map for the v velocity fluctuations reveals much lower correlation levels.

Some insight into the mechanisms leading to high correlations over large spatial extents was obtained in a study using heat as a passive scalar to follow the space-time history of the boundary layer dynamics [11]. The wall under a zero-pressure-gradient turbulent boundary layer was heated to 12°C above the free-stream temperature, and then the temperature and velocity fields were measured downstream. Figure 4.34 shows the principal results. The solid line is the mean location of an internal temperature front with the colder fluid on the upstream side and the hotter fluid downstream. Also shown are projections of the average velocity vectors on the x–y plane in a frame of reference moving with the local mean velocity at each measurement location above the wall. Clearly, the hot fluid that originates near the wall moves, on average, slower than its local mean velocity and is directed away from the wall. By contrast, the cold fluid upstream of the front moves, on average, faster than the local mean velocity and is directed toward the wall. The size of the front is striking: It extends from the probe measurement location closest to the wall in the fully turbulent layer at $y^+ = 35$ ($y/\delta \approx 0.03$) to the outer edge of the boundary layer. Based on indirect evidence, it is believed [11] that the shear layers extend even closer to the wall. Instantaneously, such fronts are highly three-dimensional, so their full extent may not be observed in a single plane, in contrast to the average front shown in Fig. 4.34. It seems likely that the fronts are the result of the juxtaposition of irrotational fluid that has penetrated

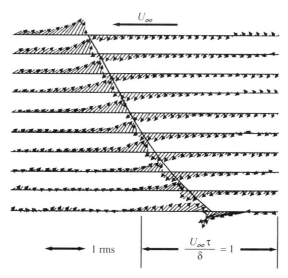

Fig. 4.34 *Temperature visualization of boundary layer structure. (From [11]. Reprinted with the permission of Cambridge University Press.)*

deep into the boundary layer from above, with fluid ejected from near the wall. Just as in Fig. 4.25, where both the wall layer and the irrotational fluid are marked with smoke, the transport both to and away from the wall occurs over almost the entire boundary layer.

These experiments are consistent with and help explain the existence of high correlations over large spatial extents. They suggest the presence of coherent motions of a scale filling a substantial part of the boundary layer. Later flow-visualization studies, discussed next, aimed at giving more precise and detailed information about the structures near the wall that are associated with the region of maximum Reynolds stress production.

4.5.3 Experimental Visualization of Boundary Layer Structure

In an experiment using combinations of up to four hydrogen bubble wires oriented parallel and normal to the wall in a low-Reynolds-number water channel [14], counterrotating streamwise vortices were identified in the wall region, $7 \leq y^+ \leq 70$, inclined at small angles (≈ 3 to $7°$) downstream and away from the wall. While convecting downstream an average distance of $\Delta x^+ \approx 450$, the vortices grew to have average radii of 30 to 40 viscous lengths. Spanwise-oriented vortices were observed downstream of the streamwise vortices, and there was some, but not conclusive evidence that they were connected.

Some additional clues about the nature of the boundary layer structure were extracted from reexamining earlier films of particles in a pipe flow [18,35], particularly at higher Reynolds numbers, where the depth of field is largest. These revealed that some of the ejection and sweep activity could be attributed to vortices oriented principally in the streamwise direction. Other experiments using hydrogen bubble wires [34,59] gave further evidence of streamwise vortices. Both counterrotating pairs and single vortices have been seen. Generally, these are in the wall region $20 \leq y^+ \leq 50$, with the most clearly discernible and energetic in the buffer layer, $14 \leq y^+ \leq 25$. The inference was also drawn in these studies that streamwise vortices are responsible for creating the low-speed streaks and lifting them away from the wall.

A multisensor hot-wire probe has been used [52] to infer the most probable orientation of vorticity in the wall region. Figure 4.35 shows the measured joint probability density function of vorticity components, $P(\Omega_1, \Omega_2)$, in the x–y plane for a position in the buffer layer. There is clearly a strong correlation between Ω_1 and Ω_2 of like sign; moreover, plots of the covariance integrand, $\Omega_1 \Omega_2 P(\Omega_1, \Omega_2)$, at different wall-normal locations reflect the presence of quasi-streamwise vortices that are lifted up at their downstream ends.

The nature of structures in the outer region, where (at least for boundary layers) the flow is intermittent, has also been studied. One technique is to acquire flow measurements in the turbulent and nonturbulent flow states separately. This has shown that in a frame of reference convecting with the interface between the turbulent and nonturbulent regions, the conditionally averaged bulges observed are spanwise vortices with a streamwise scale on the order of the boundary layer thickness. The irrotational

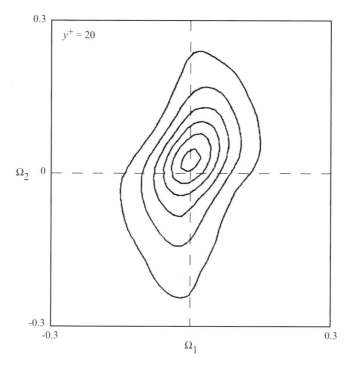

Fig. 4.35 *Joint* pdf *of vorticity components in the x–y plane measured in the buffer layer of a turbulent boundary layer at* $R_\tau = 1135$. *(From [52]. Reprinted with the permission of Cambridge University Press.)*

flow above the boundary layer moves over and around the bulges, sometimes penetrating between them deep into the boundary layer, as seen in the photograph in Fig. 4.25.

The kind of investigations cited in this section reach a reasonably consistent view of the presence of well-organized vortices in the turbulent boundary layer. In parallel with these observations have been many attempts at creating a theoretical picture of the boundary layer that includes vortical structures [62,63]. Most often these have been taken to be vortices in the shape of horseshoes or hairpins, which would cause ejections of low-speed fluid from the region between the counterrotating "legs," while the "head" or "arch" vortices (i.e., spanwise segments connecting the legs) represent bulges at the outer edge of the boundary layer when they arrive there. Although many features of such models are plausible, the experimental observations on which they are based have been too limited to allow precise conclusions to be drawn about the constitution of the boundary layer.

Obtaining detailed information about such elusive objects as turbulent vortices via flow visualization, probe measurements and conditional averaging has proven to be difficult. Thus it has fallen upon full-field numerical simulations, when they became feasible, to provide a more complete picture of these vortices and their life cycles.

4.5.4 General Observations about the Structures

To this point it should be clear that turbulent flow is highly rotational: the presence of a strong and active vorticity field is an essential aspect of turbulence. That the boundary layer has vortical objects in it is abundantly evident from direct visualization or from indirect measurements of correlation functions and other properties. Furthermore, it is not a difficult matter to infer that the other structural elements that have been touched upon earlier, such as bursts, low-speed streaks, and shear layers, can be seen as occurring because of the presence and behavior of vortices in the boundary layer. This viewpoint is now widely accepted, and it will be taken as a given here that the question of the turbulent boundary layer structure can be framed exclusively in terms of vortices. In particular, understanding of the occurrence of bursts and related events will not be sought as manifestations of localized flow instabilities, as has sometimes been considered (e.g., [27,42,57]).

Having accepted the primary importance of vortices to the boundary layer dynamics, it remains to describe them kinematically as far as is possible, and then to attempt to provide the outlines of an answer to our questions concerning where the structures come from and how they are maintained in the flow.

Even though one may have some confidence that vortices are at the heart of the structure of the boundary layer, this is no guarantee that it is easy to identify conclusively what form they take. Indeed, there continues to be debate surrounding the characteristics of such structures. Nonetheless, it may be argued that the differences between interpretations are of a secondary nature and that, in fact, there is a fair degree of consensus as to what is meant by vortical *structure* in the boundary layer.

One source of controversy in identifying structure is the unanswered question as to whether or not the boundary layer structure changes with Reynolds number. Since much of what is known about structures comes from relatively low Reynolds number simulations and experiments, it is not clear if these results are valid at Reynolds numbers in the range where many practical engineering flows occur. Although a few attempts have been made to explore such questions by examining high-Reynolds-number boundary layers found in nature or created in the laboratory, the results are far from conclusive. Thus, it is not yet certain whether or not the boundary layer structure changes with Reynolds number. Hence it must be cautioned that whatever conclusions are reached here about structures using DNS techniques are certain to apply only to the relatively low Reynolds number regime. It will have to be left to future developments to see if current results apply to all Reynolds numbers.

4.5.5 Finding Structure

The tools used for finding structure should not prejudice the finding (i.e., should not predetermine the type of structure that is found). Thus, search methods are needed that can find vortices in the flow in whatever form they might have. Moreover, the method used should not be subjected to false positives or miss structures. It cannot exclude structures according to strength, size, orientation, or any number of other characteristics. Furthermore, the field of view should be large enough to capture entire groups of structures. Ultimately, since there is an inevitable subjectivity to the concept of "vortical structure," it is unlikely that a purely objective search scheme can be found or does exist. Nonetheless, even without being precise, it is still possible to develop reasonable schemes for revealing the presence of structure, and from them draw conclusions about the structural composition of the boundary region. Hopefully, different approaches will yield similar results.

To make sense of the search for vortices, it is commonly argued that a definition of a vortex is required. This has led to some considerable debate [22] because whereas vorticity has a precise mathematical definition as the curl of the velocity field, this is not the case for a vortex. Note that the seemingly obvious choice of defining a vortex as a region of concentrated vorticity fails to take into account the fact that there can be vorticity without observable rotation, as in a laminar shear flow. It is also tempting to propose that a vortex is a region of clearly rotational motion (e.g., as in a tornado). This has to be considered carefully, since different observers might see things differently. For example, to someone translating fast enough with respect to a tornado, it might not look like it has rotational motion. A good illustration of the effect of the observer is seen in visual images of a turbulent channel flow taken by cameras translating at different speeds with respect to the flow (see [56], p. 524). Depending on the translational speed of the camera, rotational motions are revealed in different local regions of the flow field.

To be able to make progress in this discussion, it is convenient to take the definition of a vortex to be that of a region of clearly rotational flow as seen by any "reasonable" observer, namely those observers traveling in a reasonable way with respect to the flow of interest (e.g., at a translational velocity matching local mean conditions). Although this definition is not precise, it will, nonetheless, not prevent the development of a relatively unbiased idea as to what structures are in the boundary layer.

As mentioned previously, streamwise vortices are likely to be an important feature of boundary layer flow. A noteworthy property of such structures is that their rotation is the same for all observers traveling in the axial (streamwise) direction. For example, traveling up or down in a tornado does not remove the view of swirling motion, although it might in the other directions. Thus it is possible to be quite precise about the presence of streamwise vortices, merely by examining end-on velocity vector plots such as shown in Fig. 4.36 from a channel flow simulation. One can develop software for tracking the centers and sizes of the vortices and from that make the reconstruction shown in Fig. 4.37, which is the complete set of vortices found by this method in the region $0 \leq y^+ \leq 100$ of a channel flow at a single instant of time. An important aspect of this approach is that it requires no decision be made as to the strength of rotation of the vortices it finds. A different view of Fig. 4.37 is given in Fig. 4.38, a plan view of the local vorticity vectors at the centers of the structures. The thicknesses of the arrows are keyed to the distance of the vortices from the wall. Thus it is seen that the structures tend to be tilted upward away from the wall in the downstream direction. Whenever the vorticity vector points directly along the axis of the structure identified, then in this local region it is a true vortex and may be perceived of as being tornadolike. In contrast, the vorticity vectors at

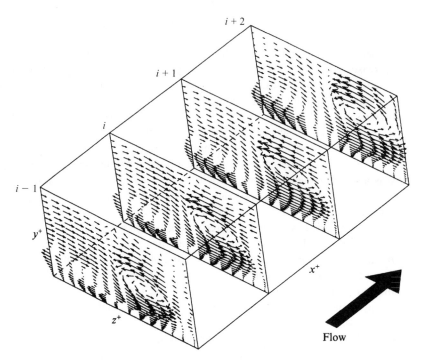

Fig. 4.36   *End-on velocity vector plots showing vortices in a channel flow at $R_\tau = 125$. (From [7]. Reprinted with the permission of Cambridge University Press.)*

the upstream ends of many structures point in the negative z direction, reflecting the fact that they are within the viscous sublayer, where large amounts of negative spanwise vorticity exist as a result of the nonslip boundary condition. Such structures may not fit our picture of a vortex, although it is an interesting part of the boundary layer physics that they generally have, or later will develop, downstream parts that are clearly vortices.

According to Fig. 4.38, vortices come in roughly equal numbers rotating in the clockwise and counterclockwise directions. The lateral distance between structures is of the same physical scale as the low-speed-streak spacing (i.e., $\Delta z^+ \approx 100$). The size and position of the vortices above the wall also appear to be consistent with physical experiments and with other numerical studies which have used similar ideas to find vortices [10]. The streamwise extent of the vortices, varying in the range from $\Delta x^+ = 200$ to $\Delta x^+ = 600$ is also consistent with related observations. It has been observed [7,10] that the quasi-streamwise structures found by this technique come in three forms: long vortices with upstream ends attached to the wall and downstream ends rising up and away from the boundary; short vortices fully attached to the wall; and moderately long vortices completely detached from the boundary. Presumably, an explanation for the occurrence of each of these manifestations of the vortices can be found when their dynamics are fully understood.

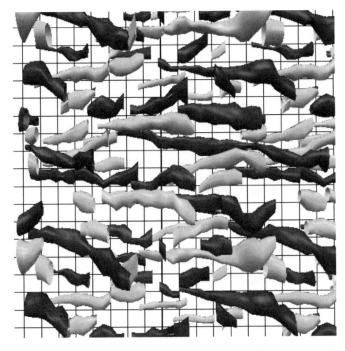

Fig. 4.37 *Instantaneous view of quasi-streamwise vortices in channel flow with sense of rotation indicated by light or dark shading. (From [7]. Reprinted with the permission of Cambridge University Press.)*

The success of the method illustrated in Fig. 4.36 for finding vortices requires that vortices have at least some nonnegligible streamwise orientation. For example, even vortices at a significant angle to the streamwise direction are visible: their flow pattern appearing elliptical rather than circular. However, vortices purely in the spanwise or wall-normal directions cannot be seen by this technique. This shortcoming can be overcome to some extent by combining the end-on view with side or top views [10,28]. Through such a procedure some vortices which are streamwise near the wall have been observed to turn into the spanwise direction at their downstream ends. Those with positive streamwise vorticity turn toward the left, and those with negative streamwise vorticity turn toward the right. Thus in each case the vortex turns in such a way that it has negative spanwise vorticity. This behavior is consistent with a vortical structure having the form of a half-horseshoe or half-hairpin vortex. In other words, two of such objects joined at their downstream ends would have the appearance of a horseshoe or hairpin vortex.

As suggested above, one of the earliest ideas about the boundary layer structure was that it might have horseshoe- or hairpinlike vortices. In the face of the evidence just cited deriving from direct examination of three-dimensional velocity vector fields, it is evident that this viewpoint has to be amended to the idea that the most prevalent structures are quasi-streamwise vortices, and that, of these, some have

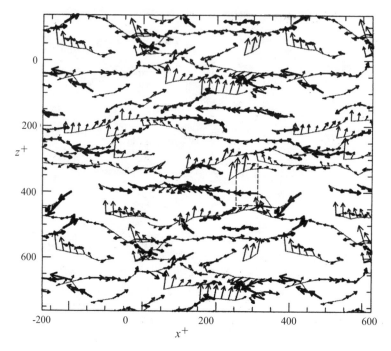

Fig. 4.38 *Vorticity vectors along axis of vortical structures in channel flow. The region inside dashes corresponds to the view in Fig. 4.36. (From [7]. Reprinted with the permission of Cambridge University Press.)*

the character of half-hairpins or half-horseshoes. This conclusion is sustained by numerical investigations of boundary layer structure in which vortices have been identified using plots of three-dimensional iso-contours of various scalar fields that are believed to "mark" the presence of rotational motion (i.e., spatial functions that take on special properties only in the positions occupied by structures, so that they may be distinguished from the background flow). Such schemes can be particularly adept at resolving vortices outside the immediate wall layer, although close to the wall, the intense shearing and high vorticity generally cause these to lose precision.

An obvious marker field is the vorticity itself since the structures of interest have significant vorticity. Some success with this method has been had under highly controlled situations where the properties of specific structures are to be investigated. The streamwise vorticity has been used [46] to find and analyze the dynamics of groups of quasi-streamwise vortices equivalent to those in Fig. 4.37. In this, attention is focused on the regions above the viscous layer where the background vorticity will not obscure the vorticity of individual structures.

In one of the earliest studies searching for structures, pressure was used as a marker [55] since it is expected to be a minimum in the core of strong vortices. The resulting three-dimensional images were found to be strongly sensitive to the chosen pressure magnitude. Thus a considerable degree of subjectivity enters the analysis in the choice of pressure threshold. Too low a threshold and the structures are indistinct; too high, and few survive to be visualized. By careful choice of an intermediate threshold value, the method successfully displays some of the strong vortices in the intermediate layer and above. Many of these are quasi-streamwise vortices, and occasionally some can be classified as half-horseshoe or even asymmetric horseshoe vortices. The latter are horseshoe vortices where one side is much less well developed than the other side. The downstream, mostly spanwise part of these vortices have been described as "arch-like", and on occasion such vortices may exist as detached structures in their own right. These results illustrate one of the principal advantages of this and related methods in the fact that the orientation of the vortices is immaterial to the success of the method.

Subsequent studies have looked for scalar quantities that are better at finding the complete range of coherent vortices in the flow. The idea is to find scalar fields whose values can be associated with rotational motions as viewed by all observers. Several such quantities have been identified and extensively studied. Perhaps the most popular is based on analyzing the types of local motion that are possible in three-dimensional flow. In this, consider the linearized velocity field in the neighborhood of an observer traveling with the fluid, namely,

$$U_i = A_{ij} x_j, \qquad (4.78)$$

where $A_{ij} \equiv \partial U_i / \partial x_j$ and x_j are the local coordinates. The eigenvalues of A satisfy

$$\det (A - \lambda I) = 0, \qquad (4.79)$$

from which the cubic equation

$$\lambda^3 + P\lambda^2 + Q\lambda + R = 0 \qquad (4.80)$$

may be derived [12]. Here $P = -A_{ii}$, $Q = \frac{1}{2}(A_{ii}^2 - A_{ij}A_{ji})$, and $R = -\det A$ are the invariants of the tensor A. In other words, P, Q, and R remain the same under arbitrary rotations and translations of the coordinate system. In incompressible flow such as considered here, $P = 0$ and $Q = -\frac{1}{2}A_{ij}A_{ji}$. It can be shown [12] that if the cubic discriminant of (4.80), namely,

$$D = \left(\frac{R}{2}\right)^2 + \left(\frac{Q}{3}\right)^3, \qquad (4.81)$$

is positive, the local streamlines consist of swirling motion about a point. In this case, (4.80) has one real root and a pair of complex roots. Corresponding to these, A has one real and two complex eigenvectors.

There are several ways of taking advantage of the properties of (4.81) in visualizing structures. One possibility is to plot iso-contours of D equal to some (small) positive number. To achieve a more refined view, this can be done in conjunction with selecting a threshold in the magnitude of vorticity [12].

Another possibility is to plot contours of Q in regions where $D > 0$ [72]. Alternatively, the magnitude of the imaginary part of the complex eigenvalue, which naturally is within the region where $D > 0$, can be used as a marker. This is a particularly effective choice since it is related directly to the amplitude of the swirling motion around the direction associated with the real eigenvector.

Yet another alternative receiving some attention [32] is based on the symmetric tensor

$$B \equiv S^2 + W^2, \tag{4.82}$$

where $S = (A + A^t)/2$ is the rate-of-strain tensor and $W = (A - A^t)/2$ is the rotation tensor. Because it is real and symmetric, B has three real eigenvalues, which can be listed as $\lambda_1 > \lambda_2 > \lambda_3$. It can be shown that regions for which $\lambda_2 < 0$ can be characterized as vortical structures.

The markers based on (4.81) and (4.82) tend to give clearer views of structure than less specialized quantities such as the pressure field. Iso-surfaces based on the magnitude of the imaginary eigenvalue of A are particularly effective, as shown in Fig. 4.39 for a DNS simulation at $R_\tau = 300$. This view includes the flow from the wall to the centerline at $y^+ = 300$ and a region of extent 950×475 viscous lengths in the streamwise and spanwise directions, respectively. Visible in the figure are many quasi-streamwise vortices consistent with Fig. 4.37. Moreover, a number of these in the outer flow fit the description of half-hairpin or half-horseshoe vortices.

It should be mentioned that all methods for finding vortices generally encounter an "end" to the structure. The end is a place where the search mechanism fails to see a continuation of the vortex. If the technique in Fig. 4.36 is used, it means that from one x position to the next, the vortex is not observed. For three-dimensional contour methods, a vortex can end because the threshold of the marker field is too high or too low. In physical space, vortices may end for any one of a number of reasons, such as their being subsumed by other vortices, or they may expand and fade away due to viscous spreading.

It is natural to expect that quasi-streamwise vortices existent at any one time cause the development of new quasi-streamwise vortices. In this sense it is possible to regard groups of structures as related to one another, where the connection has formed sequentially from an original parent vortex. This idea has crystallized recently [13] in the development of a *hairpin packet model* of the boundary layer structure, wherein a packet consists of a group of interrelated quasi-streamwise vortices that develop over time in the streamwise direction. Some of the vortices are likely to be half-hairpin or asymmetric hairpin vortices. Such a packet is thought to exist in Fig. 4.39, centered on the sequence of larger half-hairpin vortices, which it may be noticed are lined up in the streamwise direction. The large-scale coherency of packets may help explain the long-range correlations observed in earlier experiments (e.g., Fig. 4.32). It also suggests that the appearance of hairpin vortices at the outer edge of the boundary layer [29] marks the passage of a packet of vortices that have developed through a growth process begun at the boundary. Evidence of this can also be seen in the smoke-marked inner layer in Fig. 4.25.

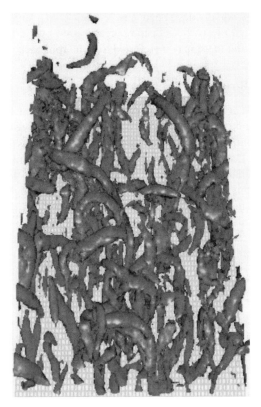

Fig. 4.39 *Iso-contour visualization of vortical structures in channel flow. (From [43].)*

To the extent that this picture is true, it may be argued that hairpin packets, rather than individual quasi-streamwise vortices, form the fundamental structure of the boundary layer. In practical terms, both points of view may be useful in characterizing the boundary layer. Thus, the idea of a self-replication mechanism for quasi-streamwise vortices is to be regarded as a fundamental dynamical aspect of hairpin packets. Moreover, the study of hairpin packets ties in with studies that have examined the detailed nature of the self-replication mechanisms. Subsequent discussion of the vortex regeneration process is devoted primarily to the near-wall region, where the process by which a single vortex generates a new vortex is the center of attention. The picture that emerges can be expected to apply to packets in the region close to the boundary where new offspring develop.

To summarize the findings thus far, it can be stated that the principal coherent vortical structures of the turbulent boundary layer are elongated tubelike vortices having a pronounced streamwise orientation and increasing wall-normal tilt toward their downstream ends. Some of these turn significantly into the spanwise direction, as well, giving the appearance of half-horseshoe or half-hairpin vortices. On some occasions, completely formed hairpin or horseshoe vortices are observed, although they tend not to have equally developed legs. Viewed from a larger perspective, there is some evidence that the vortices display organization over a scale encompassing groups of vortices, and these may be referred to as hairpin packets.

4.5.6 How Are Structures Created and Maintained?

Understanding how coherent vortices are formed in turbulent flows is critical for explaining the dynamics of the turbulent boundary layer and for suggesting strategies for modifying or controlling turbulence. Consequently, there has been much attention paid to this process [53], including many extended analyses combining insights from experiment, computations, and theory. It cannot be said that a single, universally agreed upon mechanism has emerged from this research, although there are common features to many of the viewpoints. For the present purposes the discussion is limited primarily to those studies that have attempted to directly observe the birthing process in situ (i.e., in an actual simulation or experimental realization of a turbulent flow), and for which an unrestricted view is available of the full three-dimensional field. Not touched upon will be many other potentially important theories of the boundary layer dynamics based on model computations and limited experimental views of the flow field. These tend to be somewhat technical in nature, more speculative, and difficult to establish as factual.

The goal here is to provide an explanation for how it is that the boundary layer has the structure described in Section 4.5.5. The focus is on the individual replication mechanism which it may be imagined is part of the events witnessed in the hairpin packet analysis. The question of interest here is analogous to the situation posed to visitors to Earth who wished to discover from whence human beings come. If they selected a large enough segment of the population, they would see representative examples of the entire process: pregnant women, newborn infants, teenagers, adult men. Similarly, it can be assumed that an image of the boundary layer, such as in Figs. 4.37 and 4.39 at a fixed time, shows all or most of the stages of the process by which new vortices are created. The trick is to reconstruct the typical process that creates the objects in the image. This is a difficult undertaking, since the cause-and-effect relationships are not self-evident. However, by viewing the process over time and a sufficiently large spatial extent, the dynamics may be revealed and sense can be made of the instantaneous views.

A relatively small number of studies (e.g., [7,10,30,46]) have attempted to track the dynamics of structures, and from that, explain their origin in the turbulent boundary layer. In these investigations, turbulent structure is identified and then followed in time. New structures appear, and the circumstances surrounding their birth and subsequent growth are analyzed to gain insight into the life cycle of the vortices.

The weight of evidence from these studies suggests the following picture, some aspects of which are illustrated in several accompanying figures.

1. Parent vortices (see Fig. 4.40) tend to be quasi-streamwise vortices whose upstream ends are situated in the viscous wall layer, "attached" to the wall, and their downstream ends "detached" from the wall at an increasing angle with

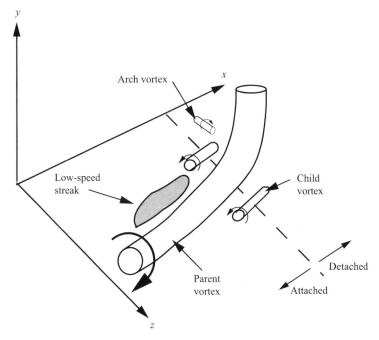

Fig. 4.40 *Parent vortex showing where new streamwise and spanwise offspring are likely to arise.*

distance downstream, sometimes turning into the spanwise direction. Child vortices are first seen in the form of streamwise vortices (with rotation opposite to that of the parent) close to the wall near the detachment point of the vortex, or spanwise arch vortices over low-speed streaks on the ejection side of the parent.

2. There is a drop-off in spanwise velocity near the wall beneath the vortex as one proceeds from the attached to the detached regions of a parent vortex, as shown in Fig. 4.41. This is tantamount to generating wall-normal vorticity (from $\partial w/\partial x$), which reorients into the streamwise direction under the action of the streamwise shearing. This mechanism has been linked [10] to the appearance of new quasi-streamwise vortices near the detachment point of a parent vortex. New wall-normal vorticity can also be expected to form by the action of a parent vortex in reorienting the spanwise vorticity in the viscous sublayer as shown in Fig. 4.42. Subsequent shearing also leads to new streamwise vortices. By either mechanism the offspring are counterrotating to the parent.

3. When new archlike vortices appear, they roll up out of spanwise wall vorticity ejected by a parent vortex over the low-speed streaks. Shearing promotes reorientation of the new vortices into the streamwise direction with opposite rotation from the parent vortex. Approximately 30% of the new vortices can be explained by this mechanism [30].

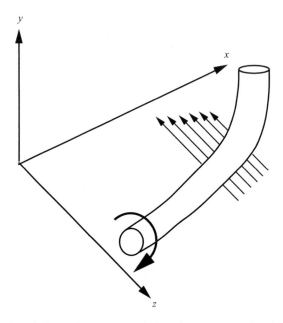

Fig. 4.41 *Spanwise velocity gradients are created where the parent vortex detaches from the wall.*

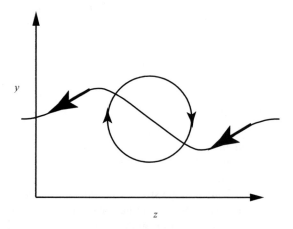

Fig. 4.42 *Reorientation of spanwise vortex lines into the wall-normal direction by a parent vortex.*

4. Parent and child vortices convect and stretch in tandem. The detached part of the parent vortex convects faster than the attached part, leading to a strengthening of the parent by stretching. New parts are adding to the child vortex at its upstream end in the neighborhood of the detachment point of the parent. At the same time, the downstream end of the newly created vortex rises away from the wall, causing stretching and then strengthening of the child vortex.

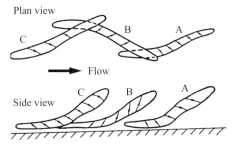

Fig. 4.43   *Conceptual model of the quasi-streamwise vortex structure.* $\omega_1 > 0$ *for* A, C; $\omega_1 < 0$ *for* B. A *is the parent of* B, *which is the parent of* C. *(From [46].)*

5. The last stage of the process has the downstream end of the child vortices situated above the upstream end of the parents [46], as shown in Fig. 4.43. This may possibly be explained by the unequal strengths of parent and child, leading to induction of the child away from the wall.

Any one vortex in the turbulent boundary layer experiences the effect of numerous other nearby structures in an ever-changing panorama, as is clear in Figs. 4.37 and 4.39. Moreover, vortices that are attached to the sublayer must ultimately succumb to the different convection velocities of the attached and detached part to evolve eventually into separate entities. This might consist of separate detached and attached vortices, either of which may perhaps evolve into parent vortices. Thus the complexity of the vortex life cycle is such that it is likely that no two vortex regeneration events need be exactly alike, even if the underlying mechanisms are quite similar and fit the foregoing description. In some sense one may view the appearance of new vortices as opportunistic; when the local arrangement of vortices allows, and a parent is available, new structures might appear, with geometrical placement assuming one of many possibilities. In this way, both the similarities and differences of the various studies can be explained.

Whether as part of the hairpin packet model or not, the idea of a series of self-replicating vortices can be linked to the occurrence of coherent features of the boundary layer such as low-speed streaks and bursts. In the later case, the observed tendency of a single burst sequence to be filled with multiple ejections conforms to the idea of a sequence of vortices. From the point of view of a fixed observer, each vortex as it convects past may cause motion with the appearance of an ejection. The collection of ejections formed from a packet may be taken to be a burst. Low-speed streaks may form by the collective action of vortices in focusing slow-moving fluid into the region between them.

There are clearly many unanswered questions about the dynamics of the boundary layer structure which need to be addressed. For example, the circumstances that lead to the creation of new quasi-streamwise vortices has yet to be mapped out fully. Presumably, future studies that track structures over greater time periods and with greater detail will be able to achieve a greater degree of closure on this and related issues.

4.5.7   Near-Wall Pressure Field

The pressure force acting on solid boundaries immersed in turbulent flow is one of its most critical features. Not only is it often the most significant force acting on bodies (e.g., in bluff body flows), but it has a direct bearing on the generation of noise and other phenomena, such as cavitation. It was remarked previously that quasi-streamwise vortices in the boundary layer can sometimes be clearly marked by their effect on lowering the magnitude of the local pressure field. This suggests that pressure fluctuations felt on solid surfaces may have a connection to coherent vortices passing by overhead. In fact, much research has explored this relationship, with such goals as determining the structural origin of the surface pressure signal. For example, at which level above the wall do the vortices contribute most to pressure fluctuations?

Despite the use of pressure transducers that could resolve only large-scale pressure fluctuations, early studies of the coupled pressure/velocity field [68,69] showed that the wall pressure perturbations correlate with velocity perturbations at various distances from the wall. The pressure perturbations were seen to convect downstream, with the local mean velocity at the wall distance corresponding to the location of the velocity perturbations with which they are correlated. Moreover, the convection speeds increase with the scale of the flow disturbances, as is to be expected since the scale of disturbances are generally related to their mean distance from the wall.

With the availability of DNS it has become possible to establish a more specific connection between vortices and the wall pressure signal. Figure 4.44 shows pressure maxima moving along the surface of a turbulent channel flow which are caused directly by the passage of coherent quasi-streamwise vortices overhead. In fact, it is evident that a series of peak pressure disturbances come and go, all driven by the same vortex. Thus, wall pressure disturbances are caused primarily by vortices close to the wall rather than those in the outer regions of the boundary layer. These observations apply strictly to the low-Reynolds-number flow attainable by DNS. It is not clear whether this is the same story in the high-Reynolds-number case.

4.6   SUMMARY

In this chapter we have examined the nature of wall-bounded flows from the perspective of force, energy, and dissipation balances, its intrinsic similarity behavior, and its structural makeup. The most complex part of bounded flows is the near-wall region, where the strong effect of viscosity gives way to fully turbulent processes involving the dynamics of vorticity. Here the comparatively simple physics of turbulence production matching dissipation, which governs the outer flow, does not apply. Instead, a balance of many physical processes determines the magnitude and distribution of the turbulent energy.

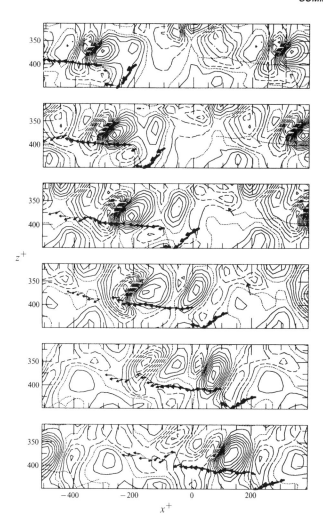

Fig. 4.44 *Wall pressure maxima generated by coherent vortices. (From [7]. Reprinted with the permission of Cambridge University Press.)*

For canonical flows such as in a channel, pipe, and boundary layer, similarity solutions of one kind or another were found to exist close to the wall in the viscosity-dominated viscous sublayer, farther from the wall in the buffer layer, and in the outer flow. It was seen in the example of pipe flow that a Reynolds-number-dependent power law may account for a wider extent of the flow than the traditional log law. Some evidence for a power law behavior also appears to exist for the boundary layer, although a number of questions remain to be answered, including proper interpretation of the implied Reynolds number.

The complex behavior of the near-wall flow takes on an entirely new meaning when considered from the perspective of the dynamics of vortical structures. In this case it is noted that quasi-streamwise vortices appear to be the dominant structure in the near-wall region. They have the capacity for self-replication, and possibly, organize into groups of structures. The dynamical importance of the structures is made clearer in Chapter 6, where their role in causing transport to the solid surface is described in detail. Before this is considered, however, the next order of business is to examine some of the important properties of turbulence in free shear flows, where the direct influence of boundaries on the development of turbulence is minimal.

REFERENCES

1. Alfredsson, P. H. and Johansson, A. W. (1984) "On the detection of turbulence-generating events," *J. Fluid Mech.* **139**, 325–345.

2. Alfredsson, P. H., Johansson, A. V., Haritonidis, J. H. and Eckelmann, H. (1988) "The fluctuating wall-shear stress and the velocity field in the viscous sublayer," *Phys. Fluids* **31**, 1026–1033.

3. Barenblatt, G. I. (1999) "Scaling laws for turbulent wall-bounded shear flows at very large Reynolds numbers," *J. Eng. Math.* **36**, 361–384.

4. Barenblatt, G. I., Chorin, A. J., Hald, O. H. and Prostokishin, V. M. (1997) "Structure of the zero-pressure-gradient turbulent boundary layer," *Proc. Natl. Acad. Sci.* **94**, 7817–7819.

5. Barenblatt, G. I., Chorin, A. J. and Prostokishin, V. M. (1997) "Scaling laws for fully developed turbulent flow in pipes," *Appl. Mech. Rev.* **50**, 413–429.

6. Barenblatt, G. I., Chorin, A. J. and Prostokishin, V. M. (2000) "Characteristic length scale of the intermediate structure in zero-pressure-gradient boundary layer flow," *Proc. Natl. Acad. Sci.* **97**, 3799–3802.

7. Bernard, P. S., Thomas, J. M. and Handler, R. A. (1993) "Vortex dynamics and the production of Reynolds stress," *J. Fluid Mech.* **253**, 385–419.

8. Blackwelder, R. F. and Kaplan, R. E. (1976) "On the wall structure of the turbulent boundary layer," *J. Fluid Mech.* **76**, 89–112.

9. Blackwelder, R. F. and Kovasznay, L.S.G. (1972) "Time scales and correlations in a turbulent boundary layer," *Phys. Fluids* **15**, 1545–1554.

10. Brooke, J. W., and Hanratty, T. J. (1993) "Origin of turbulence-producing eddies in a channel flow," *Phys. Fluids A* **5**, 1011–1022.

11. Chen, C.-H. P. and Blackwelder, R. P. (1978) "Large-scale motion in a turbulent boundary layer: a study using temperature contamination," *J. Fluid Mech.* **89**, 1–31.

12. Chong, M. S., Perry, A. E. and Cantwell, B. J. (1990) "A general classification of three-dimensional flow fields," *Phys. Fluids A* **2**, 765–777.

13. Christensen, K. T. and Adrian, R. J. (2001) "Statistical evidence of hairpin vortex packets in wall turbulence," *J. Fluid Mech.* **431**, 433–443.

14. Clark, J. A. and Markland, E.. (1971) "Flow visualization in turbulent boundary layers," *J. Hydraul. Div. ASCE* **HY19**, 1653–1664.

15. Clauser, F. H. (1954) "The turbulent boundary layer," *Adv. Appl. Mech.* **4**, 1–51.

16. Clauser, F. H. (1954) "Turbulent boundary layers in adverse pressure gradients," *J. Aerosp. Sci.* **21**, 91–108.

17. Coles, D. E. (1956) "The law of the wake in the turbulent boundary layer," *J. Fluid Mech.* **1**, 191–226.

18. Corino, E. R. and Brodkey, R. S. (1969) "A visual investigation of the wall region in turbulent flow," *J. Fluid Mech.* **37**, 1–30.

19. Favre, A., Gaviglio, J., and Dumas, R. (1957) "Space-time correlations and spectra in a turbulent boundary layer," *J. Fluid Mech.* **2**, 313–342.

20. Favre, A., Gaviglio, J., and Dumas, R. (1958) "Further space-time correlations of velocity in a turbulent boundary layer," *J. Fluid Mech.* **3**, 344–356.

21. Folz, A. B. (1997) "An experimental study of the near-surface turbulence in the atmospheric boundary layer," Ph.D. dissertation, University of Maryland.

22. Foss, J. F. (1996) "Vorticity, circulation and vortices," *FED-Vol. 238*, Fluids Engineering Conference, Vol. 3, pp. 83–95.

23. Gorski, J. J., (1993) "Application of vorticity transport analysis to the development of physically accurate turbulence models," Ph.D. dissertation, University of Maryland.

24. Grant, H. L. (1958) "The large eddies of turbulent motion," *J. Fluid Mech.* **4**, 149–190.

25. Grass, A. J. (1971) "Structural features of turbulent flow over smooth and rough boundaries," *J. Fluid Mech.* **50**, 233–255.

26. Gupta, A. K., Laufer, J. and Kaplan, R. E. (1971) "Spatial structure in the viscous sublayer," *J. Fluid Mech.* **50**, 493–512.

27. Hamilton, J. M., Kim, J. and Waleffe, F., (1995) "Regeneration mechanisms of near-wall turbulence structures," *J. Fluid Mech.* **287**, 317–348.

28. Hanratty, T. J. and Papavassilou, D. V. (1997) "The role of vortices in producing turbulence," in *Self-Sustaining Mechanisms of Wall Turbulence*, Vol. 15 (R. L. Panton, Ed.), Computational Mechanics Publications, Southampton, England, pp. 83–108.

29. Head, M. R. and Bandyopadhyay, P. (1981) "New aspects of turbulent boundary layer structure," *J. Fluid Mech.* **107**, 297–338.

30. Heist, D. K., Hanratty, T. J. and Na, Y. (2000) "Observations of the formation of streamwise vortices by rotation or arch vortices," *Phys. Fluids* **12**, 2965–2975.

31. Hinze, J. O. (1975) *Turbulence*, 2nd ed. McGraw-Hill, New York.

32. Jeong, J., Hussain, F., Schoppa, F. and Kim, J. (1997) "Coherent structures near the wall in a turbulent channel flow," *J. Fluid Mech.* **332**, 185–214.

33. Karlsson, R. I. and Johansson, T. G. (1988) "LDV measurements of higher order moments of velocity fluctuations in a turbulent boundary layer," in *Laser Anemometry in Fluid Mechanics III* (R. J. Adrian, Ed.), Ladoan-Instituto Superior Tecnico, Lisbon, pp. 273–279.

34. Kasagi, N., Hirata, M. and Nishino, K. (1986) "Streamwise pseudo-vortical structures and associated vorticity in the near-wall region of a wall-bounded turbulent boundary layer," *Exp. Fluids* **4**, 309–318.

35. Kastrinakis, E. G., Wallace, J. M., Willmarth, W. W., Ghorashi, B. and Brodkey, R. S. (1978) "On the mechanism of bounded turbulent shear flows," *Lect. Notes Phys.* **75**, 175–189.

36. Kim, H. T., Kline, S. J. and Reynolds, W. C. (1971) "The production of turbulence near a smooth wall in a turbulent boundary layer," *J. Fluid Mech.* **50**, 133–160.

37. Kim, J., Moin, P. and Moser, R. (1987) "Turbulence statistics in fully developed channel flow at low Reynolds number," *J. Fluid Mech.* **177**, 133-166.

38. Klebanoff, P.S. (1954) "Characteristics of turbulence in a boundary layer with zero pressure gradient," *NACA TN 3178*.

39. Klewicki, J. C., Metzger, M. M., Kelner, E. and Thurlow, E. M. (1995) "Viscous sublayer visualizations at $R_\theta \approx 1,500,000$," *Phys. Fluids* **7**, 857–863.

40. Kline, S. J., Reynolds, W. C., Schraub, F. A. and Runstadler, P. W. (1967) "The structure of turbulent boundary layers," *J. Fluid Mech.* **30**, 741–773.

41. Kreplin, H.-P. and Eckelmann, H. (1979) "Behavior of the three fluctuating velocity components in the wall region of a turbulent channel flow," *Phys. Fluids* **22**, 1233–1239.

42. Landahl, M. T. (1990) "On sublayer streaks," *J. Fluid Mech.* **212**, 593–614.

43. Liu, Z-C. and Adrian, R. J. (1999) "Evidence for hairpin packet structure in DNS channel flow," *Proc. First International Symposium on Turbulent Shear Flow Phenomena*, S. Bannerjee and J. Eaton, Eds., Begell House, Inc., New York, pp. 609-614.

44. Mansour, N. N., Kim, J. and Moin, P. (1988) "Reynolds-stress and dissipation-rate budgets in a turbulent channel flow," *J. Fluid Mech.* **194**, 15–44.

45. Millikan, C. B. (1938) "A critical discussion of turbulent flows in channels and circular pipes," *Proc. 5th International Conference on Applied Mechanics*, Cambridge, Mass., pp. 386–392.

46. Miyake, Y., Ushiro, R. and Morikawa, T. (1997) "The regeneration of quasi-streamwise vortices in the near-wall region," *JSME Int. J. Ser. B Fluids Thermal Eng.* **40**, 257–264.

47. Monin, A. S. and Yaglom, A. M. (1971) *Statistical Fluid Mechanics*, Vol. 1, MIT Press, Boston.

48. Moody, L. F. (1944) "Friction factors for pipe flow," *ASME Trans.* **66**, 671–684.

49. Moser, R, D., Kim, J. and Mansour, N. N. (1999) "DNS of turbulent channel flow up to $R_\tau = 590$," *Phys. Fluids* **11**, 943–945.

50. Nagib, H. and Hites, M. (1995) "High Reynolds number boundary layer measurements in the NDF," *AIAA Pap. 95-0786*.

51. Nikuradze, J. (1932) "Gesetzmässigkeiten der turbulenten Strömung in glatten Rohren," *VDI Forschungheft 356*.

52. Ong, L. and Wallace, J. M. (1998) "Joint probability density analysis of the structure and dynamics of the vorticity field of a turbulent boundary layer," *J. Fluid Mech.* **367**, 291–328.

53. Panton, R. L. (2001) "Overview of the self-sustaining mechanisms of wall turbulence," *Prog. Aerosp. Sci.* **37**, 341–383.

54. Prandtl, L. (1932) "Zur turbulenten Ströhren und längs Platten," *Ergeb. Aerodyn. Versuchanstalt* **4**, 18–29.

55. Robinson, S. K. (1991) "The kinematics of turbulent boundary layer structure," *NASA TM-103859*.

56. Schlichting, H. (1968) *Boundary Layer Theory*, 6th ed., McGraw-Hill, New York.

57. Schoppa, W. and Hussain, F. (1997) "Genesis and dynamics of coherent structures in near-wall turbulence, a new look," in *Self-Sustaining Mechanisms of Wall Turbulence*, Vol. 15 (R. L. Panton, Ed.), Computational Mechanics Publications, Southampton, England.

58. Smith, C. R. and Metzler, S. P. (1983) "The characteristics of low-speed streaks in the near-wall region of a turbulent boundary layer," *J. Fluid Mech.* **129**, 27–54.

59. Smith, C. R. and Schwartz, S. P (1983) "Observation of streamwise rotation in the near-wall region of a turbulent boundary layer," *Phys. Fluids* **26**, 641–652.

60. Spalart, P. R. (1988) "Direct simulation of a turbulent boundary layer up to $Re_\theta = 1410$," *J. Fluid Mech.* **187**, 61–98.

61. Spalding, D. B. (1961) "A single formula for the law of the wall," *Trans. ASME C J. Appl. Mech.* **28**, 455–458.

62. Theodorsen, T. (1952) "Mechanism of turbulence," in *Proc. 2nd Midwestern Conference on Fluid Mechanics*, Ohio State University, Columbus, Ohio.

63. Townsend, A. A. (1976) *The Structure of Turbulent Shear Flow*, Cambridge University Press, Cambridge, p. 120.

64. Von Kármán, Th. (1930) "Mechanische Ähnlichkeit und Turbulenz," in *Proc. 3rd International Congress on Applied Mechanics*, C. W. Oseen, and W. Weibull, Eds.), AB Sveriges Litografiska Tryckenier, pp. 85–93.

65. Wallace, J. M., Balint, J.-L., Mariaux, J.-L. and Morel, R. (1983) "Observations on the nature and mechanism of the structure of turbulent boundary layers," *Proc. 4th Engineering Mechanics Division Specialty Conference on Recent Advances in Engineering Mechanics and Their Impact on Civil Engineering Practice*, Vol. 2 (W. F. Chen and A. D. M. Lewis, Eds.), ASCE, New York, pp. 1198–1201.

66. Wallace, J. M., Eckelmann, H. and Brodkey, R. S. (1972) "The wall region in turbulent shear flow," *J. Fluid Mech.* **54**, 39–48.

67. White, F. M. (1986) *Fluid Mechanics*, 2nd ed., McGraw-Hill, New York.

68. Willmarth, W. W. (1975) "Structure of turbulence in boundary layers," *Adv. Appl. Mech.* **15**, 159–254.

69. Willmarth, W. W. and Lu, S. S. (1972) "Structure of the Reynolds stress near the wall," *J. Fluid Mech.* **55**, 65–92.

70. Zagarola, M. V. and Smits, A. J. (1998) "Mean flow scaling in turbulent pipe flow," *J. Fluid Mech.* **373**, 33–79.

71. Zagarola, M. V., Smits, A. J., Orszag, S. A., and Yakhot, V. (1996) "Experiments in high Reynolds number turbulent pipe flow," *AIAA Pap. 96-0654.*

72. Zhou, J., Adrian, R. J., Balachandar, S. and Kendall, T. M. (1999) "Mechanisms for generating coherent packets of hairpin vortices in channel flow," *J. Fluid Mech.* **387**, 353–396.

5

Properties of Turbulent Free Shear Flows

In this chapter we consider the properties of turbulent flows which are free to develop without the confining influence of solid boundaries. For the most part, such flows originate through contact with solid surfaces, as in the flow next to a body which subsequently develops into a wake downstream of the body. Thus to some extent the flows considered here represent the aftermath—without boundaries—of flows considered in Chapter 4.

The flows of most interest here have the common property of evolving within a mean shearing that may be traced to the upstream boundaries. Figure 5.1 illustrates three fundamental types of such free shear flows. Wake flows develop downstream of either streamlined or bluff bodies. In this case, turbulence is generated in the passage of the flow around the body and fills up the wake region. The mean velocity behind the body in the wake shows a momentum deficit in comparison to the region outside the wake. In a turbulent jet, fluid exits a nozzle or orifice into a larger domain. Here, an excess of velocity above that of the surrounding flow expands outward into the surrounding fluid. Turbulence can be found in the incoming jet and is created by instability during the initial stages of the jet. A third important configuration occurs in mixing layers where fluid layers traveling at different speeds are brought together. One common occurrence of such flows is behind a splitter plate. Here, boundary layers of differing mean velocities form on top and bottom splitter plate surfaces and combine downstream into a mixing layer such as that in the figure.

Flows of these basic types occur in many technologically important applications, including the flow behind ships, aircraft and surface vehicles, aircraft exhausts, and in the mixing regions of combustors. The discussion in this chapter considers idealized versions of the three basic flows: plane wakes, jets, and mixing layers. A general property of these flows is that variation of the mean field is small in the main direction of motion, x, and large in the direction across the flow, y. For plane flows the x–y plane is perpendicular to the spanwise direction, which is idealized as being infinitely

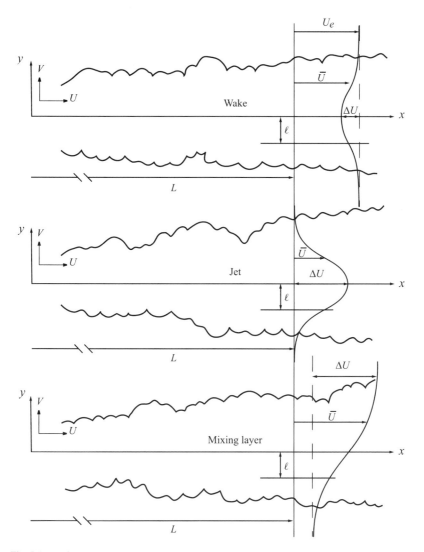

Fig. 5.1 *Defining sketch of plane turbulent wakes, jets, and mixing layers. (Adapted from [39].)*

deep. In this case, all planes parallel to the x–y plane chosen have identical two-dimensional mean fields. Closely related to the planar flows but not touched upon here are the equivalent axisymmetric flows. For these, all radial planes (i.e., those defined by a constant azimuthal angle θ), have identical two-dimensional mean fields. Many results exist for axisymmetric flows which are similar to those presented here for the planar case [16].

It is useful in the analysis of free shear flows to scale them according to characteristic velocities and lengths, ΔU and ℓ, respectively. For free shear flows, as seen in Fig. 5.1, the differences between the maximum and minimum mean velocities is a natural choice for the characteristic velocity scale, while ℓ is usually chosen as the distance from the centerline to the cross-stream location where the local mean velocity is a fixed fraction of ΔU. The exact fraction depends on how the flow is to be analyzed.

For wake flows the mean velocity at any cross-stream location behind the body is less than the mean velocity of the outer flow, U_e. The maximum velocity defect, which occurs at the centerline, is usually taken as a characteristic velocity scale (i.e., $\Delta U = U_e - \overline{U}_c$). The external velocity of a jet with no co-flow is zero, so its centerline velocity can be taken as the characteristic velocity scale, $\Delta U = \overline{U}_c$. Finally, the difference between the high- and low-speed external velocities of a mixing layer, $\Delta U = \overline{U}_h - \overline{U}_l$, can be chosen as its characteristic velocity scale. For this flow, ΔU is a constant in the downstream direction, while it changes for the jet and wake flows. It will also be seen that ℓ grows with distance downstream in all cases.

5.1 THIN, NEARLY PARALLEL FLOWS

Sufficiently far downstream of the solid surfaces that generate them, many free shear flows can be considered to be almost parallel (i.e., $d\ell/dx$ is small and the flows are "thin," meaning that $\ell/L \to 0$, where L is a measure of the downstream distance from the origin of the flow). In fact, for turbulent wakes, experiments show that the latter condition is well satisfied. On the other hand, ℓ/L remains constant with downstream distance for jets and mixing layers, but its value is small. In fact $\ell/L \approx 0.06$, small enough so that the thin shear layer hypothesis is valid.

If the Reynolds number, $R_\ell \equiv u_{\mathrm{rms}}\ell/\nu$, is also large, where, say, $u_{\mathrm{rms}} = \sqrt{\overline{u^2}}$ is a velocity scale characteristic of the turbulence, the standard boundary layer scaling of the cross-stream mean momentum equation [34,39,42] shows that the dominant balance in this direction is between convection and pressure forces, so that

$$\frac{\partial \overline{v^2}}{\partial y} = -\frac{1}{\rho}\frac{\partial \overline{P}}{\partial y}. \tag{5.1}$$

Integration across the shear layer shows that

$$\frac{\overline{P}}{\rho} + \overline{v^2} = \frac{P_e}{\rho}, \tag{5.2}$$

where P_e is the pressure in the outer flow. This result is similar to the case of channel flow [see (4.3)]. Differentiating (5.2) with respect to x gives

$$\frac{1}{\rho}\frac{\partial \overline{P}}{\partial x} + \frac{\partial \overline{v^2}}{\partial x} = 0, \tag{5.3}$$

and substituting this into the steady streamwise mean momentum equation (2.10) yields

$$\overline{U}\frac{\partial \overline{U}}{\partial x} + \overline{V}\frac{\partial \overline{U}}{\partial y} + \frac{\partial}{\partial x}(\overline{u^2} - \overline{v^2}) + \frac{\partial \overline{uv}}{\partial y} = \nu\left(\frac{\partial^2 \overline{U}}{\partial x^2} + \frac{\partial^2 \overline{U}}{\partial y^2}\right). \qquad (5.4)$$

Now applying the same thin-layer scaling analysis as done in boundary layers to this relation gives the dominant balance in the streamwise direction as

$$\overline{U}\frac{\partial \overline{U}}{\partial x} + \overline{V}\frac{\partial \overline{U}}{\partial y} + \frac{\partial \overline{uv}}{\partial y} = 0. \qquad (5.5)$$

This approximate relation is the basis for subsequent analyses of canonical free shear flows.

5.2 MOMENTUM INTEGRAL

Associated with each free shear flow is an external mean velocity, U_e, which is at most a function of x, the downstream distance. In the particular flows of interest here, U_e is constant. Thus, in a two-dimensional plane wake, it is the same on both sides; for a jet with no co-flow, it is zero; and for the mixing layer, it has two values: the constant velocities on either side.

With the help of U_e, an integral form of the momentum equation (5.5) may be derived which yields information about the global conservation properties of jet and wake flows. Thus, when U_e is constant, the relation

$$\frac{\partial}{\partial x}[\overline{U}(\overline{U} - U_e)] + \frac{\partial}{\partial y}[\overline{V}(\overline{U} - U_e)] + \frac{\partial}{\partial y}\overline{uv} = 0 \qquad (5.6)$$

may be derived from (5.5) by use of the identities $U_e(\partial \overline{U}/\partial x + \partial \overline{V}/\partial y) = 0$ and $\overline{U}\partial U_e/\partial x + \overline{V}\partial U_e/\partial y = 0$. For jets with no co-flow and wakes, $\overline{U} - U_e$ and $\overline{uv}$ go to zero for $|y|$ sufficiently large (i.e., when the external flow is reached on either side). Thus, in these two cases (5.6) can be integrated across the flow, yielding

$$\frac{d}{dx}\int_{-\infty}^{+\infty}\overline{U}(\overline{U} - U_e)\,dy = 0, \qquad (5.7)$$

which means that the total mean flux of momentum relative to the external flow does not change with downstream position. That is,

$$\rho\int_{-\infty}^{+\infty}\overline{U}(\overline{U} - U_e)\,dy = M, \qquad (5.8)$$

where M is a constant. For plane wakes it is useful to think of M as the mean streamwise momentum deficit (i.e., the momentum lost to the external flow by the presence of the wake-producing body). Finally, it should be noted that in the case of mixing layers, (5.6) cannot be made the basis for an integral relation since the integrals in (5.7) and (5.8) are unbounded in this case.

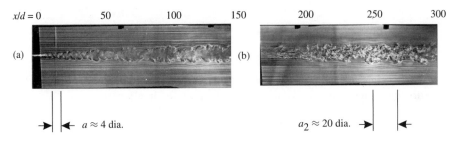

x/d = 0   50   100   150   200   250   300

(a)   (b)

↠ ↞ *a* ≈ 4 dia.   *a*₂ ≈ 20 dia. ↠ ↞

Fig. 5.2   *Circular cylinder wake at* R_e = 2200; *smoke wire at (a) x/d* = 1 *and (b) x/d* = 160. *(From [11]. Reproduced with the permission of Cambridge University Press.)*

5.3 TURBULENT WAKE

One of the best studied wake flows is that of a circular cylinder with spanwise dimension large compared to diameter. If the Reynolds number based on the far-field velocity and cylinder diameter is sufficiently large, the wake will be turbulent a short distance downstream, as illustrated in Fig. 5.2, that shows a turbulent cylinder wake at $Re \equiv U_e d/v = 2200$.

Note that very near the body (Fig 5.2a) distinct, large-scale, nearly two-dimensional laminar vortices alternately form, detach, and are convected downstream. This vortex arrangement, known as the *Kármán vortex street*, was first studied experimentally by Strouhal [37] and analyzed by von Kármán and Rubach [19]. The vortices grow in size with downstream distance, as does the width of the wake itself, by the entrainment of irrotational exterior fluid. Below $Re_d \approx 160 \rightarrow 180$, the two-dimensional Kármán vortices remain laminar and eventually decay, with downstream distance, and disappear [11,12]. The downstream location of complete decay and disappearance of the vortices moves closer to the cylinder with increasing Reynolds number. Just before transition to turbulent flow, the laminar Kármán vortices are no longer observed beyond about 75 to 100 diameters downstream. However, for $100 \lesssim Re_d \lesssim 160 - 180$, a secondary laminar vortex street, two to three times larger in scale than the primary Kármán vortex street, has been observed to develop farther downstream.

The spectra of the streamwise and cross-stream velocity components in the low-Reynolds-number laminar state have spikes of energy just at the shedding frequency, f_s, of the Kármán vortices. The nondimensional shedding frequency, called the *Strouhal number*, $St \equiv f_s d/U_e$, reaches a maximum of about 0.21 at $Re_d \approx 1300$. However, as seen in the visualization [11] in Fig. 5.2, which generally agree with the classical study of Roshko [33], for Re_d above approximately 160, the wake becomes irregular in a transition region that begins closer and closer to the body as the Reynolds number increases. For large enough Reynolds number, three-dimensional instabilities develop in the cores of the Kármán vortices and in the shear layers separating from the cylinder. This is manifested in the appearance of smaller-scale hairpinlike vortices [44,45,50], which stretch in the high-strain region between the Kármán vortices, while energy spreads to other frequencies in the spectrum.

Downstream of the transition region the wake is turbulent. Even farther downstream, in the fully turbulent wake, large-scale vortices have been observed to reappear [11,38], but it is a matter of debate as to how, or even whether, they are related to the original Kármán vortices. Such vortices are seen in Fig. 5.2b. Note that they are of considerably larger scale than the upstream Kármán vortices and are not as well organized and periodic as the large-scale secondary laminar vortices described above for the range $100 \leq \mathrm{Re}_d \leq 160 - 180$. They seem to emerge in groups or packets consistent with earlier observations in the fully turbulent far wake [26,43]. Reasonably strong evidence exists [11] that the vortices in the turbulent far wake are the product of instability associated with the turbulent mean flow. The spectra of fully turbulent wakes are broadband and do not show concentrations of energy at particular frequencies.

5.3.1 Momentum Thickness of Turbulent Wake

Similar to the momentum thickness introduced in the discussion of boundary layers, (4.64), an equivalent length scale can be defined for wake flow. In this case, θ is defined as the thickness that a layer of fluid traveling at speed U_e would have so that it had the same momentum flux as the defect that occurs in the wake region. Thus the balance is

$$- \rho U_e^2 \theta = \rho \int_{-\infty}^{+\infty} \overline{U}(\overline{U} - U_e)\, dy, \tag{5.9}$$

so that

$$\theta = \int_{-\infty}^{+\infty} \frac{\overline{U}}{U_e}\left(1 - \frac{\overline{U}}{U_e}\right) dy. \tag{5.10}$$

From the previous discussion with regard to (5.8), it may be concluded that θ is constant in wake flows.

5.3.2 Self-Preserving Far Wake

Sufficiently far downstream in a wake flow it may be imagined that conditions conducive to a similarity solution for the wake properties are achieved. Specifically, the differences between the mean velocity profile and other quantities at two x locations should be attributable to changes in scale only and not their functional form. Thus it makes sense to assume that in a wake characterized by scales $\Delta U = \Delta U(x) = U_e - \overline{U}_c(x)$ and $\ell = \ell(x)$,

$$U_e - \overline{U} = \Delta U\, f(\eta) \tag{5.11}$$

and

$$-\overline{uv} = (\Delta U)^2 g(\eta), \tag{5.12}$$

where

$$\eta \equiv \frac{y}{\ell} \tag{5.13}$$

is a similarity variable. Note that it is more reasonable to expect that the velocity defect, as in (5.11), obeys a similarity law than the mean velocity itself.

In two-dimensional symmetric wakes such as those of interest here, it follows from the definition of ΔU that

$$f(0) = 1, \tag{5.14}$$

while symmetry implies that

$$f'(0) = 0. \tag{5.15}$$

Moreover, antisymmetry in the Reynolds shear stress implies that

$$g(0) = 0. \tag{5.16}$$

The goal now is to use the momentum equation (5.5) to explore the nature of the similarity solution, if it exists, corresponding to (5.11) and (5.12) and the boundary conditions (5.14) through (5.16). To proceed, it is necessary first to develop an expression for $\overline{V}$. In fact, the functional form of $\overline{V}$ can be determined by integrating the continuity equation. Specifically, at a fixed x position, it can be asserted that

$$\overline{V} = \int_0^y \frac{\partial \overline{V}}{\partial y} dy = -\int_0^y \frac{\partial \overline{U}}{\partial x} dy, \tag{5.17}$$

since $\overline{V}$ is zero at the centerline $y = 0$. According to (5.11),

$$\frac{\partial \overline{U}}{\partial x} = -f \frac{d\Delta U}{dx} + \Delta U \, f' \frac{\eta}{\ell} \frac{d\ell}{dx}, \tag{5.18}$$

where the prime denotes differentiations with respect to η, and the relations $df/dx = (df/d\eta)(d\eta/d\ell)(d\ell/dx)$ and $d\eta/d\ell = -\eta/\ell$ have been used to obtain the second term on the right-hand side. Substituting this into (5.17) and converting the y integration into η integration gives

$$\overline{V} = \ell \frac{d\Delta U}{dx} G(\eta) - \Delta U \frac{d\ell}{dx} H(\eta), \tag{5.19}$$

where $G(\eta) \equiv \int_0^\eta f(\eta) \, d\eta$ and $H(\eta) \equiv \int_0^\eta f'(\eta)\eta \, d\eta$.

Now we are in a position to investigate the similarity properties of (5.5) for wake flow. Noting that

$$\frac{\partial \overline{U}}{\partial y} = -\frac{\Delta U}{\ell} f' \tag{5.20}$$

and

$$\frac{\partial \overline{uv}}{\partial y} = -\frac{\Delta U^2}{\ell} g', \tag{5.21}$$

and using (5.19), (5.5) becomes, after dividing through by $\Delta U^2/\ell$,

$$-\alpha^* f + \beta^* \eta f' + \alpha^* \frac{\Delta U}{U_e} \left(-f'G + f^2 \right) - \beta^* \frac{\Delta U}{U_e} \left(-f'H + \eta f f' \right) = g', \tag{5.22}$$

where

$$\alpha^* = \frac{U_e \ell}{(\Delta U)^2} \frac{d \Delta U}{dx} \tag{5.23}$$

and

$$\beta^* = \frac{U_e}{\Delta U} \frac{d\ell}{dx} \tag{5.24}$$

are dimensionless parameters. Equation (5.22) shows that a necessary condition for a similarity solution to exist is that α^* and β^* not be functions of x (i.e., they must be constant). Moreover, if attention is confined to the far wake, characterized by the condition that

$$\frac{\Delta U}{U_e} \to 0 \quad \text{as} \quad x \to \infty, \tag{5.25}$$

the similarity form of the momentum equation is

$$-\alpha^* f + \beta^* \eta f' = g', \tag{5.26}$$

where α^* and β^* are constant.

Setting the expressions in (5.23) and (5.24) to constant values leads to a coupled system of equations for ℓ and ΔU. Substituting $\Delta U/U_e$ from the second of these into the first and setting

$$n \equiv \frac{\alpha^*}{\beta^*} \tag{5.27}$$

gives

$$\frac{1}{\Delta U}\frac{d\Delta U}{dx} = n\frac{1}{\ell}\frac{d\ell}{dx}.$$ (5.28)

After integration this gives

$$\Delta U = C\ell^n,$$ (5.29)

where C is a constant. Substituting (5.29) into (5.24), defining $\alpha = (1 - n)\beta^* C / U_e$, integrating, and introducing an integration constant x_0 gives

$$\ell(x) = \alpha^m (x - x_0)^m,$$ (5.30)

where

$$m = \frac{1}{1 - n}.$$ (5.31)

Moreover, from (5.29) it then follows that

$$\Delta U(x) = C\alpha^{m-1}(x - x_0)^{m-1}.$$ (5.32)

The exponent m may be found by substituting (5.11) into (5.10), giving

$$\frac{\Delta U}{U_e}\left[\int_{-\infty}^{+\infty} f(\eta)\,d\eta - \frac{\Delta U}{U_e}\int_{-\infty}^{+\infty} f^2(\eta)\,d\eta\right] = \frac{\theta}{\ell}.$$ (5.33)

The second term on the left-hand side becomes small relative to the first in the far wake because $\Delta U / U_e \to 0$, so that (5.33) reduces to

$$\Delta U\,\ell = \frac{U_e\theta}{\int_{-\infty}^{+\infty} f(\eta)\,d\eta}.$$ (5.34)

The right-hand side of this is constant, with the implication that the product of ΔU and ℓ is constant in the far wake. Using (5.30) and (5.32), this means that $m + m - 1 = 0$ or $m = \frac{1}{2}$. It is thus concluded that

$$\ell(x) = \alpha^{1/2}(x - x_0)^{1/2}$$ (5.35)

and

$$\Delta U = C\alpha^{-1/2}(x - x_0)^{-1/2},$$ (5.36)

where x_0 is a virtual origin of the wake and α and C are constant scales. Later, they will be given specific values. It is evident from these results that the plane wake widens with the one-half power and the velocity defect decreases with the inverse one-half power of the distance downstream of the virtual origin of the wake.

Experiments show that the mean wake of a circular cylinder, for example, reaches a self-preserving state with the characteristics in (5.35) and (5.36) about 80 to 90 diameters downstream. It takes a considerably greater downstream distance for second- and higher-order properties of the velocity fluctuations to obtain a self-preserving state.

Equation (5.34), which is valid only in the far wake, also serves to establish a relationship between θ and the drag, D, associated with the wake-producing body. This requires performing a standard control volume analysis of streamwise momentum on a large rectangle exterior to the cylinder in question [16]. In this, Eq. (5.4) is integrated over the control volume. The convective terms yield the difference in streamwise momentum flux across lines upstream and downstream of the body, while the viscous and pressure terms contribute to the drag force. The result is

$$D = \rho U_e \int_{-\infty}^{\infty} \left(U_e - \overline{U} \right) dy, \tag{5.37}$$

and after changing the integration to η and using (5.11) and (5.34), this gives

$$D = \rho U_e^2 \theta. \tag{5.38}$$

5.3.3 Mean Velocity

Now that the x dependence of ℓ and ΔU has been determined, attention is turned to finding the mean velocity field, $\overline{U}$, which is tantamount to finding the similarity function $f(\eta)$. The governing equation for f, namely (5.26), is not closed owing to the presence of the function $g(\eta)$, accounting for the effect of the Reynolds shear stress. To make progress in solving for f, it is traditional to use a simple eddy viscosity model (see Sections 6.1 and 8.1) to express g in terms of f and thereby close the governing equation. A reasonably good assumption is that of a constant eddy viscosity, ν_t, in the central part of the wake where there is no intermittency. Closer to the outer part of the mean wake, where the flow is alternately turbulent and nonturbulent, this assumption likely is less accurate.

By definition of ν_t,

$$\overline{uv} = -\nu_t \frac{\partial \overline{U}}{\partial y}, \tag{5.39}$$

and using (5.12) and (5.20), it follows that

$$g = -\frac{f'}{R_t}, \tag{5.40}$$

where $R_t = \ell\,\Delta U/\nu_t$ is a constant Reynolds number, as is evident from (5.34). Substituting (5.40) into (5.26) and noting from (5.27) that $\beta^* = -\alpha^*$, since according to (5.31) $n = -1$, the following closed equation for $f(\eta)$ results:

$$f'' - R_t\alpha^*(\eta f' + f) = 0. \tag{5.41}$$

Integrating (5.41), applying the condition (5.15), and then integrating a second time and applying (5.14) gives

$$f(\eta) = e^{-U_e\alpha/4\nu_t\eta^2}, \tag{5.42}$$

where the definitions of R_t and α^* and formulas (5.35) and (5.36) have been used. It may be noticed that the constant α, appearing in ℓ, ΔU, and f, affects only the scaling of distances and may be chosen arbitrarily. One obvious choice suggested by (5.42) is to take $\alpha \sim \nu_t/U_e$. However, it is customary when considering the wake behind a cylinder of diameter d to take

$$\alpha = d. \tag{5.43}$$

Since data from this case are examined here, the scaling in (5.43) will be used. It thus follows that

$$f(\eta) = e^{-R\eta^2/4}, \tag{5.44}$$

where $R = dU_e/\nu_t$. Furthermore, with this choice of α, $\eta = y/\sqrt{d(x - x_0)}$.
    Experimental measurements of

$$\frac{U_e - \overline{U}}{\Delta U} = f(\eta) \tag{5.45}$$

in the far wake of a circular cylinder at several cross sections are shown in Fig. 5.3. For $\eta < 0.3$, (5.44) is seen to give a very good fit of the data when the empirically derived value $R = 61.04$ is used. As expected, the agreement is not as good in the outer part of the wake where the effect of intermittency becomes more prominent. However, more careful analysis, which includes an intermittency factor, $\gamma(\eta)$, in (5.39), shows good agreement of (5.45) with (5.44) throughout the wake [39].
    From the value of R it may be computed that $\nu_t = 0.0164\,U_ed$. A calculation also reveals that $\eta_{1/2} = 0.213$, where $\eta_{1/2}$ is defined as the value of η where the velocity defect is half its maximum [i.e., $f(\eta_{1/2}) = 0.5$].
    Using (5.44), it is now possible to determine the constant C appearing in equation (5.36) for ΔU. This entails carrying out the integration in (5.34) and yields

$$C = \sqrt{\frac{R}{\pi}\frac{U_e\theta}{2}} = 2.204\,U_e\theta. \tag{5.46}$$

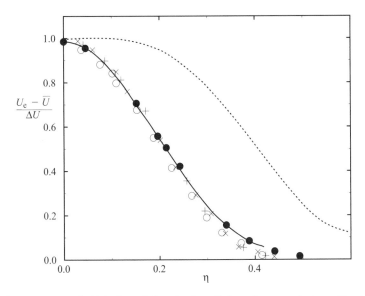

Fig. 5.3 *Comparison of self-similar turbulent circular cylinder wake mean velocity profile with physical experiments at $R_d = 1360$. •, $x/d = 500$; +, 650; ○, 800; ×, 950; —, Eq. (5.44); $\cdots$, intermittency function $\gamma(\eta)$. (From [40]. Reproduced from the Australian Journal of Scientific Research (Vol. A2, 1949), with permission of CSIRO Publishing.)*

Thus it has been determined that

$$\frac{\Delta U(x)}{U_e} = 2.204 \frac{\theta}{d} \sqrt{\frac{d}{x - x_0}}. \tag{5.47}$$

Introducing the drag coefficient

$$C_D \equiv \frac{D}{\frac{1}{2}\rho d U_e^2} \tag{5.48}$$

and noting the result (5.38), it is found that

$$\frac{\theta}{d} = \frac{1}{2}C_D, \tag{5.49}$$

which provides an alternative way of expressing $\Delta U/U_e$ in (5.47).

5.3.4 Velocity and Vorticity Fluctuations

Measurements have been made of the turbulent velocity fluctuations in the far wake [40]. For the same conditions as in Fig. 5.3, fits of the measured normal and shear

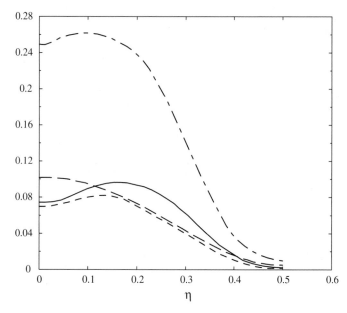

Fig. 5.4 *Normal Reynolds stress components across turbulent wake of a cylinder at $R_d = 1360$. Curve fit of measurements at $x/d = 500, 650, 800,$ and 950. —, $\overline{u^2}/(\Delta U)^2$; — —, $\overline{v^2}/(\Delta U)^2$; – – $\overline{w^2}/(\Delta U)^2$; — $\cdot$ —, $2K/(\Delta U)^2$. (From [40]. Reproduced from the Australian Journal of Scientific Research (Vol. A2, 1949), with permission of CSIRO Publishing.)*

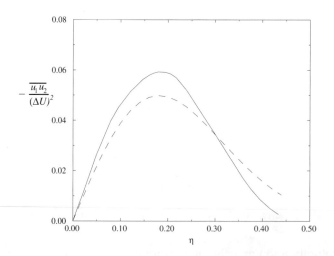

Fig. 5.5 *Reynolds shear stress across turbulent wake of a cylinder at $R_d = 1360$. — —, curve fit at locations $x/d = 500, 650,$ and 800; ——, Eq. (5.50). (From [40]. Reproduced from the Australian Journal of Scientific Research (Vol. A2, 1949), with permission of CSIRO Publishing.)*

Reynolds stress distributions are given in Figs. 5.4 and 5.5, respectively. They are plotted in the form $\overline{u_i u_j}/(\Delta U)^2$ versus η. Note that the Reynolds shear stress in the wake is predicted using (5.12), (5.35), (5.40), and (5.47), so that

$$\frac{\overline{uv}}{(\Delta U)^2} = \frac{-\eta}{2.204 C_D} e^{-R\eta^2/4},\tag{5.50}$$

and this curve is also plotted in Fig. 5.5 using the measured value, $C_D \approx 1$. The scatter of the data used in determining these stresses is greater than for the mean velocity profile, although the data collapse fairly well to the empirically fitted curves, demonstrating the general validity of similarity forms such as given in (5.50). The streamwise and spanwise velocity fluctuation variance distributions have peaks away from the centerline, but this is not the case for the cross-stream component. Equation (5.50) is seen to agree reasonably well with the data. Moreover, the Reynolds shear stress peak occurs close to the location of the maximum mean velocity gradient, lending additional support to (5.50). This is also near the cross-stream location of the peak in turbulent kinetic energy production, as will be seen below when its budget is considered.

Measurements of vorticity in the wake have been made only rarely. In one such experiment [1], two parallel X-array hot wires were used to measure vorticity in the far wake for $\text{Re}_d = 1170$ at $x/d = 420$. The root-mean-square distributions of vorticities are given in Fig. 5.6. It may be noticed that the spanwise and cross-stream components are roughly equal, but the streamwise component has a significantly

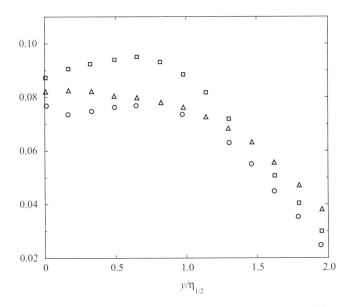

Fig. 5.6 *Root-mean-square vorticity components in turbulent wake normalized with $\overline{U}$ and $\eta_{1/2}$. $\square$, ω_1'; $\circ$, ω_2'; $\triangle$, ω_3'. (From [1]. Reprinted with the permission of Cambridge University Press.)*

larger peak value. The data have been normalized by the local mean velocity $\overline{U}$ and the cross-stream length scale $\ell = \eta_{1/2}$.

Some vorticity measurements have been made in the intermediate turbulent wake $10 \leq x/d \leq 70$ of a circular cylinder at $\text{Re}_d = 2000$ [23] and $\text{Re}_d = 3000$ [25]. These are relatively close to the bluff body where the flow is not self-similar, and the Kármán vortices are still in evidence. The character of the flow in this region can be seen in Fig. 5.2a, obtained at a similar Reynolds number. At positions near the cylinder, the rms streamwise vorticity fluctuation is only slightly higher than the spanwise and cross-stream components.

By bandpass-filtering the spanwise, ω_3, vorticity component signal about the sharp peak at the Kármán vortex shedding frequency of St $= 0.21$, one obtains phase reference to the passage of Kármán vortices. In this way, conditional averages of the flow properties can be made which reveal such interesting information as the relative location of peak Reynolds stress magnitude with respect to the passing vortices. Some results along these lines are given in Section 5.3.6.

5.3.5 Turbulent Kinetic Energy Budget

The turbulent kinetic energy budget [40] is shown in Fig. 5.7. The displayed curves represent fits to the data for the same x/d locations as in previous figures. Consistent with the earlier remarks about $\overline{uv}$, the production term has a pronounced peak at $\eta \approx 0.2$; however, the dissipation rate has its maximum at the wake centerline where flow symmetry assures that turbulence production is zero. The dissipation rate decreases gradually with distance toward the exterior of the wake. The advection and turbulent transport terms are largely equal and opposite, except near the centerline where the former is much greater in magnitude. From this, one can infer that near the wake centerline, turbulent kinetic energy is gained through advection from upstream while it is lost by dissipation and its transport outward by the turbulent velocity fluctuations. Between the interior of the wake, and the exterior flow, at about $\eta \approx 0.3$, the rate of production is roughly equal to that of dissipation. In the outer part of the wake, turbulent kinetic energy is gained through turbulent transport from the interior, and lost through a net downstream advection.

5.3.6 Structure of the Turbulent Wake

It was mentioned previously that vortical structures exist in both the turbulent far wake and in the laminar and transitional wake regions. Whether there is a dynamical connection between them is not clear, but as will now be shown, they do appear to have several similar kinematic characteristics. First, consider the transition region wake structure. Vortices with axes oriented primarily in the x–y plane appear here as a result of several types of instabilities in the laminar wake. Called *rib vortices*, they are smaller in diameter than the large-scale *roller vortices* which are oriented primarily in the spanwise direction. The relative position of the ribs and rollers can be deduced indirectly from conditional averaging of the vorticity field, measured according to the passage of vortex rollers. From the vorticity field the joint pdf

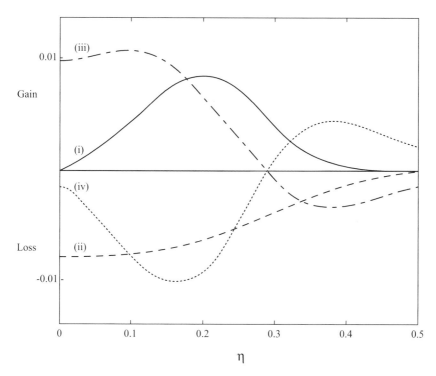

Fig. 5.7 *Experimentally determined turbulent kinetic energy balance in turbulent wake of a cylinder*
($R_e = 1360, x/d \geq 500$): (i) production, $\overline{uv} \, \partial \overline{U}/\partial y$; (ii) dissipation obtained from isotropic approxi-
mation, $15\nu \overline{(\partial u/\partial x)^2}$ [see (7.40)]; (iii) advection, $-\overline{U} \, \partial K/\partial x$ (note $\overline{V} \, \partial K/\partial y$ is negligible in a wake);
(iv) turbulent transport, $-\partial(1/2 \, \overline{u_i^2 u} + \overline{pv}/\rho)/\partial y$. Terms made nondimensional by U_e^3 and d. (From [40].
Reproduced from the Australian Journal of Scientific Research (Vol. A2, 1949), with permission of CSIRO
Publishing.)

of ω_1 and ω_2 vorticity, $P(\omega_1, \omega_2)$, can also be computed. The identity $\overline{\omega_1 \omega_2} =$
$\int \int \omega_1 \omega_2 P(\omega_1, \omega_2) \, d\omega_1 \, d\omega_2$ shows that the largest contributions to $\overline{\omega_1 \omega_2}$ occur at
the maximum of $\omega_1 \omega_2 P(\omega_1, \omega_2)$. Figure 5.8 shows qualitatively the end result of
reconstructing the position and orientation of rib vorticity filaments that correspond
to the maximum of this integrand. These measurements were made in the intermediate
transitional wake at $x/d = 20$.

Roller vortices in the transitional wake region occur at the same frequency as in
the laminar region, indicating that they are the evolving remains of the Kármán vortices. However, in the transition region they contain smaller-scale turbulent motions
within them. The rib vortices appear to wrap over the top of the downstream and
the bottom of the upstream roller vortices on each side of the transitional wake [17].
Measurements in the transitional region [10,17,28] show that the position of highest
turbulent kinetic energy production is located in the region of high strain rate, often
called the *braid region*, between the rollers. There is also evidence suggesting that
the dissipation rate is largest near the roller centers [28].

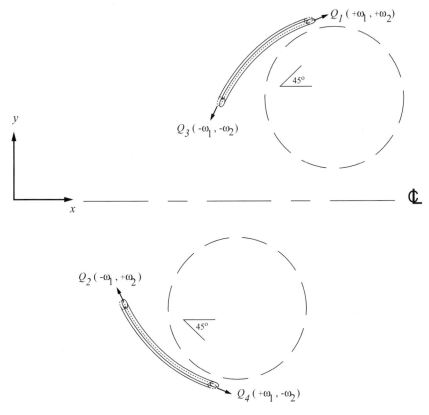

Fig. 5.8 *Orientation of projection on the x–y plane of rib vorticity filaments that contribute most to the $\overline{\omega_1 \omega_2}$ covariance. Dashed circles are locations of large spanwise roller vortices. (From [28].)*

In the fully turbulent far wake it is more difficult to educe the flow structure because the motion does not exhibit an unambiguous periodicity. Some early two-point correlation measurements of the streamwise velocity fluctuations in the far wake [41] showed that the cross-stream separation correlations are always positive but that the spanwise separation correlations are negative for large separations. This behavior has been attributed to the presence of groups of large eddies with axes in the streamwise direction.

A more extensive set of measurements of the entire two-point correlation tensor, $\mathcal{R}^*_{ij}(\mathbf{x}, \mathbf{x} + \mathbf{r}, t)$, were later carried out [14] in the far turbulent wake of a circular cylinder at $\mathrm{Re}_d = 1300$, with the fixed probe located at $x/d = 533$. Figure 5.9 shows measurements of the nine correlations making up the two-point correlation tensor. It is seen that $\mathcal{R}^*_{22}(r, 0, 0)$ and $\mathcal{R}^*_{33}(r, 0, 0)$ change sign with streamwise separation. This suggests, on average, a change in signs of the v and w velocity components over a distance less than it takes for the $\mathcal{R}^*_{11}(r, 0, 0)$ correlation to go to zero. Such behavior may be explained by the vortical makeup of the far wake region. In particular,

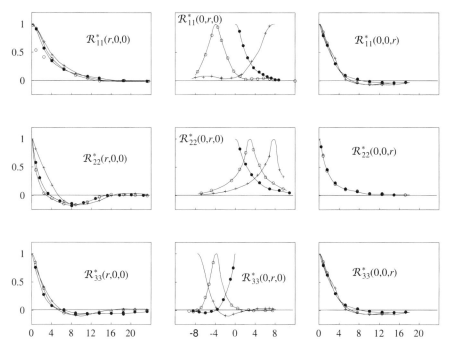

Fig. 5.9 *The nine correlations in the turbulent wake of a cylinder. The measurements were made with the fixed probe at three values of y:* ●, *y/d* = 0; ○, *y/d* = 4; +, *y/d* = 7.6 except for $\mathcal{R}_{22}^*(0, r, 0)$, *where the positions are 0, 2.8, and 7.6, respectively. (From [14]. Reproduced with the permission of Cambridge University Press.)*

a possible cause is the presence of pairs of counterrotating vortices inclined downstream, away from the wake centerline, at an angle of approximately 45° (i.e., roughly in the direction of the principal rate of strain). The vortices in each pair are separated from each other in the spanwise direction, and their circulation is confined entirely to the x–z plane (i.e., the fluid associated with motions of the rollers is contained within the x–z plane at any y position). This double-roller model [14] was in the main reaffirmed [27] later via a pattern-matching algorithm which searched through various types and combinations of structures to find those that are consistent with measurements of the two-point correlation.

Analysis of Grant's data [30] using a *proper orthogonal decomposition,*[1] [35] associated the vortices with the eigenfunctions of the correlation tensor $\mathcal{R}_{ij}$. In this, the eddy structure inferred from the first and most energetic eigenfunction is similar in many respects to that of Grant except that the circulation is not confined to the

---

[1] Consider a region of flow V and define $\phi_i^{(n)}$ to be eigenvectors corresponding to the integral operator over V with kernel $\mathcal{R}_{ij}(\mathbf{x}, \mathbf{y})$ [i.e., $\phi_i^{(n)}$ for $n = 1, 2, 3, \ldots$ satisfy $\int_V \mathcal{R}_{ij}(\mathbf{x}, \mathbf{y})\phi_j^{(n)}(\mathbf{y})\, d\mathbf{y} = \lambda^{(n)}\phi_i^{(n)}(\mathbf{x})$, where $\lambda^{(n)}$ are the eigenvalues]. The expansion $u_i^N(\mathbf{x}, t) = \sum_{n=1}^N a_n \phi_i^{(n)}(\mathbf{x})$, where $a_n = \int_V u_i(\mathbf{y}, t)\phi_i^{(n)}(\mathbf{y})\, d\mathbf{y}$ minimizes the error $\epsilon_N \equiv [\int_V (u_i - u_i^N)^2 d\mathbf{y}]^{1/2}$ and is referred to as the *Karhunen–Loève expansion* or proper orthogonal decomposition in the turbulence literature.

x–z plane, thus accounting for some of the strong cross-stream motion between the vortices on each side of the centerline. This cross-stream motion had been attributed [14] to mixing jets which expel fluid from the interior to the outer reaches of the wake.

Data from an array of eight pairs of hot-wire sensors oriented so as to measure the streamwise and cross-stream vorticity components at $x/d = 170$ of a $\text{Re}_d = 8000$ turbulent wake [43] has been used in a pattern-matching algorithm to educe groups of nearly periodic vortices. Each group has three to five equally spaced vortices, with the spacing varying considerably from group to group. The vortices appear to resemble the Kármán vortices in the manner in which they contribute to two-point correlations. It has also been observed [27] that double rollers frequently occur in groups and do not always occur in two pairs that are directly opposite each other on the two sides of the wake. The groups of vortices seem to be the same ones that have been observed [11] in the far wake downstream of the region where the Kármán vortices have disappeared. Flow visualization in two planes indicates that the double-roller vortices are actually manifestations of three-dimensional hairpin vortices.

Measurements of the streamwise and cross-stream velocity components using a linear array of pairs of hot wires [5] yields conditionally averaged patterns that clearly reveal vortex structures in the far wake at $x/d = 420$ for $R_d = 1170$. Figure 5.10 shows instantaneous velocity vectors, approximate stream functions, and sectional streamlines viewed in a frame of reference moving with approximate mean speed at the half-velocity deficit location in the wake. Here the centers of the vortices and the high-strain-rate or *saddle regions* between them are quite apparent. In this convecting frame of reference, the saddle region contains a stagnation point. Although these patterns are sensitive to the convection speed chosen for the frame of reference in which they are viewed, similar results have been obtained using measurements of the large-scale spanwise vorticity, a method that is frame of reference (Galilean) invariant. Analysis of many data segments shows that an alternating arrangement of vortices on the two sides of the wake occurs roughly twice as frequently as one in which the vortices are opposite each other.

5.4 TURBULENT JET

A turbulent axisymmetric round jet discharging into a motionless surrounding fluid was shown in Fig. 1.1. Near the nozzle two single-stream mixing layers form on either side because of the velocity discontinuities between the essentially uniform velocity of the discharging flow and the surrounding motionless fluid. Some of the characteristics of mixing layers are considered in Section 5.5. Here it may be noticed that vortices develop at the juncture between the fast- and slow-moving fluid.

Downstream the mixing layers grow outward into the surrounding fluid and inward toward the centerline and each other. As they grow, the region of essentially inviscid potential flow at the core of the jet shrinks. As the figure makes abundantly clear, the region of orderly vortex motion is short-lived since a complete breakdown is evident just a few diameters downstream of the orifice. The turbulent region that forms is filled with vortical motions of many scales. Far enough downstream it may be expected

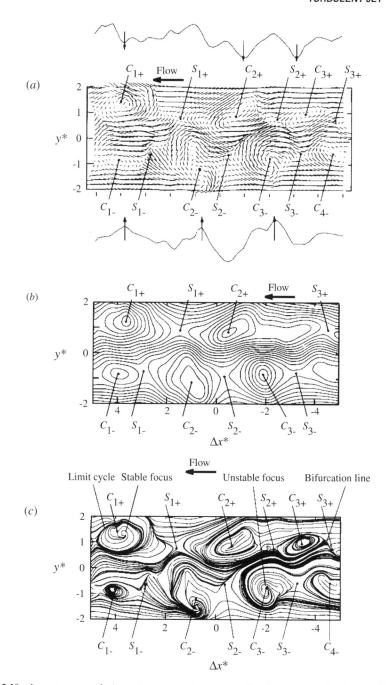

Fig. 5.10 *Instantaneous velocity vectors, approximate stream functions, and sectional streamlines observed in a frame of reference moving at the half deficit velocity from right to left. C are centers/foci, S are saddles, subscripts + and − refer to positive and negative y positions, respectively. (a) Velocity vectors with traces of instantaneous large-scale vorticity centered at $y^* \equiv y/\eta_{1/2} = 0.88$ (above) and $y^* = -0.85$ (below); (b) equally spaced contours of two-dimensional approximate stream functions; (c) sectional streamlines. (From [5]. Reprinted with the permission of Cambridge University Press.)*

that the turbulent jet develops under the auspices of a similarity law with some of the same characteristics as in the previous wake flow analysis, but with some significant differences as well. This is now considered for a plane jet, followed by a description of some of its measured statistical properties.

5.4.1 Self-Preserving Jet

A sufficient distance downstream from the point where the mixing layers that formed at each side of the jet orifice meet at the centerline, a fully developed jet is established. From here onward, experiments show that the cross-stream distribution of the mean velocity acquires the self-preserving form

$$\frac{\overline{U}}{\Delta U} = f(\eta), \tag{5.51}$$

where $\eta = y/\ell$, ΔU and ℓ are x dependent, and $\Delta U = \overline{U}_c$ is the centerline velocity, as discussed previously. Generally, (5.51) is found to be true starting at 50 or more jet nozzle widths downstream of the exit plane. The Reynolds stresses and higher-order velocity statistics of the turbulent field reach a self-preserving state somewhat farther downstream.

It proves to be convenient in the analysis of the plane jet to introduce a mean streamfunction $\Psi(x, y)$ satisfying

$$\Psi = \ell \, \Delta U \, F(\eta), \tag{5.52}$$

with $F(\eta)$ a similarity function. The coefficient $\ell \, \Delta U$ is chosen for dimensional consistency. Moreover, it is a simple matter to see from the definitions

$$\overline{U} \equiv \frac{\partial \Psi}{\partial y}, \quad \overline{V} \equiv -\frac{\partial \Psi}{\partial x} \tag{5.53}$$

that (5.51) is recovered as long as

$$F'(\eta) = f(\eta). \tag{5.54}$$

From (5.53) it is also seen that

$$\overline{V} = -\frac{d(\ell \, \Delta U)}{dx} F + \Delta U \frac{d\ell}{dx} \eta F', \tag{5.55}$$

and substitution of these results into (5.5) gives

$$\frac{\ell}{\Delta U} \frac{d(\Delta U)}{dx} \left(F'^2 - F F'' \right) - \frac{d\ell}{dx} F F'' = g', \tag{5.56}$$

where once again it is assumed that

$$-\overline{uv} = \Delta U^2 \, g(\eta). \tag{5.57}$$

Equation (5.56) differs from (5.22) due primarily to the different velocity scalings. Self-similarity can be achieved in the present case if it happens that

$$\frac{d\ell}{dx} = \alpha \tag{5.58}$$

and

$$\frac{\ell}{\Delta U} \frac{d(\Delta U)}{dx} = \beta, \tag{5.59}$$

where α and β are constants. It follows from (5.58) that

$$\ell = \alpha(x - x_0) \tag{5.60}$$

and from (5.59) that

$$\Delta U = C(x - x_0)^m, \tag{5.61}$$

where C is a constant and the exponent m is still to be determined.

The information provided by (5.7) has yet to be used in the present circumstances. In fact, after a change of integration variable and use of (5.51) and (5.54) and the fact that $U_e = 0$ for a jet with no co-flow, (5.7) becomes

$$\frac{d}{dx}\left(\ell \, \Delta U^2 \int_{-\infty}^{+\infty} F'^2 \, d\eta\right) = 0. \tag{5.62}$$

After substituting for ℓ and ΔU from (5.60) and (5.61), respectively, it is found that $1 + 2m = 0$ (i.e., $m = -1/2$). Thus, in addition to (5.60), it follows that

$$\Delta U = C(x - x_0)^{-1/2}. \tag{5.63}$$

Thus the important results have been found that ℓ grows linearly with downstream distance as the jet develops after it becomes self-preserving, while the mean jet centerline velocity decays according to a $x^{-1/2}$ power law. These results show that the Reynolds number, $\mathrm{Re}_\ell = \Delta U \ell / \nu$, increases as $(x - x_0)^{1/2}$ with x, so that neglecting the viscous terms to obtain (5.5) is more and more justified with increasing downstream distance.

5.4.2 Mean Velocity

Similar to the case of wake flow, (5.56) provides an equation with which to determine the similarity form of the mean velocity field once a model is proposed that allows g'

to be related to F. Once again it is simplest to assume that $\overline{uv}$ can be modeled by the gradient law (5.39), in which case

$$g' = \frac{1}{R_t} F''', \tag{5.64}$$

where

$$R_t = \frac{\Delta U \ell}{v_t}. \tag{5.65}$$

Substituting (5.58), (5.60), (5.63), and (5.64) into (5.56) yields

$$\frac{\alpha}{2}\left(FF'' + F'^2\right) + \frac{1}{R_t} F''' = 0. \tag{5.66}$$

For this to have a similarity solution, it must be that R_t is constant. For this to occur, (5.65) implies that $v_t \sim \sqrt{x - x_0}$.

Accompanying (5.66) are the boundary conditions $F(0) = 0$, which forces the symmetry line to be a streamline, and $F'(0) = 1$ from the definition of ΔU. A last condition is that $\lim_{\eta \to \infty} F'(\eta) = 0$ since the velocity is zero far from the jet centerline. Integrating (5.66) twice and applying the boundary conditions gives

$$F^2 + \frac{4}{\alpha R_t}(F' - 1) = 0. \tag{5.67}$$

This is an example of a Riccati equation [3] the solution of which is

$$F(\eta) = \frac{2}{\sqrt{\alpha R_t}} \tanh\left(\frac{\sqrt{\alpha R_t}}{2}\eta\right). \tag{5.68}$$

Taking a derivative and substituting into (5.51) gives

$$\overline{U} = \Delta U \left[1 - \tanh^2\left(\eta \frac{\sqrt{\alpha R_t}}{2}\right)\right], \tag{5.69}$$

a result first obtained by Görtler [13]. The similarity variable, η, depends on the arbitrary parameter α through ℓ. For convenience it may be assumed that $\alpha = 4/R_t$, so that (5.69) becomes

$$\overline{U} = \Delta U(1 - \tanh^2\eta), \tag{5.70}$$

where $\eta = y R_t / 4(x - x_0)$. Equation (5.70) shows, via (5.63), that $\overline{U}$ ultimately depends on the two parameters R_t and C. The latter can be expressed in terms of the momentum flux, M, by substituting (5.70) into (5.8) and integrating. The result is $C = (3M R_t / 16\rho)^{1/2}$. Finally, R_t is determined with reference to experimental data.

Generally, it is found that a good fit occurs with $R_t = 25.7$. With these parameter values it may be observed from (5.70) that when $\eta = 1$, $y = \ell$ and $\overline{U}$ is approximately 42% of the centerline velocity.

Measurements of the mean and fluctuating properties of a plane jet with no co-flow [15] support the results of the similarity argument and are in good qualitative agreement with measurements in an axisymmetric jet [47]. For example, Fig. 5.11 shows the development of the centerline velocity and the width of the jet, ℓ, with distance downstream. In this experiment, the jet has Reynolds number $\mathrm{Re}_d = 3.4 \times 10^4$, d is the jet width at the exit plane, ℓ is the cross-stream distance from the jet centerline to where the mean velocity is $\Delta U/2$, and the velocity scale is the jet exit velocity.

The evident linear growth of $1/(\Delta U)^2$ for $x/d \geq 45$ confirms the validity of (5.63). Also evident is the linear growth of the normalized jet half-width, ℓ/d, for $x/d \geq 65$, confirming (5.60). The intercept of the two lines in the figure is taken to be the virtual origin, x_0.

Figure 5.12 shows profiles of the jet mean velocity normalized by ΔU, at six locations between $x/d = 47$ and 155 downstream of the exit plane of the nozzle. The cross-stream similarity variable is taken as $\eta = y/(x - x_0)$ (i.e., the choice $\alpha = 1$ is made in this instance). Also plotted is the predicted similarity form (5.70). The measured profiles clearly collapse on one another, displaying self-preservation, and the theoretical curve agrees well with the measured mean profile except near the exterior of the jet, where the intermittency of the turbulence plays a role.

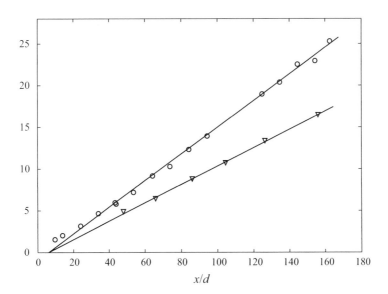

Fig. 5.11   *Centerline mean velocity and jet width development of turbulent plane jet at* $\mathrm{Re}_d = 3.4 \times 10^4$. *o,* $1/(\Delta U)^2 \times 10^4 (\mathrm{ft/s})^{-2}$; ▽, ℓ/d. *(From [15]. Reprinted with permission of ASME.)*

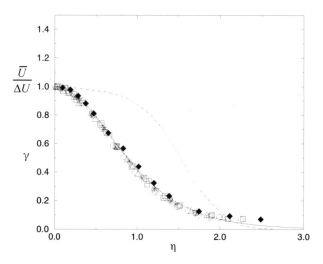

Fig. 5.12   *Mean streamwise velocity profiles of turbulent plane jet at* $\mathrm{Re}_d = 3.4 \times 10^4$ *for* ♦ *, x/d = 47;* ○*, 65;* □*, 85;* ▽*, 103;* △*, 125;* ∗*, 155; and* —*, Eq. (5.70). Also plotted as a dashed line is the intermittency function* $\gamma(\eta)$ *at x/d = 102. (From [15]. Reprinted with permission of ASME.)*

5.4.3 Velocity and Vorticity Fluctuations

The growth of the root-mean-square streamwise velocity fluctuation u_{rms} along the jet centerline, shown in Fig. 5.13, is also linear for $x/d \geq 45$. Furthermore it is observed in experiments that the cross-stream distribution of the variance of the fluctuation in the streamwise direction displays approximate self-preservation for $x/d \geq 45$. In Fig. 5.14, this and the other normal Reynolds stresses in the laboratory coordinate system directions are plotted in the self-preserving region at $x/d = 101$.

The variance of u has a peak more than twice that of v and about twice that of w. The variances of all three components drop to negligible values compared to ΔU for $\eta \gtrsim 0.3$, which is located at about 2.5ℓ from the centerline of the jet. At this location the jet intermittency function, γ, has decreased to nearly zero, as shown in Fig. 5.12. For $\eta \leq 0.15$ (within a cross-stream distance of approximately 1.3ℓ from the jet centerline), $\gamma \geq 0.8$, and the flow is nearly always turbulent, although incursions of irrotational fluid occasionally penetrate nearly to the jet centerline.

The Reynolds shear stress peaks at about $\eta \approx 0.07$ (i.e., at a cross-stream distance of about 0.6ℓ from the jet centerline), as shown in Fig. 5.15. It is zero at the centerline, due to symmetry of the mean flow. For comparison sake the predicted similarity solution for $\overline{uv}$ is also shown. This is computed from

$$g = -\sqrt{\frac{\alpha}{R_t}} \tanh\left(\eta \frac{\sqrt{\alpha R_t}}{2}\right) \left[1 - \tanh^2\left(\eta \frac{\sqrt{\alpha R_t}}{2}\right)\right], \qquad (5.71)$$

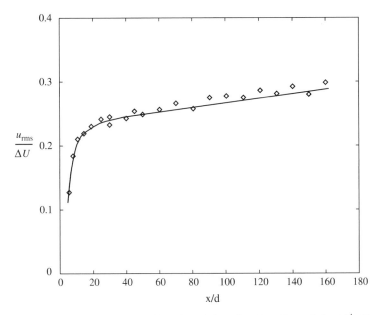

Fig. 5.13   *Growth of u_{rms} along the centerline of a turbulent plane jet at* $Re_d = 3.4 \times 10^4$. *(From [15]. Reprinted with permission of ASME.)*

which is obtained from the fact that $g = F''/R_t$. As in earlier figures, $\alpha = 1$ is assumed. It may be noted that with increasing distance from the jet centerline, the calculated values of $\overline{uv}$ are systematically greater than the measured values. Within about 0.7ℓ ($\eta \approx 0.08$) of the centerline, the agreement is reasonably good. One possible reason for the discrepancy is that at larger distances the turbulence intensities become very large as the mean velocity gets smaller, and many of the assumptions used in justifying hot-wire measurements become less valid.

To date there are no reported measurements of the statistical properties of the vorticity field in the self-preserving region of the turbulent plane jet. It should be noted, however, that such measurements have been made [24] very close to the nozzle, within three jet nozzle diameters downstream of the jet exit plane.

5.4.4 Turbulent Kinetic Energy Budget

Figure 5.16 presents the measured distribution of the terms in the turbulent kinetic energy budget of a planar jet. This balance does not achieve an approximate self-preserving state until at least $x/d = 100$, the location where these measurements were made. Over the inner part of the jet, Reynolds shear stress production is roughly balanced by transport, while advection is balanced by turbulent dissipation. For $\eta \geq 0.07$ the balance of terms is more complex. These results are in relatively good qualitative agreement with other measurements in jet flow (e.g., [6]). Overall, the

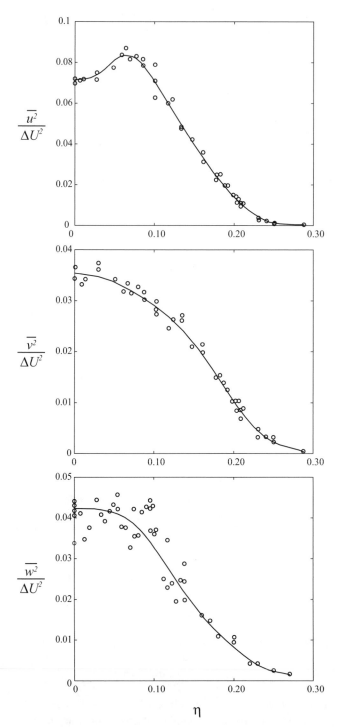

Fig. 5.14 *Variances of fluctuating velocity components for turbulent plane jet at* $\mathrm{Re}_d = 3.4 \times 10^4$ *and* $x/d = 101;$ —, *is best fit of data. (From [15]. Reprinted with permission of ASME.)*

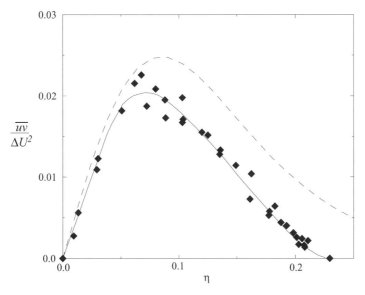

Fig. 5.15   *Reynolds shear stress distribution for turbulent plane jet at* $\mathrm{Re}_d = 3.4 \times 10^4$ *and* $x/d = 101;$ *—, is best fit of data. Also shown: — —, Eq. (5.71). (From [15]. Reprinted with permission of ASME.)*

balance has much the same character as in the two-dimensional wake shown in Fig. 5.7. For example, energy produced near the centerline from upstream diffuses outward, supplying turbulent energy to the outer jet edge.

5.4.5   Structure of the Turbulent Jet

The structural properties of the plane turbulent jet, particularly in the self-preserving region, are less well known than the equivalent region in wakes or mixing layers. In one study [26], the space-time velocity component correlations, $\mathcal{R}_{ii}^*(y^*, r^*, t^*)$, were measured in planes parallel to the symmetry plane of the jet and in the cross-stream plane using a rake of eight hot-wire probes. Here, $y^* \equiv y/\ell$ is the normalized cross-stream location of the fixed probe, ℓ is the jet half-width, $r^* \equiv r/\ell$ is the scaled separation between the fixed and second probes, and $t^* \equiv t\,\Delta U/\ell$ is the normalized time delay. An approximate reconstruction of the sort of structure that is thought to exist in the flow to cause the two-point correlations to have the appearance they have is shown in Fig. 5.17. Parts (b) and (c) show a sketch of vortices, alone and in pairs, with axes oriented in the spanwise and strainwise directions and ideal correlation coefficient curves corresponding to the vortices for fixed separations of the measurement locations. In part (a) of the figure the actual measured correlation coefficient curves are shown. From a pattern-matching technique [26] which attempts to extract more precise information about the eddy structure of the flow, it was determined that the spanwise and strainwise vortices are probably joined together.

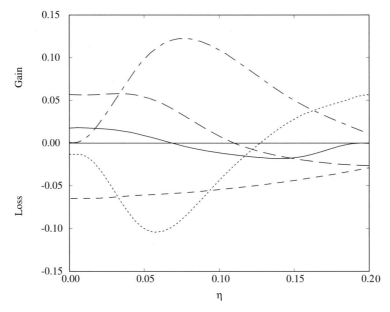

Fig. 5.16 *Turbulent kinetic energy balance in a turbulent plane jet at* $\mathrm{Re}_d = 3.4 \times 10^4$ *and* $x/d = 101$. — · —, *production by shear stresses* $\overline{uv}\partial\overline{U}/\partial y$; —, *production by normal stresses* $(\overline{u^2} - \overline{v^2})\partial\overline{U}/\partial x$; — —, *advection* $\overline{U}\partial K/\partial x + \overline{V}\partial K/\partial y$; · · ·, *turbulent transport* $\partial v(\overline{u_i^2/2 + p})/\partial y$; - - -, *dissipation rate using isotropic approximation* $15\nu\overline{(\partial u/\partial x)^2}$. *Terms made nondimensional by* $(\Delta U)^3$ *and* x. *(From [15]. Reprinted with permission of ASME.)*

The uniqueness of this construction cannot be guaranteed, but that some structure is present is evident in the two-point correlations. The rudimentary model in Fig. 5.17 is a plausible start at an explanation of the structure of the self-similar planar jet flow.

5.5 TURBULENT MIXING LAYER

Turbulent mixing layers occur whenever two parallel streams with different mean velocities flow next to one another at sufficiently high Reynolds numbers. It has already been seen how they appear as building blocks of wake and jet flows in their initial stages. They are also found in many engineering applications, such as in turbomachinery. For these reasons mixing layers have been studied widely. Figure 5.18 shows shadowgraph photos of the x–z and x–y planes of a turbulent mixing layer developing downstream of a thin plate separating two streams of different fluids with the same densities, a velocity ratio $\overline{U}_\ell/\overline{U}_h = 0.38$, and a Reynolds number of 6500 based on the vorticity thickness where the mixing layer becomes turbulent, (i.e., $\Delta U/(dU/dy)_{\max}$).

The high shear in the interfacial region between the two streams gives rise to a Kelvin–Helmholtz type of instability, resulting in the characteristic spanwise vortices sometimes called *rollers*, shown in the lower part of the figure. The vortices grow

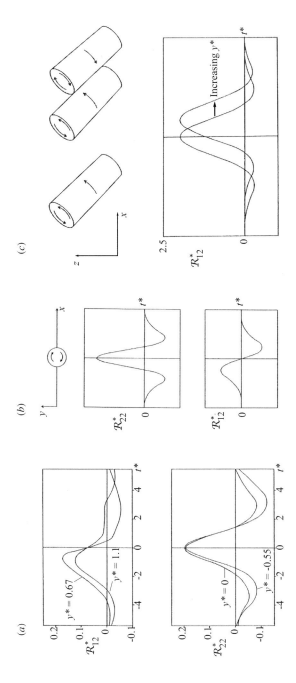

Fig. 5.17 *Two-point correlations in turbulent plane jet at* $Re_d \approx 10^4$: *(a)* $\mathcal{R}_{22}^*$ *and* $\mathcal{R}_{12}^*$ *measured space-time correlation functions; (b)* $\mathcal{R}_{22}^*$ *and* $\mathcal{R}_{12}^*$ *modeled for single spanwise-oriented roller vortex; (c)* $\mathcal{R}_{12}^*$ *modeled for a combination of single and paired inclined vortices. (From [26]. Reprinted with the permission of Cambridge University Press.)*

as they convect downstream by ingesting irrotational fluid from the exterior flow. Although the vortices are initially laminar, small-scale perturbations within them amplify so that they quickly transition to a turbulent state that can be seen as the irregular motions appearing within the vortices not far downstream of the splitter plate. It has often been observed (e.g., [46]) that roller vortices such as these undergo pairing (i.e., that two adjacent vortices merge into a single larger vortex). The large-scale spanwise vortices that form are strikingly coherent, even considerable distances downstream. How far they persist is a matter of some debate; in particular, it is not certain if they persist into the self-preserving region, which like that of the wake and jet, is imagined to occur sufficiently far downstream from the mixing layer source. How the pairing mechanism affects transverse growth of the mixing layer is an important issue.

The upper part of Fig. 5.18 gives a view of the x–z plane looking down on the mixing layer from above. It is apparent that the small-scale turbulence within and between the rollers also exhibits some coherency. Initially, streamwise streaky structures appear, out of which form thin "braid" regions between the spanwise vortices at the downstream side of the photo.

5.5.1 Self-Preserving Mixing Layer

The analysis of the mixing layer proceeds along the same lines as for the wake and jet. As in the case of those flows, the nearly parallel, thin-flow form of the

Fig. 5.18 *Growth of a turbulent mixing layer viewed in x–z (top) and x–y (bottom) planes. (From [20].)*

streamwise momentum equation (5.5) is used to determine the similarity conditions and predict the resulting mean flow characteristics. For a mixing layer with $\overline{U}_l$ and $\overline{U}_h$ as the uniform mean velocities on the low- and high-speed sides, respectively, it is convenient to postulate a self-preserving form for the mean streamfunction as

$$\Psi = \ell U_m F(\eta), \tag{5.72}$$

where $U_m = (\overline{U}_h + \overline{U}_l)/2$ and $\eta = y/\ell$ is the similarity variable. In this case

$$\overline{U} = U_m F'(\eta), \tag{5.73}$$

and assuming that

$$\overline{uv} = -U_m^2 g(\eta), \tag{5.74}$$

it follows from (5.5) that

$$\frac{d\ell}{dx} F F'' + g' = 0. \tag{5.75}$$

As in the case of the jet, (5.75) implies that for a similarity solution to exist, ℓ should depend linearly on x, that is

$$\ell(x) = \alpha(x - x_0), \tag{5.76}$$

where α is a constant.

Assuming a constant-eddy viscosity model as in (5.64) so as to get a closed system, (5.75) becomes

$$F''' + R_t \alpha F F'' = 0, \tag{5.77}$$

where

$$R_t = \frac{U_m \ell}{\nu_t}. \tag{5.78}$$

Since R_t must be constant to maintain similarity, it is evident that ν_t varies with x; in fact, it must increase linearly with x since U_m is constant.

The boundary conditions to be applied to (5.77) are

$$F'(\pm\infty) = 1 \pm \lambda \tag{5.79}$$

where $\lambda = \Delta U/2U_m$ and $\Delta U = \overline{U}_h - \overline{U}_l$. Note that (5.77) is identical in form to the classical Blasius boundary layer equation for a zero-pressure-gradient boundary layer [34]. The only difference is in the boundary conditions.

5.5.2 Mean Velocity in Mixing Layer

As in the case of boundary layer flow, (5.77) does not have a closed-form solution. Nonetheless, a reasonably good approximate solution can be obtained when λ is small. This entails assuming that F obeys a power series in the small parameter, λ [13], as in

$$F(\eta) = \eta + \lambda F_1(\eta) + \lambda^2 F_2(\eta) + \cdots, \tag{5.80}$$

where the functions $F_1, F_2, \ldots$ need to be determined. The leading term in (5.80) is chosen specifically so that using (5.73) it follows that

$$\overline{U} = U_m + U_m \lambda F_1'(\eta) + U_m \lambda^2 F_2'(\eta) + \cdots. \tag{5.81}$$

Substituting (5.80) into (5.77) and collecting together terms of like powers of λ gives equations from which F_1, F_2, and so on, can be determined. The lowest-order equation yields

$$F_1''' + \alpha R_t \eta F_1'' = 0, \tag{5.82}$$

which is a closed equation for F_1. One integration yields

$$F_1'' = C e^{-\alpha R_t \eta^2 / 2}, \tag{5.83}$$

where C is a constant. The arbitrary constant, α, may be picked conveniently so that $\alpha R_t = 2$. In this case,

$$\eta = \frac{R_t}{2} \frac{y}{x - x_0} \tag{5.84}$$

and

$$F_1'' = C e^{-\eta^2}. \tag{5.85}$$

If (5.80) is truncated at the level of F_1 so that $F = \eta + \lambda F_1$, the boundary condition (5.79) implies that

$$F_1'(\pm\infty) = \pm 1. \tag{5.86}$$

Now integrating (5.85) and applying (5.86), it is found that

$$F_1' = \mathrm{erf}\,\eta, \tag{5.87}$$

where

$$\operatorname{erf}\eta = \frac{2}{\sqrt{\pi}} \int_0^{\eta} e^{-\xi^2}\, d\xi \tag{5.88}$$

is the error function. Finally, it is seen from (5.81) that

$$\overline{U} \approx U_m \left(1 + \frac{\Delta U}{2U_m}\operatorname{erf}\eta \right) \tag{5.89}$$

where higher-order terms are neglected.

There remains the parameter R_t, which needs to be evaluated from experiments. The latter show a dependence of R_t on $\overline{U}_h/\overline{U}_l$ [48]. A typical value [31] is $R_t = 27$.

The momentum thickness defined for wakes and jets in (5.10) is not suitable for the mixing layer since the integral appearing in it is unbounded. However, it is not hard to come up with a useful variant of θ appropriate to mixing layers. One possibility is based on the definition

$$\rho\theta(\Delta U)^2 = \rho \int_{-\infty}^{\infty} (\overline{U}_h - \overline{U})(\overline{U} - \overline{U}_l)\, dy, \tag{5.90}$$

which is well defined since the integral goes to zero for $y \to \pm\infty$. The right-hand side may be viewed as the momentum deficit (in relationship to $\overline{U}_h$) in the flux of momentum measured relative to $\overline{U}_l$. θ, then, is the thickness of a layer that one would need to account for this momentum defect if the fluid were all traveling at speed $\overline{U}_h$ with its momentum flux measured relative to $\overline{U}_\ell$.

Substituting (5.89) into (5.90) reveals that

$$\frac{\theta}{l} = \frac{1}{4} \int_{-\infty}^{\infty} (1 - \operatorname{erf}^2\eta)\, d\eta \tag{5.91}$$

or $\theta \sim \ell$. Thus, in view of (5.76), the rate of growth of θ with x is constant. It is observed empirically [49] that the growth-rate parameter,

$$r_\theta = \frac{U_m}{\Delta U}\frac{d\theta}{dx}, \tag{5.92}$$

is a universal constant (i.e., this definition has taken properly into account the dependence of $d\theta/dx$ on the velocities of the mixing streams).

To illustrate the accuracy of (5.89), it is compared in Fig. 5.19 to mean velocity measurements in two-stream mixing layers with a velocity ratio of about 0.6. Predictions from a temporally varying DNS [32] are also indicated. All three of these comparison mixing layers were initiated with turbulent boundary layers on the two sides of the upstream dividing plane. Görtler's theoretical self-preserving mean profile is seen to represent the experimental and DNS data quite well. In this, R_t has been

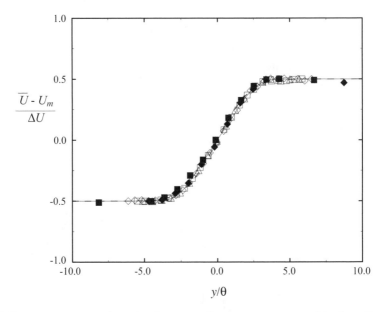

Fig. 5.19 *Mean streamwise velocity distributions in self-preserving two-stream mixing layer. Experiment [22]:* ◆, $R_\theta = 1792$; ■, $R_\theta = 2483$. Experiment [2]: □, $R_\theta \approx 3820$; ◇, $R_\theta \approx 4380$; △, $R_\theta \approx 6365$. DNS [32]: —, averaged for $R_\theta = 1500 - 2000$. Dashed line is Eq. (5.89).

chosen so that the slope of (5.89) matches that of the experimental and DNS profiles at the inflection point.

5.5.3 Velocity and Vorticity Fluctuations

The normal Reynolds stresses scaled by $(\Delta U)^2$ as measured in physical experiments [2,22] and computed by DNS [32] are compared in Fig. 5.20. The cross-stream similarity variable in this figure is take to be $\eta = y/\theta$. The maxima of the velocity fluctuation variances are seen to occur at $\eta = 0$, where the mean velocity profile has its inflection point and maximum gradient. The streamwise variance is largest and the cross-stream variance the smallest. It should be noted that earlier investigations [29,36,49] show considerable variation in the peak values, a fact that is often attributed to differences in initial conditions of the mixing layer experiments. In general, the experimental values and DNS curves in Fig. 5.20 are in good agreement and show that self-preservation has been reached.

Figure 5.21 shows the distribution of the Reynolds shear stress together with the implied turbulent viscosity $\nu_t = -\overline{uv}/(d\overline{U}/dy)$. The experimental results agree quite well with a similar curve from the DNS suggesting that $\overline{uv}$ exhibits self-preservation. In the center portion of the mixing layer, between the limits of $\eta = \pm 3$, a constant turbulent viscosity is not a bad assumption, but this gets worse on either side as the free stream is approached.

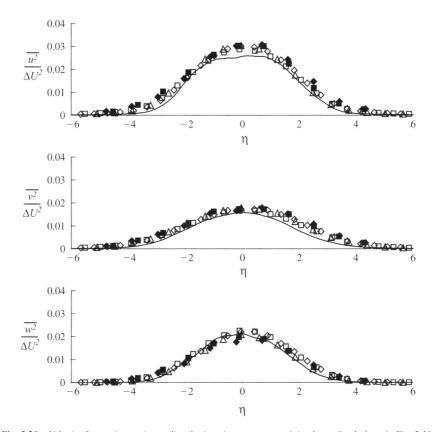

Fig. 5.20 *Velocity fluctuation variance distributions in two-stream mixing layer. Symbols as in Fig. 5.19.*

The distribution of the vorticity covariance $\overline{\omega_1 \omega_2}$ from a physical experiment [22] and computed in the time-evolving DNS [32] is shown in Fig. 5.22. The large discrepancies between the two are possibly due to differences in upstream conditions of the mixing layer. Although they are both initiated with turbulent boundary layers, the sign of the mean vorticity is different in these boundary layers for the spatially developing physical experiment compared to the temporally developing DNS. Nevertheless, it is clear that there is a preference for simultaneous occurrences of like sign values of ω_1 and ω_2, a fact that reflects the presence of a structure underling the vorticity field. Some aspects of this are examined further in Section 5.5.5.

5.5.4 Turbulent Kinetic Energy Budget

Knowledge of the turbulent kinetic energy budget comes only through temporal DNS results [32], which are presented in Fig. 5.23. The production and dissipation rate terms are dominant and vary symmetrically about the center of the mixing layer. The peak production is larger than dissipation, so the sum of these effects leaves a net gain

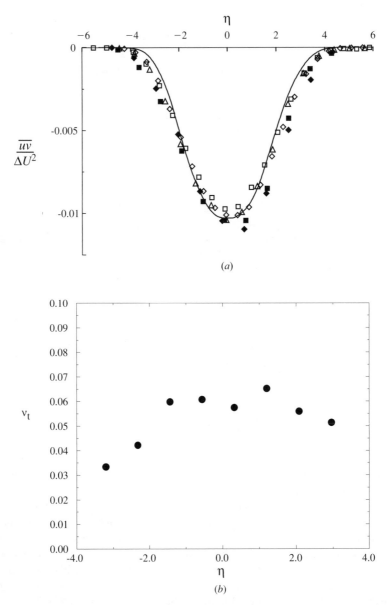

Fig. 5.21 (a) *Reynolds shear stress distributions in two-stream mixing layer. Symbols as in Fig. 5.19. (b) Implied turbulent viscosity ν_t.*

of turbulent energy. This coincides with the behavior observed for the transport term, which indicates that energy is being transported to the outer edge of the shear layer from the central region. In the temporally evolving simulation considered here, the unsteady term in the kinetic energy budget plays the same role as that of the advection term in a spatially developing flow. Its effect is similar to that of the transport term.

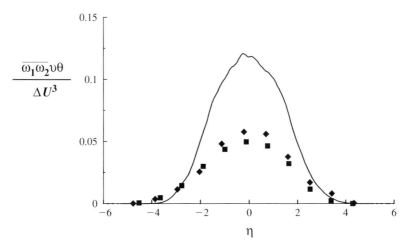

Fig. 5.22 *Vorticity fluctuation covariance* $\overline{\omega_1 \omega_2}$ *distribution in two-stream mixing layer. Symbols as in Fig. 5.19.*

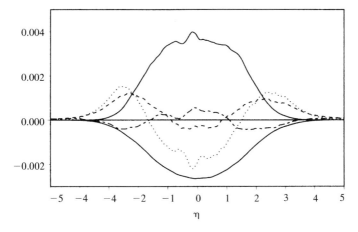

Fig. 5.23 *Turbulent kinetic energy balance in two-stream mixing layer for* $R_\theta = 1500 - 2000$. —, *production (positive curve) and dissipation rate (negative curve);* − −, *time derivative;* ···, *turbulent diffusion* — · —, *pressure diffusion. Terms made nondimensional by* ΔU^3 *and* θ. *(From [32].)*

The viscous diffusion term is not shown because it is small in the self-preserving region.

5.5.5 Structure of the Turbulent Mixing Layer

Some indication of the presence of large-scale structure in the two-stream mixing layer can be seen in two-point correlation measurements [48]. Shown in Fig. 5.24 is $\mathcal{R}_{11}^*(0, r, 0)$, plotted for three fixed probe positions located at the cross-stream

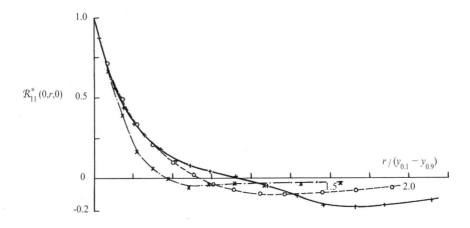

Fig. 5.24 *Spatial correlations $\mathcal{R}_{11}^{*}(0, \Delta y, 0)$ in two-stream mixing layer. Fixed probe positions with $(\overline{U} - U_{\ell})/\Delta U$ values: +, 0.1; ×, 0.47; ○, 0.9. (From [48].)*

positions where $(\overline{U} - \overline{U}_{l})/\Delta U = 0.1$, 0.47, and 0.9, and $r = \Delta y$. The cross-stream separation distance between the two probes is normalized by the mixing-layer width, which is defined in this experiment as the distance between the location where $(\overline{U} - \overline{U}_{l})/\Delta U = 0.1$ and 0.9. For the three fixed probe locations, the correlations become negative at separations of about 0.5 to 1.0 layer widths. This could be viewed as an indication of the presence of spanwise vortices convecting in the x direction. The low amplitude of negative correlation can be explained by the fact that the vortices pass the probes only intermittently. To be more definitive as to the nature of the mixing layer structure, alternative analyses are necessary.

One analysis technique discussed previously in the context of bounded flow is quadrant analysis of the Reynolds shear stress. Figure 5.25 shows the result of this decomposition at the point where $R_{\theta} = 1792$ in a self-preserving mixing layer. As in the case of wall-bounded flows, Q2 and Q4 provide the dominant contributions to the Reynolds shear stress. Their relative importance changes as the splitter plate plane is crossed. Q2 motions (i.e., fluid motions with a deficit of streamwise velocity moving outward toward the high-speed freestream) dominate on the high-speed side of the mixing layer. Conversely, Q4 motions (i.e., fluid motions with an excess of streamwise velocity moving outward toward the low-speed freestream) dominate on the low-speed side. These observations are consistent with earlier studies (e.g., [8]).

The discovery of large-scale roller structures in the original mixing-layer visualizations of Brown and Roshko [9] was unexpected since the flow exhibited all the random characteristics of turbulence. Moreover, the structure clearly persisted with increasing Reynolds numbers, despite the fact that energy is distributed to higher wavenumbers. Subsequent investigations [7,20] revealed the presence of other coherent structures in the splitter plate plane of the mixing layer. These appeared to

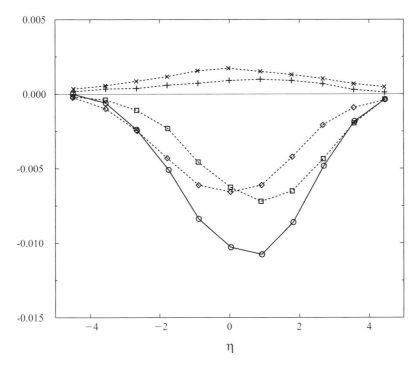

Fig. 5.25   *Quadrant analysis of u and v velocity fluctuations in a two-stream mixing layer at $R_\theta = 1792$.*
$\circ$, $\overline{uv}$; $\times$, Q1($+u$, $+v$); $\square$, Q2($-u$, $+v$); $+$, Q3($-u$, $-v$); $\diamond$, Q4($+u$, $-v$). *(From [22].)*

be streamwise vortices of small diameter compared to their length in the streamwise direction. They are evident near the right-hand side of Fig. 5.18*a*. By examining the two parts of the figure, it appears that the secondary streamwise vortices are located on the "braid" between the adjacent roller vortices and wrap over the high-speed side of the downstream vortex and under the low-speed side of the upstream vortex. Subsequent observations [21] showed the streamwise vortices to occur in counterrotating pairs in the early transition stage to three-dimensionality of the mixing layer soon after the appearance of the spanwise rollers. Moreover, the streamwise vortices appear first on the braids, but then extend into the cores of the rollers. Sometimes the streamwise vortices are seen to pair. Direct evidence for the vortices is provided in the cross-stream *y–z* plane laser-induced fluorescence photographs in Fig. 5.26 through (*a*) the braid region and (*b*) the roller core for a Reynolds number of 3000 based on the vorticity thickness. It is speculated that the counterrotating vortices are joined in a hairpin configuration as they wrap around the spanwise rollers.

Recent investigations [32] using DNS to simulate a temporally evolving mixing layer have called into question the belief that the roller pairings and rib vortices are a fundamental component of turbulent mixing layers. In fact, DNS simulations in which two unperturbed turbulent boundary layers are joined reveal the presence of roller and rib vortices only during the initial flow development, before self-similarity

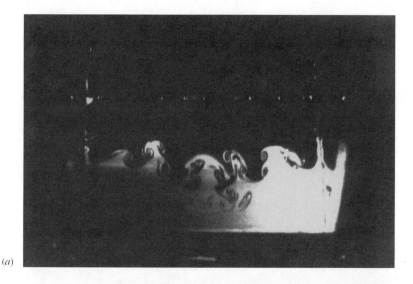

(a)

(b)

Fig. 5.26 *Counterrotating streamwise vortices (a) in the braid region between spanwise roller vortices and (b) wrapped over the roller core in a two-stream mixing layer. (From [4]. Reprinted with the permission of Cambridge University Press.)*

is achieved. Moreover, the pairings occur in limited spanwise regions, not across the entire spanwise simulation domain. It is suggested that the experimental observations may, in fact, be the result of uncontrolled upstream disturbances in the experiments, conditions that are absent in the DNS. This viewpoint is based on the observation that by applying a strong additional perturbation to the initial turbulent boundary layers in the numerical simulations, a self-similar region containing roller vortices

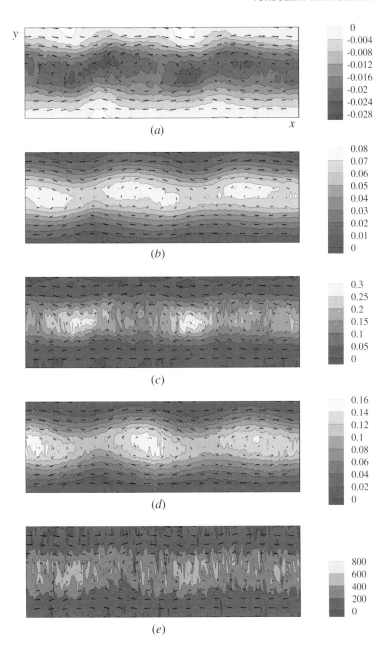

Fig. 5.27 *Conditionally averaged properties of the two-stream mixing layer at $R_\theta = 2483$. Vectors indicate the spanwise roller vortices; superimposed gray-scale levels are (a) Reynolds shear stress (m^2/s^2), (b) turbulent kinetic energy (m^2/s^2), (c) turbulent kinetic energy production rate (m^2/s^3), (d) turbulent kinetic energy dissipation rate (m^2/s^3), and (e) covariance $\overline{\omega_1\omega_2}$, $(1/s^2)$. (From [22].)*

and a distinct braid region is observed. Furthermore, pairings and rib vortices are also observed. This is in strong contrast to the characteristics of structures with no amplification.

Regardless of the origin of the rollers and braid regions, if present, they play a significant part in the dynamics of the mixing layer. The nature of this connection can be elucidated, in part, via conditional averaging of the flow statistics predicated on the passage of roller vortices. Experimentally, their passage is detected through the use of a single-sensor hot-wire probe in the freestream just outside the high-speed side of the mixing layer [18,22]. Figure 5.27 displays the conditionally averaged $x-y$ planar field of several flow variables superimposed on the projection of the velocity vector field on this plane for $R_\theta = 2483$ in the self-preserving region. In (a) the Reynolds shear stress is seen to peak throughout the braid region and where the braid and the rollers intersect. The turbulent kinetic energy is broadly distributed throughout the layer but with its largest values within the rollers themselves toward their outer edges, as shown in (b). The turbulent kinetic energy production shown in (c) has a similar pattern, which is not surprising given the shape of the mean velocity profile. From (d) it is clear that the dissipation rate also peaks within the rollers. Finally, it is observed in (e) that the conditional vorticity covariance $\langle \omega_1 \omega_2 \rangle$ peaks in the region where the braid intersects the rollers on their downstream side. This suggests that the positivity of the $\overline{\omega_1 \omega_2}$ correlation is tied in with the structure of the vorticity field found in the organized braid regions. These results are in good general agreement with conditionally average mixing-layer characteristics observed by Hussain and Zaman [18].

5.6   SUMMARY

Our consideration of wakes, jets, and mixing layers in this chapter reveals the distinctive character of turbulence free of the direct influence of boundaries. Despite the technical differences between the mechanisms causing the appearance of these different kinds of free shear flows, there is a high degree of commonality between them, including the fact that they each develop a self-similar region far enough downstream from their source.

Spanwise vortical structures, perhaps representing the legacy of upstream structures, appear systematically in these shear layers. As the flows develop in their initial stages, they often appear to be subject to instabilities, leading to the appearance of streamwise vortices. Coincidentally with the growth of vortex structures and turbulence is a tendency for free shear flows to grow laterally. This property is also readily understood from the turbulent kinetic energy budgets that were examined for these flows.

REFERENCES

1. Antonia, R. A., Browne, L. W. B. and Shah, D. A. (1988) "Characteristics of vorticity fluctuations in a turbulent wake," *J. Fluid Mech.* **189**, 349–365.

2. Bell, J. H. and Mehta, R. D. (1990) "Development of a two-stream mixing layer from tripped and untripped boundary layers," *AIAA J.* **28**, 2034–2042.

3. Bender, C. M. and Orszag, S. A. (1978) *Advanced Mathematical Methods for Scientists and Engineers*, McGraw-Hill, New York.

4. Bernal, L. P. and Roshko, A. (1986) "Streamwise vortex structure in plane mixing layers," *J. Fluid Mech.* **170**, 499–525.

5. Bisset, D. K., Antonia, R. A. and Browne, L. W. B. (1990) "Spatial organization of large structures in the turbulent far wake of a cylinder," *J. Fluid Mech.* **218**, 439–461.

6. Bradbury, L. J. S. (1965) "The structure of the self-preserving turbulent plane jet," *J. Fluid Mech.* **23**, 31–64.

7. Breidenthal, R. (1982) "Structure in turbulent mixing layers and wakes using a chemical reaction," *J. Fluid Mech.* **109**, 1–24.

8. Browand, F. K. and Ho, C.-M. (1983) "The mixing layer: an example of quasi two-dimensional turbulence," *J. Mech. Theor. Appl. Spec. Publ. 0750-7240*, pp. 99–120.

9. Brown, G. L. and Roshko, A. (1974) "On density effects and large structure in turbulent mixing layers," *J. Fluid Mech.* **64**, 775–816.

10. Cantwell, B. and Coles, D. (1983) "An experimental study of entrainment and transport in the turbulent near wake of a circular cylinder," *J. Fluid Mech.* **136**, 321–374.

11. Cimbala, J., Nagib, H. and Roshko, A. (1988) "Large structure in the far wakes of two-dimensional bluff bodies," *J. Fluid Mech.* **190**, 265–298.

12. Fey, U., König, M. and Eckelmann, H. (1998) "A new Strouhal–Reynolds-number relationship for the circular cylinder in the range $47 < Re < 2 \times 10^5$", *Phys. Fluids* **10**, 1547–1549.

13. Görtler, H. (1942) "Berechnung von Aufgaben der freien Turbulenz auf Grund eines neuen Nahrungsansatzes," *Z. Angew. Math. Mech.* **22**, 244–254.

14. Grant, H. L. (1958) "The large eddies of turbulent motion," *J. Fluid Mech.* **4**, 149–190.

15. Heskestad, G. (1965) "Hot-wire measurements in a plane turbulent jet," *Trans. ASME Ser. E J. Appl. Mech.* **32**, 721–734.

16. Hinze, J. O. (1975) *Turbulence*, 2nd ed., McGraw-Hill, New York.

17. Hussain, A. K. M. F. and Hayakawa, M. (1987) "Eduction of large scale organized structures in a turbulent plane wake," *J. Fluid Mech.* **180**, 193–229.

18. Hussain, A. K. M. F. and Zaman, K. B. M. Q. (1985) "An experimental study of organized motions in the turbulent plane mixing layer," *J. Fluid Mech.* **159**, 85–104.

19. Kármán, T. von and Rubach, H. (1912) "Uber den Mechanismus de Flüssigkeits und Luftwiderstandes," *Phys. Z.* **13**, 49–59.

20. Konrad, J. H. (1976) "An experimental investigation of mixing in two-dimensional turbulent shear flows with applications to diffusion-limited chemical reactions," Ph.D. dissertation, California Institute of Technology.

21. Lasheras, J. C., Cho, J. S. and Maxworthy, T. (1986) "On the origin and evolution of streamwise vortical structures in a plane, free shear-layer," *J. Fluid Mech.* **172**, 231–258.

22. Loucks, R. (1998) "An experimental examination of the velocity and vorticity fields in a plane mixing layer," Ph.D. dissertation, University of Maryland.

23. Marasli, B., Nguyen, P. and Wallace, J. M. (1993) "A calibration technique for multiple-sensor hot-wire probes and its application to vorticity measurements," *Exp. Fluids* **15**, 209–218.

24. Menon. S (1993) "A study of velocity and vorticity statistics in an axisymmetric jet," M.S. thesis, University of Houston.

25. Mi, J. and Antonia, R. A. (1996) "Vorticity characteristics of the turbulent intermediate wake," *Exp. Fluids* **20**, 383–392.

26. Mumford, J. C. (1982) "The structure of the large eddies in fully-developed turbulent shear flows. 1. The plane jet," *J. Fluid Mech.* **118**, 241–268.

27. Mumford, J. C. (1983) "The structure of the large eddies in fully-developed turbulent shear flows. 2. The plane wake," *J. Fluid Mech.* **137**, 447–456.

28. Nguyen, P. N. (1993) "Simultaneous measurements of the velocity and vorticity vector fields in the turbulent near wake of a circular cylinder," Ph.D. dissertation, University of Maryland.

29. Oster, D. and Wygnanski, I. (1982) "The forced mixing layer between parallel streams," *J. Fluid Mech.* **123**, 91–130.

30. Payne, F. and Lumley, J. L. (1967) "Large eddy structure of the turbulent wake behind a circular cylinder," *Phys. Fluids Suppl.* **10**, S194–S196.

31. Reichardt, H. (1942) "Gesetzmässigkeiten der freien Turbulenz," VDI-Forschungsh. **414**.

32. Rogers, M. M. and Moser, R. D. (1994) "Direct simulation of a self-similar turbulent mixing layer," *Phys. Fluids* **6**, 903–923.

33. Roshko, A. (1953) "On the development of turbulent wakes from vortex streets," *NACA TN 291*.

34. Schlichting, H. (1968) *Boundary Layer Theory*, 6th ed., McGraw-Hill, New York.

35. Sirovich, L. (1991) "Empirical eigenfunctions and low dimensional systems," in *New Perspectives in Turbulence* (L. Sirovich, Ed.), Springer-Verlag, New York, pp. 139–164.

36. Spencer, B. W. (1970) "Statistical investigation of turbulent velocity and pressure fields in a two-stream mixing layer," Ph.D. dissertation, University of Illinois.

37. Strouhal, V. (1878) "Über eine besondere Art der Tonerregung," *Ann. Phys. Chem.* **5**, 216–251.

38. Taneda, S. (1959) "Downstream development of wakes behind cylinders," *J. Phys. Soc. Jpn.* **14**, 843–848.

39. Tennekes, H. and Lumley, J. L. (1972) *A First Course in Turbulence*, MIT Press, Cambridge.

40. Townsend, A. A. (1949) "The fully developed turbulent wake of a circular cylinder," *Aust. J. Sci. Res. Ser. A* **2**, 451–468.

41. Townsend, A. A. (1956) *The Structure of Turbulent Shear Flow*, Cambridge University Press, Cambridge.

42. Townsend, A. A. (1976) *The Structure of Turbulent Shear Flow*, 2nd ed., Cambridge University Press, Cambridge, p. 120.

43. Townsend, A. A. (1979) "Flow patterns of large eddies in a wake and in a boundary layer," *J. Fluid Mech.* **95**, 515–537.

44. Wei, T. and Smith, C. R. (1986) "Secondary vortices in the wake of circular cylinders," *J. Fluid Mech.* **169**, 513–533.

45. Williamson, C. H. K. (1988) "The existence of 2 stages in the transition to 3-dimensionality of a cylinder wake," *Phys. Fluids* **31**, 3165–3168.

46. Winant, C. D. and Browand, F. K. (1974) "Vortex pairing: mechanism of turbulent mixing layer growth at moderate Reynolds number," *J. Fluid Mech.* **63**, 237–255.

47. Wygnanski, I. and Fiedler, H. (1969) "Some measurements in self-preserving jet," *J. Fluid Mech.* **38**, 577–612.

48. Yule, A. J. (1971) "Two-dimensional self-preserving turbulent mixing layers at different free stream velocity ratios," *Aeronautical Res. Counc. Rep. Mem. 3683.*

49. Yule, A. J. (1972) "Spreading of turbulent mixing layers," *AIAA J.* **10**, 686–687.

50. Zhang, H.-Q., Fey, U., Noack, B. R. and Eckelmann, H. (1995) "On the transition of the cylinder wake," *Phys. Fluids* **7**, 779–794.

6

$$\overline{}$$

Turbulent Transport

The importance of turbulent transport phenomena to flow prediction has been made clear in previous chapters. Here the physics of transport as it relates to the Reynolds stress and the vorticity flux correlation are considered; further discussion of passive scalar transport is postponed until Chapter 11. Our immediate goal is to explain what the physical processes are that go on inside turbulent flow to cause transport, while a discussion of how these are modeled is given in Chapter 8. The present discussion should provide a useful background with which to assess critically the legitimacy of modeling assumptions.

6.1 REYNOLDS STRESS

It was suggested in Section 2.1 that the principal avenue by which transport correlations such as R_{ij}, defined in (1.6), have traditionally been modeled is by invoking a molecular transport analogy. In this, the random eddying motion of turbulence is assumed to have the same phenomenological relationship to the mean momentum transport rate (i.e., Reynolds stress) as random molecular motions do to the molecular transport rate (i.e., the shear stress tensor). By this reasoning one expects that analogous to (2.21), one should have for turbulent flow

$$\sigma_{ij}^T = -\frac{2}{3}\rho K \delta_{ij} + \mu_t \left(\frac{\partial \overline{U}_i}{\partial x_j} + \frac{\partial \overline{U}_j}{\partial x_i} \right), \tag{6.1}$$

where, in place of pressure, $\frac{2}{3}\rho K$ appears to guarantee consistency of (6.1) after contraction of the tensor (i.e., setting $i = j$ and summing $i = 1, 2, 3$). Note that an assumed eddy viscosity μ_t appears in (6.1), in contrast to the real molecular viscosity μ appearing in (2.21).

The assumption inherent in (6.1) is much more than the introduction of an unknown eddy viscosity. In fact, (6.1) assumes that the tensor $\sigma_{ij}^T + \frac{2}{3}\rho K \delta_{ij}$, which has six independent components, is proportional to the mean rate of strain tensor, with the same proportionality constant, μ_t, for all components. If (6.1) turns out in practice to be in error, it can be either because the entire structure of the approximation is wrong (i.e., μ_t should change from one component to the next), or because the choice of μ_t is incorrect. Turbulence modeling strategies tend to place more weight on the second possibility [i.e., (6.1) is taken for granted and consideration is given to finding the best possible values of μ_t to use in an application]. Our primary concern here, however, is with the legitimacy of (6.1). In Chapter 8 we consider strategies for finding optimal forms of μ_t.

To get at the validity of (6.1), it is helpful first to review the classical phenomenological argument used in justifying (2.21) for the case of nondense gases. For simplicity, consider the particular case of unidirectional flow in which the fluid velocity varies only in a direction transverse to the flow direction [i.e., it is assumed that $\mathbf{U}(\mathbf{x}) = (U(y), 0, 0)$].

It was seen in Section 2.1 that $\rho < c_1 c_2 >$ represents the flux of streamwise momentum in the y direction and $c_2 \, d\mathcal{A} \, dt$ is the volume of space containing molecules with velocities near c_2 that crosses through a small area element $d\mathcal{A}$ in time dt. Here, once again, $\mathbf{C} = < \mathbf{C} > + \mathbf{c}$ is the molecular velocity and $\mathbf{U} = < \mathbf{C} >$, so in this case $< C_2 > = 0$. Now, molecules in a nondense gas can be expected to travel a distance on the order of magnitude of the mean free path λ between collisions. During this time it can be assumed that molecules maintain their momentum. Moreover, it can be hypothesized that due to the high molecular speed and the frequency of collisions, a given molecule shares its momentum rapidly with neighboring molecules. Thus the time between collisions may be interpreted as a *mixing time*, the time over which a molecule exchanges its momentum successfully with the surrounding molecules.

Consider molecules crossing a surface $y = y_0$ in the positive y direction. A standard argument from the kinetic theory of gases [6] shows that the rate at which molecules cross the surface (in units of molecules per area per second) is given by $\frac{1}{4}nc$, where $c = < \mathbf{c} \cdot \mathbf{c} >^{1/2}$ and n, the number density, is the number of molecules per unit volume. On average, as shown in Fig. 6.1, those molecules crossing the surface from below carry streamwise momentum from the location $y_0 - \alpha\lambda$ [i.e., $mU(y_0 - \alpha\lambda)$], where α is an empirical constant which turns out to be smaller than 1, and m is the mass of a molecule. In fact, this is the momentum acquired by the molecules from their last collision and which will be maintained until they cross the surface to give it up in a collision on the other side. By this reckoning, the rate at which momentum is carried upward across the surface is $\frac{1}{4}\rho c U(y_0 - \alpha\lambda)$ since $\rho = nm$. Conversely, molecules traveling across the surface from above carry streamwise momentum at the rate $\frac{1}{4}\rho c U(y_0 + \alpha\lambda)$. The net flux of momentum per area per second is the difference between the momentum going up and that going down, so that

$$\rho < c_1 c_2 > \approx \tfrac{1}{4}\rho c \left[U(y_0 - \alpha\lambda) - U(y_0 + \alpha\lambda) \right]. \tag{6.2}$$

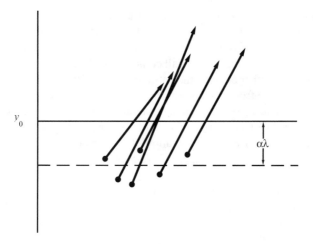

Fig. 6.1   *Molecules crossing the surface $y = y_0$ in the positive y direction, on average, carry unchanged, the momentum from the location $y_o - \alpha\lambda$.*

Since $\alpha\lambda$ is very small, in all but the most exceptional circumstances (e.g., in a shock wave where gas properties can change dramatically on the scale of λ), the approximation $U(y_0 \pm \alpha\lambda) = U(y_0) \pm \alpha\lambda dU/dy(y_0)$ is very well justified. In this case, (6.2) gives

$$\rho < c_1 c_2 > \approx -\frac{1}{2}\alpha\lambda\rho c\frac{dU}{dy}, \qquad (6.3)$$

while from (2.21),

$$\rho < c_1 c_2 > \approx -\mu\frac{dU}{dy}, \qquad (6.4)$$

from which it may be deduced that

$$\mu = \frac{1}{2}\alpha\lambda\rho c \qquad (6.5)$$

(i.e., the viscosity is proportional to the density, mean speed of the velocity fluctuation, and the mean free path). Extension of this argument to the general case yields the stress rate-of-strain law (2.21).

In summary, it may be concluded that the validity of (6.3), and by extension (2.21), depends on the confluence of three essential properties of molecular momentum transport: (1) mixing occurs after a well-defined mixing time; (2) momentum is preserved between collisions; and (3) the mean free path is exceedingly small in comparison to the scale of greater-than-linear variation of the velocity field. Since (6.1) is developed in analogy to (2.21), its legitimacy should depend on whether or not the equivalent set of conditions is satisfied for turbulent transport. In other words, it is required that (1) a mixing time representing the interval over which fluid particles exchange momentum in turbulent flow can be defined; (2) parcels of fluid in turbulent motion can be shown to preserve their momentum over the mixing time; and (3) to a reasonable degree of approximation, the change in $\overline{U}$ is linear over the distance traveled by fluid particles during the mixing time, a distance referred to as the *mixing length*.

To examine whether or not these conditions are met so that (6.1) is justified, it is again convenient to consider the case of a unidirectional mean flow, $\overline{\mathbf{U}} = (\overline{U}(y), 0, 0)$, in which case (6.1) asserts, in analogy to (6.4), that the turbulent flux of x momentum in the y direction may be modeled by

$$\sigma_{12}^T = -\rho\overline{uv} = \mu_t \frac{d\overline{U}}{dy}. \tag{6.6}$$

At the outset, it is not self-evident what the mixing time should be, since fluid particles, unlike molecules, are not well-defined coherent entities. Thus it is difficult to judge when and after what time interval, mixedness has occurred. Second, momentum is not preserved along fluid particle paths even for short times, a fact that was pointed out very early by Taylor [11]. In fact, the continuous action of pressure and viscous forces in accelerating fluid particles makes it unlikely that their momentum will not change. Third, there is little reason to expect that the length scale over which mixing takes place will be much smaller than that at which the mean field has greater-than-linear variation. Thus the validity of the local linear approximation to the mean velocity field implied in assuming (6.6) is called into question. It is thus seen that on all three counts there are significant a priori conceptual obstacles to be overcome before (6.1) can be accepted as an appropriate description of turbulent momentum transport.

It also needs to be pointed out there are a number of well-known flows that are direct counterexamples to (6.6). These have in common the fact that the zeros of $\overline{uv}$ on the left-hand side do not coincide with the zeros of the mean velocity derivative on the right-hand side. The result is a prediction of a negative and/or infinite eddy viscosity, as the case may be. One example occurs in the case of a channel with unequally roughened walls, where the maximum mean velocity is shifted toward the rough wall without a corresponding shift in the zero of the Reynolds shear stress. As a consequence, (6.6) predicts an infinite eddy viscosity near the rough wall. A similar phenomenon is observed in two-dimensional turbulent wall jets [10], in which a stream of fast-moving fluid is injected into a flow parallel to the boundary. Here, $\overline{uv}$ changes sign at a point closer to the wall than the peak mean velocity, once again implying that (6.6) is unphysical.

If the validity of (6.6) is not known with certainty, what would constitute a physically correct model of the Reynolds stress? This issue can be explored by analyzing the turbulent fluid motions that cause u and v to become correlated. Specifically, this can be done by using a DNS of channel flow to generate an ensemble of fluid particle paths arriving at a given point in the flow, and then interrogating them to find out what went on in the recent history of the fluid particles to cause u and v to have a favored (i.e., correlated) relationship so that $\overline{uv} \neq 0$ at that point. Formal application of this methodology reveals much about the physics of the Reynolds shear stress correlation, the conditions under which (6.6) may be valid, and some insight into how $\overline{uv}$ should be modeled.

To begin, consider the set of fluid particles arriving at a point $\mathbf{x}$ at time t, one particle and its path for each realization of the flow. Paths differ between realizations except for having a common endpoint. The paths are denoted by $\mathbf{X}(s)$, s denoting time, so that by definition $\mathbf{X}(t) = \mathbf{x}$. Time $s < t$ denotes motion prior to arriving at $\mathbf{x}$, while $s > t$ is future motion.

Now select a time interval τ, and denote the position of the fluid particle at $t - \tau$ as $\mathbf{b}$ so that $\mathbf{X}(t - \tau) = \mathbf{b}$. Clearly, in contrast to the point $\mathbf{x}$, which is a common destination point for all paths, $\mathbf{b}$ is random (i.e., it varies from one realization to the next), as illustrated in Fig. 6.2.

The velocity of the fluid particle having path $\mathbf{X}(s)$, say $\mathbf{V}(s) \equiv d\mathbf{X}(s)/ds$, must everywhere equal the local velocity of the fluid at the location of the fluid particle. In other words, it follows that

$$\frac{d\mathbf{X}}{ds}(s) = \mathbf{U}(\mathbf{X}(s), s), \tag{6.7}$$

linking the Lagrangian viewpoint on the left-hand side with the combined Eulerian–Lagrangian expression on the right-hand side, in which the Eulerian field $\mathbf{U}$ is evaluated at the Lagrangian point $\mathbf{X}(s)$.

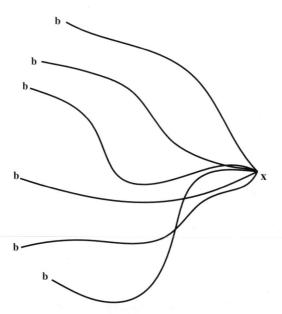

Fig. 6.2 *Ensemble of paths arriving at* **x**.

$U(X(s), s)$ is random both through the velocity field U, which varies from real-ization to realization of a turbulent flow, and from the difference in $X(s)$ from one realization to the next. If it is imagined that the ensemble averaging needed in com-puting the nonrandom mean field $\overline{U}$ has been done a priori, the velocity of each fluid particle can be decomposed according to

$$U(X(s), s) = \overline{U}(X(s), s) + u(x(s), s). \tag{6.8}$$

$\overline{U}(x(s), s)$ in this expression is random only because the mean (deterministic) velocity field is evaluated at a point that varies between realizations. At time $t - \tau$, (6.8) gives $U^b = \overline{U}^b + u^b$, with the superscript b referring to the spatial point $\mathbf{b}$. A similar decomposition for the destination point is $U = \overline{U} + u$, where evaluation at (x, t) is understood. Note that $\overline{U}$ in this expression is not random.

With this notation in place, the aim now is to decipher the cause of nonzero correlation in the Reynolds shear stress, $\overline{uv}$, evaluated at (x, t). Using the previous definitions, one can write the identity

$$u \equiv u^b + (\overline{U}^b - \overline{U}) + (U - U^b), \tag{6.9}$$

which comes from the desire to express the velocity fluctuation u at x in terms of what it was earlier in time at $\mathbf{b}$, namely u^b, plus the factors that have caused it to change since then. Equation (6.9) shows that u differs from u^b because of $\overline{U}^b - \overline{U}$ representing the change in the mean velocity field along the particle path, and $U - U^b$, which is the overall change in the fluid particle velocity component. The first of these expressions records how the local velocity fluctuation changes merely because the fluid particle has changed position in a nonuniform mean field, while the latter reflects changes due to acceleration or deceleration of the fluid particle. An important point to note is that even if fluid particles travel at constant velocity so that $U = U^b$, nevertheless, the Eulerian velocity fluctuation along their paths may change, due to the middle term in (6.9).

To see how the Reynolds shear stress comes about, substitute for u using (6.9) yielding the following exact Lagrangian Reynolds stress decomposition:

$$\overline{uv} = \overline{u^b v} + \overline{v(\overline{U}^b - \overline{U})} + \overline{v(U - U^b)}. \tag{6.10}$$

From this perspective the physical origin of the Reynolds stress is contained in the correlations on the right-hand side of the equation. The first of these, $\overline{u^b v}$, is between Eulerian velocity fluctuation components at different times (and positions). When τ is small, $\overline{u^b v}$ should be as significant as $\overline{uv}$ itself. On the other hand, as τ gets larger, there is less and less reason for u^b and v to be correlated, reflecting what—for lack of a better expression—may be called the essential randomizing nature of turbulence. This is an intrinsic property of turbulent flow presumably associated with the limitations of eddy size and their lifetimes. Thus it is safe to say that $\overline{u^b v} = 0$ when τ is large enough.

The conclusion may be drawn that for τ sufficiently large, it is justified to assert that

$$\overline{uv} = \overline{v(\overline{U}^b - \overline{U})} + \overline{v(U - U^b)}, \tag{6.11}$$

so that the flow processes causing Reynolds stress are represented by just the two terms on the right-hand side. The first of these involves transport coming about primarily by the movement of fluid particles through a Eulerian mean field, so it will be referred to as *displacement transport*. The second term is clearly connected with accelerations of fluid particles and will subsequently be referred to in terms of this property.

Before considering the displacement and acceleration processes in detail, note that the time at which the condition $\overline{u^b v} \approx 0$ is first reached is a natural candidate for the role of mixing time, which, it was noted previously, is otherwise lacking a definition. In particular, if $\overline{uv} \neq 0$ at a point, the minimum time until $\overline{u^b v} = 0$ measures the interval over which events conspire in the flow to cause a correlation between u and v which was formerly lacking between u^b and v. The explanation for nonzero turbulent transport lies in events occurring in the flow over this mixing time.

For the present discussion the turbulent mixing time is defined formally as the minimum time interval at which u^b and v become decorrelated (i.e., $\overline{u^b v} = 0$). It should also be understood that closely related mixing times can be defined in a similar way for any of a range of flow variables (e.g., between v and v^b or v and w^b or velocities with vorticities, and so on). Thus it does not appear to be possible to select a single definition of mixing time covering all correlations, although one may expect that similarly defined mixing times are of the same order of magnitude.

Figure 6.3 shows an evaluation of (6.10) as a function of τ for a fixed point near the lower wall of a turbulent channel flow where $d\overline{U}/dy > 0$. Both backward and forward times are considered. The former clearly shows the mixedness property in that the correlation $\overline{u^b v}$ drops rapidly to zero. Moreover, once the mixing time is achieved, the displacement and acceleration terms are essentially time independent, and their sum equals $\overline{uv}$, as in (6.11). The relative magnitudes of the two terms shows which physical effect is most important at this location.

The forward time correlations in Fig. 6.3 are seen to behave very differently from the backward ones. In fact, somewhat surprisingly, $\overline{u^b v}$ initially increases in magnitude before eventually approaching zero (i.e., v correlates more strongly with u^b in the immediate future than it does with u at the present). An explanation for this may be found in the presence of significant mean shear as occurs in channel flow and as illustrated in Fig. 6.4. In this part of a channel, u and v are negatively correlated, so $v < 0$ tends to be associated with $u > 0$, and vice versa. If $v < 0$, a fluid particle is traveling into a region where $\overline{U}^b < \overline{U}$, and thus u^b will tend to be more positive than u, with the result that $|u^b v| > |uv|$. A similar observation applies for particles for which $v > 0$, $u < 0$. Thus the net result is to enhance the correlation $\overline{u^b v}$. For longer times, randomization of the motion becomes more dominant than the correlation mechanism and $\overline{u^b v} \to 0$. Note that in the case of the backward time

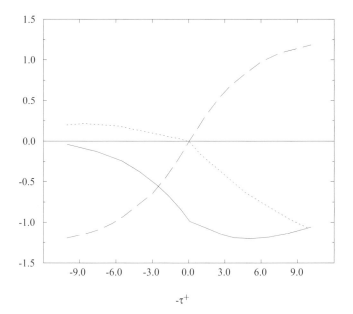

Fig. 6.3 *Effect of time delay on Reynolds stress decomposition at $y^+ = 15.8$ in a channel flow. —, $\overline{u^b v}$;*
– –, $\overline{v(\overline{U}^b - \overline{U})}$; $\cdots$, $\overline{v(U - U^b)}$. All quantities scaled by $|\overline{uv}|$. (From [2]. Copyright © 1989 by the
American Institute of Aeronautics and Astronautics, Inc. Reprinted with permission.)

analysis, the comparable effect of the mechanisms shown in Fig. 6.4 is to enhance the rate at which decorrelation takes place.

Another interesting result implied by Fig. 6.3 when $-\tau > 0$ (i.e., a forward time analysis) is sufficiently large so that $\overline{u^b v} = 0$, is that the resulting decomposition of $\overline{uv}$ in terms of displacement and acceleration effects is meaningless. In fact, the displacement term in this case is opposite in sign to the Reynolds stress, and in compensation, the acceleration term becomes very large. These results are devoid of physical interest and therefore not useful for deciphering the cause of transport correlations. Evidently, a future analysis of the Reynolds stress correlation can only tell the consequences of knowing that u and v are correlated, but it cannot tell why they are correlated. To get insight into the latter question, one must look backward in time.

The terms in the backward time decomposition (6.11), have been evaluated using ensembles of backward particle paths computed in a DNS of turbulent channel flow [3]. The result is shown in Fig. 6.5. Several interesting conclusions may be drawn. Thus, while the Reynolds shear stress is primarily the result of displacement transport, the acceleration term is also significant, both in countering Reynolds stress near the boundary and in enhancing it farther from the wall. More insight into these trends can be had by taking a closer look at how different types of fluid motions contribute to each of the terms in (6.11).

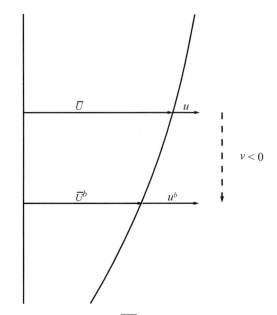

Fig. 6.4   *Explanation for large $\overline{u^b v}$ correlation in a forward time analysis.*

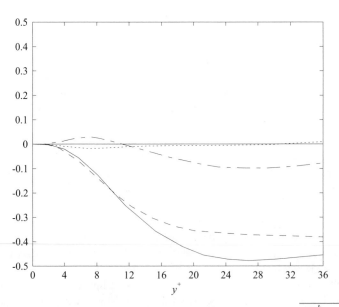

Fig. 6.5   *Reynolds stress decomposition in channel flow.* —, $\overline{uv}$; — —, $\overline{v(\overline{U}^b - \overline{U})}$; — · —, $\overline{v(U - U^b)}$; · · · , $\overline{u^b v}$. *(From [3]. Reprinted with the permission of Cambridge University Press.)*

First, consider the acceleration term. An integration of the x component of the Navier–Stokes equation (2.5) along a particle path yields the identity

$$U - U^b = -\int_{t-\tau}^{t} \frac{\partial P}{\partial x}(s)\, ds + \int_{t-\tau}^{t} \nabla^2 U(s)\, ds, \qquad (6.12)$$

where, for example, $\partial P(s)/\partial x$ is shorthand for $\partial P(\mathbf{x}(s), s)/\partial x$. From (6.12) it follows that

$$\overline{v(U - U^b)} = -\int_{t-\tau}^{t} \overline{v\frac{\partial P}{\partial x}(s)}\, ds + \int_{t-\tau}^{t} \overline{v\nabla^2 U(s)}\, ds. \qquad (6.13)$$

The two terms on the right-hand side account for the effect of pressure and viscous forces, respectively, in accelerating and decelerating fluid particles over the mixing time. For these to contribute to transport, the forces must be related systematically to the speed and direction of the fluid particles, as exemplified by v. A numerical evaluation of (6.13) is presented in Fig. 6.6, where it is seen that viscous forces underlie acceleration transport near the wall, while pressure forces begin to dominate beyond $y^+ \approx 40$. A detailed look at the particle paths used in evaluating (6.13) shows that the reduction in Reynolds stress near the boundary is caused by the action of viscous forces in slowing down the streamwise momentum of fluid particles as they move toward the wall (i.e., there is a strong connection between $v < 0$ and $U - U^b < 0$). The mechanisms at work farther from the boundary, where

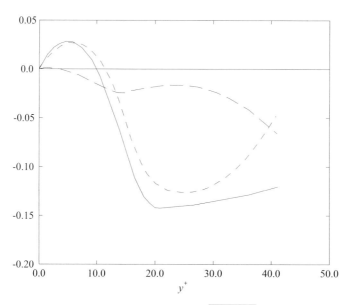

Fig. 6.6 *Decomposition of acceleration transport terms.* —, $\overline{v(U - U^b)}$; –, *viscous term;* — —, *pressure term. (From [7].)*

the acceleration term enhances the Reynolds stress, involves the action of vortical structures and is described in Section 6.1.2, when the dynamical connection between vortices and transport is discussed.

Displacement transport depends on how fluctuations in u arise solely from the movement of fluid particles within a nonuniform mean field. In this, imagine fluid particles leaving from **b** with the typical fluid velocity at that point, namely, $\overline{U}^b$, and traveling with constant velocity to destination point **x**. In a nonuniform mean field, the direction and distance that fluid particles travel control the sign and magnitude of $\overline{U}^b - \overline{U}$. For example, consider Fig. 6.7, which illustrates a case in which the movement of a fluid particle for which $v > 0$ generates a negative contribution, $\overline{U}^b - \overline{U}$, to u. Motion in the opposite direction is correlated with $\overline{U}^b - \overline{U} > 0$. Either way the product $v(\overline{U}^b - \overline{U}) < 0$ and the displacement term will contribute to $\overline{uv}$. This is the analog for turbulent transport of the argument used in justifying (6.2) as a model of molecular transport in nondense gases, with the exception that nowhere here is an assumption of linearity in $\overline{U}$ used or implied.

If a linear stress rate-of-strain law for the Reynolds shear stress such as (6.6) is to be valid, it should fall out of the displacement term and not the acceleration term (i.e., only the displacement transport term has the appropriate physics). Indeed, substituting for $\overline{U}^b$ its Taylor series about **x**, namely,

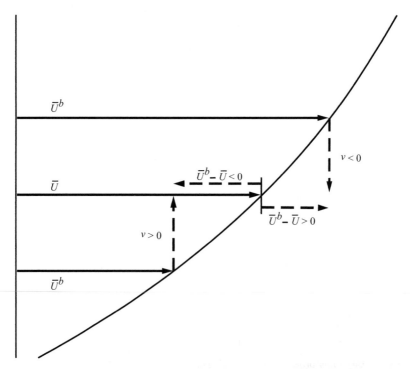

Fig. 6.7   *Source of displacement transport correlation.*

$$\overline{U}^b = \overline{U} - L_2 \frac{d\overline{U}}{dy} + \frac{L_2^2}{2} \frac{d^2\overline{U}}{dy^2} + \cdots, \tag{6.14}$$

where

$$\mathbf{L} \equiv \mathbf{x} - \mathbf{b} = \int_{t-\tau}^{t} \mathbf{U}(\mathbf{X}(s), s) \, ds, \tag{6.15}$$

denotes the change in position recorded by a fluid particle moving between $\mathbf{b}$ and $\mathbf{x}$. Note that the second equality in (6.15) comes from integrating (6.7) from $t - \tau$ to t. From (6.14) the displacement term then becomes

$$\overline{v(U^b - \overline{U})} = -\overline{vL_2}\frac{d\overline{U}}{dy} + \frac{\overline{vL_2^2}}{2}\frac{d^2\overline{U}}{dy^2} + \cdots, \tag{6.16}$$

where the first term on the right-hand side is a gradient law with eddy viscosity $\overline{vL_2}$. Substituting (6.16) into (6.11) gives

$$\overline{uv} = -\overline{vL_2}\frac{d\overline{U}}{dy} + \Phi_{12} + \overline{v(U - U^b)}, \tag{6.17}$$

where Φ_{12} represents the higher-order terms in (6.16). Equation (6.17) reveals the precise conditions under which a gradient law can be justified: (1) the higher-order terms in (6.16) must be zero; and (2) the acceleration transport term must vanish. The first of these conditions depends on how much and in what way $\overline{U}$ varies over the length L_2, while the second concerns the extent to which U changes on particle paths.

Note that one case where the gradient model may be fully justified is a linear mean flow, $\overline{U}(y) = d\overline{U}/dy \, y$, where $d\overline{U}/dy$ is constant. In these circumstances, $\overline{U}^b = \overline{U} - L_2 \, d\overline{U}/dy$ is exact, and (6.17) reduces to

$$\overline{uv} = -\overline{vL_2}\frac{d\overline{U}}{dy} + \overline{v(U - U^b)}. \tag{6.18}$$

Thus, the only reason that a gradient transport law would not be exact in this case is if the as yet unknown effect of the acceleration transport term happened to be significant.

Equation (6.16) implies that the true "physical" eddy viscosity in the simple shear flow considered here is given by $\overline{vL_2}$. Using (6.15) and the fact that $\overline{V} = 0$ gives

$$\overline{vL_2}(\mathbf{x}, t) = \int_{t-\tau}^{t} \overline{v(\mathbf{x}, t)v(\mathbf{X}(s), s)} \, ds. \tag{6.19}$$

This can be placed into a more suggestive form by introducing the concept of a *Lagrangian integral time scale*. In this, it is convenient to make a general definition suitable for arbitrary pairs of velocity components. Thus a Lagrangian correlation coefficient $\mathcal{T}_{\alpha\beta}(\tau)$ may be defined via

$$\overline{u_\alpha u_\beta}\ \mathcal{T}_{\alpha\beta}(\tau) \equiv \int_{t-\tau}^{t} \overline{u_\alpha(\mathbf{x}, t)u_\beta(\mathbf{X}(s), s)}\, ds, \tag{6.20}$$

where according to our previous convention, Greek letter indices are not summed. The integrand in (6.20) is a correlation between Eulerian velocity fluctuations, one of which is evaluated at a Lagrangian point. Only in the absence of a mean velocity field would this correlation be a true Lagrangian correlation function involving the average of the velocities of actual fluid particles.

Note that $\mathcal{T}_{\alpha\beta}(\tau)$ is not a tensor (i.e., it will not transform as a tensor between coordinate systems). When τ is greater than a mixing time, $\mathcal{T}_{\alpha\beta}(\tau)$ is independent of τ since the integrand in (6.20) has become zero. In this case, for each index pair $\mathcal{T}_{\alpha\beta}$ may be referred to as a Lagrangian integral time scale. This may be contrasted with the Eulerian integral scale defined using Eulerian correlation functions in (1.26).

Proceeding now to use (6.20) with (6.19) gives

$$\overline{vL_2} = \mathcal{T}_{22}\overline{v^2}, \tag{6.21}$$

so that the Reynolds shear stress relation (6.17) is

$$\overline{uv} = -\mathcal{T}_{22}\overline{v_2}\frac{d\overline{U}}{dy} + \Phi_{12} + \overline{v(U - U^b)}. \tag{6.22}$$

Thus, a gradient transport law—with eddy viscosity $\mathcal{T}_{22}\overline{v_2}$—is fully justified if $\Phi_{12} = 0$ and acceleration effects are small. It has already been seen in Fig. 6.5 that the latter term is nonnegligible in channel flow, although it is not the dominant effect.

It is interesting to consider what role Φ_{12} might play in these circumstances. Figure 6.8 shows the contribution that the gradient term makes to the displacement term. Clearly, it is significant, but the truncated term, Φ_{12}, is not negligible, suggesting that a model based on a local linear mean velocity field cannot be physically correct. Some insight into how the gradient model fails can be had by looking at the properties of the fluid particle paths making up the Lagrangian ensemble. Consider Fig. 6.9, containing a plot of $\overline{U}$, superimposed local linear approximations to it at points near and far from the wall, and at the particular point $y^+ = 17.8$, where also is plotted the pdf of initial fluid particle locations for paths arriving at $y^+ = 17.8$ after a mixing time. The paths are the same ones that have been used to evaluate (6.10). It is seen that many of the fluid particles arriving at $y^+ = 17.8$ have traveled from locations well beyond the validity of the linear approximation to $\overline{U}$. The figure establishes the interesting fact that for fluid particles traveling toward the wall, the linear model *overpredicts* $\overline{U}^b - \overline{U}$, while the opposite happens for particles traveling outward. Thus the errors derived from the linear approximation appear to be somewhat canceling, and indeed, it may be noted that there is one point at $y^+ \approx 30$ where $\Phi_{12} = 0$. For points nearer and farther from the wall, however, the errors do not cancel, as is made clear by examining

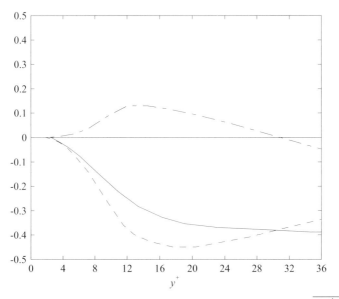

Fig. 6.8 *Evaluation of displacement transport decomposition in channel flow.* —, $\overline{v(\overline{U}^b - \overline{U})}$; – –, *gradient term;* — · —, Φ_{12}. *(From [3]. Reprinted with the permission of Cambridge University Press.)*

the properties of the linear models for $\overline{U}$ at these locations, as shown in Fig. 6.9. In fact, near the wall, outward-moving particles are well served by the linear model, while inbound ones are not. The opposite is true far from the wall where a linear approximation to $\overline{U}$ is good for inbound particles but not for outbound particles. This explains why $\Phi_{12} > 0$ near the wall and $\Phi_{12} < 0$ away from the wall.

Figure 6.9 suggests that the distances traveled by the particles and the size of the resultant errors are too large to be repaired simply by carrying through the analysis, including another few terms in the Taylor series (6.14). In fact, the only recourse is to design a nonlocal closure scheme (i.e., one that takes into account the action of the fluid over the entire region associated with the movement of fluid particles during the mixing time). Whatever technique is chosen to do this falls within the realm of turbulence modeling, and further discussion of this point is deferred until Chapter 8.

It is tempting to develop a Reynolds stress model from just the gradient term in (6.22), so for example in channel flow, (4.1) becomes

$$0 = -\frac{1}{\rho}\frac{d\overline{P}}{dy} + \frac{d}{dy}\left((\nu + \mathcal{T}_{22}\overline{v^2})\frac{d\overline{U}}{dy}\right). \tag{6.23}$$

However, besides the obvious difficulty of obtaining values for $\mathcal{T}_{22}\overline{v^2}$, it is also clear that neglect of the other terms in (6.22) will be a cause of error. One way to sidestep these difficulties is to alter the eddy viscosity in (6.23) away from the physical value $\mathcal{T}_{22}\overline{v^2}$, to a modeled value ν_t, in such a way that the nongradient terms are accounted

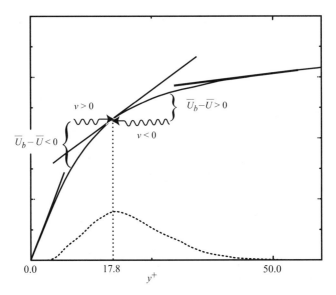

Fig. 6.9 *Local analysis of displacement transport term. —, $\overline{U}$; $\cdots$, pdf. (From [3]. Reprinted with the permission of Cambridge University Press.)*

for. In fact, for a channel flow it is possible to pick ν_t in such a way as to make (6.6) exact. This is only because $\overline{uv}$ and $d\overline{U}/dy$ have common zeros in the channel. In many other flows, such fortuitous circumstances do not exist and (6.6), or more generally (6.1), must fail.

6.1.1 Momentum Transport in General Mean Flows

The momentum transport analysis in Section 6.1 generalizes to encompass the full Reynolds stress tensor. In particular, it is straightforward to derive

$$\overline{u_i u_j} = \overline{u_j u_i^b} + \overline{u_j (\overline{U}_i^b - \overline{U}_i)} + \overline{u_j (U_i - U_i^b)} \tag{6.24}$$

and then conclude that as long as τ is large enough so that each of the conditions $\overline{u_i^b u_j} = 0$ is satisfied, the Reynolds stress obeys

$$\overline{u_i u_j} = \overline{u_j (\overline{U}_i^b - \overline{U}_i)} + \int_{t-\tau}^{t} \overline{u_j \frac{\partial P}{\partial x_i}(s)} \, ds + \int_{t-\tau}^{t} \overline{u_j \nabla^2 U_i(s)} \, ds. \tag{6.25}$$

Here the acceleration term has been expanded in a manner similar to what was done in (6.13). The goal now is to extract a first-order gradient term from the displacement correlation. This is accomplished via a suitable generalization of the procedure used in Section 6.1. The limitation of steady mean fields may be imposed to simplify the analysis, although the same result is achieved for the nonsteady case.

As before, the Taylor series expansion

$$\overline{U}_i^b = \overline{U}_i - L_j \frac{\partial \overline{U}_i}{\partial x_j} + \frac{L_j L_k}{2} \frac{\partial^2 \overline{U}_i}{\partial x_j \partial x_k} + \cdots \tag{6.26}$$

is substituted into the displacement term, yielding

$$\overline{u_j(\overline{U}_i^b - \overline{U}_i)} = -\overline{u_j L_k} \frac{\partial \overline{U}_i}{\partial x_k} + \Phi_{ij}, \tag{6.27}$$

where Φ_{ij} represents all terms with second or higher derivatives in $\overline{U}_i$. The truncated terms are nominally $O(\tau^2)$, although as occurred in Section 6.1, they may very well be significant. According to (6.27), $\overline{u_i L_j}$ is the natural eddy viscosity governing turbulent diffusion in the general case. From (6.8) and (6.15) it follows that

$$\overline{u_i L_j} = \int_{t-\tau}^t \overline{u_i(\mathbf{x}, t) \overline{U}_j(\mathbf{X}(s), s)} \, ds + \int_{t-\tau}^t \overline{u_i(\mathbf{x}, t) u_j(\mathbf{X}(s), s)} \, ds. \tag{6.28}$$

The first term on the right-hand side—after replacing $\overline{U}_j(\mathbf{X}(s), s)$ by its own Taylor series expansion—is found to be $O(\tau^2)$ (i.e., the same order as Φ_{ij}). Therefore, the part of the transport law corresponding to this part of (6.28) may now be assumed to be included in Φ_{ij}. Consequently, after applying (6.20) to the second term in (6.28), the following Reynolds stress decomposition is derived:

$$\overline{u_\alpha u_j} \approx -\mathcal{T}_{\alpha k} \overline{u_\alpha u_k} \frac{\partial \overline{U}_j}{\partial x_k}. \tag{6.29}$$

Note that this relation is of limited value in predicting the Reynolds stresses, not only because of the difficulty of estimating $\mathcal{T}_{\alpha k}$, but because the gradient term involves the very same Reynolds stresses appearing on the left-hand side.

6.1.2 Vortical Structures and Transport

In Section 6.1 we described in general terms how fluid particle motions can be associated with the generation of Reynolds stress through either displacement or acceleration transport. Left unexamined is how these Reynolds stress-producing motions come about in the first place. In Chapter 4 evidence was presented for the existence of self-replicating quasi-streamwise vortices in the wall region. If these are to be dynamically significant, it must be as the source of Reynolds stress-producing motions. It will now be shown in fact, that the two are strongly connected by visualizing fluid particle motions and evolving vortical structures simultaneously.

The ensemble of fluid particle paths arriving at a given point in the boundary region may be ordered according to the size of their respective contributions to the terms in the Lagrangian decomposition of the Reynolds stress (6.11). For example, shown in Fig. 6.10 are the contributions to displacement transport from the ensemble of fluid

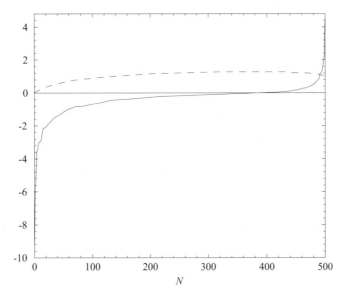

Fig. 6.10 *Ranking of contributions to displacement transport from 500 particles in ensemble of paths arriving at* $y^+ = 24.6$. —, *individual contributions; − −, cumulative contributions. (From [4]. Reprinted with permission of ASME International.)*

particles arriving at $y^+ = 24.6$. Also shown is the cumulative fractional contribution to displacement transport at this location, which reveals that 100% of the Reynolds stress is accounted for by the action of approximately one-fifth of the particles. The remaining particles may be considered to cancel, plus and minus.

It is furthermore revealed in Figs. 6.11 and 6.12 that almost the entire net Reynolds stress exists because of a small number of events transpiring in the turbulent flow. For example, Fig. 6.11 shows the instantaneous Reynolds stress contours with respect to the mean, namely, $uv - \overline{uv}$, at $y^+ = 7.3$, together with the particle paths belonging to the top 10% of contributors to displacement transport projected on and ending at this plane. The paths are as they would appear to an observer translating through the flow at the mean velocity at $y^+ = 7.3$. A similar plot for paths from the ensemble that end at $y^+ = 24.6$ is shown in Fig. 6.12. The large circles denote the position of the particles when they arrive at the given end plane; open circles are for those heading away from wall, and filled circles for those traveling toward it. From this perspective, fast-moving particles appear as starting downstream and arriving upstream, while slow-moving particles look as if they have come from upstream. Two important observations may be made: in agreement with the prior discussion, sweeps dominate Reynolds stress production near the wall, since the paths in Fig. 6.11 belong predominantly to fast-moving particles traveling to the wall. Conversely, at $y^+ = 24.6$, ejections of slow-moving fluid away from the boundary are most evident. Second, the massing together of the paths into just a few main events shows that the Reynolds stress is primarily the result of relatively small numbers of localized events throughout the flow region.

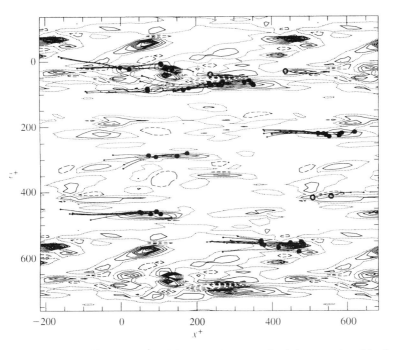

Fig. 6.11 *Source of Reynolds stress at* $y^+ = 7.3$. *(From [5]. Reprinted with the permission of Cambridge University Press.)*

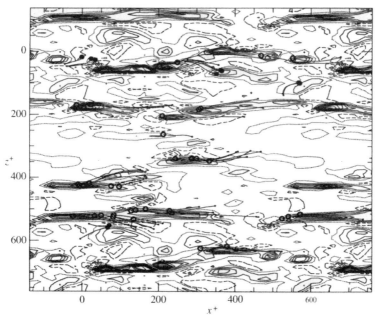

Fig. 6.12 *Source of Reynolds stress at* $y^+ = 24.6$. *(From [5]. Reprinted with the permission of Cambridge University Press.)*

The flow events causing Reynolds stress are driven by vortices, as may be inferred from Figs. 6.13 and 6.14, in which paths most associated with Reynolds stress production are plotted simultaneously with depictions of nearby vortical structures. Figure 6.13 is an example of the motions associated with the sweep events in Fig. 6.11. The direct influence of a quasi-streamwise vortex in causing this motion is unmistakable. A similar conclusion is clear from Fig. 6.14 for the case of an ejection event in Fig. 6.12. The spatial scale over which the paths extend in these figures is determined completely by the coherent vortices that control the motion of fluid particles over the mixing time. Conversely, the mixing time may be thought of as the time over which coherent vortices can exert influence over the motions of fluid particles. During this time the net momentum flux associated with the Reynolds stress is created.

Vortical structures are also implicated as a cause of the acceleration transport phenomenon. This is illustrated in Fig. 6.15, showing the path of a typical fluid particle that makes a large contribution to acceleration transport. In this case the fluid particle travels around the vortex, apparently due to its influence. Reynolds stress is produced in this example because transit of the fluid particle over the vortex and then down toward the wall is accompanied by a considerable enhancement of streamwise velocity. This occurs because vortices tend to be tilted so that their rotational motion is projected into the streamwise direction. On the sweep side this increases the streamwise velocity, and on the ejection side it decreases it. Thus fluid particles caught up in traveling over a vortex, as in Fig. 6.15, experience an acceleration $U - U^b > 0$, while those traveling under a vortex are decelerated (i.e., $U - U^b < 0$). In the former case, fluid particles end up heading toward the wall, so that $v(U - U^b) < 0$. The same kind of contribution to this correlation happens for fluid particles traveling under the vortex as they are slowed and then flung outward. Either way a net enhancement of Reynolds stress results. The degree to which the particle in Fig. 6.15 accelerates is made clear in Fig. 6.16, where it is seen that the fluid particle velocity changes much more than that of the local mean field along its path.

The transport processes described here are specific to canonical nonseparating turbulent flow near solid walls. In more complex flows (e.g., where separation takes place), it still may be presumed that vortical structures underlie the observed Reynolds stress distribution, although how they do so is an open question. In view of the primacy of vortical structures as a cause of Reynolds stress, it appears unlikely that the Reynolds stress can be successfully modeled without taking their presence into account. Examples of this have already been seen in the fact that vortical structures are responsible for the large fluid particle displacements, which render gradient models of $\overline{uv}$ in channel flow inappropriate. Vortices were also seen to have a critical role in causing acceleration transport. While vortex-driven transport may be problematical in developing successful gradient Reynolds stress models, the opposite may be true when it comes to the development of LES techniques. In fact, because the Reynolds stress appears to result from the action of just a relatively few principal vortical structures, it may be reasonably well accounted for in the context of a LES. Thus, in such schemes, the scales of motion causing the Reynolds stress would be directly represented in the computation even if smaller scales are not. Moreover, in a vortex

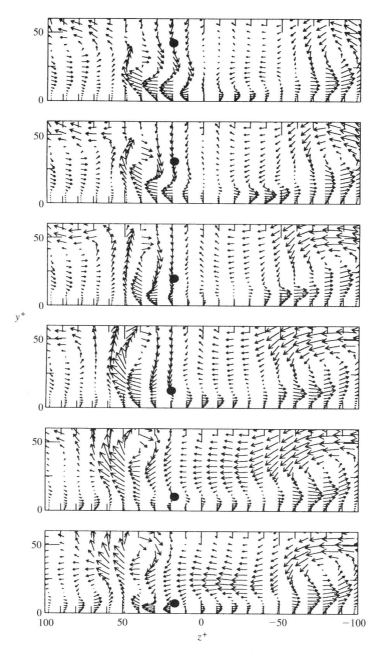

Fig. 6.13 *Vortex motion leading to a sweep. Time advances from top to bottom in increments* $\Delta t^+ = 6.4$. *(From [5]. Reprinted with the permission of Cambridge University Press.)*

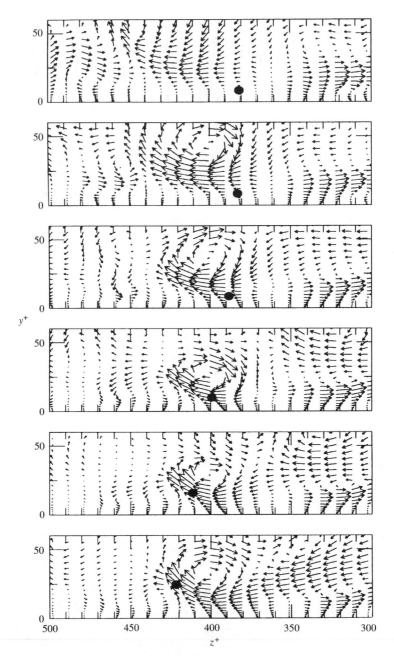

Fig. 6.14 *Vortex motion causing an ejection. Time advances from top to bottom in increments* $\Delta t^+ = 6.4$. *(From [5]. Reprinted with the permission of Cambridge University Press.)*

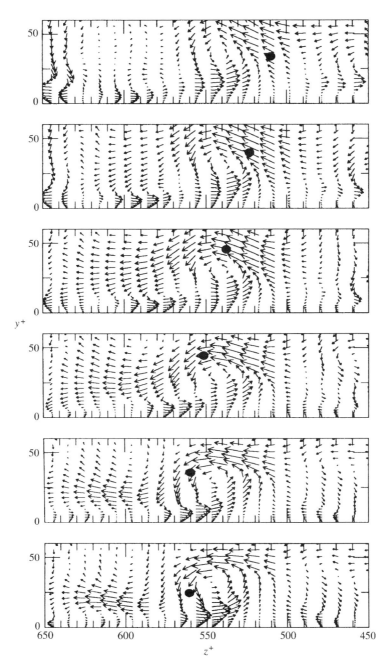

Fig. 6.15 *Vortex motion causing fluid particle accelerations. Time advances from top to bottom in increments* $\Delta t^+ = 6.4$. *(From [5]. Reprinted with the permission of Cambridge University Press.)*

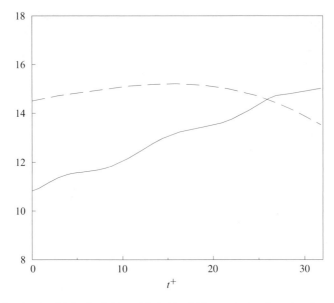

Fig. 6.16 *Acceleration felt by the fluid particle in Fig. 6.15. —, velocity of accelerating particle; − −, mean velocity at fluid particle location. (From [5]. Reprinted with the permission of Cambridge University Press.)*

method implementation of LES, the Reynolds stress-producing vortices are computed as part of the numerical scheme.

Finally, it should be mentioned that some turbulence modeling schemes have gone so far to synthesize a boundary layer as a collection of vortical objects (e.g. Λ vortices [8,9], which are meant to be model horseshoe or hairpin vortices). There is much empiricism in such methodologies, and they may be inappropriate in general flow configurations. Thus their application has been limited.

6.2 VORTICITY TRANSPORT

It is interesting to note that Taylor's pursuit of a model for the vorticity flux [11] predates the systematic development of closure models for the Reynolds stresses. Taylor preferred the vorticity formulation since he suspected that pressure forces would inevitably change momentum on fluid particle paths, thus calling into question the validity of arguments purporting to justify a gradient law. In contrast, at least in some circumstances, the vorticity field is preserved on paths (e.g., two-dimensional inviscid flow,) thus perhaps giving greater justification for a vorticity transport model. The previous sections showed that Taylor's reservations about a momentum transport model were justified. As it turns out, however, similar objections exist for vorticity transport once three-dimensional flows are considered, although it will be seen that it is technically easier to model the effects of vorticity changes during transport than it is momentum changes (i.e., accelerations).

A starting point for a vorticity transport analysis of turbulent flow is the rotational form of the momentum equation (2.107). For a unidirectional flow as in a channel, where $\overline{U} = (\overline{U}, 0, 0)$ and $\overline{\Omega} = (0, 0, \overline{\Omega}_3)$, this reduces to

$$0 = -\frac{1}{\rho}\frac{\partial \overline{P}}{\partial x} + v\frac{d^2\overline{U}}{dy^2} + \overline{v\omega_3} - \overline{w\omega_2}. \tag{6.30}$$

Taylor's earliest work on the vorticity transport theory was limited to purely two-dimensional flow (i.e., flow confined to the x–y plane). In this case $\overline{w\omega_2} = 0$ in (6.30) since $w = 0$ as are ω_1 and ω_2.

Taylor derived a gradient transport law in which

$$\overline{v\omega_3} = -\overline{vL_2}\frac{d\overline{\Omega}_3}{dy}, \tag{6.31}$$

where it should be noted that the diffusivity, $\overline{vL_2} = \mathcal{T}_{22}\overline{v^2}$, is the same as it is for momentum transport in (6.17). This is expected since the action of the turbulent eddies which leads to the turbulent diffusivity $\mathcal{T}_{22}\overline{v^2}$ is independent of the quantity being transported.

The closure resulting from (6.31) and a gradient Reynolds shear stress model are dissimilar. For example, in channel flow the later gives (6.23), while (6.31) gives, after substituting into (6.30),

$$0 = -\frac{1}{\rho}\frac{\partial \overline{P}}{\partial x} + (v + \mathcal{T}_{22}\overline{v^2})\frac{d^2\overline{U}}{dy^2}, \tag{6.32}$$

since $\overline{w\omega_2}$ vanishes in a gradient transport model when $\overline{\Omega}_2 = 0$. Thus it is seen that unlike (6.23), the eddy diffusivity in (6.32) is undifferentiated. Taylor found that (6.32) cannot be applied successfully near boundaries [12]. One reason for this surfaced later when DNS calculations revealed the situation in Fig. 6.17: the vorticity flux is actually countergradient very close to the wall in a bounded flow, so (6.31) is untenable in this flow region.

Taylor's effort to derive a three-dimensional vorticity transport law containing the effects of vortex stretching did not succeed in the sense that the additional terms required in this case were not in a tractable form. After neglecting these, vorticity transport theory reduced to a gradient model that is no different than that which appears in (6.31). It may be noted that the Lagrangian transport analysis used previously to analyze momentum transport bears a resemblance to the Lagrangian analysis technique used in Taylor's modeling of vorticity transport. However, unlike the latter approach it leads to useful model expressions for the effect of vortex stretching on vorticity transport, as will now be demonstrated.

To apply the methodology of Section 6.1 to vorticity transport, one starts with the exact decomposition

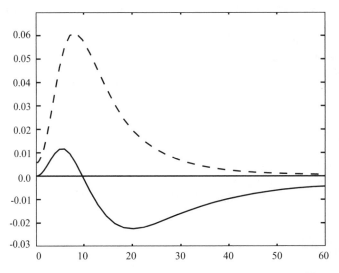

Fig. 6.17 *Wall-normal vorticity flux in channel flow.* —, $\overline{v\omega_3}$; – –, $d\overline{\Omega}/dy$.

$$\omega_j = \omega_j^b + (\overline{\Omega}_j^b - \overline{\Omega}_j) + (\Omega_j - \Omega_j^b). \qquad (6.33)$$

Moreover, an integration of (2.108) along a fluid particle path gives for the last term in (6.33),

$$\Omega_j - \Omega_j^b = \int_{t-\tau}^t \Omega_k(s) \frac{\partial U_j}{\partial x_k}(s)\, ds + \int_{t-\tau}^t \nu\, \nabla^2 \Omega_j(s)\, ds. \qquad (6.34)$$

The first integral in (6.34) accounts for changes in vorticity due to vortex stretching and the second from viscous effects. Substituting (6.33) and (6.34) into the flux correlation yields

$$\overline{u_i \omega_j} = \overline{u_i(\overline{\Omega}_j^b - \overline{\Omega}_j)} + \int_{t-\tau}^t \overline{u_i \Omega_k(s) \frac{\partial U_j}{\partial x_k}(s)}\, ds + \int_{t-\tau}^t \overline{\nu u_i\, \nabla^2 \Omega_j(s)}\, ds, \qquad (6.35)$$

where it has been assumed that τ is sufficiently large so that the mixing condition $\overline{u_i \omega_j^b} = 0$ is satisfied. Vortex stretching is accommodated in (6.35) by the second term on the right-hand side.

The exact formula (6.35) retains all nonlocal transport effects. In the same vein as the earlier analysis of momentum transport, one can extract only first-order effects by substituting a Taylor series expansion for $\overline{\Omega}_j^b$ in the displacement term and $\Omega_k(s)$ and $\partial U_j(s)/\partial x_k$ in the stretching term. For simplicity, the viscous term is omitted. After making these substitutions and retaining just the principal first-order contributions to the vorticity flux, it is found that

$$\overline{u_i \omega_j} = -\int_{t-\tau}^{t} \overline{u_i u_k(s)}\, ds \, \frac{\partial \overline{\Omega}_j}{\partial x_k} + \int_{t-\tau}^{t} \overline{u_i \frac{\partial u_j}{\partial x_k}(s)}\, ds \, \overline{\Omega}_k. \tag{6.36}$$

The first term on the right-hand side is the expected gradient term. This is now accompanied by a first-order stretching term.

To write (6.36) in a more useful form, it is helpful to define—in addition to (6.20)—a Lagrangian correlation coefficient $Q_{\alpha\beta\gamma}(\tau)$ via

$$\overline{u_\alpha \frac{\partial u_\beta}{\partial x_\gamma}}\, Q_{\alpha\beta\gamma}(\tau) = \int_{t-\tau}^{t} \overline{u_\alpha \frac{\partial u_\beta}{\partial x_\gamma}(s)}\, ds. \tag{6.37}$$

For τ large enough, so that the integrand on the right-hand side reaches zero, $Q_{\alpha\beta\gamma}(\tau)$ is independent of τ and may thus be regarded as a family of Lagrangian integral scales, $Q_{\alpha\beta\gamma}$. With these definitions, (6.36) becomes

$$\overline{u_\alpha \omega_\beta} = -\mathcal{T}_{\alpha k} \overline{u_\alpha u_k} \frac{\partial \overline{\Omega}_\beta}{\partial x_k} + Q_{\alpha\beta k}\, \overline{u_\alpha \frac{\partial u_\beta}{\partial x_k}}\, \overline{\Omega}_k, \tag{6.38}$$

which is a general vorticity transport law for three-dimensional flow in which gradient transport is included together with a first-order effect of vortex stretching. Of course, knowledge of the many time scales appearing in (6.38) is necessary to make this relation effective in turbulent flow predictions.

6.2.1   Vorticity Transport in Channel Flow

For channel flow (6.38) gives for the two correlations in (6.30),

$$\overline{v \omega_3} = -\mathcal{T}_{22} \overline{v^2} \frac{d\overline{\Omega}_3}{dy} + Q_{233}\, \overline{v \frac{\partial w}{\partial z}}\, \overline{\Omega}_3 \tag{6.39}$$

and

$$\overline{w \omega_2} = Q_{323}\, \overline{w \frac{\partial v}{\partial z}}\, \overline{\Omega}_3, \tag{6.40}$$

where (6.31) is now modified by the presence of a contribution from vortex stretching, while $\overline{w \omega_2}$, which was formerly zero, is now nonzero, due to a stretching effect. Apart from the Lagrangian scales, the terms in (6.39) and (6.40) are readily obtained from DNS, and it is natural to wonder if the inclusion of stretching effects on transport is able to overcome the fundamental limitations of the gradient law (6.31) in accounting for the behavior of $\overline{v \omega_3}$, as implied by Fig. 6.17.

To analyze the validity of (6.38) more fully, it is also helpful to consider its implications for the remainder of the vorticity fluxes. In fact, (6.38) yields

$$\overline{u \omega_3} = -\mathcal{T}_{12} \overline{uv} \frac{d\overline{\Omega}_3}{dy} + Q_{133}\, \overline{u \frac{\partial w}{\partial z}}\, \overline{\Omega}_3 \tag{6.41}$$

and

$$\overline{w\omega_1} = \mathcal{Q}_{313}\, \overline{w\frac{\partial u}{\partial z}}\, \overline{\Omega}_3. \tag{6.42}$$

In addition, the remaining five vorticity fluxes, $\overline{u\omega_1}$, $\overline{v\omega_2}$, $\overline{w\omega_3}$, $\overline{u\omega_2}$, and $\overline{v\omega_1}$, are predicted to be zero. This result is in complete agreement with the exact behavior of these correlations as predicted by symmetry and homogeneity conditions.

To make a complete test of (6.39) through (6.42) requires having values of the time scales upon which they depend. Unfortunately, such information is not yet available. However, it is the case that the scales must be nonnegative, and a reasonably stringent test of (6.39) through (6.42) can be had by seeing if positive, constant values of the time scales can be found to cause the predicted fluxes to agree closely with their DNS values. The results of such an exercise are shown in Fig. 6.18 for the values (in wall units) $\mathcal{T}_{22}^+ = 4.8$, $\mathcal{T}_{12}^+ = 12.3$, $\mathcal{Q}_{233}^+ = 5.5$, $\mathcal{Q}_{323}^+ = 9.5$, $\mathcal{Q}_{133}^+ = 16.3$, and $\mathcal{Q}_{313}^+ = 0.95$. Evidently, the figures suggest that by including the first-order vortex stretching model, the essentials of the turbulent vorticity flux can be accounted for. In particular, the gradient terms in (6.39) and (6.41) are satisfactory away from the boundary, while near the wall the stretching terms account successfully for nongradient conditions. Furthermore, the stretching terms in (6.40) and (6.42) provide a satisfactory explanation as to how these correlations come about. Note that in all four cases the stretching terms have the correct sign and functional behavior near the wall.

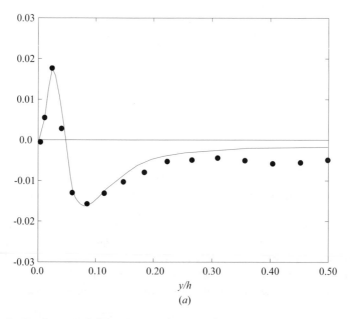

Fig. 6.18 *Predicted vorticity fluxes:* •, *DNS results;* —, *prediction from Eqs. (6.39) through (6.42). (a)* $\overline{v\omega_3}$. *(From [1]. Copyright © Springer-Verlag.)*

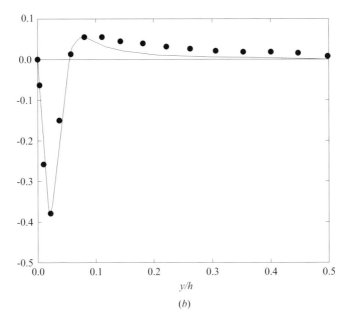

Fig. 6.18 *Predicted vorticity fluxes: •, DNS results; —, prediction from Eqs. (6.39) through (6.42). (b)*
$\overline{u\omega_3}$. *(From [1]. Copyright © Springer-Verlag.)*

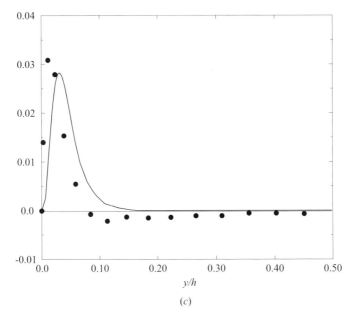

Fig. 6.18 *Predicted vorticity fluxes: •, DNS results; —, prediction from Eqs. (6.39) through (6.42). (c)*
$\overline{w\omega_1}$. *(From [1]. Copyright © Springer-Verlag.)*

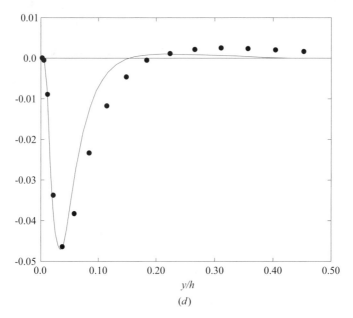

Fig. 6.18   *Predicted vorticity fluxes: •, DNS results; —, prediction from Eqs. (6.39) through (6.42). (d)*
$\overline{w\omega_2}$. *(From [1]. Copyright © Springer-Verlag.)*

These results suggest that (6.38) captures a large measure of the physics of vorticity transport. The conclusion may be reached that with the addition of stretching effects, the vorticity transport theory may be viewed as a potentially fruitful avenue to pursue in developing turbulence models. This question is examined further in Chapter 8.

6.3   SUMMARY

The focus of this chapter has been on the physics of turbulent transport, which is, perhaps, the single most important property of turbulent flows. It is not surprising, then, that modeling of turbulent transport rates form a major part of the effort at deriving accurate turbulent flow prediction schemes, as described in Chapter 8. Here it was shown that the requirements necessary to justify a gradient Reynolds stress model are typically not satisfied in turbulent flow. The problem stems from the large distances covered by fluid particles in the mixing time, which invalidates a local linear approximation to $\overline{U}$. In addition, the fact that momentum is very likely to change in a systematic way on fluid particle paths leads to another nongradient source of Reynolds stress. Similar considerations for vorticity transport suggest that by the inclusion of a term accounting for the effect of vortex stretching on fluid particle motions, a workable vorticity transport model can be developed.

Finally, a dynamical link was established between the Reynolds shear stress on the one hand, and the presence of quasi-streamwise vortices in the turbulent boundary layer on the other. A particularly favorable outcome of this consideration is that the presence of a net Reynolds shear stress appears to be the result of a relatively small number of significant vortical-driven events near the wall. This offers encouragement to the idea that such processes can be modeled effectively via LES techniques, at least if they are able to resolve this vortical activity.

REFERENCES

1. Bernard, P. S. (1990) "Turbulent vorticity transport in three dimensions," *Theor. Comput. Fluid Dyn.* **2**, 165–183.
2. Bernard, P. S., Ashmawey, M. F. and Handler, R. A. (1989) "Evaluation of the gradient model of turbulent transport through direct Lagrangian simulation," *AIAA J.* **27**, 1290–1292.
3. Bernard, P. S. and Handler, R. A. (1990) "Reynolds stress and the physics of turbulent momentum transport," *J. Fluid Mech.* **220**, 99–124.
4. Bernard, P. S. and Handler, R. A. (1990) "Dynamical significance of wall layer streaks," *Appl. Mech. Rev.* **43** No. 5 pt. 2, S219–S226.
5. Bernard, P. S., Thomas, J. M. and Handler, R. A. (1993) "Vortex dynamics and the production of Reynolds stress," *J. Fluid Mech.* **253**, 385–419.
6. Chapman, S. and Cowling, T. G. (1952), *The Mathematical Theory of Non-uniform Gases*, 2nd ed., Cambridge University Press, London.
7. Handler, R. A., Bernard, P. S., Rovelstad A., and Swearingen, J. (1992) "On the role of accelerating particles in the generation of Reynolds stress," *Phys. Fluids A* **4**, 1317–1319.
8. Perry, A. E. and Chong, M. S. (1982) "On the mechanism of wall turbulence," *J. Fluid Mech.* **119**, 173–217.
9. Perry, A. E., Henbest, S. and Chong, M. S. (1986) "A theoretical and experimental study of wall turbulence," *J. Fluid Mech.* **165**, 163–199.
10. Tangemann, R. and Gretler, W. (2001) "The computation of a two-dimensional turbulent wall jet in an external stream," *ASME J. Fluids Eng.* **123**, 154–157.
11. Taylor, G. I. (1915) "Eddy motion in the atmosphere," *Philos. Trans. R. Soc. London* **215**, 1–26.
12. Taylor, G. I. (1932) "The transport of vorticity and heat through fluids in turbulent motion," *Proc. R. Soc. Ser. A* **135**, 685–705.

7

Theory of Idealized Turbulent Flows

7.1 INTRODUCTION

For the most part, the turbulent flows considered in earlier chapters involve the interplay between a range of physical phenomena, including turbulent transport, production, dissipation, pressure work, and so on. For example, even for such relatively "simple" flows as in a channel or boundary layer, the energy budget, particularly near the wall, is a delicate balance of many effects. Clearly, the presence of multiple flow phenomena complicates the attempt to develop models representing the characteristics of individual flow processes. For this reason, among others, there has been much interest over the years in investigating idealized flows in which some of the most important physical processes of turbulent motion can be viewed in a setting free of extraneous factors.

In this chapter we consider in detail two particular flows in which the variety of physical phenomena is kept to a minimum. The first is the decay of turbulent energy in homogeneous, isotropic turbulence in the absence of turbulence production. Second, a homogeneous shear flow in which a uniform rate of turbulent production acts throughout the flow volume is examined. As will be seen, the isotropy assumption is invaluable for reducing the mathematical complexities to the point where it becomes possible to learn a considerable amount about turbulent physics, including the decay process. In fact, one important application of this knowledge is in providing a basis for modeling energy dissipation in more general circumstances.

7.2 IMPLICATIONS OF ISOTROPY

As mentioned in Chapter 1, the assumption of isotropy in turbulent flow presupposes that it is homogeneous as well, since otherwise there would be directional preferences to the motion. Thus the discussion in this chapter can generally avoid the need to specify spatial position. Among the obvious consequences of isotropy is that each of the components of the normal Reynolds stress are equal (i.e., $\overline{u^2} = \overline{v^2} = \overline{w^2} = \frac{2}{3} K$), and similarly for the diagonal components of the vorticity covariance: $\overline{\omega_1^2} = \overline{\omega_2^2} = \overline{\omega_3^2} = \zeta/3$. Less obvious is that the isotropy condition is deeply felt in the structure of turbulence, where multipoint tensor correlation functions must take on very specialized forms. From these it will be seen that such quantities as the dissipation rate, stretching correlation, and other flow statistics simplify considerably.

The properties of isotropic correlation tensors were first derived by Taylor [19] through consideration of their rotation and reflection properties. For example, as indicated in Fig. 7.1, where just planar coordinates are indicated, in the absence of a preferred direction, such correlations as $\mathcal{R}_{11}(r\mathbf{e}_1)$, $\mathcal{R}_{22}(r\mathbf{e}_2)$, and $\mathcal{R}_{11}(-r\mathbf{e}_1)$ [see (1.14)] must be equal since they can be obtained from each other merely by rotating the coordinate axes. In particular, note from Fig. 7.1c that $\overline{[-u(0,0)][-u(-r,0)]} = \overline{u(0,0)u(-r,0)} = \mathcal{R}_{11}(-r\mathbf{e}_1)$. This also means, according to (1.22), that $f(r) = f(-r)$ (i.e., f is an even function in homogeneous, isotropic turbulence). The same fact is true for $g(r)$, defined in (1.23).

In a similar vein, reflection of $\mathcal{R}_{12}(r\mathbf{e}_2)$ through the y axis implies that $\mathcal{R}_{12}(r\mathbf{e}_2) = -\mathcal{R}_{12}(r\mathbf{e}_2)$, so that $\mathcal{R}_{12}(r\mathbf{e}_2) = 0$ (see Fig. 7.2). Systematic application of these principles leads to general formulas for the multipoint correlation tensors. A more direct

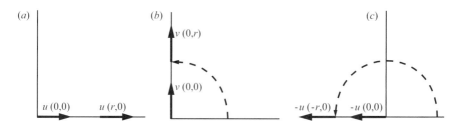

Fig. 7.1 *Rotational invariance in isotropic turbulence.*

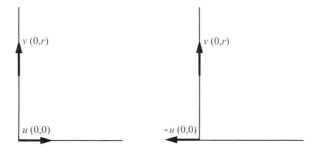

Fig. 7.2 *Antisymmetry of $\mathcal{R}_{12}(r\mathbf{e}_2)$ under reflection.*

route to the same end can be had through application of the formal mathematical theory of isotropic tensors [13]. From either approach it is found that the most general isotropic form of the two-point double (1.14) and triple (1.15) velocity correlation tensors are, respectively,

$$\mathcal{R}_{ij}(\mathbf{r}) = \overline{u^2}[R_1(r)r_i r_j + R_2(r)\delta_{ij}] \tag{7.1}$$

and

$$S_{ij,l}(\mathbf{r}) = S_1(r)r_i r_j r_l + S_2(r)r_l\delta_{ij} + S_3(r)(r_j\delta_{il} + r_i\delta_{jl}), \tag{7.2}$$

where $R_1(r)$, $R_2(r)$, $S_1(r)$, $S_2(r)$, and $S_3(r)$ are scalar functions of $r = |\mathbf{r}|$, and δ_{ij} is the Kronecker delta function.

From the definitions of $f(r)$ and $g(r)$ in (1.22) and (1.23), (7.1) implies that $f = R_1 r^2 + R_2$ and $g = R_2$, so that, after replacing R_1 and R_2, (7.1) becomes

$$\mathcal{R}_{ij}(\mathbf{r}) = \overline{u^2}\left[(f-g)\frac{r_i r_j}{r^2} + g\delta_{ij}\right]. \tag{7.3}$$

Note that in the interest of notational simplicity, the dependence of f and g on r is not indicated explicitly. Moreover, even though it is not indicated, f, g and $\overline{u^2}$ may very well be time dependent.

In the case of $S_{ij,l}$ scalar correlation functions, $k(r)$, $h(r)$, and $q(r)$ are defined via

$$S_{11,1}(r\mathbf{e}_1) = u_{\text{rms}}^3 k(r), \tag{7.4}$$

$$S_{22,1}(r\mathbf{e}_1) = u_{\text{rms}}^3 h(r), \tag{7.5}$$

$$S_{21,2}(r\mathbf{e}_1) = u_{\text{rms}}^3 q(r) \tag{7.6}$$

(see Fig. 7.3), so that from (7.2) it follows that

$$u_{\text{rms}}^3 k = S_1 r^3 + S_2 r + 2r S_3 \tag{7.7}$$

$$u_{\text{rms}}^3 h = S_2 r \tag{7.8}$$

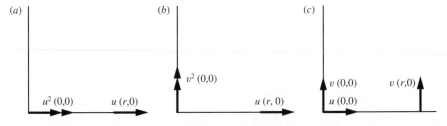

Fig. 7.3   *Definition of two-point triple velocity correlations: (a) k(r); (b) h(r); (c) q(r).*

and

$$u_{\text{rms}}^3 q = S_3 r \tag{7.9}$$

where $u_{\text{rms}} = \sqrt{\overline{u^2}}$. Solving (7.7) through (7.9) for S_1, S_2, and S_3 and substituting into (7.2) gives

$$S_{ij,l}(\mathbf{r}) = u_{\text{rms}}^3 \left[(k - h - 2q) \frac{r_i r_j r_l}{r^3} + \delta_{ij} h \frac{r_l}{r} + q \left(\delta_{il} \frac{r_j}{r} + \delta_{jl} \frac{r_i}{r} \right) \right]. \tag{7.10}$$

Note that the appearance of the factor u_{rms}^3 in (7.4) through (7.6) is by choice, not by necessity. The intention is to make k, h, and q dimensionless in the simplest way. In fact, other choices are possible. The need to select a scaling factor in this case has arisen from the fact that $S_{ij,l}(0) = 0$. In contrast, $\mathcal{R}_{ij}(0) = \overline{u^2}\delta_{ij}$, so the coefficient $\overline{u^2}$ in (7.1) is predetermined.

The two-point longitudinal triple correlation coefficient $k(r)$ is antisymmetric [i.e., $k(r) = -k(-r)$], as illustrated in Fig. 7.4, where it is seen that $\overline{u(0,0)^2 u(r,0)} = \overline{(-u(0,0))^2 (-u(-r,0))} = -\overline{u(0,0)^2 u(-r,0)}$. This means that $k(0) = 0$, as are all its even derivatives. It turns out as well that $k'(0) = 0$, by the following argument. Since $u_{\text{rms}}^3 k(r) = \overline{u^2(x)u(x+r)}$, after taking an r derivative and setting $r = 0$, then $u_{\text{rms}}^3 dk/dr(0) = \overline{u^2 \, \partial u/\partial x} = \frac{1}{3}\overline{\partial u^3/\partial x} = 0$ in homogeneous turbulence. These results imply that if $k(r)$ were to be expanded in a Taylor series about $r = 0$, its leading-order term would be $\sim r^3 d^3 k/dr^3(0)$ for small r.

Further simplification of (7.3) and (7.10) follows from demanding that they be consistent with the incompressibility condition. For example, consider the identity

$$0 = \overline{u_i(\mathbf{x}) \frac{\partial u_j}{\partial x_j}(\mathbf{y})} = \frac{\partial \mathcal{R}_{ij}}{\partial r_j}(\mathbf{r}) \tag{7.11}$$

[see the discussion in Chapter 2, centered on (2.74)]. Substituting (7.3) into (7.11) for $\mathcal{R}_{ij}$, and making use of identities such as

$$\frac{\partial}{\partial r_j}(r) = \frac{r_j}{r} \tag{7.12}$$

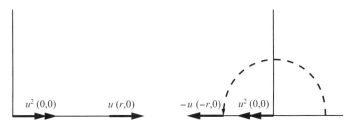

Fig. 7.4 *Antisymmetry of the two-point longitudinal triple velocity correlation.*

and

$$\frac{\partial}{\partial r_j}(r_i) = \delta_{ij},$$ (7.13)

it turns out that

$$g = f + \frac{r}{2}\frac{df}{dr}.$$ (7.14)

Experimental confirmation of this result is shown in Fig. 7.5, where it interesting to note the presence of a region where $g(r)$ is negative. Equation (7.14) can be used to eliminate $g(r)$ from (7.3), which then gives

$$\mathcal{R}_{ij}(\mathbf{r}) = \overline{u^2}\left[\left(f + \frac{r}{2}\frac{df}{dr}\right)\delta_{ij} - \frac{r_i r_j}{r^2}\frac{r}{2}\frac{df}{dr}\right].$$ (7.15)

Thus $\mathcal{R}_{ij}$ depends on a single, as yet undetermined scalar function, either $f(r)$ or $g(r)$.

A comparable result also obtains for $S_{ij,l}$. Using a relation similar to (2.77), incompressibility implies that

$$\frac{\partial S_{ij,l}}{\partial r_l}(\mathbf{r}) = 0.$$ (7.16)

Substituting for $S_{ij,l}$ via (7.10) and performing a lengthy calculation gives

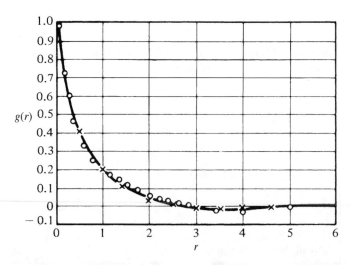

Fig. 7.5 *Confirmation of the isotropic identity (7.14). ×, measured $g(r)$; ○, evaluation of (7.14). (From [8]. Reproduced with permission of The McGraw-Hill Companies.)*

$$q = \frac{1}{4r} \frac{d(kr^2)}{dr} \tag{7.17}$$

and

$$h = -\frac{k}{2}. \tag{7.18}$$

Thus it may be concluded that $S_{ij,l}$ depends on just one scalar function [e.g., $k(r)$], through the relation

$$S_{ij,l}(\mathbf{r}) = u_{\mathrm{rms}}^3 \left[\left(k - r\frac{dk}{dr} \right) \frac{r_i r_j r_l}{2r^3} - \frac{k}{2} \delta_{ij} \frac{r_l}{r} + \frac{1}{4r} \frac{d(kr^2)}{dr} \left(\delta_{il} \frac{r_j}{r} + \delta_{jl} \frac{r_i}{r} \right) \right]. \tag{7.19}$$

This result, together with (7.15), proves to be quite useful in simplifying the equations governing the decay of turbulent kinetic energy in isotropic turbulence.

7.3 ENERGY DECAY

In Section 4.2.4 the turbulent kinetic energy equation was examined, including a consideration of what DNS could tell about its budget in channel flow. The particular importance of the dissipation rate ϵ in setting the overall balance of K was noted and the behavior of the terms in its own conservation equation were considered. Away from the wall the energy balance was between production and dissipation only and thus, in relative terms, a particularly simple form with which to consider dissipation. Here we go one step further to consider the decay of homogeneous, isotropic turbulence without turbulence production. The physical flow domain of interest can be thought of as either a large region far away from boundaries, or as considered in Section 3.6.2, turbulent flow in a box assuming periodic boundary conditions. In the latter case, the box is assumed to be of sufficient size so that periodicity has no influence on the evolution of the flow. Presumably, the minimal requirement for this to occur is that correlations such as f and k have fallen to zero within the domain.

For the ideal flows of interest in this section it is assumed that at an initial time the fluid in the domain is uniformly stirred, creating a homogeneous turbulent field whose mean statistics obey all isotropic conditions. The problem, then, is to determine how the flow proceeds through time back to a quiescent state, which it must do in the absence of production.

As mentioned previously, complete statistical specification of the turbulent velocity field requires having knowledge of all its multipoint correlations. For example, in the present case, besides specifying K, ϵ, and other one-point statistics at the initial time, it appears to be necessary to specify the initial state of $f(r)$, $k(r)$, and an infinite number of higher-order moments. Even if there were a reasonable way of obtaining initial conditions for all these statistics, it is not clear how a practical solution method could be devised that used the information. Consequently, it is necessary to limit the equations and initial conditions to a manageable number, even if in so doing it becomes impossible to specify the isotropic state exactly. This is but another manifestation of the closure problem in turbulence.

At the very minimum, the initial state of the turbulent fluid is proscribed by its turbulent kinetic energy, K, which, according to (2.40), satisfies the ordinary differential equation

$$\frac{dK}{dt} = -\epsilon \tag{7.20}$$

in homogeneous conditions. Clearly, the history of $K(t)$ after the stirring force is removed depends intimately on the behavior of ϵ. Consequently, to solve for K it is necessary to specify the initial value of ϵ, as well as an equation governing its history. The latter was considered in Section 2.2.2, [see (2.42)] and in the present circumstances has the form

$$\frac{d\epsilon}{dt} = P_\epsilon^4 - \Upsilon_\epsilon = -2v\overline{\frac{\partial u_i}{\partial x_l}\frac{\partial u_i}{\partial x_j}\frac{\partial u_l}{\partial x_j}} - 2v^2\overline{\left(\frac{\partial^2 u_i}{\partial x_j\, \partial x_l}\right)^2}, \tag{7.21}$$

in which the rate of change of ϵ depends on the balance between stretching, given by P_ϵ^4, and its dissipation rate, given by Υ_ϵ. Although (7.21) is formidable in appearance, the isotropy condition has the effect of reducing this relation to a much more manageable form.

The first step is to express the velocity derivative correlations in (7.21) in terms of $\mathcal{R}_{ij}$ and $S_{il,i}$, given by (7.15) and (7.19). This entails taking their appropriate derivatives as was done in (7.11). In the case of the last correlation in (7.21), a straightforward calculation yields

$$\overline{\frac{\partial^2 u_i}{\partial x_j\, \partial x_l}(\mathbf{x})\frac{\partial^2 u_i}{\partial y_j\, \partial y_l}(\mathbf{y})} = \frac{\partial^4 \mathcal{R}_{ii}}{\partial r_j^2\, \partial r_l^2}(\mathbf{y} - \mathbf{x}), \tag{7.22}$$

where $\mathbf{r} = \mathbf{y} - \mathbf{x}$. Taking the limit as $\mathbf{y} \to \mathbf{x}$ then gives

$$\overline{\left(\frac{\partial^2 u_i}{\partial x_j \partial x_l}\right)^2} = \frac{\partial^4 \mathcal{R}_{ii}}{\partial r_j^2\, \partial r_l^2}(0). \tag{7.23}$$

To find a similar simplification of the triple velocity derivative correlation in (7.21), $S_{il,i}$ in (1.15) and (2.69) is differentiated according to $\partial^3/\partial x_j\, \partial x_l\, \partial y_j$, yielding

$$\overline{u_l(\mathbf{x})\frac{\partial^2 u_i}{\partial x_j\, \partial x_l}(\mathbf{x})\frac{\partial u_i}{\partial y_j}(\mathbf{y})} + \overline{\frac{\partial u_i}{\partial x_l}(\mathbf{x})\frac{\partial u_l}{\partial x_j}(\mathbf{x})\frac{\partial u_i}{\partial y_j}(\mathbf{y})} = \frac{\partial^3 S_{il,i}}{\partial r_l\, \partial r_j^2}(\mathbf{y} - \mathbf{x}), \tag{7.24}$$

where the continuity equation (2.12) has been used. Letting $\mathbf{y} \to \mathbf{x}$, gives

$$\overline{\frac{\partial u_i}{\partial x_l} \frac{\partial u_l}{\partial x_j} \frac{\partial u_i}{\partial x_j}} = \frac{\partial^3 S_{il,i}}{\partial r_l \, \partial r_j^2}(0), \tag{7.25}$$

because

$$\overline{u_l \frac{\partial^2 u_i}{\partial x_j \partial x_l} \frac{\partial u_i}{\partial x_j}} = \frac{1}{2} \frac{\partial}{\partial x_l} \left[\overline{u_l \left(\frac{\partial u_i}{\partial x_j} \right)^2} \right] = 0 \tag{7.26}$$

in homogeneous turbulence.

To summarize the results thus far, according to (7.23) and (7.25), (7.21) may be written as:

$$\frac{d\epsilon}{dt} = -2\nu \frac{\partial^3 S_{il,i}}{\partial r_l \, \partial r_j^2}(0) - 2\nu^2 \frac{\partial^4 \mathcal{R}_{ii}}{\partial r_j^2 \, \partial r_l^2}(0). \tag{7.27}$$

The next step is to apply (7.15) and (7.19) to take full advantage of the assumed isotropy.

First consider the last term in (7.27). After contracting indices, (7.15) gives

$$\mathcal{R}_{ii} = \overline{u^2}(3f + rf'), \tag{7.28}$$

where r derivatives of f are henceforth denoted as in $f' = df/dr$. Differentiating (7.28) according to $\partial/\partial r_j$ yields

$$\frac{\partial \mathcal{R}_{ii}}{\partial r_j} = \overline{u^2} \left(4\frac{r_j}{r} f' + r_j f'' \right), \tag{7.29}$$

and after taking the remaining derivatives, it is found that

$$\frac{\partial^4 \mathcal{R}_{ii}}{\partial r_j^2 \, \partial r_l^2}(\mathbf{r}) = \overline{u^2} \left[\frac{24}{r} f'''(r) + 11 f^{iv}(r) + r f^{v}(r) \right]. \tag{7.30}$$

To evaluate this at $r = 0$, as required by (7.27), note that a Taylor series expansion of $f'''(r)$ gives

$$f'''(r) = r f^{iv}(0) + \frac{r^3}{3!} f^{vi}(0) + \cdots \tag{7.31}$$

since f is an even function of r. Consequently,

$$\lim_{r \to 0} \frac{f'''(r)}{r} = f^{iv}(0), \tag{7.32}$$

and finally from (7.32) and (7.30), it follows that

$$\frac{\partial^4 R_{ii}}{\partial r_j^2 \, \partial r_l^2}(0) = 35\overline{u^2}\, f^{\text{iv}}(0). \tag{7.33}$$

A similar calculation may be done using (7.19) to simplify the first term on the right-hand side of (7.27). The result is

$$\frac{\partial^3 S_{il,i}}{\partial r_l \, \partial r_j^2}(0) = \frac{35}{2}u_{\text{rms}}^3 k'''(0). \tag{7.34}$$

Applying this and the previous result to (7.27) gives

$$\frac{d\epsilon}{dt} = -35\nu u_{\text{rms}}^3 k'''(0) - 70\nu^2\overline{u^2}\, f^{\text{iv}}(0). \tag{7.35}$$

Thus, despite the complexity of (7.21), in point of fact the ϵ equation in isotropic turbulence depends on only the two time-dependent scalars $k'''(0)$ and $f^{\text{iv}}(0)$ besides K and ϵ (where, of course, $u_{\text{rms}} = \sqrt{\frac{2}{3}K}$).

A form of (7.35) that is more useful for later analysis can be had by replacing $k'''(0)$ and $f^{\text{iv}}(0)$ by nondimensional parameters. Convenient in this regard are the skewness of the velocity derivative field

$$S_K \equiv -\frac{\overline{(\partial u/\partial x)^3}}{\overline{(\partial u/\partial x)^2}^{\frac{3}{2}}} \tag{7.36}$$

and the *palenstrophy* coefficient,

$$G = \frac{\overline{(\partial^2 u/\partial x^2)^2}}{\overline{(\partial u/\partial x)^2}^2}. \tag{7.37}$$

These can be incorporated into (7.35) by making use of some additional identities. First, consider the correlation $\overline{(\partial u/\partial x)^2}$. By a calculation similar to that leading to (7.33), it may be shown that

$$\overline{\left(\frac{\partial u}{\partial x}\right)^2} = -\overline{u^2}\, f''(0), \tag{7.38}$$

and in fact, more generally that

$$\overline{\left(\frac{\partial u_i}{\partial x_j}\right)^2} = -15\overline{u^2}\, f''(0). \tag{7.39}$$

In view of (2.36), (7.38), and (7.39), it follows that

$$\epsilon = -15\nu\overline{u^2}f''(0) = 15\nu\overline{\left(\frac{\partial u}{\partial x}\right)^2}, \tag{7.40}$$

a result first obtained by Taylor [19]. An interesting sidelight of these calculations is that it may be shown that

$$\overline{\left(\frac{\partial u}{\partial y}\right)^2} = 2\overline{\left(\frac{\partial u}{\partial x}\right)^2} \tag{7.41}$$

(i.e., in a loose sense the mean-square shearing in isotropic turbulence is twice the mean-square stretching).

Similar to (7.38) it may be computed that

$$\overline{\left(\frac{\partial u}{\partial x}\right)^3} = u_{\text{rms}}^3 k'''(0) \tag{7.42}$$

and

$$\overline{\left(\frac{\partial^2 u}{\partial x^2}\right)^2} = \overline{u^2}f^{\text{iv}}(0). \tag{7.43}$$

Using (7.40) and (7.42) in (7.36) then gives

$$k'''(0) = -S_K\left(\frac{\epsilon}{15\overline{u^2}\nu}\right)^{3/2}, \tag{7.44}$$

while (7.40) and (7.43) in (7.37) yield

$$f^{\text{iv}}(0) = G\left(\frac{\epsilon}{15\overline{u^2}\nu}\right)^2. \tag{7.45}$$

Substituting (7.44) and (7.45) into (7.35) and using the fact that $K = \frac{3}{2}\overline{u^2}$ gives the standard form of the ϵ equation for homogeneous isotropic turbulence:

$$\frac{d\epsilon}{dt} = S_K^* R_T^{1/2}\frac{\epsilon^2}{K} - G^*\frac{\epsilon^2}{K}, \tag{7.46}$$

where for convenience we have defined

$$S_K^* = \frac{7}{3\sqrt{15}}S_K, \tag{7.47}$$

$$G^* = \frac{7}{15}G, \tag{7.48}$$

and

$$R_T = \frac{K^2}{\nu\epsilon}. \tag{7.49}$$

R_T is a dimensionless parameter that may be interpreted as a Reynolds number, as will be seen in the next section.

The coupled system of equations (7.20) and (7.46) represent two equations in the four unknowns, K, ϵ, S_K^*, and G^*, and thus is not closed. The initial state of the turbulence is specified by assigning values $K_0, \epsilon_0, S_{K_0}^*$, and G_0^* to these quantities. Alternatively, in view of (7.44) and (7.45), initial forms for $f(r)$ and $k(r)$ can be specified from which S_{K_0} and G_0 can be obtained.

7.3.1   Turbulent Reynolds Number

It is helpful to regard R_T in (7.49) as a turbulent Reynolds number formed from the velocity scale $\sqrt{K}$ and length scale, $K^{3/2}/\epsilon = T_t\sqrt{K}$, where the time scale $T_t = K/\epsilon$ is referred to as the *eddy turnover time*. In fact, according to (7.20),

$$\frac{1}{T_t} = \frac{\epsilon}{K} = -\frac{1}{K}\frac{dK}{dt} \tag{7.50}$$

is the fractional rate at which energy is being dissipated. Consequently, T_t is the time scale of that dissipation (i.e., the time over which a significant fraction of the turbulent kinetic energy dissipation might occur). For example, if K were decaying exponentially according to $K(t) = K(0)e^{-t/\tau}$, a calculation with (7.50) gives $T_t = \tau$ (i.e., in this case, T_t is the time over which K falls by the factor $1/e$). T_t can also be conceptualized as the time over which a turbulent eddy loses a measurable amount of its energy and might be replaced by another more energetic eddy; hence there is a "turnover" in the eddies.

Since R_T can be written as the quotient $(K/\epsilon)/(\nu/K)$, it can also be interpreted as the ratio of turbulent and viscous time scales, T_t/T_μ, where $T_\mu = \nu/K$ is characteristic of small-scale dissipative motions (i.e., it is the time scale over which viscous dissipation can be expected to act). Large R_T means that the turbulence is very energetic and far from being dissipated, since the time scale over which significant energy is being lost, T_t, is much larger than that over which the smallest dissipation scales can act. In this case, many small, dissipative time units must come and go before there is a significant change in the energy. When R_T is small, the energy must be mainly in the dissipation range, since the energy of the flow itself drops as fast as it is being dissipated. This is a sign that the turbulence may be considered to be weak. This interpretation helps make clear why R_T decreases to zero during the decay of isotropic turbulence. The appearance of R_T in (7.46) suggests that the stretching term, which depends on it, will tend to be large when the flow is energetic and turbulence has not yet filled out the dissipative range of scales.

Another useful turbulence Reynolds number is traditionally defined using the microscale λ, computed from (1.25) using $g(r)$ in place of $f(r)$. In this case

$$R_\lambda = \frac{\lambda u_{\rm rms}}{\nu}, \tag{7.51}$$

which, in fact, can be related to R_T as follows. Since $g(r)$ is an even function in isotropic turbulence, $g'(0) = 0$. Thus (1.25) implies that

$$\lambda^2 = \frac{-2}{g''(0)} = -\frac{1}{f''(0)}, \tag{7.52}$$

with the second equality a consequence of (7.41). It follows from (7.40) that

$$\epsilon = 15\nu \frac{\overline{u^2}}{\lambda^2}. \tag{7.53}$$

A calculation then gives

$$R_T = \frac{3}{20} R_\lambda^2. \tag{7.54}$$

R_T and R_λ may thus be used interchangeably to characterize the degree to which a homogeneous field is turbulent. $R_\lambda > 100$ is a useful measure of strong turbulence, while $R_\lambda < 1$ signifies weak turbulence. In fact, flow with $R_\lambda < 1$ is referred to as the *final period* (i.e., the last stage of turbulence before the flow relaminarizes). It has been studied in some detail by Batchelor and Townsend [3] and others. Our interest in the decay process centers on following the history of a turbulent field as it changes from an initial state with large R_T to the final period where $R_T < 1$.

Before considering the decay process further, it is useful to note that equations (7.20) and (7.46) can be combined into a single equation for R_T. This is derived by first computing, from (7.49),

$$\frac{dR_T}{dt} = \frac{2K}{\nu\epsilon} \frac{dK}{dt} - \frac{K^2}{\nu\epsilon^2} \frac{d\epsilon}{dt} \tag{7.55}$$

and then substituting (7.20) and (7.46) to obtain

$$\frac{dR_T}{dt} = -\frac{2K}{\nu} - S_K^* \sqrt{R_T} \frac{K}{\nu} + G^* \frac{K}{\nu}. \tag{7.56}$$

Since K and ϵ are always positive, a dimensionless time can be defined unambiguously via

$$\tau(t) = \int_0^t \frac{\epsilon(t')}{K(t')} \, dt', \tag{7.57}$$

which, in view of (7.20), integrates exactly to yield

$$\tau(t) = \ln(K(0)/K(t)), \tag{7.58}$$

if it is assumed for convenience that $\tau(0) = 0$. Since (7.20) implies that K decays monotonically to zero, $\tau(t)$ is well defined and, in particular, $\tau \to \infty$ as $t \to \infty$.

Equation (7.58) is a mapping from t to τ. Let $t(\tau)$ be the inverse relation mapping τ to t. Thus $\tau(t(\tau)) = \tau$. Defining

$$R_T^*(\tau) = R_T(t(\tau)), \tag{7.59}$$

or equivalently,

$$R_T^*(\tau(t)) = R_T(t), \tag{7.60}$$

it is not hard to show from (7.56) that

$$\frac{dR_T^*}{d\tau} = R_T^* \left(G^* - 2 - S_K^* \sqrt{R_T^*} \right). \tag{7.61}$$

Thus, as an alternative to solving the decay problem via the coupled system (7.20) and (7.46), there is the option of solving the single differential equation (7.61). Of course, G^* and S_K^* must be found by additional considerations. Using either way of setting up the decay problem, further progress requires that additional assumptions be made so that a closed system of equations can be deduced.

7.3.2 Self-Similarity

Since our formulation of the isotropic decay problem takes place in free space without an externally imposed length scale, it is natural to inquire if a similarity solution might exist (i.e., one in which just the scale of the turbulence changes with time and not its structural features, such as the mathematical form of its correlation functions). In the present case this means that we are enquiring as to whether or not there may exist functions $\tilde{f}(s)$ and $\tilde{k}(s)$ and a length scale $L(t)$ such that

$$f(r, t) = \tilde{f}(r/L(t)) \tag{7.62}$$

and

$$k(r, t) = \tilde{k}(r/L(t)). \tag{7.63}$$

If (7.62) and (7.63) hold for all r, the turbulent decay would have complete similarity, or be *self-similar* or *self-preserving*. If they are valid for only a limited range of r, the flow has only "partial" or "incomplete" similarity. Both of these cases have been treated in the literature. Here, the implications of just the former assumption (i.e., complete similarity) is considered, and it will be shown that this has the remarkable consequence of rendering the isotropic decay problem fully closed.

An immediate consequence of complete similarity is found by substituting (7.62) into (7.52), to yield

$$\frac{\lambda^2}{L^2} = \frac{-1}{\tilde{f}''(0)},$$
(7.64)

where $\tilde{f}''(0)$ is a constant. This means that $L \sim \lambda$, so the choice of L is clearly not arbitrary. Without loss of generality, it is convenient to take

$$L = \lambda.$$
(7.65)

Applying (7.53), (7.63), and (7.65) to (7.44) gives

$$- S_K = \tilde{k}'''(0)$$
(7.66)

and similarly, (7.45) yields

$$G = \tilde{f}^{\mathrm{iv}}(0)$$
(7.67)

(i.e., both S_K and G are constants during the decay). In this case, the self-similar decay of isotropic turbulence is governed by the *closed* system of equations

$$\frac{dK}{dt} = -\epsilon$$
(7.68)

$$\frac{d\epsilon}{dt} = S_{K_0}^* R_T^{1/2} \frac{\epsilon^2}{K} - G_0^* \frac{\epsilon^2}{K},$$
(7.69)

where $S_{K_0}^*$ and G_0^* can be determined from (7.66) and (7.67) as constant values during the decay. Note that it makes little practical difference whether $S_{K_0}^*$ and G_0^* are assigned or if they are deduced from given forms of $\tilde{f}$ and $\tilde{k}$.

In the case of complete similarity, (7.61) becomes the single solvable equation for $R_T^*(\tau)$:

$$\frac{dR_T^*}{d\tau} = R_T^* \left(G_0^* - 2 - S_{K_0}^* \sqrt{R_T^*} \right).$$
(7.70)

Our focus now is on investigating the properties of the solution to (7.70), or equivalently, (7.68) and (7.69), from an initial state with $R_T^*(0) \gg 1$ to the point when $R_T^* \ll 1$.

7.3.3 Fixed-Point Analysis

A useful means [18] for discovering the general properties of solutions to (7.70) is by considering its *fixed points*, those values of R_T^* where the right-hand side is zero and hence R_T^* does not change in time. Such points often correspond to *attracting solutions*, those toward which all other solutions travel as $t \to \infty$. For this reason, the fixed points of (7.70) will be denoted as $R_{T_\infty}^*$.

From (7.70) a calculation gives

$$R_{T_\infty}^* \left(G_0^* - 2 - S_{K_0}^* \sqrt{R_{T_\infty}^*} \right) = 0,$$ (7.71)

which is satisfied by either

$$R_{T_\infty}^* = 0$$ (7.72)

or

$$R_{T_\infty}^* = \left(\frac{G_0^* - 2}{S_{K_0}^*} \right)^2.$$ (7.73)

In the present case these are *attracting*, that is, solutions starting from arbitrary R_T^* will move toward one or the other of the fixed points, depending on the value of G_0^*. For example, (7.72) is reached for all initial states if $G_0^* < 2$, while (7.73) is reached for all initial conditions if $G_0^* > 2$. These behaviors are evident from examination of (7.70). Thus, if $G_0^* < 2$, the term in parentheses on the right-hand side is always negative and R_T^* must decay to zero. For $G_0^* > 2$, the right-hand side of (7.70) will be positive if $R_{T_0} < R_{T_\infty}^*$ and negative if $R_{T_0} > R_{T_\infty}^*$. Either way the tendency is to raise or lower R_T^* until it converges to $R_{T_\infty}^*$ given by (7.73).

The two fixed-point solutions represent, in essence, different equilibrium states of turbulence during isotropic decay. Assuming only that self-similarity is maintained during the decay, one or the other of these is achieved depending on the given value of G_0^*. Before considering some of the properties of these equilibrium states and what they suggest about the nature of the decay process, it should be noted that between the two cases (7.72) and (7.73), only the former satisfies the expectation that at the end of the decay process, $R_T^* = 0$ [i.e., the asymptotic limit of case (7.73) has nonzero R_T^*]. At the outset this suggests that if the decay is similar and it must end up in the final period solution, where $R_T^* \to 0$, then $G_0^* < 2$. As it turns out, this viewpoint cannot be supported, if for no other reason than that experiments often reveal the occurrence of values of $G_0^* > 2$. A more reasonable scenario (suggested in the next sections) is that the two equilibrium solutions correspond to different stages of the decay process. In the high-Reynolds-number range, an equilibrium associated with (7.73) occurs, while a second equilibrium, associated with (7.72), occurs in the final period. This suggests that complete self-preservation is not physically possible over the entire decay from an initial large value of R_T^* until the final period, although aspects of self-similarity are relevant to the dynamics.

7.3.4   Final Period of Isotropic Decay

Now the nature of isotropic decay at small R_T^* is considered. Rewriting equation (7.69) in the form

$$\frac{d\epsilon}{dt} = \left(\frac{S_{K_0}^* R_T^{1/2}}{G_0^*} - 1 \right) G_0^* \frac{\epsilon^2}{K}, \tag{7.74}$$

it is clear that the term containing $R_T^{1/2}$, representing vortex stretching, can be neglected when

$$\frac{S_{K_0}^* R_T^{1/2}}{G_0^*} << 1. \tag{7.75}$$

For typical values of $S_K \approx 0.5$ and $G \approx 3$ found in low-Reynolds-number experiments, so that $S_K^* \approx 0.3$ and $G_0^* \approx 1.4$, (7.75) is reasonably well satisfied when $R_T < 0.1$. This implies that in the vicinity of $R_T = 0$, the coupled equations for K and ϵ reduce to

$$\frac{dK}{dt} = -\epsilon \tag{7.76}$$

and

$$\frac{d\epsilon}{dt} = -G_0^* \frac{\epsilon^2}{K}. \tag{7.77}$$

These have the exact solution

$$\frac{K}{K_0} = \left(1 + \frac{t}{\alpha T_{t_0}} \right)^{-\alpha} \tag{7.78}$$

$$\frac{\epsilon}{\epsilon_0} = \left(1 + \frac{t}{\alpha T_{t_0}} \right)^{-1-\alpha}, \tag{7.79}$$

where

$$\alpha = \frac{1}{G_0^* - 1}, \tag{7.80}$$

and K_0, ϵ_0, and $T_{t_0} = \epsilon_0 / K_0$ are initial values of the flow properties. After t advances several multiples of T_{t_0}, so that $t/(\alpha T_{t_0}) >> 1$, (7.78) and (7.79) become simple power laws:

$$K \sim t^{-\alpha} \tag{7.81}$$

and

$$\epsilon \sim t^{-1-\alpha}, \tag{7.82}$$

respectively. Thus for $G_0^* < 2$, so that $\alpha > 1$, and R_T small, the self-similar solution for K and ϵ consists of power laws with exponents depending on G_0^*.

It is found in experiments [3] of isotropic decay with R_T small (i.e., in the "final period") that f is given by the Gaussian function

$$f(r, t) = e^{-r^2/2\lambda^2} \tag{7.83}$$

as illustrated in Fig. 7.6. Substituting this into (7.67) yields $G = 3$, so that $G_0^* = 7/5$ according to (7.48), and by (7.80), $\alpha = 5/2$. According to (7.81), this means that K obeys a $-5/2$ decay law in the final period if the decay is self-preserving. Experimental results are shown in Fig. 7.7 which establish that this is indeed the case. In this experiment, turbulence is generated downstream of a mesh of bars with spacing M. The decay time is associated with distance along the wind tunnel x_1, via the mapping $t = x_1/\overline{U}$, where $\overline{U}$ is the mean velocity of the uniform flow. The linear behavior of $(\overline{U}^2/\overline{u^2})^{2/5}$ in the figure implies that $\overline{u^2}$ satisfies a $-5/2$ decay law. It can be concluded from this that the equilibrium solution associated with (7.72) when $G_0^* = 7/5$, is fully consistent with the measured final period behavior.

Substituting (7.81) and (7.82) into (7.53) and solving for λ gives

$$\lambda^2 \sim t \tag{7.84}$$

in the final period, so that, in fact, the microscale increases as small scales dissipate energy faster than the large scales. This behavior is confirmed experimentally, as

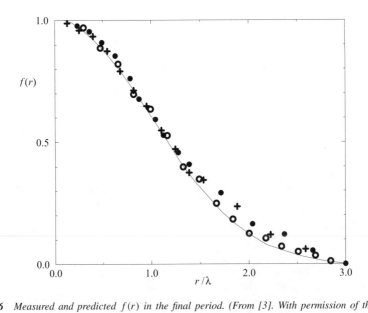

Fig. 7.6 *Measured and predicted $f(r)$ in the final period. (From [3]. With permission of the Royal Society.)*

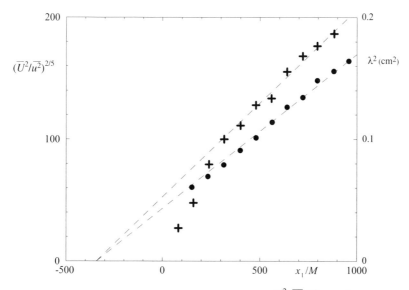

Fig. 7.7   *Confirmation of the* $-5/2$ *decay law in the final period.* •, $(\overline{U^2}/\overline{u^2})^{2/5}$; +, λ^2. *(From [3]. With permission of the Royal Society.)*

shown in Fig. 7.7, where λ^2 varies linearly with x_1. Another interesting facet of the final period is that, according to (7.14) and (7.83),

$$g(r, t) = \left(1 - \frac{r^2}{2\lambda^2} \right) e^{-r^2/2\lambda^2}, \tag{7.85}$$

a function which, similar to the measured values in Fig. 7.5, is also negative when r is sufficiently large.

Now consider the applicability of the fixed-point solution (7.72) to the case when $G_0^* < 2$, yet $R_{T_0} >> 1$, a situation that is expected to be unphysical. To explore this possibility, note that (7.70) admits an exact solution of the form

$$R_T^* = R_{T_0}^* \left[\frac{(G_0^* - 2) \exp\left[(G_0^* - 2)\tau/2 \right]}{G_0^* - 2 - S_{K_0}^* \sqrt{R_{T_0}^*} \left\{ 1 - \exp\left[(G_0^* - 2)\tau/2 \right] \right\}} \right]^2 \tag{7.86}$$

when $G_0^* \neq 2$, and

$$R_T^* = R_{T_0}^* \left[\frac{1}{1 + S_{K_0}^* \sqrt{R_{T_0}^*} \, \tau/2} \right]^2 \tag{7.87}$$

when $G_0^* = 2$. When $R_{T_0}^* >> 1$, $G_0^* < 2$ and τ is large enough, (7.86) gives

$$R_T^* \sim e^{(G_0^*-2)\tau}. \tag{7.88}$$

This means that R_T^* is small only when τ is large. The latter condition, in view of (7.58), happens when $K \ll K_0$ (i.e., most of the energy is lost). An example of this solution is given in Fig. 7.8, where it is seen that the outset of the decay consists of an unphysical precipitous drop in K. Time is scaled here by the initial eddy turnover time, T_{t_0}; thus, in fact, K falls five orders of magnitude in one such time unit. Also shown is that R_T falls swiftly to values representative of the final period. Taken together, it appears evident that the fixed-point solution associated with (7.72) cannot apply to the initial, high-Reynolds-number stage of isotropic decay.

7.3.5 High-Reynolds-Number Equilibrium

Now consider the decay associated with the second equilibrium solution, (7.73). When R_T reaches this value during decay, which it would if $G_0^* > 2$ (and the physical decay is self-similar), it does not subsequently change, in which case (7.70) reduces to

$$\frac{d\epsilon}{dt} = (G_0^* - 2)\frac{\epsilon^2}{K} - G_0^*\frac{\epsilon^2}{K} = -2\frac{\epsilon^2}{K}. \tag{7.89}$$

According to (7.73), if $R_{T_\infty} \gg 1$, then G_0^* is generally large (assuming that S_{K_0} never strays significantly from its typical measured values). As an example, $R_{T_\infty} =$

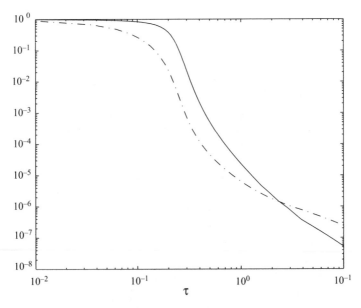

Fig. 7.8 *Self-similar decay corresponding to $G_0^* = 7/5$ and $R_{T_0}^* = 1000$. —, K/K_0; — · —, R_T/R_{T_0}.*

10^4 coincides with $G_0^* \approx 25$. In this case the middle expression in (7.89) consists of a near balance between vortex stretching, with coefficient $G_0^* - 2$, and viscous dissipation, with coefficient $-G_0^*$. The slight edge toward the latter leaves a net dissipation with coefficient -2. The important point here is that the high-Reynolds-number equilibrium decay represents one with nearly, but not quite exactly, equal contributions from vortex stretching and dissipation of dissipation. In this, the net dissipation rate assumes a universal value independent of initial conditions, including the value of G_0^*.

Similar to the $R_{T_\infty} = 0$ case, (7.68) and (7.89) can be solved, this time yielding an asymptotic power law decay of the form [18]

$$K \sim t^{-1} \tag{7.90}$$

$$\epsilon \sim t^{-2}, \tag{7.91}$$

which, as expected because of (7.89), is independent of the initial condition. In general, $R_{T_0} \neq R_{T_\infty}$, so a transient period occurs, during which time the coefficient $S_{K_0}\sqrt{R_T}$ in (7.69) changes to G_0^*-2. If R_{T_0} is large, then the larger G_0^* is, the closer the solution is to equilibrium and the shorter the time interval until equilibrium is reached. Thus, if $G_0^* - 2$ is large, so is $e^{(G_0^*-2)\tau/2}$ for relatively small τ. The exponential terms then dominate the numerator and denominator in (7.86) and thus cancel, so that in lieu of (7.73), it follows from (7.86) that $R_T \approx R_{T_\infty}$ in a relatively short time (i.e., just a few eddy turnover times).

To the extent that (7.46) and (7.69) have similar behavior, the high-Reynolds-number equilibrium and its associated flow properties can be expected to be relevant to actual isotropic turbulence. However, the fixed points (7.72) and (7.73) are stable nodes of the coupled dynamical system. This means [6] that the character of the solution to the K and ϵ equations is unaffected by small changes in G^* and S_K^* during the decay. Therefore, the results of this section are likely to provide a reasonably good description of isotropic decay at high Reynolds numbers. In particular, it is expected that G_0^* will be relatively large and a t^{-1} decay law will be reached in a matter of a few eddy turnover times.

To illustrate these results, Fig. 7.9 shows the behavior of K, ϵ, and R_T displayed in a log-log plot for the relatively high value $G_0^* = 5$, where $R_{T_0} = 1000$ is purposefully taken to be large and a typical value of $S_K^* = 0.3$ is used. In contrast to the low G_0^* solution, the solution for K converges to the t^{-1} decay law and it does so relatively quickly, within which time only a modest energy loss has occurred. R_T has an initial rapid drop, followed by convergence to the nonzero value given by (7.73). While R_T is falling, it may be noted that K is relatively constant, while ϵ increases. Thus, whether through consideration of (7.53) or (7.54), it is expected that λ should decrease during this time period, as may be seen to be the case in Fig. 7.10. Physically, this is the natural outcome of having insufficient energy at the outset in the dissipation range scales. The turbulence makes a relatively rapid adjustment in which vortex stretching brings energy to small scales so that λ decreases and ϵ rises. After just a few eddy

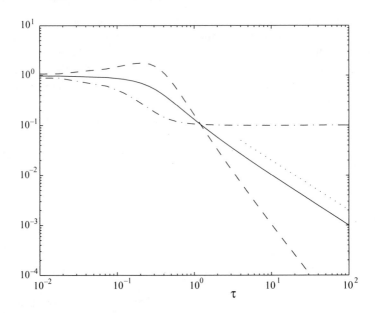

Fig. 7.9 *Self-similar decay corresponding to $G_0^* = 5$ and $R_{T_0}^* = 1000$. —, K/K_0; − −, ϵ/ϵ_0; − · −, R_T/R_{T_0}; · · ·, line with slope −1.*

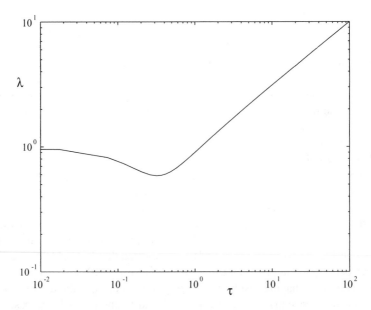

Fig. 7.10 *λ in isotropic decay corresponding to the conditions in Fig. 7.9.*

turnover times, the equilibrium between dissipation and stretching given in (7.73) is achieved and the energy begins to decay according to (7.90). By this time, as the turbulence weakens, λ is rising and ϵ is falling. Eventually, the cumulative energy decay is such that the high-Reynolds-number equilibrium can no longer be tenable and the decay must shift toward that of the final period. At this time, self-similarity must be lost and G_0^* decreases to 7/5.

While Fig. 7.7 shows that a $-5/2$ decay law in the final period is a true physical effect, the experimental record for decay at high Reynolds number provides a less clearcut view of what occurs. In part, this is due to the difficulty of attaining high values of R_T in physical experiments. In addition, the duration of the experiments may not be long enough to witness the appearance of a full-fledged t^{-1} law. Those few experiments that have either been performed at relatively high Reynolds numbers or that covered an unusually long elapsed time (i.e., the length of a wind tunnel), in fact do show the presence of the t^{-1} decay law. Other experiments, at lower Reynolds numbers, show power law decays with exponents on average approximately equal to 1.25. It may be surmised that these results correspond to the approach toward the t^{-1} state. Indeed, this behavior is observed in the similarity solution for many choices of $R_{T_0}^*$ and G_0^*. Fig. 7.11 illustrates that decay laws with exponents representative of those seen in experiments approximate very well the solutions to the self-similar decay equations for time intervals typically seen in experiments (i.e., just a few eddy turnover times).

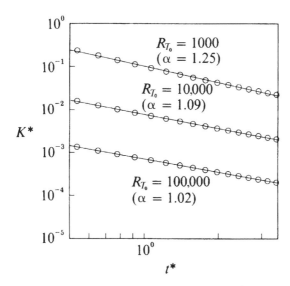

Fig. 7.11 *Comparison of self-similar decay with typical experimental observations.* ○, *self-preserving solution;* —, $K \sim t^{-\alpha}$. *(From [18]. Reprinted with the permission of Cambridge University Press.)*

7.4 SELF-PRESERVATION AND THE KÁRMÁN–HOWARTH EQUATION

It is in the nature of self-similarity as it has been defined in (7.62) and (7.63) that $\tilde{f}$ and $\tilde{k}$ are fixed throughout the decay. According to (7.66) and (7.67), the third and fourth derivatives of $\tilde{k}$ and $\tilde{f}$ at the origin, respectively, set the scales S_K and G and hence subsequent decay of K and ϵ. It is clear from Section 7.3 that at least G must change in time during isotropic decay if $R_{T_0} \neq 0$, with the consequence that $\tilde{f}$ must change as well. Exactly what kind of behavior to expect from $\tilde{f}$ is an interesting question that will now be considered.

The starting point for this analysis is the Kármán–Howarth equation [9] relating f, k, and $\overline{u^2}$, which is derived after a fairly long calculation, by substituting (7.15) and (7.19) into (2.82):

$$\frac{\partial(\overline{u^2}f)}{\partial t} = (\overline{u^2})^{3/2}\left(\frac{\partial k}{\partial r} + \frac{4}{r}k\right) + 2\nu\overline{u^2}\left(\frac{\partial^2 f}{\partial r^2} + \frac{4}{r}\frac{\partial f}{\partial r}\right). \tag{7.92}$$

In fact, equations (7.20) and (7.46) can be derived from (7.92) by substituting Taylor series for f and k and gathering terms depending on like powers of r.

As it stands, (7.92) is intractable, but by adding the additional constraint that the decay be self-similar so that (7.62) and (7.63) hold, the Kármán–Howarth equation becomes

$$2\eta^{-4}\frac{d}{d\eta}\left(\eta^4\frac{d\tilde{f}}{d\eta}\right) + \eta\frac{d\tilde{f}}{d\eta}\left(\frac{7}{3}G_0 - 5\right) + 10\tilde{f}$$

$$= R_\lambda\left(\frac{7}{6}S_{K_0}\eta\frac{d\tilde{f}}{d\eta} - \eta^{-4}\frac{d(\eta^4\tilde{k})}{d\eta}\right), \tag{7.93}$$

where $\eta = r/\lambda$ is a similarity variable. This equation constitutes a single ordinary differential equation for the two unknown functions $\tilde{f}(\eta)$ and $\tilde{k}(\eta)$, with R_λ as time-dependent parameter.

If completely self-similar decay were possible, (7.93) would have to be satisfied at all times during the decay, regardless of how R_λ varies. The only way this could occur is if each side of the equation vanished independently of the other, that is,

$$2\eta^{-4}\frac{d}{d\eta}\left(\eta^4\frac{d\tilde{f}}{d\eta}\right) + \eta\frac{d\tilde{f}}{d\eta}\left(\frac{7}{3}G_0 - 5\right) + 10\tilde{f} = 0 \tag{7.94}$$

and

$$\frac{7}{6}S_{K_0}\eta\frac{d\tilde{f}}{d\eta} - \eta^{-4}\frac{d(\eta^4\tilde{k})}{d\eta} = 0, \tag{7.95}$$

assuming that $R_\lambda \neq 0$. If not, the left side of (7.93) would be multivalued as R_λ changed.

Equation (7.94) is an example of the confluent hypergeometric equation [10,16,17] and has solution

$$\tilde{f}(\eta) = M\left(\frac{1}{G_0^* - 1}, \frac{5}{2}, -\frac{5(G_0^* - 1)}{4}\eta^2\right), \tag{7.96}$$

where M is the confluent hypergeometric function. Since $\tilde{f}$ is known, (7.95) can be integrated to give

$$\tilde{k}(\eta) = \frac{7}{6}S_{K_0}\frac{1}{\eta^4}\int_0^{\eta} s^5 \frac{d\tilde{f}}{ds}\, ds, \tag{7.97}$$

with $\tilde{f}$ given by (7.96). It is interesting to note from (7.96) and (7.97) that $\tilde{f}$ and $\tilde{k}$ are entirely specified once G_0 and S_{K_0} are chosen. Thus, within the constraint of complete self-similarity, a unique pair of functions $\tilde{f}$ and $\tilde{k}$ are associated with constant values of G_0 and S_{K_0}, and vice versa.

It has already been shown that complete similarity is unattainable in isotropic decay, so it is clear that $\tilde{f}$ and $\tilde{k}$ must change in time in a real flow. Nonetheless, it is still useful to consider the meaning of (7.93) in the neighborhood of the high- and low-Reynolds-number equilibrium solutions. First consider the case of small R_T in the neighborhood of $R_{T_\infty} = 0$. Here the right-hand side of (7.93) is small, so (7.94) is well satisfied regardless of the separability condition. Since $G_0^* = 7/5$ for the final period, the solution to (7.94) given by (7.96) is [4]

$$\tilde{f}(\eta) = M\left(\frac{5}{2}, \frac{5}{2}, -\frac{\eta^2}{2}\right) = e^{-\eta^2/2}, \tag{7.98}$$

in complete agreement with (7.83).

Assuming that the decay is self-similar in the neighborhood of $R_T = 0$, then (7.97) holds and, after carrying out the integration using (7.98), it is found that

$$\tilde{k}(\eta) = \frac{7}{6}S_{K_0}\frac{1}{\eta^4}\left[(\eta^5 + 5\eta^3 + 15\eta)e^{-\eta^2/2} - 15\sqrt{\frac{\pi}{2}}\,\mathrm{erf}\left(\frac{\eta}{\sqrt{2}}\right)\right], \tag{7.99}$$

where erf is the error function [7]. A plot of (7.99) is given in Fig. 7.12, where $\tilde{k}$ is seen to have a much slower decay for large η than the Gaussian form of $\tilde{f}$. In fact, (7.97) and (7.98) make clear that $\tilde{k}(\eta) \sim \eta^{-4}$ as $\eta \to \infty$.

In the case of the high-Reynolds-number equilibrium, R_T is a nonzero constant, as is R_λ. Thus, in the equilibrium state, (7.93) is indeterminate [i.e., it does not have the separability requirement leading to (7.94) and (7.95)]. In fact, solutions to (7.94) for $G_0 > 3$ are unphysical in the sense that their corresponding energy spectrum functions are singular at the origin, as shown in Section 7.6.1.

It may be concluded from this that, unlike the system of self-similar K and ϵ equations which are expected to be a reasonable model of the exact equations, the self-similar Kármán–Howarth equation is not useful for analyzing high-Reynolds-number flow which is *close* to similarity, but not exactly self-similar.

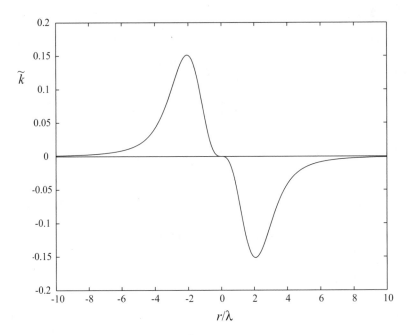

Fig. 7.12 $\tilde{k}$ *in the final period.*

7.5 IMPLICATIONS FOR TURBULENCE MODELING

Looking back to (7.21) and (7.46), it is evident that we have derived the exact isotropic formulas

$$P_\epsilon^4 = S_K^* R_T^{1/2} \frac{\epsilon^2}{K} \tag{7.100}$$

and

$$\Upsilon_\epsilon = G^* \frac{\epsilon^2}{K}. \tag{7.101}$$

When the modeling of the ϵ equation is considered in Chapter 8, it will be seen that traditional RANS modeling assumes that

$$P_\epsilon^4 - \Upsilon_\epsilon = (S_K^* R_T^{1/2} - G^*) \frac{\epsilon^2}{K} = -C_{\epsilon_2} \frac{\epsilon^2}{K}, \tag{7.102}$$

where C_{ϵ_2} is a constant. In other words, it is assumed that $S_K^* R_T^{1/2} - G^* = -C_{\epsilon_2}$, which is similar to the high-Reynolds-number equilibrium result in (7.89), except that a constant C_{ϵ_2} which is not necessarily 2 is assumed. The rationale for this assumption is based on a scaling argument that is interesting to examine in the light of the previous results.

The first step is to note from (2.50) and (2.126) that $\Upsilon_\epsilon = \Upsilon_\zeta$ in isotropic turbulence by virtue of the identity

$$\overline{\left(\frac{\partial^2 u_i}{\partial x_j\,\partial x_l}\right)^2} = \overline{\left(\frac{\partial \omega_i}{\partial x_j}\right)^2}, \tag{7.103}$$

which is easily proven after substituting the definition of the vorticity on the right-hand side. Then from (7.101) and the fact that $\epsilon = \nu\zeta$, it follows that

$$G^* \sim \frac{\Upsilon_\zeta/\zeta}{\epsilon/K}, \tag{7.104}$$

which says that G^* is the ratio of the fractional rate of change of enstrophy by its dissipation (i.e., Υ_ζ/ζ) to the fractional rate of change of energy (i.e., ϵ/K). In effect, both numerator and denominator are inverse time scales.

Since enstrophy dissipation is influenced most strongly by small-scale phenomena, it can be hypothesized that the numerator in (7.104) scales with the Kolmogorov dissipation time scale $(\nu/\epsilon)^{1/2}$ [see (2.93)]. Substituting this into (7.104) gives

$$G^* \sim \sqrt{R_T} \tag{7.105}$$

for large Reynolds numbers. This result is consistent with (7.73). To justify (7.102), it is necessary to go beyond (7.105) to assert more specifically that

$$G^* = S_K^* \sqrt{R_T} + C_{\epsilon_2}, \tag{7.106}$$

so that (7.69) becomes

$$\frac{d\epsilon}{dt} = \frac{\epsilon^2}{K}\left[S_K^*\sqrt{R_T} - (S_K^*\sqrt{R_T} + C_{\epsilon_2})\right] = -C_{\epsilon_2}\frac{\epsilon^2}{K}. \tag{7.107}$$

In other words, the coefficient of $\sqrt{R_T}$ in (7.106) is chosen to be precisely what is needed to cancel the effect of vortex stretching from the ϵ equation.

In view of our previous discussion, it is clear that (7.107) is tantamount to imposing an equilibrium structure to the turbulent decay process. This in and of itself would appear to be a major assumption that may conflict with the circumstances of many turbulent flows. Moreover, if the constant C_{ϵ_2} is not taken to be 2, there can also be expected to be conflict with the equilibrium determined naturally as part of the physics of the high-Reynolds-number self-preserving flow. For example, the solution of (7.68) and (7.107) gives an imposed decay law of the form

$$K \sim t^{-1/(C_{\epsilon_2}-1)}, \tag{7.108}$$

wherein C_{ϵ_2} must take on different values to satisfy different experiments.

Another way to view the questionable nature of (7.107) is to consider the case of vanishing viscosity. Here (7.107) reduces to

$$\frac{d\epsilon}{dt} = 0 \tag{7.109}$$

as $\nu \to 0$, implying that ϵ, and therefore ζ, are constant. This is an unphysical result since it is expected that ζ must grow rapidly due to vortex stretching in the absence of dissipation. Such behavior has been observed in inviscid simulations of turbulent flow [11]. Moreover, in the inviscid limit, it can be shown that (7.46) becomes

$$\frac{d\zeta}{dt} = S^*_{K_\infty} \zeta^{3/2}, \tag{7.110}$$

where $S^*_{K_\infty}$ is the infinite Reynolds number limit of the skewness factor. The solution to (7.110) is

$$\zeta(t) = \zeta(0) \left(1 - \frac{S^*_{K_\infty} \sqrt{\zeta(0)}t}{2} \right)^{-2}, \tag{7.111}$$

which says that in the absence of viscosity, enstrophy is expected to blow up at the time $t = 2/S^*_{K_\infty} \sqrt{\zeta(0)}$. Although these arguments do not constitute a proof about the behavior of ζ in turbulent flow, nonetheless it can be concluded that (7.107) unnecessarily limits the range of physical phenomena that can be captured through dissipation modeling. More realism is attained by assuming a modification of (7.106) that allows for the residual effect of vortex stretching. We return to this point in our discussion of homogeneous shear flow in Section 7.8.

7.6   ENERGY SPECTRUM AND ISOTROPIC DECAY

The isotropy constraint limits the allowable form of the energy spectrum tensor, $E_{ij}(\mathbf{k}, t)$, similar to the way that it forces $\mathcal{R}_{ij}(\mathbf{r}, t)$ to be as given in (7.1). Thus it must be the case that

$$E_{ij}(\mathbf{k}, t) = E_1(k, t)k_i k_j + E_2(k, t)\delta_{ij}, \tag{7.112}$$

where E_1 and E_2 are unknown scalar functions of $k = |\mathbf{k}|$. From (1.17) and the incompressibility condition, it follows that

$$0 = \frac{\partial \mathcal{R}_{ij}}{\partial r_j}(\mathbf{r}, t) = -\int_{\mathfrak{R}^3} \iota k_j E_{ij}(\mathbf{k}, t)e^{-\iota \mathbf{k}\cdot\mathbf{r}} \, d\mathbf{k} \tag{7.113}$$

for all $\mathbf{r}$, which can only be true if

$$k_j E_{ij}(\mathbf{k}, t) = 0. \tag{7.114}$$

Applying this condition to (7.112) yields

$$E_2(k, t) = -k^2 E_1(k, t), \tag{7.115}$$

so that

$$E_{ij}(\mathbf{k}, t) = E_1(k, t)(k_i k_j - k^2 \delta_{ij}). \tag{7.116}$$

Contracting indices gives

$$E_{ii}(\mathbf{k}, t) = -2E_1(k, t)k^2. \tag{7.117}$$

In other words, $E_{ii}(\mathbf{k}, t)$ depends only on k.

A consequence of (7.117) is that the integration in (1.20) can be carried out, yielding

$$E(k, t) = 2\pi k^2 E_{ii}(\mathbf{k}, t), \tag{7.118}$$

and in view of (7.117),

$$E(k, t) = -4\pi k^4 E_1(k, t). \tag{7.119}$$

Thus, (7.116) and (7.119) give

$$E_{ij}(\mathbf{k}, t) = \frac{E(k, t)}{4\pi k^2} \left(\delta_{ij} - \frac{k_i k_j}{k^2} \right). \tag{7.120}$$

Assuming that $E_1(k, t)$ is a reasonably well behaved function near $k = 0$ (e.g., it approaches a finite constant there), so will E_{ij}, according to (7.116). In this case, (7.119) implies that

$$E(k, t) \sim k^4 \tag{7.121}$$

as $k \to 0$, and when used in (7.120), the analyticity of E_{ij} at $k = 0$ is assured.

7.6.1   Relation between f and E

In isotropic turbulence the relationship between R_{ij} and E_{ij} reduces to one between f and E, which is now considered. Shifting to spherical coordinates and carrying out the θ and ϕ integrations in the defining equation for E_{ii}, namely,

$$E_{ii}(k, t) = \frac{1}{(2\pi)^3} \int_{\Re^3} R_{ii}(r, t) e^{i\mathbf{k}\cdot\mathbf{r}} \, d\mathcal{V}, \tag{7.122}$$

yields

$$E_{ii}(k, t) = \frac{1}{2\pi^2} \int_0^\infty \mathcal{R}_{ii}(r, t) r^2 \frac{\sin kr}{kr} \, dr. \tag{7.123}$$

Here $k = |\mathbf{k}|$, which is not to be confused with the correlation function $k(r, t)$. In the following, which definition of k is implied will be clear from the context in which it appears.

Now, from (7.28), (7.120), and (7.123), it follows that

$$E(k, t) = \frac{1}{\pi} \int_0^\infty \mathcal{R}_{ii}(r, t) kr \sin kr \, dr$$

$$= \frac{\overline{u^2}}{\pi} \int_0^\infty \left[3f(r, t) + rf'(r, t) \right] kr \sin kr \, dr. \tag{7.124}$$

Integrating the second term by parts and collecting terms gives

$$E(k, t) = \frac{\overline{u^2}}{\pi} \int_0^\infty f(r, t)(kr \sin kr - k^2 r^2 \cos kr) \, dr \tag{7.125}$$

where it has been assumed that $\lim_{r \to \infty} r^2 f(r, t) = 0$. This condition is satisfied as long as (7.121) holds, as may be seen as follows: Substitute a Taylor series for $\sin kr$ in (7.124) giving

$$E(k, t) = \left[\frac{\overline{u^2}}{\pi} \int_0^\infty (3f + rf') r^2 \, dr \right] k^2$$

$$+ \left[\frac{\overline{u^2}}{6\pi} \int_0^\infty (3f + rf') r^4 dr \right] k^4 + \cdots . \tag{7.126}$$

If it is accepted that (7.121) is true, it must be the case that

$$\int_0^\infty (3f + rf') r^2 dr = 0. \tag{7.127}$$

But

$$3r^2 f + r^3 f' = \frac{d(r^3 f)}{dr}, \tag{7.128}$$

so (7.127) implies that

$$\lim_{r \to \infty} r^3 f(r, t) = 0. \tag{7.129}$$

If this is satisfied, so too is $\lim_{r \to \infty} r^2 f(r, t) = 0$, so (7.125) is proven.

The inverse equation to (7.125) can be derived by switching to spherical coordinates in (1.17) and performing the θ and ϕ integrations since $E_{ii}(k,t)$ depends on k and not $\mathbf{k}$. The result is

$$\mathcal{R}_{ii}(r,t) = 2\int_0^\infty \frac{\sin kr}{kr} E(k,t)\, dk, \tag{7.130}$$

and using (7.28) and (7.128) with (7.130) gives

$$\overline{u^2}\frac{d(r^3 f)}{dr} = 2\int_0^\infty \frac{r\,\sin kr}{k} E(k,t)\, dk. \tag{7.131}$$

Integrating (7.131) from $0 \to r$ yields

$$\overline{u^2} f(r,t) = 2\int_0^\infty E(k,t)\left(\frac{\sin kr}{k^3 r^3} - \frac{\cos kr}{k^2 r^2}\right) dk. \tag{7.132}$$

Thus (7.125) and (7.132) represent transforms between f and E.

Equation (7.125) is helpful for investigating the physicality of solutions to (7.94). In particular, if it is assumed that $\tilde{f}(\eta) \sim \eta^\gamma$ for large η, this function will be consistent with (7.94) if $\gamma = -2/(G^* - 1)$, as is found by direct substitution. Transforming (7.125) to the similarity variable η yields

$$E(k^*,t) = \frac{\lambda\overline{u^2}}{\pi}\int_0^\infty \tilde{f}(\eta,t)(k^*\eta\sin k^*\eta - k^{*^2}\eta^2\cos k^*\eta)\, d\eta, \tag{7.133}$$

where $k^* = k\lambda$. Now consider the boundedness of (7.133) in the situation when $\tilde{f}(\eta) \sim \eta^\gamma$. For the integral in (7.133) to be convergent for large η, it must be the case that $\eta^2\tilde{f} \sim \eta^{2+\gamma}$ decays faster than does η^{-1} (i.e., $2 + \gamma < 1$ or $G^* < 5/3$). Thus, as suggested previously, the high-Reynolds-number solution of (7.94), which has $G^* > 2$, can be dismissed as being unphysical.

7.6.2 Energy Spectrum Equation

A dynamical equation for E is obtained from (2.83) by contracting indices and integrating over a spherical shell so as to convert E_{ii} to E according to the definition (1.20). For the same reason as in (2.82), the pressure terms drop out on contraction, so it is found that

$$\frac{\partial E}{\partial t}(k,t) = T(k,t) - 2\nu k^2 E(k,t), \tag{7.134}$$

where

$$T(k,t) = \frac{1}{2}\int_{|\mathbf{k}|=k} T_{ii}(\mathbf{k},t)\, d\Omega \tag{7.135}$$

is the transfer term and $T_{ii}(\mathbf{k},t)$ is given in (2.84) as the Fourier transform of $S_{ii}(\mathbf{r},t)$.

A calculation using (7.19) in (2.85) gives

$$S_{ii}(r, t) = u_{rms}^3 \frac{1}{r^2} \frac{d}{dr} \left(r^3 \frac{dk}{dr} + 4r^2 k \right), \tag{7.136}$$

which shows that S_{ii} is a function of r only in isotropic turbulence. In exactly the same way that the definition (7.122) leads to (7.123) because $\mathcal{R}_{ii}$ depends only on r, so does the Fourier transform of $S_{ii}(\mathbf{r}, t)$ simplify to

$$T_{ii}(k, t) = \frac{1}{2\pi^2} \int_0^\infty S_{ii}(r, t) r^2 \frac{\sin kr}{kr} \, dr \tag{7.137}$$

(i.e., T_{ii} is a function of k and not $\mathbf{k}$). Putting (7.137) into (7.135) after substituting for $S_{ii}(r, t)$, using (7.136) and integrating by parts gives

$$T(k, t) = \frac{2u_{rms}^3}{\pi} \int_0^\infty \frac{1}{k} \left[(3 - k^2 r^2) \sin kr - 3kr \cos kr \right] k(r, t) \, dr. \tag{7.138}$$

In this way it is seen that the two-point triple velocity correlation $k(r, t)$ determines the rate of transfer of energy between scales. Finally, evaluating the inverse transform to (2.84), namely,

$$S_{ii}(r, t) = \int_{\mathfrak{R}^3} T_{ii}(\mathbf{k}, t) e^{-\imath \mathbf{k} \cdot \mathbf{r}} \, dV \tag{7.139}$$

at $r = 0$, remembering that $k'(0) = 0$, and using the definition (7.135) gives

$$\int_0^\infty T(k, t) \, dk = 0. \tag{7.140}$$

In other words, the net amount of energy transfer between scales is zero: just as much as is gained by transfer is lost.

Whenever $f(r, t)$ is known, $E(k, t)$ can be obtained from (7.125). Then $T(k, t)$ can be gotten directly from (7.134) without a need for (7.138). As an important example, consider the final decay period where f is given by (7.83). Substituting this into (7.125) and carrying out the integration gives

$$E(k, t) = \frac{\overline{u^2} \lambda}{\sqrt{2\pi}} (k\lambda)^4 e^{-(1/2)(k\lambda)^2}. \tag{7.141}$$

It is apparent that this formula is consistent with the condition (7.121). It is a simple matter to substitute (7.141) into (7.134) and calculate

$$T(k, t) = E(k, t) \frac{u_{rms}}{\lambda} \left[(k\lambda)^2 - 5 \right]. \tag{7.142}$$

Figure 7.13 contains a budget of the energy spectrum equation (7.134) in the final period. It is seen that energy given up at the large scales is passed to smaller scales.

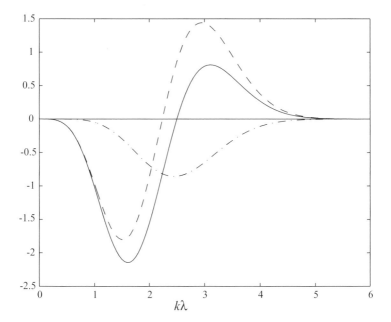

Fig. 7.13   *Energy spectrum budget in final period.* —, $\partial E/\partial t$; $- \cdot -$, *dissipation term*; $- -$, *transfer term. The dimensionless parameter* $R_\lambda (7/6) S_K = 10$. *(See [4].)*

The peak in energy is at $k\lambda = 2$, while that of the dissipation, as seen in the figure, is at $k\lambda = \sqrt{6}$. In this solution, the energy and dissipation range scales are not well separated such as would occur at high Reynolds numbers. In particular, there is no inertial subrange associated with (7.141). As λ rises during the final period, the balance in Fig. 7.13 applies to smaller wavenumbers (i.e., higher wavenumbers lose all their energy before lower wavenumbers).

7.6.3   One-Dimensional Energy Spectrum

In Section 1.2 it was mentioned that physical experiments can be used to estimate the energy spectrum $E(k, t)$ from measured values of the one-dimensional energy spectrum, $E_{11}(\omega)$. The relationship between these quantities is approximate and depends on using Taylor's hypothesis (Section 3.2.3.3) together with the relationships between isotropic tensors. To see how this goes, first note that under the conditions necessary for Taylor's hypothesis and the definitions (1.22) and (1.27), it may be assumed that

$$f(x) = \mathcal{R}_E(x/\overline{U}) \tag{7.143}$$

where the time correlation at zero time delay is taken to correspond to the spatial correlation at zero separation. Equation (7.143) implies that $f(x)$ and $E_{11}(\omega)$ are related through a cosine transformation as deduced from (1.32) and (1.34):

$$E_{11}(\omega) = \frac{4\overline{u^2}}{\overline{U}} \int_0^\infty \cos\frac{2\pi\omega x}{\overline{U}} f(x)\,dx \tag{7.144}$$

and then that

$$f(x) = \frac{1}{\overline{u^2}} \int_0^\infty \cos\frac{2\pi\omega x}{\overline{U}} E_{11}(\omega)\,d\omega. \tag{7.145}$$

Taylor's hypothesis can also be used to connect wavenumber k_1 with angular frequency ω using

$$k_1\overline{U} = 2\pi\omega, \tag{7.146}$$

in which case a one-dimensional wavenumber energy spectrum, $E_{11}^*(k_1)$, can be defined via

$$E_{11}^*(k_1) \equiv \frac{\overline{U}}{2\pi} E_{11}\left(\frac{k_1\overline{U}}{2\pi}\right) \tag{7.147}$$

or, conversely,

$$E_{11}(\omega) \equiv \frac{2\pi}{\overline{U}} E_{11}^*\left(\frac{2\pi\omega}{\overline{U}}\right). \tag{7.148}$$

Applying these relations to (7.145) and taking advantage of symmetry to convert the cosine integral to an exponential gives

$$f(x) = \frac{1}{2\overline{u^2}} \int_{-\infty}^\infty e^{\iota x k_1} E_{11}^*(k_1)\,dk_1. \tag{7.149}$$

A similar relation can be derived for $g(x)$, namely,

$$g(x) = \frac{1}{2\overline{u^2}} \int_{-\infty}^\infty e^{\iota x k_1} E_{22}^*(k_1)\,dk_1. \tag{7.150}$$

Now, using (7.149) and (7.150) in (7.3), it is found that

$$R_{ii}(x,0,0) = \frac{1}{2} \int_{-\infty}^\infty e^{\iota x k_1} [E_{11}^*(k_1) + 2E_{22}^*(k_1)]\,dk_1 \tag{7.151}$$

and taking the inverse transform gives

$$E_{11}^*(k_1) + 2E_{22}^*(k_1) = \frac{1}{\pi} \int_{-\infty}^\infty e^{-\iota x k_1} R_{ii}(x,0,0)\,dx. \tag{7.152}$$

Differentiating both sides with respect to k_1 and then applying the symmetry relations to simplify the integral on the right-hand side, it is found that

$$\frac{\partial E_{11}^*}{\partial k_1}(k_1) + 2\frac{\partial E_{22}^*}{\partial k_1}(k_1) = -\frac{2}{\pi}\int_0^\infty x\sin xk_1\, R_{ii}(x,0,0)\, dx, \qquad (7.153)$$

and then using (7.124), it follows that

$$\frac{\partial E_{11}^*}{\partial k_1}(k_1) + 2\frac{\partial E_{22}^*}{\partial k_1}(k_1) = -\frac{2E(k_1,t)}{k_1}. \qquad (7.154)$$

Now, a relation between E_{22}^* and E_{11}^* can be obtained by taking a Fourier transform of (7.14) and using the inverse transforms to (7.149) and (7.150). The result is

$$E_{22}^*(k_1) = \frac{1}{2}\left(E_{11}^*(k_1) - k_1\frac{\partial E_{11}^*}{\partial k_1} \right), \qquad (7.155)$$

and finally, substituting this into (7.154) gives

$$E(k_1,t) = \frac{1}{2}k_1^2\frac{\partial^2 E_{11}^*}{\partial k_1^2} - \frac{1}{2}k_1\frac{\partial E_{11}^*}{\partial k_1}, \qquad (7.156)$$

which is our desired relationship between the one-dimensional energy spectrum and the three-dimensional spectrum. Of course, Taylor's hypothesis and the isotropy assumption play big parts in the derivation of this equation. Thus, it must be understood as being an approximation.

The inverse relationship to (7.156) is readily derived by solving the differential equation for E_{11} for a given forcing E. The result is

$$E_{11}^*(k_1,t) = \int_{k_1}^\infty dk\frac{E(k,t)}{k}\left(1 - \frac{k_1^2}{k^2} \right), \qquad (7.157)$$

in which the condition that $\lim_{k_1\to\infty} E_{11}(k_1,t) = 0$ is used to determine the constants of integration [8]. A particularly important aspect of (7.157) is the fact that it shows that E_{11} and E share the same power law behavior. This justifies previous conclusions that were reached about the occurrence of Kolmogorov's $-5/3$ spectrum based on measurements of one-dimensional spectra (e.g., as shown in Fig. 2.3).

Another example of the usefulness of one-dimensional spectra is in displaying wavenumber contributions to the enstrophy and kinetic energy as shown in Fig. 7.14 for a boundary layer at $y^+ = 18$ and $R_\theta \approx 2685$. As described above, Taylor's hypothesis is used to transform frequency spectra to wavenumber spectra. The figure shows that the peak in the kinetic energy spectrum is at half an order of magnitude lower wavenumber than the peak in the enstrophy spectrum. Thus enstrophy is associated with smaller scales of motion than the kinetic energy. This separation of scales can be expected to become larger with increasing Reynolds number.

7.7   FOURIER ANALYSIS OF THE VELOCITY FIELD

The numerical scheme for periodic flow in a box discussed in Section 3.6.2 incorporated the Fourier decomposition of the velocity field:

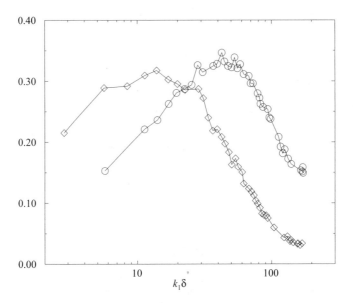

Fig. 7.14   *k_1 spectra of the enstrophy ($\circ$) and the kinetic energy ($\diamond$) are compared for a boundary layer at $y^+ = 18$ for $R_\theta \approx 2685$. (From [1]. Reprinted with the permission of Cambridge University Press.)*

$$u_i(\mathbf{x}, t) = \sum_{\mathbf{k}} \widehat{u}_i(\mathbf{k}, t) e^{\iota \mathbf{k} \cdot \mathbf{x}}, \qquad (7.158)$$

where the wavenumber vector $\mathbf{k} = 2\pi \mathbf{n}/L$ and $\mathbf{n}$ denotes the set of integer triples (n_1, n_2, n_3) running from $-\infty \rightarrow +\infty$. The goal now is to use (7.158) as the basis for deriving an alternative form of the transfer term in the energy spectrum equation (7.134) for isotropic turbulence, one that provides additional insights into the meaning of the terms and is often used as a starting point for theoretical analyses of turbulent motion, as shown in Chapter 12.

By taking a Fourier transform of the Navier–Stokes equation, a dynamical equation is derived for the Fourier coefficients $\widehat{u}_i(\mathbf{k}, t)$. How to transform the diffusion and advection terms was discussed in (3.92) and (3.95), respectively. Fourier transform of the pressure gradient term is along the lines of (3.91):

$$\widehat{\frac{\partial p}{\partial x_i}}(\mathbf{k}, t) = \iota k_i \widehat{p}(\mathbf{k}, t), \qquad (7.159)$$

where pressure has the Fourier series

$$p(\mathbf{x}, t) = \sum_{\mathbf{k}} \widehat{p}(\mathbf{k}, t) e^{\iota \mathbf{k} \cdot \mathbf{x}} \qquad (7.160)$$

and inverse transform

$$\widehat{p}(\mathbf{k}, t) = \frac{1}{L^3} \int_{\mathcal{V}_L} p(\mathbf{x}, t) e^{-\iota \mathbf{k} \cdot \mathbf{x}} \, d\mathbf{x}, \tag{7.161}$$

where $\mathcal{V}_L$ denotes a cube of side L. Using (7.159) together with (3.92) and (3.95) gives the Fourier-transformed Navier–Stokes equation in the form

$$\frac{\partial \widehat{u}_i}{\partial t}(\mathbf{k}, t) = -\iota \sum_{\mathbf{l}} l_j \widehat{u}_i(\mathbf{l}, t) \widehat{u}_j(\mathbf{k} - \mathbf{l}, t) + \iota k_i \widehat{p}(\mathbf{k}, t) - \nu k^2 \widehat{u}_i(\mathbf{k}, t). \tag{7.162}$$

A Fourier transform of the continuity equation (2.12) yields

$$k_i \widehat{u}_i(\mathbf{k}, t) = 0, \tag{7.163}$$

meaning that in incompressible flow, $\widehat{u}_i(\mathbf{k}, t)$ is perpendicular to $\mathbf{k}$. In other words, for any value of $\mathbf{k}$, $\widehat{u}_i(\mathbf{k}, t)$ is oriented tangent to the surface of the sphere of radius $|\mathbf{k}|$ centered at the origin.

Equation (7.163) can be used to eliminate the pressure from (7.162): take a dot product over i, yielding

$$\widehat{p}(\mathbf{k}, t) = \frac{k_i}{k^2} \sum_{\mathbf{l}} l_j \widehat{u}_i(\mathbf{l}, t) \widehat{u}_j(\mathbf{k} - \mathbf{l}, t). \tag{7.164}$$

In view of (7.163),

$$l_j \widehat{u}_j(\mathbf{k} - \mathbf{l}, t) = k_j \widehat{u}_j(\mathbf{k} - \mathbf{l}, t). \tag{7.165}$$

Using this in (7.164) and substituting for $\widehat{p}$ in (7.162) yields a final form of the $\widehat{u}_i$ equation:

$$\frac{\partial \widehat{u}_i}{\partial t}(\mathbf{k}, t) + \nu k^2 \widehat{u}_i(\mathbf{k}, t) = -\iota P_{ij}(\mathbf{k}) k_m \sum_{\mathbf{l}} \widehat{u}_j(\mathbf{l}, t) \widehat{u}_m(\mathbf{k} - \mathbf{l}, t), \tag{7.166}$$

where

$$P_{ij}(\mathbf{k}) = \delta_{ij} - \frac{k_i k_j}{k^2} \tag{7.167}$$

is a projection operator. P_{ij} maps any vector into its component perpendicular to the $\mathbf{k}$ direction. For example, for an arbitrary vector $\mathbf{V}$, $P_{ij}(\mathbf{k}) V_j = V_i - (V_j k_j) k_i / k^2$ is perpendicular to k_i, since $(P_{ij}(\mathbf{k}) V_j)(k_i) = 0$. The appearance of P_{ij} in (7.166) is consistent with the fact that the two other terms in the equation are perpendicular to $\mathbf{k}$, as may be deduced from (7.163). In other words, if the left-hand side of the equation is normal to $\mathbf{k}$, so must be the right-hand side.

An equivalent form of (7.166), which is often used in theoretical analyses, is given by

$$\frac{\partial \widehat{u}_i}{\partial t}(\mathbf{k}, t) + \nu k^2 \widehat{u}_i(\mathbf{k}, t) = M_{ijm}(\mathbf{k}) \sum_{\mathbf{l}} \widehat{u}_j(\mathbf{l}, t) \widehat{u}_m(\mathbf{k} - \mathbf{l}, t), \qquad (7.168)$$

where

$$M_{ijm} = -\frac{\iota}{2} \left[k_m P_{ij}(\mathbf{k}) + k_j P_{im}(\mathbf{k}) \right]. \qquad (7.169)$$

This is derived by noting that the right-hand side of (7.166) is left unchanged if the dummy indices j and m are switched, and summation on $\mathbf{l}$ is replaced by the equivalent summation on $\mathbf{l}' \equiv \mathbf{k} - \mathbf{l}$. The right-hand side of (7.168) represents the average of these two equivalent expressions for the right-hand side of (7.166).

The energy spectrum tensor, $E_{ij}(\mathbf{k}, t)$, was defined previously as the Fourier transform of the two-point velocity correlation tensor, $\mathcal{R}_{ij}(\mathbf{r}, t)$, and its governing dynamical equation, (2.83), was derived by taking a Fourier transform of (2.80). Specialization to isotropic turbulence led to the energy spectrum equation, (7.134). Here a different form of the spectral equation is derived by working directly with (7.166). The analysis is first done for the case of a finite cubic box in which $\mathbf{k}$ is discrete, and then the limit as $L \to \infty$ is taken to get an alternative form of (7.134).

First note from (3.85) that

$$\overline{\widehat{u}_i(\mathbf{k}, t) \widehat{u}_j(\mathbf{l}, t)} = \frac{1}{L^6} \int_{V_L} \int_{V_L} \overline{u_i(\mathbf{x}, t) u_j(\mathbf{y}, t)} e^{-\iota(\mathbf{k} \cdot \mathbf{x} + \mathbf{l} \cdot \mathbf{y})} \, d\mathbf{x} \, d\mathbf{y}. \qquad (7.170)$$

Under the restriction of homogeneous turbulence, one can set $\mathbf{y} = \mathbf{x} + \mathbf{r}$ in (7.170), transform the $\mathbf{y}$ integration to $\mathbf{r}$ integration, note that $\overline{u_i(\mathbf{x}, t) u_j(\mathbf{y}, t)} = \mathcal{R}_{ij}(\mathbf{r}, t)$ is independent of $\mathbf{x}$, and carry out the $\mathbf{x}$ integration. This yields

$$\overline{\widehat{u}_i(\mathbf{k}, t) \widehat{u}_j(\mathbf{l}, t)} = \delta_{\mathbf{k}+\mathbf{l}\,\mathbf{0}} \frac{1}{L^3} \int_{V_L} \mathcal{R}_{ij}(\mathbf{r}, t) e^{-\iota \mathbf{l} \cdot \mathbf{r}} \, d\mathbf{r} \qquad (7.171)$$

where (3.96) has been used. The right-hand side of (7.171) is nonzero only when $\mathbf{l} = -\mathbf{k}$, so, in fact, this relation proves that

$$\overline{\widehat{u}_i(\mathbf{k}, t) \widehat{u}_j(\mathbf{l}, t)} = 0 \quad \text{unless} \quad \mathbf{k} + \mathbf{l} = 0. \qquad (7.172)$$

Although it is not presented here, a straightforward generalization of this argument gives

$$\overline{\widehat{u}_i(\mathbf{k}, t) \widehat{u}_j(\mathbf{l}, t) \widehat{u}_n(\mathbf{m}, t)} = 0 \quad \text{unless} \quad \mathbf{k} + \mathbf{l} + \mathbf{m} = 0, \qquad (7.173)$$

and so on, for higher-order moments.

From (7.171) it follows that

$$\overline{\widehat{u}_i(\mathbf{k}, t) \widehat{u}_j(-\mathbf{k}, t)} = \frac{1}{L^3} \int_{V_L} \mathcal{R}_{ij}(\mathbf{r}, t) e^{\iota \mathbf{k} \cdot \mathbf{r}} \, d\mathbf{r}. \qquad (7.174)$$

This relation is appropriate for a discrete set of $\mathbf{k}$ values, in contrast to (1.16), which is valid for all real vector wavenumbers. The inverse relation to (7.174) gives the Fourier transform of $\mathcal{R}_{ij}(\mathbf{r}, t)$ as

$$\mathcal{R}_{ij}(\mathbf{r}, t) = \sum_{\mathbf{k}} \overline{\widehat{u}_i(\mathbf{k}, t)\widehat{u}_j(-\mathbf{k}, t)} e^{-i\mathbf{k}\cdot\mathbf{r}}, \tag{7.175}$$

which is the discrete analog of (1.17).

For $\mathbf{r} = 0$ and $i = j$, after using (3.88), (7.175) gives

$$K = \frac{1}{2}\sum_{\mathbf{k}} \overline{|\widehat{u}_i(\mathbf{k}, t)|^2}, \tag{7.176}$$

since for any complex number z, $|z|^2 = zz^*$. Defining $E_{\mathbf{k}} = \frac{1}{2}\overline{|\widehat{u}_i(\mathbf{k}, t)|^2}$, then (7.176) gives

$$K = \sum_{\mathbf{k}} E_{\mathbf{k}} \tag{7.177}$$

as the discrete analog of (1.18). This shows how different wavenumber modes contribute to turbulent kinetic energy.

A dynamical equation for $E_{\mathbf{k}}$ may be derived by taking the average of the sum of (7.166) times $\widehat{u}_i^*(\mathbf{k}, t)$ and the complex conjugate of (7.166) times $\widehat{u}_i(\mathbf{k}, t)$. The result is

$$\frac{\partial E_{\mathbf{k}}}{\partial t}(t) + 2\nu k^2 E_{\mathbf{k}}(t)$$

$$= \frac{1}{2}M_{ijm}(\mathbf{k})\sum_{\mathbf{l}}[\overline{\widehat{u}_i(-\mathbf{k})\widehat{u}_j(\mathbf{l})\widehat{u}_m(\mathbf{k}-\mathbf{l})} - \overline{\widehat{u}_i(\mathbf{k})\widehat{u}_j(\mathbf{l})\widehat{u}_m(-\mathbf{k}-\mathbf{l})}], \tag{7.178}$$

where $\mathbf{l}$ has been replaced with $-\mathbf{l}$ in the second term for later convenience. The goal now is to find the limiting form of this equation as $L \to \infty$. This provides an alternative statement to (7.134) and (7.138).

7.7.1 Limit of Infinite Space

With the goal of taking the limit as $L \to \infty$, define

$$E_{ij}^L(\mathbf{k}, t) = \left(\frac{L}{2\pi}\right)^3 \overline{\widehat{u}_i(\mathbf{k})\widehat{u}_j(-\mathbf{k})}, \tag{7.179}$$

which becomes, after substituting (7.174),

$$E_{ij}^L(\mathbf{k}, t) = \left(\frac{1}{2\pi}\right)^3 \int_{\mathcal{V}_L} \mathcal{R}_{ij}(\mathbf{r}, t)e^{i\mathbf{k}\cdot\mathbf{r}} \, d\mathbf{r}. \tag{7.180}$$

In the limit as $L \to \infty$, the right-hand side becomes the Fourier transform of $\mathcal{R}_{ij}(\mathbf{r}, t)$, so it has been established that

$$\lim_{L \to \infty} E_{ij}^L(\mathbf{k}, t) = E_{ij}(\mathbf{k}, t). \tag{7.181}$$

In this limit the values of $\mathbf{k}$ become closer and closer together, and at the end of the limiting process, $\mathbf{k}$ covers all real vectors, not just the discrete set where $E_{ij}^L(\mathbf{k}, t)$ is defined at a particular value of L.

This same argument needs to be repeated in the case of triple velocity correlations. Thus, define

$$T_{ijn}^L(\mathbf{k}, \mathbf{l}, t) = \left(\frac{L}{2\pi} \right)^6 \overline{\widehat{u}_i(\mathbf{k}) \widehat{u}_j(\mathbf{l}) \widehat{u}_n(-\mathbf{k} - \mathbf{l})}, \tag{7.182}$$

where (7.173) has been taken into account. Substituting for the Fourier components using (3.85) then gives

$$T_{ijn}^L(\mathbf{k}, \mathbf{l}, t) \tag{7.183}$$

$$= \left(\frac{L}{2\pi} \right)^6 \frac{1}{L^9} \int_{V_L} \int_{V_L} \int_{V_L} \overline{u_i(\mathbf{x}, t) u_j(\mathbf{y}, t) u_n(\mathbf{z}, t)} e^{-\imath \mathbf{k} \cdot (\mathbf{x} - \mathbf{z}) - \imath \mathbf{l} \cdot (\mathbf{y} - \mathbf{z})} \, d\mathbf{x} \, d\mathbf{y} \, d\mathbf{z}.$$

In homogeneous turbulence the correlation within the integral depends only on the two vectors $\mathbf{r} = \mathbf{x} - \mathbf{z}$ and $\mathbf{s} = \mathbf{y} - \mathbf{z}$, so a three-point triple velocity correlation function is defined via

$$S_{ijn}(\mathbf{r}, \mathbf{s}, t) \equiv \overline{u_i(\mathbf{z} + \mathbf{r}, t) u_j(\mathbf{z} + \mathbf{s}, t) u_n(\mathbf{z}, t)}, \tag{7.184}$$

which is to be contrasted with the two-point triple correlation $S_{ij,n}$ in (1.15). Changing the $\mathbf{x}$ and $\mathbf{y}$ variables in (7.183) to $\mathbf{z} + \mathbf{r}$ and $\mathbf{z} + \mathbf{s}$, respectively, using (7.184), and carrying out the $\mathbf{z}$ integration yields

$$T_{ijn}^L(\mathbf{k}, \mathbf{l}, t) = \left(\frac{1}{2\pi} \right)^6 \int_{V_L} \int_{V_L} S_{ijn}(\mathbf{r}, \mathbf{s}, t) e^{-\imath \mathbf{k} \cdot \mathbf{r} - \imath \mathbf{l} \cdot \mathbf{s}} \, d\mathbf{r} \, d\mathbf{s}. \tag{7.185}$$

In the limit as $L \to \infty$ this leads to

$$T_{ijn}(\mathbf{k}, \mathbf{l}, t) = \left(\frac{1}{2\pi} \right)^6 \int_{\Re^3} \int_{\Re^3} S_{ijn}(\mathbf{r}, \mathbf{s}, t) e^{-\imath \mathbf{k} \cdot \mathbf{r} - \imath \mathbf{l} \cdot \mathbf{s}} \, d\mathbf{r} \, d\mathbf{s} \tag{7.186}$$

as the Fourier transform of $S_{ijn}(\mathbf{r}, \mathbf{s}, t)$.

Now we are prepared to consider the limit of (7.178) as $L \to \infty$. Multiply this equation by $(L/2\pi)^3$ and note through (7.179), (7.181), and (7.118) that

$$\lim_{L\to\infty} \left(\frac{L}{2\pi}\right)^3 E_{\mathbf{k}} = \frac{1}{2} E_{ii}(\mathbf{k}, t) = \frac{E(k,t)}{4\pi k^2}. \tag{7.187}$$

Moreover, it may be calculated using (7.182) that

$$\lim_{L\to\infty} \left(\frac{L}{2\pi}\right)^3 \sum_{\mathbf{l}} \overline{\widehat{u}_j(\mathbf{l})\widehat{u}_m(\mathbf{k}-\mathbf{l})\widehat{u}_i(-\mathbf{k})} = \lim_{L\to\infty} \sum_{\mathbf{l}} \left(\frac{2\pi}{L}\right)^3 T_{jmi}^L(\mathbf{l}, \mathbf{k}-\mathbf{l}, t)$$

$$= \int_{\Re^3} T_{jmi}(\mathbf{l}, \mathbf{k}-\mathbf{l}, t)\, d\mathbf{l}, \tag{7.188}$$

where the second equality depends on the fact, implied by (7.185) and (7.186), that $\lim_{L\to\infty} T_{jmi}^L = T_{jmi}$ and that $(2\pi/L)^3$ is the volume surrounding each of the wavenumber vectors in the sum [see (3.83)]. The integral appears in the last term in (7.188) as the limit of the Riemann sums as $L \to \infty$. Collecting these results together, it has been established that

$$\frac{\partial E(k,t)}{\partial t} + 2\nu k^2 E(k,t)$$

$$= 2\pi k^2 M_{ijm}(\mathbf{k}) \int_{\Re^3} (T_{jmi}(\mathbf{l}, \mathbf{k}-\mathbf{l}, t) - T_{jmi}(\mathbf{l}, -\mathbf{k}-\mathbf{l}, t))\, d\mathbf{l}. \tag{7.189}$$

Using the homogeneity properties of $T_{jmi}(\mathbf{l}, -\mathbf{k}-\mathbf{l}, t)$, it can be shown that the second term in (7.189) contributes the same as the first, so that the equation simplifies to

$$\frac{\partial E(k,t)}{\partial t} + 2\nu k^2 E(k,t) = 4\pi k^2 M_{ijm}(\mathbf{k}) \int_{\Re^3} T_{jmi}(\mathbf{l}, \mathbf{k}-\mathbf{l}, t)\, d\mathbf{l}. \tag{7.190}$$

The form of the stretching term in (7.189) or (7.190) differs significantly from that in (7.138), and is generally preferable for the theoretical analyses of isotropic decay that are considered in Chapter 12.

7.8 HOMOGENEOUS SHEAR FLOW

The ways in which the decay process described in the first part of this chapter is modified by the presence of a turbulence producing mean shear is of considerable theoretical and practical interest. Fortunately, this fundamental question can be explored within the context of homogeneous shear flow, where the simplification of homogeneity is still available. Thus consider the case of an initially homogeneous, isotropic turbulence subjected to a constant mean shearing, say $S \equiv d\overline{U}/dy > 0$. Under these circumstances, according to (2.40), (7.20) is modified by the presence of a production term, $P = -\overline{uv}\, d\overline{U}/dy > 0$, which is positive since $\overline{uv} < 0$. Thus the K equation is given by

$$\frac{dK}{dt} = P - \epsilon. \tag{7.191}$$

The presence of a mean shear leads to additional production terms in the ϵ equation, which now simplifies from (2.42) to the form

$$\frac{d\epsilon}{dt} = P_\epsilon^1 + P_\epsilon^2 + P_\epsilon^4 - \Upsilon_\epsilon. \tag{7.192}$$

A number of experimental and theoretical studies have attempted to predict the short- and long-term behavior of K and ϵ subjected to homogeneous shear. For the most part these have observed exponential growth in K and ϵ at approximately the same rate over the times that have been examined, generally, St < 30, as shown in Fig. 7.15. If K and ϵ grow at the same rate, their ratio should become constant, and such a trend has been seen in experiments. Generally, the following asymptotic relations are observed:

$$\frac{SK}{\epsilon} \approx 6.0 \tag{7.193}$$

and

$$\frac{P}{\epsilon} \approx 1.80, \tag{7.194}$$

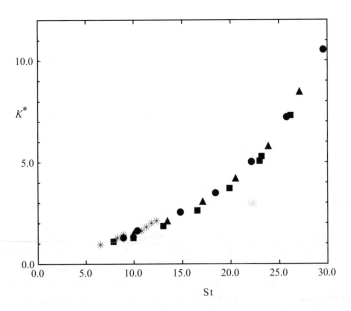

Fig. 7.15 *Measured $K/K(0)$ in homogeneous shear flow for* St < 30. *(From [15]. Reprinted with permission of Cambridge University Press.)*

although in some calculations using LES (e.g., [14]), even by time $St = 30$, an equilibrium state has yet to be reached.

In view of the limited observations of homogeneous shear flow, it is not certain what occurs at large times, but two principal scenarios have been put forward. Townsend [20] hypothesized that given enough time, a production-equals-dissipation equilibrium results in which the terms on the right-hand sides of (7.191) and (7.192) achieve balances so that K and ϵ asymptote to constant finite values. An alternative viewpoint is that the exponential growth in K and ϵ observed for relatively short times, continues unabated for long times as well.

Taking $t^* = St$ as a nondimensional time, then (7.191) may be rewritten as

$$\frac{dK}{dt^*} = \frac{\epsilon}{SK} \left(\frac{P}{\epsilon} - 1 \right) K. \tag{7.195}$$

It is clear from (7.195) that for the production-equals-dissipation equilibrium to hold, (7.194) must be valid only for relatively short times. On the other hand, if it and (7.193) were to hold indefinitely, the right-hand side of (7.195) would asymptote to $\approx 0.13K$. The long time behavior of K in the time scale t^* would thus be given by

$$K(t^*) = e^{0.13t^*}, \tag{7.196}$$

with ϵ having a similar growth rate. Note that the latter scenario is relevant only to the ideal case, since, clearly, unlimited growth in K is not physical.

More insight into what is likely to be the most physically correct characterization of the long-time behavior of K and ϵ in homogeneous shear flow can be had by solving modeled equations for K and ϵ. This anticipates the results in Chapter 8. The value of this approach in the present context is enhanced by the fact that relatively few modeling assumptions are required to obtain a closed system of equations.

Closure to (7.191) requires a model for the Reynolds shear stress $\overline{uv}$. It was shown in Section 6.1 that the gradient law $\overline{uv} = -\nu_t \, d\overline{U}/dy$ can be justified in a uniform shear flow if only the acceleration transport effect is not significant. For the present purposes it is assumed that this is so and, furthermore, that $\nu_t = C_\mu K^2/\epsilon$. As shown in Chapter 8, this is a standard model used in the most popular closure schemes. It is an obvious choice here since K and ϵ are the only two quantities available to characterize turbulent mixing scales in this problem. With these assumptions, (7.191) takes the closed form

$$\frac{dK}{dt^*} = C_\mu \frac{K^2}{\epsilon} S^2 - \epsilon. \tag{7.197}$$

As for the ϵ equation modeling, it may be shown that in homogeneous turbulence,

$$P_\epsilon^1 + P_\epsilon^2 = 2\nu \, \overline{\omega_1 \omega_2} \, S, \tag{7.198}$$

after using the isotropic identity

$$\overline{\frac{\partial u_i}{\partial x_m}\frac{\partial u_j}{\partial x_n}} = \overline{\frac{\partial u_i}{\partial x_n}\frac{\partial u_j}{\partial x_m}}. \tag{7.199}$$

Moreover, the approximation

$$\frac{\overline{\omega_1 \omega_2}}{\zeta} \sim -\frac{\overline{uv}}{2K} \tag{7.200}$$

is excellent in simple shear flows such as a channel, as shown in Fig. 7.16, where the ratio of the two sides of the equation is plotted and seen to be approximately constant. Taking the constant to be C_{ϵ_1} and noting that $\epsilon = \nu\zeta$, leads to the model

$$P_{\epsilon}^1 + P_{\epsilon}^2 = C_{\epsilon_1} C_{\mu} K S^2. \tag{7.201}$$

For the terms P_{ϵ}^4 and Υ_{ϵ} in (7.192), the expressions (7.100) and (7.101), respectively, are also available for the present application, so that

$$P_{\epsilon}^4 - \Upsilon_{\epsilon} = S_K^* R_T^{1/2}\frac{\epsilon^2}{K} - G^*\frac{\epsilon^2}{K}, \tag{7.202}$$

where S_K^* and G^* are as yet unknown. As discussed in Section 7.5, the standard assumption (7.106), if applied to (7.202), excludes a role for vortex stretching. Instead, to maintain generality, the assumption is made here [5] that

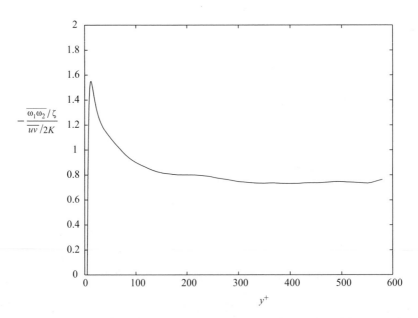

Fig. 7.16 *Demonstration of the near constancy of* $-(\overline{\omega_1\omega_2}/\zeta)/(\overline{uv}/2K)$ *in channel flow. (From data in [12].)*

$$G^* = (S_K^* - C_{\epsilon_3})\sqrt{R_T} + C_{\epsilon_2}, \tag{7.203}$$

where the constant C_{ϵ_3} is used to enable the possibility that the vortex stretching effect is not canceled completely by the diffusion term. Thus, only when $C_{\epsilon_3} = 0$ is the standard model recovered. Substituting (7.201), (7.202), and (7.203) into (7.192) gives the ϵ equation model

$$\frac{d\epsilon}{dt} = C_{\epsilon_1} C_\mu K S^2 + C_{\epsilon_3} R_T^{1/2} \frac{\epsilon^2}{K} - C_{\epsilon_2} \frac{\epsilon^2}{K}, \tag{7.204}$$

which is to be solved in conjunction with (7.197). In these relations the calibrated values of the constants are given as $C_\mu = 0.09$, $C_{\epsilon_1} = 1.45$, and $C_{\epsilon_2} = 1.90$ (see Section 8.4). C_{ϵ_3} is taken to be either zero, corresponding to its traditional modeling, or 0.1, which allows for a residual effect of vortex stretching to be felt independently of the dissipation term.

Numerical solutions to the system of equations (7.197) and (7.204) are shown in Figs. 7.17 and 7.18, where some results from a LES are also included. It is seen that strikingly different long-term behavior is observed depending on whether or not vortex stretching is present or absent in (7.204). In the latter case, unbounded exponential growth at the same rate for both K and ϵ occurs with an asymptotic equilibrium state equivalent to (7.196).

For nonzero C_{ϵ_3}, a calculation reveals that a constant equilibrium state results in which K and ϵ asymptote to

$$K_\infty = \frac{135}{49} \frac{\sqrt{C_\mu}(C_{\epsilon_2} - C_{\epsilon_1})^2}{C_{\epsilon_3}^2} \nu S \tag{7.205}$$

and

$$\epsilon_\infty = \frac{135}{49} \frac{C_\mu(C_{\epsilon_2} - C_{\epsilon_1})^2}{C_{\epsilon_3}^2} \nu S^2, \tag{7.206}$$

respectively. Thus the slightly added ϵ production caused by the stretching term causes a saturation in K and then, ultimately, a saturation in ϵ so that equilibrium is reached.

Equations (7.205) and (7.206) make clear that for small C_{ϵ_3}, the equilibrium states K_∞ and ϵ_∞ will be large, and the solution will generally pass through an initial growth phase which for all intents and purposes has the appearance of unbounded exponential growth. Thus it can be argued that current experimental and numerical simulations of homogeneous shear flow, which are limited to $St < 30$, do not contradict either of the alternatives.

For a number of reasons the production-equals-dissipation scenario is the more likely of the two outcomes. In particular, despite the fact that model equations were used in the analysis, it is apparent that the issue is equivalent to the question of whether or not vortex stretching retains an independent role with respect to dissipation in the

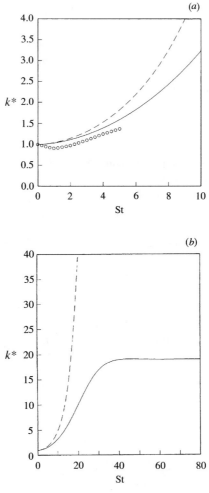

Fig. 7.17 *Computed solution for K in homogeneous shear flow. —, with vortex stretching; – –, without vortex stretching; o, LES calculation [2]. (a) Short time solution, and (b) long time solution. (From [5]. Reprinted with permission of ASME.)*

ϵ (or ζ) equation at high Reynolds numbers. An affirmative answer in the case of isotropic decay in Section 7.3 led to a richer understanding of the decay process. A similar conclusion holds here, particularly when it is recognized that the alternative depends on the exact annihilation of vortex stretching as an independent physical process by dissipation.

7.9 SUMMARY

The ideal flows considered in this chapter provide the best opportunity there is for obtaining complete closure to a turbulent flow problem. Indeed, with the condition of

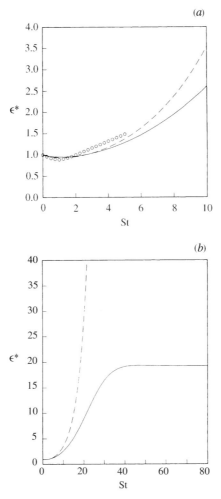

Fig. 7.18 *Computed solution for ϵ in homogeneous shear flow: —, with vortex stretching; — —without vortex stretching; ○, data from LES calculation [2]. (a) Short time solution, and (b) long time solution. (From [5]. Reprinted with permission of ASME.).*

self-similarity, a closure to the isotropic decay problem was achieved. This, however, cannot explain the entire decay process beginning from a high-Reynolds-number state to the final decay period at small Reynolds number, although it appears to give significant insight into how this process occurs. In particular, a decay scenario ensues in which within a few eddy turnover times the turbulence achieves a balance between vortex stretching and dissipation consistent with a t^{-1} decay law, independent of the initial state. As time progresses, and after sufficient energy is lost, the flow departs from the t^{-1} law to evolve toward the final period in which $K \sim t^{-5/2}$. Here the turbulence is weak, the longitudinal correlation coefficient is Gaussian, and the turbulent length scales increase because of the more efficient elimination of small scales via dissipation.

In this chapter we also examined the isotropic decay problem from the point of view of spectral space. The transfer function, which expresses the tendency of turbulent motion to distribute energy between scales due to vortex stretching, was shown to depend on triadic interactions between wavenumbers. This is the starting point for many theoretical analyses of turbulent fluid motion (e.g., by making statistical hypothesis concerning energy transfer).

In the last part of the chapter we examined the situation where uniform production is placed on an isotropic turbulence. This immediately prevents exact treatment of the problem, but with certain approximations, a credible job of accounting for the ensuing flow development was achieved. The modeling necessary to obtain closure in this case is also valuable in the prediction of general flows. In fact, it is standard practice to expropriate the isotropic models for use in general turbulence closure schemes, as will be shown in subsequent chapters.

REFERENCES

1. Balint, J.-L., Wallace, J. M. and Vukoslavčević, P. (1991) "The velocity and vorticity vector fields of a turbulent boundary layer. 2. Statistical properties," *J. Fluid Mech.* **228**, 53–86.

2. Bardina, J., Ferziger, J. H. and Reynolds, W. C. (1983) "Improved turbulence models based on large-eddy simulation of homogeneous, incompressible turbulent flows," *Stanford Univ. Rep. TF-19*.

3. Batchelor, G. K. and Townsend, A. A. (1948) "Decay of isotropic turbulence in the final period," *Proc. R. Soc. London Ser. A*, **194**, 527–543.

4. Bernard, P. S. (1985) "Energy and vorticity dynamics in decaying isotropic turbulence," *Int. J. Eng. Sci.* **23**, 1037–1057.

5. Bernard, P. S. and Speziale, C. G. (1992) "Bounded energy states in homogeneous turbulent shear flow–an alternative view," *ASME J. Fluids Eng.* **114**, 29–39.

6. Guckenheimer, J. and Holmes, P. J. (1986) *Nonlinear Oscillations, Dynamical Systems and Bifurcations of Vector Fields*, Springer-Verlag, New York.

7. Hildebrand, F. B. (1976) *Advanced Calculus for Applications*, 2nd ed., Prentice-Hall, Englewood Cliffs, N.J.

8. Hinze, J. O. (1975) *Turbulence*, McGraw-Hill, New York.

9. Kármán. T. von and Howarth, L. (1938) "On the statistical theory of isotropic turbulence," *Proc. R. Soc. London Ser. A* **164**, 192–215.

10. Korneyev, A. I. and Sedov, L. I, (1976) "Theory of isotropic turbulence and its comparison with experimental data," *Fluid Mech. Sov. Res.* **5**, 37–48.

11. Lesieur, M. (1997) *Turbulence in Fluids*, 3rd ed., Kluwer Academic, Dordrecht, The Netherlands.

12. Moser, R. D., Kim, J. and Mansour, N. N. (1999) "DNS of turbulent channel flow up to $R_\tau = 590$," *Phys. Fluids* **11**, 943–945.

13. Robertson, H. P. (1940) "The invariant theory of isotropic turbulence," *Cambridge Philos. Soc.* **36**, 209–223.

14. Rogers, M. M., Moin, P. and Reynolds, W. C. (1986) "The structure and modeling of the hydrodynamic and passive scalar fields in homogeneous turbulent shear flow," *Stanford Univ. Rep. TF-25*.

15. Rohr, J. J., Itsweire, E. C., Helland, K. N. and Van Atta, C. W. (1988) "An investigation of the growth of turbulence in a uniform mean shear flow," *J. Fluid Mech.* **187**, 1–33.

16. Sedov, L. I. (1944), "Decay of isotropic turbulent motions of an incompressible fluid," *Dokl. Akad. Nauk. SSSR* **42**, 116–119.

17. Slater, L. J. (1960) *Confluent Hypergeometric Functions*, Cambridge University Press, Cambridge.

18. Speziale, C. G. and Bernard, P. S. (1992) "The energy decay in self-preserving isotropic turbulence revisited," *J. Fluid Mech.* **241**, 645–667.

19. Taylor, G. I. (1935) "Statistical theory of turbulence, Part I," *Proc. R. Soc. Ser. A* **151**, 421–444.

20. Townsend, A. A. (1956) *The Structure of Turbulent Shear Flow*, Cambridge University Press, New York.

8

Turbulence Modeling

Turbulence modeling comprises the various strategies that have been developed for predicting mean flow quantities other than by directly averaging solutions to the Navier–Stokes equation. The most popular approaches in this regard seek to close one or more of the RANS equations, which, for the present purposes, are considered to be the equations derived in Chapter 2 for mean velocity and vorticity, kinetic energy, Reynolds stresses, dissipation, and enstrophy. These are all one-point equations, in contrast to the two-point equations derived in Section 2.4, whose solutions were touched upon in Chapter 7 and are considered further in Chapter 12.

To effect closure, models for the unclosed correlations appearing in the governing equations have to be developed. Often, these make use of modeling constants that are a priori unknown. A calibration process is then used to find their values in such a way as to achieve the best overall prediction of one or more particular flows. The latter usually consist of canonical flows such as a zero-pressure-gradient turbulent boundary layer on a smooth wall, channel flow, homogeneous shear flow, and decaying isotropic turbulence.

The ideal is to find a universally applicable closure model, one that encompasses the full range of turbulent physics and applies to flows beyond those used in its calibration. With this aim in mind, turbulence models have tended to become more ambitious with each succeeding decade. The first closures were for just the $\overline{U}_i$ equation, (2.10), but dissatisfaction with their predictive capability led to closures to the K (2.40), ϵ (2.42), and Reynolds stress equation (2.52). These changes have echoed the continual increase in computer capabilities which now permit economical solutions to the large sets of coupled, nonlinear partial differential equations associated with advanced closure models.

Ultimately, the capabilities of RANS models, no matter how complex, remains limited, and alternative modeling approaches continue to be sought. The most popular of these is LES, in which the "filtered" Navier–Stokes equation is solved with the aid of simplified models for small-scale phenomena, called *subgrid-scale stress models.* Such methods reduce the dependence on modeling at the cost of greater computer resources. With time, the range of flows to which LES can usefully be applied can be expected to grow. However, currently and into the foreseeable future, LES is not sufficiently accurate or cost-effective to replace the methods described in this chapter, so RANS models continue to be the dominant force in turbulent flow prediction today.

8.1 TYPES OF RANS MODELS

RANS models must wrestle with the problem of predicting the Reynolds stress, $R_{ij} = \overline{u_i u_j}$, whose presence in the averaged momentum equation prevents closure. The most common approach to predicting R_{ij} is to assume a constitutive model that mimics the molecular momentum transport model described at length in Sections 2.1 and 6.1. Thus this class of models obeys the transport law, (6.1):

$$R_{ij} = \frac{2}{3} K \delta_{ij} - \nu_t \left(\frac{\partial \overline{U}_i}{\partial x_j} + \frac{\partial \overline{U}_j}{\partial x_i} \right), \tag{8.1}$$

where, as before, $\nu_t > 0$ is termed the *eddy viscosity*. The wide range of models of this type are referred to as eddy viscosity models (EVMs). After substitution into (2.10), (8.1) leads to the closed $\overline{U}_i$ equation in the form

$$\frac{\partial \overline{U}_i}{\partial t} + \overline{U}_j \frac{\partial \overline{U}_i}{\partial x_j} = -\frac{\partial}{\partial x_i} \left(\frac{\overline{P}}{\rho} + \frac{2}{3} K \right) + \nu \, \nabla^2 \overline{U}_i$$

$$+ \frac{\partial}{\partial x_j} \left[\nu_t \left(\frac{\partial \overline{U}_i}{\partial x_j} + \frac{\partial \overline{U}_j}{\partial x_i} \right) \right]. \tag{8.2}$$

It is interesting to note that K only appears in (8.2) combined with $\overline{P}$. Thus the usual numerical procedures that solve (8.2), which are equivalent to those used to solve the Navier–Stokes equation, yield values for the combined variable $\overline{P}/\rho + \frac{2}{3}K$. Since $K = 0$ at solid boundaries, surface pressures are thus readily obtained. To get $\overline{P}$ elsewhere, however, requires a separate means for predicting K.

Models adopting (8.1) are distinguished according to how they plan to estimate ν_t. For the simplest models, the functional form of eddy viscosity is complete and self-contained in the sense that it can be evaluated from just $\overline{U}_i$ and parameters related to the local flow geometry (e.g., distance from a solid boundary). These are referred to as *zero-equation models*. Clearly, such methods cannot be used to find $\overline{P}$ except at solid boundaries, where $K = 0$.

In other approaches, ν_t is made to depend on additional turbulence fields, most often K and/or ϵ, so that additional theories are necessary to provide information about these quantities. Part of the motivation for developing such models is an attempt to bring "history effects" into the determination of ν_t. Thus, instead of having ν_t

depend on the instantaneous mean velocity field, a dependence on K, for example, allows the eddy viscosity to better reflect evolutionary processes within the flow. The usual practice is to use the structure of the governing differential equations for K and ϵ as the starting point for developing the desired relations. One- and two-equation models opt to introduce one and two more, respectively, additional differential equations from which to calculate K and/or ϵ. The result is then a system of coupled differential equations, including (8.2).

Some examples where the simple eddy viscosity model of the Reynolds shear stress must fail were considered previously (e.g., in the case of flow in a rough wall channel; see Section 6.1). Another way in which a formula such as (8.1) can fail is in its incompatibility with flow anisotropy. For example, in a turbulent channel flow, $d\overline{U}/dy$ is the only nonzero mean velocity derivative, so that (8.1) can be seen to predict the equality of $\overline{u^2}$, $\overline{v^2}$, and $\overline{w^2}$. In fact, each of these is predicted to be $\frac{2}{3}K$. This is a clear impossibility in view of the experimental observations shown in Fig. 4.7. The important conclusion that may be drawn from this is that although (8.1) can fail because a model cannot compute a correct value for ν_t, it can also fail because the entire relation is inappropriate (i.e., no value of eddy viscosity will make the relation accurate). This happens, for example, when the principal axes of the matrices on either side are not in approximate alignment. As was just demonstrated, failure of the eddy viscosity law is a real possibility and has encouraged the development of closure schemes not depending on (8.1).

There are three principal avenues for improving the prediction of Reynolds stress beyond (8.1). One is to search directly for alternatives to (8.1) that better model the true physics of turbulent transport. Here, attention is paid to understanding the physical mechanisms underlying transport and devising constitutive forms which reflect them. Such models tend to generalize the linear stress rate-of-strain law to include nonlinear effects. These are usually developed formally as truncated series expansions in increasing powers of various tensor quantities (e.g., the rate-of-strain tensor). Such models are referred to as *nonlinear eddy viscosity models* (NLEVMs).

The second approach is to determine the Reynolds stress as the solution of a suitably closed form of its own governing differential equation. This requires modeling physical processes affecting the conservation of Reynolds stress. Since the Reynolds stresses are second moments of the fluctuating velocity field, closures that pursue the second strategy (i.e., modeling the Reynolds stress equation) are commonly referred to as *second moment closures* (SMCs). This notation is adopted in the present context.

Closely related to the first two approaches is a third methodology, which depends on extracting a constitutive law for R_{ij} out of each SMC by specializing the latter to equilibrium flow conditions (in the sense of Section 7.8). In particular, at equilibrium a set of coupled algebraic equations for R_{ij} are derived whose solution is a generalization of (8.1). These are known as *algebraic Reynolds stress models* (ARSMs, ASMs, or ARMs). These are in essence very similar to the NLEVMs; in fact, they unite constitutive equation modeling with Reynolds stress equation modeling. In the following, ARMs are considered in Section 8.8 after first presenting Reynolds stress equation models in Section 8.7; NLEVMs are treated in Section 8.6.

The NLEVM, SMC, and ARM methodologies have been developed extensively with many competing formalisms. The goal here is to present the fundamental ideas behind these schemes, without any pretense at making an exhaustive survey of the possibilities. To help explain the nature of the various closures, some specific models are presented in detail, although the intention is not to imply that they are necessarily the "best" of the closure models available.

8.2 GENERAL CONSIDERATIONS IN RANS MODELING

The various modeling strategies considered here do not have the force of rigorous theory [e.g., as in the stress rate-of-strain relation (2.2) underlying the Navier–Stokes equation]. Rather, turbulence models usually aim only toward "engineering accuracy" (i.e., sufficient accuracy to be useful in a given context). For example, often it is sufficient if a model can accurately reflect trends in the data as the Reynolds number or other parameters are changed. Or, a closure may be adequate for an application if only the model can obtain a good estimate of $\overline{U}_i$ or K, and it is of no concern that other aspects of the modeling may be seriously in error. This helps explain why lower-order models based on (8.1) are widely used today and have value despite their innate fallacies.

It will become clear below that no particular approach holds the position as the "best" closure model, although it makes some sense to talk about the best model for a given problem. This viewpoint is mirrored in two competing philosophies of turbulence modeling. In the first, turbulence models are regarded as approximations to universal laws, in the sense that their mathematical forms and calibrated constants should be inviolable (i.e., they should be kept the same for all applications). The second philosophy is that it is fully legitimate to alter both the calibrated constants and even the actual mathematical forms of closure models in order to derive the variant that is best for a particular flow or class of flows. There are many examples where the latter approach is successful, even if the revised model acquires diminished capabilities in other areas. Ultimately, the decision as to whether or not it makes sense to modify a closure model to achieve necessary predictive goals depends on the perceived benefit. The price of forfeiting generality might or might not be too high a price to pay.

It was mentioned in Section 8.1, for example, that in some cases EVMs are impervious to the presence of flow anisotropy. In fact, similar fundamental limitations can be found for closures at any given level of complexity, in the sense that no matter how they are modified, certain physical phenomena cannot be accommodated. As a general rule, it makes sense to utilize models for a given flow that are not prevented fundamentally from representing the relevant physics. However, the use of a closure that is not prevented a priori from representing a wide range of physical phenomena does not necessarily guarantee that it gives better predictions than a less general model. For example, in flows containing rotation, a closure that can accommodate rotation should be superior to those that cannot, but may not be superior in other contexts.

Finding the best model for a particular flow—when, in fact, there are literally hundreds of possible models—may seem to be a daunting task. Moreover, for those who do not object to changing "coefficient" values, this can add even greater dimensionality to the modeling process. Fortunately, in practical terms, much of the variability of models occurs within just a few kinds of basic approaches, and the improvements to be had within each approach are often not significant. Consequently, it is sufficient for the present purposes to confine the discussion to the best known models of each type. Some of these are incorporated in commercial computational fluid dynamics (CFD) codes and may easily be compared to one another for specific applications to see which is the best basic approach to take. Many others mentioned here are not available in commercial codes. The performance of a selection of approaches of both types is explored in Chapter 9. Ultimately, the decision as to how much investment needs to be made in devising an effective closure depends on the application.

Also coloring the decision as to which model to use, particularly in industrial applications of CFD, is the relative cost of implementing different solution techniques. Since it is generally true that simpler models require less computational resources for their implementation than more advanced models, they tend to be preferred over the latter approaches, unless the payoff in accuracy is notable. If computational costs are important to a user, it is reasonable to initiate any investigation with simple models, proceeding to more advanced ones only if performance is unsatisfactory. However, judging by the overwhelming preference for EVMs in industrial applications today, it is apparent that there is not a widespread belief that higher-order closures offer added value. The trend, however, is changing if for no other reason than the fact that advances in computer speed and affordability lessen the economic incentive for EVMs.

The task of deciding which closure ought to be used in a given situation is also simplified by the recognition that a relatively small number of closure schemes have emerged as de facto standard models with widespread application. For example, in the category of two equation models, the $K-\epsilon$ closure is dominant. This has a standard high-Reynolds-number form as well as a family of low-Reynolds-number forms which extend its applicability to near-wall flows. Similarly, SMCs are most commonly encountered in the form of a few relatively closely related models.

It is often remarked that closure models tend to perform well in the flows for which they have been tuned (i.e., calibrated). Thus, a well-tuned model can be of significant use in studying flows which are close to that used in the calibration. Much more challenging is having models perform predictably well for flows unrelated to the calibration flows. This level of accomplishment is still being awaited.

Finally, it should be noted that the discussion in this chapter first considers the high-Reynolds-number form of closures, as against the form they take near boundaries where the turbulent Reynolds number generally goes to zero. In most cases, the high-Reynolds-number closures will require modification before they can be applied in the near-wall region. Some of the ways in which this is done for various closures is taken up in Section 8.9.

8.3 EDDY VISCOSITY MODELS

The eddy viscosity in (8.1) has units of length times velocity, as in any diffusivity coefficient. Turbulence models exploit this property by proposing that

$$\nu_t = \mathcal{U}\mathcal{L}, \tag{8.3}$$

where $\mathcal{U}$ and $\mathcal{L}$ are appropriately chosen turbulence velocity and length scales, respectively. The similarity of (8.3) in comparison to (6.5) for the molecular viscosity is intentional. Great latitude exists for the selection of the scales appearing in (8.3), and in this way the different eddy viscosity models are distinguished one from another.

For a unidirectional flow, such as in a channel or pipe, (8.1) gives

$$\overline{uv} = -\nu_t \frac{d\overline{U}}{dy}, \tag{8.4}$$

where y is distance from the wall. The first workable closure model, the well-known mixing-length theory of Ludwig Prandtl [41], was developed specifically for this case and assumes that

$$\mathcal{L} = l_m \tag{8.5}$$

and

$$\mathcal{U} = l_m \left| \frac{d\overline{U}}{dy} \right|, \tag{8.6}$$

where l_m is known as the *mixing length*. Using (8.5) and (8.6), (8.4) becomes

$$\overline{uv} = -l_m^2 \left| \frac{d\overline{U}}{dy} \right| \frac{d\overline{U}}{dy}. \tag{8.7}$$

l_m is intended to represent the distance over which turbulent eddies "carry" momentum before mixing takes place. It should be clear from the discussion in Chapter 6 that (8.7) should only be taken at face value [i.e., the physical arguments justifying the selection of $\mathcal{L}$ and $\mathcal{U}$, and even (8.4) for that matter, have at best a tenuous connection to the real physics of turbulent flows].

An essential difficulty in applying (8.7) is deciding a value for l_m. For flow either adjacent to or far from a flat surface, simple dimensional considerations give useful estimates of l_m. For example, next to a wall it is reasonable to assume that $l_m \sim y$. Far from walls, l_m may be assumed to vary slowly enough to be taken as constant. With these choices, mixing-length theory is consistent with some of the basic properties of turbulence mentioned in Chapters 4 and 5. For example, in the case of channel flow, the integrated mean momentum equation (4.9) becomes, after the use of (8.7),

$$\left(v + l_m^2 \left| \frac{d\overline{U}}{dy} \right| \right) \frac{d\overline{U}}{dy} = \frac{\tau_w}{\rho} \left(1 - \frac{2y}{h} \right). \tag{8.8}$$

In the constant-stress layer near the wall (see Section 4.2), v can be neglected in comparison to the eddy viscosity to which it is being added on the left-hand side. Assuming that

$$l_m = \kappa y, \tag{8.9}$$

where κ is a constant, (8.8) reduces to

$$\kappa^2 y^2 \left(\frac{d\overline{U}}{dy} \right)^2 = \frac{\tau_w}{\rho}. \tag{8.10}$$

After integration and transformation into wall variables, (8.10) gives

$$\overline{U}^+ = \frac{1}{\kappa} \log y^+ + B \tag{8.11}$$

(i.e., the log law discussed in Section 4.3.2).

Very close to the wall, (8.7) incorporating (8.9) conflicts with the requirement that $\overline{uv} \sim y^3$, a condition that follows from the product of Taylor series for u and v. To be consistent with this condition, the mixing-length form of the eddy viscosity must be specially modified in the sublayer. This is but one example of many where high-Reynolds-number closure relations may conflict with near-wall conditions.

Several generalizations of (8.7) have been developed to accommodate flows other than the restricted class of unidirectional flows for which it is valid. The first, due to Smagorinsky [51], postulates that

$$v_t = l_m^2 (2\overline{S}_{ij} \overline{S}_{ij})^{1/2}, \tag{8.12}$$

where

$$\overline{S}_{ij} = \frac{1}{2} \left(\frac{\partial \overline{U}_j}{\partial x_i} + \frac{\partial \overline{U}_j}{\partial x_i} \right) \tag{8.13}$$

is the mean rate of strain tensor. Equation (8.12) is most often used in the context of subgrid models for LES and will be revisited in Chapter 10. More commonly used in RANS computations are the Cebeci and Smith model [8], wherein

$$v_t = l_m^2 \left(\frac{\partial \overline{U}_i}{\partial x_j} \frac{\partial \overline{U}_i}{\partial x_j} \right)^{1/2}, \tag{8.14}$$

and the popular Baldwin–Lomax model [3], which has

$$\nu_t = l_m^2 (\overline{\Omega_i}\,\overline{\Omega_i})^{1/2}, \tag{8.15}$$

where $\overline{\Omega_i}$ is the mean vorticity vector. The three relations (8.12), (8.14), and (8.15) reduce to (8.7) in unidirectional flow.

Although there may be a rationale for the selection of l_m in a typical zero-pressure-gradient boundary layer or pipe flow, in more general flows there is little basis for its determination. For example, distance to the body has little meaning near the front and trailing edges of a slender body, or near cavities and other highly curved or sharp edges, as illustrated in Fig. 8.1.

As a practical matter, turbulence closures based on models such as (8.14) and (8.15) give reasonable performance only for flows that are not too far removed from those for which (8.7) is appropriate. This includes, for example, three-dimensional boundary layers with mild cross-flow. However, if such phenomena as curvature, separation, large pressure gradients, rotation, secondary flows, transient effects, sudden changes in shear, and so on, are present, the performance of the mixing-length model is degraded to the point where an alternative modeling scheme is required.

If it is reasonable to assume that (8.1) is appropriate for a particular flow, then failure of the mixing-length model, per se, might be due to a poor choice for l_m or perhaps because (8.6) is unacceptable. The latter possibility is not unexpected since, for example, (8.6) is insensitive to the local turbulence level, which according to (6.29), has a major influence on setting the diffusion rate.

A convenient measure of the local turbulence level is provided by K and so it has been pursued [38] as an alternative to (8.6). Within the framework of (8.1) and (8.3), it is thus assumed that

$$\mathcal{U} \sim \sqrt{K}, \tag{8.16}$$

which introduces the need to develop a means for predicting K. If this is to be done by closing the K equation, (2.40), then among other problems, it is necessary to find a means for computing ϵ. If a differential equation for ϵ is to be avoided, an alternative route to determining ϵ is necessary. One possibility is to use a dimensional argument [as in (2.96)] to justify the expression

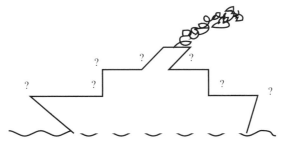

Fig. 8.1   *There is little to guide the selection of a mixing length in some regions of complex flows.*

$$\epsilon \sim \frac{K^{3/2}}{\mathcal{L}}, \tag{8.17}$$

where $\mathcal{L}$ is assumed to be available as part of the mixing-length assumption. Models of this type, which require just a single extra equation to be solved in conjunction with the mean velocity equation, are referred to as *one-equation models*.

Another approach along these lines, which has been effective in treating two-dimensional boundary layers with large pressure gradients and flow separation, such as occurs in diffuser flows, was developed by Johnson and King [25,26]. This abandons (8.6) in favor of a velocity scale formed from the maximum Reynolds shear stress. The latter is computed along the boundary surface by solving a modeled differential equation developed out of the K equation. An externally supplied length scale is also necessary in setting up the auxiliary differential equation.

On the whole, one-equation models suffer from the same limitations as zero-equation models in having to determine a length scale. Partly for this reason they are convenient and effective only for restricted classes of flows. This has led to the pursuit of alternative EVMs that are entirely self-contained (i.e., no special parameters or scale functions need to be supplied externally). In particular, such models incorporate a means for selecting the length scale automatically.

The most popular approach toward accomplishing this goal is to continue to assume that (8.16) is legitimate, but depart from (8.17) by incorporating a differential equation with which to solve for ϵ. The result is an example of a *two-equation model*. In this case, the use of (8.17) is turned around so that it provides the estimate

$$\mathcal{L} \sim \frac{K^{3/2}}{\epsilon}, \tag{8.18}$$

and thus (8.3) gives

$$\nu_t = C_\mu \frac{K^2}{\epsilon}, \tag{8.19}$$

where C_μ is a constant. Recalling the discussion in Section 7.3.1, it is clear that a physical interpretation may be assigned to (8.18) as the approximate distance over which fluid particles move during the eddy turnover time.

When (8.19) is made the basis for closure by using it in (8.2), additional, modeled equations for K and ϵ must be solved. The resulting system of equations constitutes the K–ϵ closure [27], perhaps the most commonly used closure today. Modeling the K and ϵ equations is taken up in the next section.

Other two-equation models have been developed, most notably the K–ω model [65], which differs from the K–ϵ closure in using a model equation for the inverse time scale $\omega = \epsilon/K$ (with units the same as vorticity). In fact, ω expresses the fractional rate at which energy is dissipated in the flow. The K–ω model assumes that $\nu_t = \gamma^* K/\omega$, where γ^* is a constant. The modeling of the ω and ϵ equations differs primarily in the transport term, and the consequences for turbulent flow prediction are generally not significant. The greatest practical difference between the K–ϵ and

$K-\omega$ closures lies in their near-wall behaviors, where it is sometimes argued that the $K-\omega$ closure has better numerical properties.

In recent years there has been renewed interest in one-equation models, not as a relation for K, but rather, as a single model equation for the eddy viscosity. Since the eddy viscosity does not naturally have a governing conservation equation, the model ν_t equations are constructed artificially as a balance of production, dissipation, and transport. Models of this type [2,56] have been used successfully in predicting a variety of airfoil flows, although the technique does not appear to be suitable for modeling arbitrary three-dimensional flows.

8.4 $K-\epsilon$ CLOSURE

The closed system of equations composing the standard, high-Reynolds-number form of the $K-\epsilon$ closure is now considered. Recalling (2.40), the exact K equation can be written in the form

$$\frac{\partial K}{\partial t} + \overline{U}_j \frac{\partial K}{\partial x_j} = \mathcal{P} - \epsilon + \nu \nabla^2 K - \frac{\partial}{\partial x_i} \left(\frac{\overline{p u_i}}{\rho} + \overline{u_i (u_j^2/2)} \right), \tag{8.20}$$

where

$$\mathcal{P} = -R_{ij} \frac{\partial \overline{U}_i}{\partial x_j} \tag{8.21}$$

is the turbulent kinetic energy production term, and it will be noted that the pressure work and kinetic energy flux terms have been combined into a single fluxlike term. Assuming that (8.1) holds, it follows that

$$\mathcal{P} = \nu_t \frac{\partial \overline{U}_i}{\partial x_j} \left(\frac{\partial \overline{U}_i}{\partial x_j} + \frac{\partial \overline{U}_j}{\partial x_i} \right), \tag{8.22}$$

where ν_t is given by (8.19).

Since ϵ is to be determined from its own closed equation, only the last term in (8.20) needs to be modeled. For wont of a formal means for analyzing the relevant physics, it is traditional to assume that this term obeys a gradient transport law of the form

$$\frac{1}{\rho} \overline{p u_i} + \overline{u_i (u_j^2/2)} = -\frac{\nu_t}{\sigma_K} \frac{\partial K}{\partial x_i}, \tag{8.23}$$

where the constant σ_K may be considered to be a turbulent Prandtl number. Putting these results together, the closed K equation is

$$\frac{\partial K}{\partial t} + \overline{U}_j \frac{\partial K}{\partial x_j} = \mathcal{P} - \epsilon + \frac{\partial}{\partial x_i} \left[\left(\nu + \frac{\nu_t}{\sigma_K} \right) \frac{\partial K}{\partial x_i} \right]. \tag{8.24}$$

In this, K is determined by the balance between convection on the left-hand side, and production, dissipation, and transport on the right-hand side.

The ϵ equation (2.42) was previously given in the form

$$\frac{D\epsilon}{Dt} = P_\epsilon^1 + P_\epsilon^2 + P_\epsilon^3 + P_\epsilon^4 + \Pi_\epsilon + T_\epsilon + D_\epsilon - \Upsilon_\epsilon, \tag{8.25}$$

where the correlations on the right-hand side are defined in (2.43) through (2.50). With the sole exception of the diffusion term, D_ϵ, these terms require modeling. Clearly, the ϵ equation represents a significantly greater modeling challenge than the K equation.

The modeling of the stretching and dissipation terms, P_ϵ^4 and Υ_ϵ, was discussed in Sections 7.5 and 7.8, in the context of isotropic and homogeneous shear flow turbulence. The results of that analysis are typically adopted for the general case also, since there is, as yet, no rigorous means for predicting how inhomogeneities and anisotropies might affect these correlations. Thus (7.202) and (7.203) may be invoked to construct the model

$$P_\epsilon^4 - \Upsilon_\epsilon = C_{\epsilon_3} R_T^{1/2} \frac{\epsilon^2}{K} - C_{\epsilon_2} \frac{\epsilon^2}{K}, \tag{8.26}$$

where $R_T = K^2/\nu\epsilon$. As mentioned previously, the traditional viewpoint assumes that $C_{\epsilon_3} = 0$, in which case vortex stretching makes no independent contribution to the dissipation rate balance.

The modeling of P_ϵ^1 and P_ϵ^2 was also considered in Section 7.8. The approach taken there under homogeneous shear flow conditions is now expanded to accommodate the general case. Introducing what is sometimes called the complementary dissipation tensor,

$$\epsilon_{ij}^c = 2\nu \overline{\frac{\partial u_k}{\partial x_i} \frac{\partial u_k}{\partial x_j}}, \tag{8.27}$$

(2.43), (2.44), (2.51), and (8.27) yield the expressions

$$P_\epsilon^1 = -\epsilon_{ij}^c \frac{\partial \overline{U}_i}{\partial x_j} \tag{8.28}$$

and

$$P_\epsilon^2 = -\epsilon_{ij} \frac{\partial \overline{U}_i}{\partial x_j}. \tag{8.29}$$

Note that $\epsilon_{ii} = \epsilon_{ii}^c = 2\epsilon$, and, moreover, in isotropic turbulence,

$$\epsilon_{ij} = \epsilon_{ij}^c = \delta_{ij} \tfrac{2}{3}\epsilon. \tag{8.30}$$

When turbulence is anisotropic, it is difficult to predict how the individual components of ϵ_{ij} and ϵ_{ij}^c will behave. Rather than assume isotropy so that (8.30) holds, with the implausible consequence that $P_\epsilon^1 = P_\epsilon^2 = 0$, it is usual to assume that the anisotropy of ϵ_{ij} and ϵ_{ij}^c matches that of the Reynolds stress tensor. A measure of the degree to which the flow may be anisotropic is given by the anisotropy tensor

$$b_{ij} \equiv \frac{R_{ij} - \frac{2}{3}K\delta_{ij}}{2K}, \tag{8.31}$$

which is identically zero in isotropic turbulence and has the convenient property that $|b_{ij}| \le 1$.

The formal assumption is now made that the deviatoric parts of ϵ_{ij} and ϵ_{ij}^c are proportional to b_{ij}, that is,

$$\frac{\epsilon_{ij} - \frac{2}{3}\epsilon\delta_{ij}}{2\epsilon} \sim b_{ij} \tag{8.32}$$

and a similar relation for ϵ_{ij}^c. Note that this is the same idea as in (7.200). Substituting these results into (8.28) and (8.29), combining the equations, and introducing a proportionality constant C_{ϵ_1} yields

$$P_\epsilon^1 + P_\epsilon^2 = C_{\epsilon_1}\frac{\epsilon}{K}\mathcal{P}. \tag{8.33}$$

This reduces to (7.201) in homogeneous shear flow.

The current practice is not to model explicitly the production term, P_ϵ^3. Rather, its contribution is imagined to be contained within that of the other production terms in (8.25). Finally, the transport terms, T_ϵ and Π_ϵ, are normally given a gradient law treatment:

$$T_\epsilon + \Pi_\epsilon = \frac{\partial}{\partial x_i}\left(\frac{\nu_t}{\sigma_\epsilon}\frac{\partial \epsilon}{\partial x_i}\right), \tag{8.34}$$

where σ_ϵ is a constant, so that the final modeled form of the ϵ equation is

$$\frac{\partial \epsilon}{\partial t} + \overline{U}_j\frac{\partial \epsilon}{\partial x_j} = C_{\epsilon_1}\frac{\epsilon}{K}\mathcal{P} - C_{\epsilon_2}\frac{\epsilon^2}{K} + \frac{\partial}{\partial x_i}\left[\left(\nu + \frac{\nu_t}{\sigma_\epsilon}\right)\frac{\partial \epsilon}{\partial x_i}\right]. \tag{8.35}$$

This has a similar structure to the K equation: convection and transient terms on the left-hand side are balanced by the combined effect of production, dissipation, and transport on the right-hand side.

A common step used in calibrating (8.35) is to ensure its compatibility with turbulent flow in the constant-stress layer in a simple flat plate boundary layer or channel. In this circumstance it is assumed that

$$-\overline{uv} = \frac{\tau_w}{\rho} = U_\tau^2 \tag{8.36}$$

as well as that the turbulence production and dissipation terms in the K equation are in balance:

$$\mathcal{P} = -\overline{uv}\frac{d\overline{U}}{dy} = \epsilon. \tag{8.37}$$

Replacing $\overline{uv}$ in (8.4) using (8.36) yields an equation for the mean velocity derivative. Substituting this into (8.37), replacing $\overline{uv}$ using (8.36) once again, and substituting for ν_t from (8.19) gives the approximation

$$K = \frac{U_\tau^2}{\sqrt{C_\mu}}. \tag{8.38}$$

Thus, in this modeling scenario, K is constant in the equilibrium, constant-stress layer. Note that (8.38) exactly solves (8.24) under the present circumstances.

Our interest in these results is for what constraints they may place on the constants in the ϵ equation (8.35) in the constant-stress layer. According to (8.4), (8.36), and (8.37),

$$\epsilon = U_\tau^2 \frac{d\overline{U}}{dy}, \tag{8.39}$$

and substituting for $d\overline{U}/dy$ using the log law (4.42), this becomes

$$\epsilon = \frac{U_\tau^3}{\kappa y}. \tag{8.40}$$

Thus it is seen that ϵ increases in this part of the boundary layer as the wall is approached. Incorporating (8.38) and (8.40) into (8.35), and omitting the viscous diffusion term yields the condition

$$\frac{\sqrt{C_\mu}\,\sigma_\epsilon}{\kappa^2}(C_{\epsilon_2} - C_{\epsilon_1}) = 1. \tag{8.41}$$

Satisfaction of (8.41) by model constants is one step toward ensuring that the K–ϵ closure behaves properly near walls.

To summarize, the K–ϵ closure consists of the coupled system of equations (8.2), (8.24), and (8.35) with eddy viscosity (8.19). The standard values of the constants appearing in these equations are $C_\mu = 0.09$, which guarantees consistency with the log law of the wall, $C_{\epsilon_1} = 1.44$, $C_{\epsilon_2} = 1.92$, $\sigma_K = 1$, and $\sigma_\epsilon = 1.3$. The choice of C_{ϵ_2} is made with a view toward guaranteeing a decay law in isotropic turbulence in the range of experimental observations, as discussed in Section 7.5. Values for the other constants are ascertained after applying the closure to standard boundary layer flows and with the aim of satisfying (8.41). It should be noted that (8.24) and (8.35) must be modified before they can be applied to low-Reynolds-number flow near boundaries. This question is taken up in Section 8.9.2.

8.5 MODELING CONSTRAINTS

Although it has not been mentioned previously, there are a number of tensorial and transformation properties which all constitutive models of the Reynolds stress tensor must satisfy if they are to be considered physically valid. The fact that Eq. (8.1) satisfies all such constraints is one of the compelling reasons for its use despite its many known drawbacks. Since the goal now is to consider Reynolds stress models that depart from this standard linear stress rate-of-strain law, it is generally a good idea, where possible, to make sure that the new models also satisfy the constitutive constraints. In fact, the use of such constraints has played a significant role in shaping the form of many of the alternative formulas used in Reynolds stress modeling.

Thus, consider a turbulent flow that is being studied by two different observers that are moving relative to each other. Suppose that each of them has the same physical idea for modeling the Reynolds stress tensor. A fundamental question is: what constraints must be imposed on the mathematical models they propose so that they are consistent with each other? Specifically, how can it be guaranteed that the mathematical forms of the models given by each observer will naturally transform into each other if the appropriate coordinate transformation is made? In fact, if a model does not satisfy this transformation property, the proposed physical law is observer dependent and open to the criticism of not being physical.

Since $R_{ij} = \overline{u_i u_j}$, it is evident that the transformation properties of R_{ij} are the same as that for the fluctuating velocity vector, u_i. The latter transforms according to rules established by its own governing equation. Thus, consider two observers, the first of which is an inertial observer (i.e., one who is not accelerating), and measures position according to $\mathbf{x}$ and velocities according to $\mathbf{U}(\mathbf{x}, t)$, while the second observer is accelerating arbitrarily and has coordinates, $\mathbf{x}^*$, and measures velocities to be $\mathbf{U}^*(\mathbf{x}^*, t)$. If $\mathbf{r}(t)$ is the position vector connecting the origin of the two coordinate systems and $Q(t)$ is a rotation tensor,[1] the most general relationship between the two observers is

$$\mathbf{x}^* = Q(t)\mathbf{x} + \mathbf{r}(t). \tag{8.42}$$

This means that the second observer sees the first translating away from himself according to $\mathbf{r}(t)$ and rotating with respect to himself according to $Q(t)$. If there is no relative rotation between the observers, so that $Q(t) = I$, (8.42) simplifies to

$$\mathbf{x}^* = \mathbf{x} + \mathbf{r}(t). \tag{8.43}$$

Furthermore, if the second observer is also inertial, (8.43) reduces to

$$\mathbf{x}^* = \mathbf{x} + \mathbf{U}^c t, \tag{8.44}$$

where $\mathbf{U}^c$ is a constant, translational velocity.

---

[1]A rotation tensor Q satisfies $Q^{-1} = Q^t$ and has $|Q| = 1$. For any vector $\mathbf{v}$, $|Q\mathbf{v}| = |\mathbf{v}|$ and $Q\mathbf{v}$ is rotated with respect to $\mathbf{v}$.

It may readily be shown that the Navier–Stokes equation for a general noninertial observer is [4,21]

$$\frac{\partial U_i^*}{\partial t} + U_j^* \frac{\partial U_i^*}{\partial x_j^*} = -\frac{1}{\rho}\frac{\partial P^*}{\partial x_i^*} + \nu \nabla^2 U_i^* - \ddot{\mathbf{r}} - 2(\mathbf{\Omega}^o \times \mathbf{U}^*)_i$$

$$- [\mathbf{\Omega}^o \times (\mathbf{\Omega}^o \times \mathbf{x}^*)]_i - (\dot{\mathbf{\Omega}}^o \times \mathbf{x}^*)_i, \tag{8.45}$$

where the last four terms on the right-hand side, respectively, account for rectilinear accelerations of the reference frame, the Coriolis force, the centrifugal force, and the effect of changes in rotation rate. $\mathbf{\Omega}^o$, which is associated with Q, is the angular velocity or rotation rate of the noninertial coordinate system.[2]

By subtracting its own average from itself, (8.45) yields an equation for the velocity fluctuation in the form

$$\frac{\partial u_i^*}{\partial t} + \overline{U}_j^* \frac{\partial u_i^*}{\partial x_j^*} = -u_j^* \frac{\partial u_i^*}{\partial x_j^*} - u_j^* \frac{\partial \overline{U}_i^*}{\partial x_j^*} - \frac{1}{\rho}\frac{\partial p^*}{\partial x_i^*} + \nu \nabla^2 u_i^*$$

$$- \frac{\partial R_{ij}^*}{\partial x_j^*} - 2(\mathbf{\Omega}^o \times \mathbf{u}^*)_i, \tag{8.46}$$

where the presence of the Coriolis term is all that prevents the velocity fluctuation from satisfying the identical equation for every observer. As it is, however, all observers who are undergoing only rectilinear accelerations without rotation (i.e., $\mathbf{\Omega}^o = 0$) will see the identical velocity fluctuation equation. If this is so, any proposed Reynolds stress model should satisfy the condition that it appear the same to all observers related through (8.43). This is referred to as invariance over the extended Galilean group. This is a significant constraint on the allowable tensor forms of Reynolds stress models, and thus is of considerable benefit in reducing the complexity of the general tensor models which are commonly proposed for predicting the Reynolds stress, as will be seen below.

Material frame indifference (MFI) occurs when a tensor property has the identical form for all observers, including those who are rotating. If (8.46) is specialized to two-dimensional turbulence, it may be seen that the Coriolis term vanishes. This suggests that ideally, Reynolds stress models should satisfy MFI in the limit of two-dimensional turbulence [57]. This physical condition is most nearly achieved in rapidly rotating systems where, according to the Taylor–Proudman theorem [39], the motion becomes largely two-dimensional. It is also approximated in some sense near solid boundaries, where the wall-normal Reynolds stress component is much smaller than the wall-parallel components. Since such conditions are far removed

---

[2]In fact, the vector $\mathbf{\Omega}^o$ is the axial vector corresponding to the skew-symmetric tensor $Q^{-1}\dot{Q}$. Note that a tensor W is skew-symmetric if $W = -W^t$. Such tensors have only three independent components and, in fact, have an associated axial vector $\mathbf{w}$ such that $W\mathbf{v} = \mathbf{w} \times \mathbf{v}$ for any vector $\mathbf{v}$. The fact that $Q^{-1}\dot{Q}$ is skew-symmetric follows from the property $Q^{-1} = Q^t$ of rotation tensors and taking a time derivative of the identity $QQ^{-1} = I$ (see [21]).

from the general three-dimensional state of turbulent flows, enforcing MFI in the two-dimensional limit does not necessarily provide benefit to closures destined to be used for three-dimensional turbulence. In fact, this constraint can have the harmful effect of overly restricting the forms of models and thereby severely reducing their range of applicability.

Another strategy for constraining models through (8.46) is suggested by the fact that only the Coriolis force appears. Thus, consider the identity

$$\frac{\partial \overline{U}_i^*}{\partial x_j^*} = \overline{S}_{ij}^* + \overline{W}_{ij}^*, \tag{8.47}$$

where $\overline{S}_{ij}^*$ is the mean rate of strain tensor and

$$\overline{W}_{ij}^* = \frac{1}{2}\left(\frac{\partial \overline{U}_i^*}{\partial x_j^*} - \frac{\partial \overline{U}_j^*}{\partial x_i^*} \right) \tag{8.48}$$

is the mean vorticity tensor. It is not hard to show that $\overline{S}_{ij}^*$ obeys MFI; both $\overline{S}_{ij}$ and $\overline{S}_{ij}^*$ are given by (8.13) written in their respective coordinate systems, and they transform between each other according to the mappings associated with (8.42). It may also be shown that the absolute mean vorticity tensor

$$\overline{W}_{ij}^a \equiv \overline{W}_{ij}^* + \epsilon_{kji}\Omega_k^o \tag{8.49}$$

is invariant to all observers. $\overline{W}_{ij}^a$ is a skew tensor and its associated axial vector is $\mathbf{\Omega}^a/2$, where $\mathbf{\Omega}^a$ is the absolute mean vorticity, defined by

$$\mathbf{\Omega}^a = \nabla \times \mathbf{U} + 2\mathbf{\Omega}^o. \tag{8.50}$$

Introducing these definitions into (8.46) yields

$$\frac{\partial u_i^*}{\partial t} + \overline{U}_j^* \frac{\partial u_i^*}{\partial x_j^*} - u_j^* \frac{\partial \overline{U}_i^*}{\partial x_j^*} = -u_j^* \frac{\partial u_i^*}{\partial x_j^*} - 2u_j^*(\overline{S}_{ij}^* + \overline{W}_{ij}^a) - \frac{1}{\rho}\frac{\partial p^*}{\partial x_i^*}$$
$$+ \nu\nabla^2 u_i^* + \frac{\partial R_{ij}^*}{\partial x_j^*}. \tag{8.51}$$

The left-hand side of this equation is the frame-indifferent *Oldroyd derivative* [59] of u_i^* and henceforth is denoted as

$$\frac{D_o u_i^*}{Dt} \equiv \frac{\partial u_i^*}{\partial t} + \overline{U}_j^* \frac{\partial u_i^*}{\partial x_j^*} - u_j^* \frac{\partial \overline{U}_i^*}{\partial x_j^*}. \tag{8.52}$$

In particular, the form of $D_o u_i^*/Dt$ does not change from one observer to the next. The importance of (8.51) is that it retains the form it has here for all observers.

Consequently, if Reynolds stress models are constructed using $D_o u_i^*/Dt$, $\overline{S}_{ij}$, and $\overline{W}_{ij}^a$, they can be constrained to be frame indifferent. This property has sometimes been exploited in deriving models, as shown below.

Another potentially useful constraint on closure design is the concept of *realizability*. In the case of the Reynolds stress tensor, it means that a particular closure model should not predict negative values of the normal Reynolds stress components R_{11}, R_{22}, and R_{33}, which by definition are nonnegative. Moreover, it implies that the off-diagonal components, R_{12}, R_{23}, and R_{13}, should satisfy the Schwarz inequality, to the effect that

$$|R_{\alpha\beta}| \leq \sqrt{\overline{u_\alpha^2}}\sqrt{\overline{u_\beta^2}} \tag{8.53}$$

for $\alpha \neq \beta$.

A distinction needs to be drawn between enforcing realizability for the computed solutions to turbulent flow problems, a condition that is certainly necessary, and ensuring that *all* solutions to a set of closure equations are realizable. The latter requirement is legitimate if the goal of the model calculation is to get time-accurate predictions of turbulent flow fields. But if, in fact, it is only a steady-state solution that one is interested in, as is often the case, there is a real possibility that the act of forcing realizability to the closure equations will unintentionally degrade the accuracy of the prediction (i.e., the realizability condition may get in the way of better modeling the physics of the steady-state flow balance). This viewpoint is reinforced by the fact that in practical terms, there is more than one way to enforce realizability on a given model prototype [14]. So there is a considerable degree of arbitrariness in selecting one particular realizable form over another. It is also true that the typical conditions where a model is likely to violate realizability are often far removed from the states occurring in practical flows. Forcing realizability is therefore not likely to have a beneficial effect on predictions, and in fact, may be harmful. For these several reasons, most turbulence models are not developed under the realizability constraint.

One final idea that sometimes finds application in limiting the possible form of closures is enforcing their compatibility with *rapid distortion theory* (RDT) [24]. This theory describes the type of turbulent flow behavior that results when external conditions are suddenly changed. For example, fluid in turbulent motion flowing down a pipe would experience a rapid change in circumstances when it passes through a narrowing exit nozzle. In such circumstances, if the time scale over which changes in the mean field are significant is much smaller than the time scale governing the turbulence decay, it can be argued that the nonlinear terms in the Navier–Stokes equation have such a small influence that they can be neglected. The idea in this case is that these terms account for the self-interaction of the turbulent field on itself, a process that does not become significant in a short time period.

To quantify the necessary conditions for RDT to be relevant, consider an initially homogeneous turbulence that is suddenly subjected to a mean shearing, $S = d\overline{U}/dy > 0$. Then one may pursue a RDT analysis if

$$\frac{1}{S} << \frac{K}{\epsilon}.$$ (8.54)

Without the nonlinear term, the velocity fluctuation equation becomes

$$\rho \left(\frac{\partial u_i}{\partial t} + \overline{U}_j \frac{\partial u_i}{\partial x_j} \right) = -\rho u_j \frac{\partial \overline{U}_i}{\partial x_j} - \frac{\partial p}{\partial x_i} + \mu \, \nabla^2 u_i.$$ (8.55)

This is a linear system in u_i that can be analyzed in many fruitful ways, particularly if the mean field is of a relatively simple structure. Many examples of such analyses are available in the literature. One particular situation which is often used in model calibration is that in which a mean shear is placed on an initially isotropic turbulence [11]. RDT analysis in this case leads to a specific form of the pressure-strain term. In another example [43] it may be shown that an initially isotropic turbulent flow subjected to a sudden rotation remains isotropic. The wisdom of forcing closures to satisfy such constraints is not guaranteed: as in the case of realizability, if the RDT limit is far removed from the circumstances of a flow of interest, forcing compatibility with the RDT limit might seriously hinder the performance of the method. For this reason, consistency with RDT predictions is not a universal goal of model developers.

8.6 GENERALIZED CONSTITUTIVE MODELS

The constitutive model (8.1) has its inspiration in the analysis of molecular momentum transport. This analogy, however, can only go so far before it is inappropriate, and a more comprehensive theory of turbulent transport is required. In Chapter 6 we gave some elements of a formal analysis of transport, but such techniques have had only limited use in developing practical constitutive laws. One such application, with an indirect consequence for Reynolds stress modeling, is the case of vorticity transport theory discussed in Section 8.10.

Apart from the ARM methodology considered below, the most common approach to expanding beyond (8.1) so as to develop NLEVMs is first to assume a general functional relationship between Reynolds stress, mean velocity, and other relevant parameters, and then apply a series of "calibrations" with a view toward turning the general expression into a useful prognosticator of turbulent flow behavior. For example, it may be hypothesized that

$$R_{ij}(\mathbf{x}, t) = F_{ij}(\overline{\mathbf{U}}(\mathbf{y}, s) - \overline{\mathbf{U}}(\mathbf{x}, s), K, \epsilon),$$ (8.56)

where $s \leq t$ represents times up to and including the present and $\mathbf{y}$ represents an arbitrary point in the flow domain. The velocity difference is used here instead of just the velocity since the former satisfies invariance under the extended Galilean group, as was previously shown to be a property of the Reynolds stress tensor.[3]

---

[3] By extension of (8.43) it may be shown that the mean velocity transforms between observers as $\overline{\mathbf{U}}^* = \overline{\mathbf{U}} + \dot{\mathbf{r}}$, so $\dot{\mathbf{r}}$ cancels when taking velocity differences.

To generate nonlinear models from (8.56), the usual procedure [60,62] is first to substitute a truncated Taylor series expansion for $\overline{U}(\mathbf{y}, s) - \overline{U}(\mathbf{x}, s)$ and then require form invariance under a change of observer. Keeping first-order terms in the Taylor series yields (8.1), while keeping quadratic terms yields the general form [52]

$$
R_{ij} = \frac{2}{3} K \delta_{ij} - 2C_\mu \frac{K^2}{\epsilon} \overline{S}_{ij} + A_1 \frac{K^3}{\epsilon^2} \left(\overline{S}_{ik} \overline{S}_{jk} - \frac{1}{3} \overline{S}_{kl} \overline{S}_{kl} \delta_{ij} \right)
$$

$$
+ A_2 \frac{K^3}{\epsilon^2} \left(\overline{W}_{ik} \overline{W}_{jk} - \frac{1}{3} \overline{W}_{kl} \overline{W}_{kl} \delta_{ij} \right) \tag{8.57}
$$

$$
+ A_3 \frac{K^3}{\epsilon^2} (\overline{S}_{ik} \overline{W}_{jk} + \overline{S}_{jk} \overline{W}_{ik}) + A_4 \frac{K^3}{\epsilon^2} \left(\frac{\partial \overline{S}_{ij}}{\partial t} + \overline{U}_k \frac{\partial \overline{S}_{ij}}{\partial x_k} \right),
$$

where it may be noticed that the first two terms on the right-hand side are equivalent to (8.1) with ν_t given by (8.19). The expressions containing K and ϵ in front of the various terms are inserted to ensure dimensional consistency. A number of different models derived through contrasting arguments (e.g., [66]) turn out to be special cases of this general equation.

While the coefficients $A_i, i = 1, \ldots, 4$, in (8.57) are nominally constants, in fact, it is legitimate and often helpful to assume that they are functions of tensor invariants such as $\overline{S}_{ij} \overline{S}_{ij}$ and $\overline{W}_{ij} \overline{W}_{ij}$. This adds greater generality to the constitutive model without the necessity of deriving higher-order terms in the fundamental expansion. The inclusion of tensor invariants in this way is also a natural by-product of the formal development of constitutive models via the ARM approach, which is visited in Section 8.8.

In some approaches to NLEVMs, it is considered desirable to formally include cubic and even quartic terms in the fundamental expansion (8.56). For example, in [1], cubic terms of the form

$$
A_5 \frac{K^4}{\epsilon^3} \left(\overline{W}_{ik} \overline{W}_{kl} \overline{S}_{lj} + \overline{S}_{ik} \overline{W}_{kl} \overline{W}_{lj} + \overline{W}_{kl} \overline{W}_{kl} \overline{S}_{ij} - \frac{2}{3} \overline{W}_{kl} \overline{S}_{lm} \overline{W}_{mk} \delta_{ij} \right)
$$

$$
+ A_6 \frac{K^4}{\epsilon^3} \left(\overline{W}_{ik} \overline{S}_{kl} \overline{S}_{lj} - \overline{S}_{ik} \overline{S}_{kl} \overline{W}_{lj} \right) \tag{8.58}
$$

are added to those in (8.57). Part of the rationale for adding such terms is that it is believed that they are necessary if a NLEVM is to account for effects such as mean streamline curvature and swirl. This opinion is not shared by all, however, since some view quadratic models as sufficiently general to capture the physics of general three-dimensional mean flows. Another recent NLEVM incorporating cubic and quartic terms [10] includes an independent model equation for the stress anisotropy,

$$
\text{II} \equiv b_{ij} b_{ij}, \tag{8.59}
$$

with b_{ij} defined in (8.31).

The development of cubic and quartic models often tends to be tied in with considerations of low-Reynolds-number near-wall modeling. Here, much information about the physics of the flow is known from DNS calculations, and the higher-order models provide more flexibility with which to match known trends. How the many coefficients are calibrated in models that include forms such as (8.57) and (8.58), is beyond the scope of this presentation. Nonetheless, in Chapter 9 we include a number of results from such models, which in some cases can be quite accurate.

Another alternative to (8.57), which has a relatively compact mathematical expression, has been developed using the condition of MFI in the limit of two-dimensional turbulence [58]. The starting point in this case is the functional form

$$R_{ij} = \frac{2}{3} K \delta_{ij} + F_{ij} \left(\nabla \overline{U}, \frac{D}{Dt} (\nabla \overline{U}), K, \epsilon \right),\qquad (8.60)$$

where $(\nabla \overline{U})_{ij} = \partial \overline{U}_i / \partial x_j$ is the mean velocity gradient tensor and the functional F_{ij} must be traceless. The convective derivative of $\nabla \overline{U}$ is included in (8.60) since it contains quadratic terms in the mean velocity and so is consistent with the goal of deriving a relationship that is quadratic in the mean velocity and its derivatives. The requirement that MFI be satisfied in the two-dimensional limit implies that F_{ij} should be a functional of just the invariant parts of its arguments, namely, $\overline{S}_{ij}$, coming from $\nabla \overline{U}$, and

$$\frac{D_o \overline{S}_{ij}}{Dt} = \frac{\partial \overline{S}_{ij}}{\partial t} + \overline{U}_k \frac{\partial \overline{S}_{ij}}{\partial x_k} - \frac{\partial \overline{U}_i}{\partial x_k} \overline{S}_{kj} - \frac{\partial \overline{U}_j}{\partial x_k} \overline{S}_{ki},\qquad (8.61)$$

which is the Oldroyd derivative associated with $(D/Dt)(\nabla \overline{U})$. The demand that (8.60) be invariant under rotations gives, after truncating at the level of quadratic terms in the implied tensor forms,

$$\begin{aligned}
R_{ij} = {} & \frac{2}{3} K \delta_{ij} - 2 C_\mu \frac{K^2}{\epsilon} \overline{S}_{ij} - 4 C_D C_\mu^2 \frac{K^3}{\epsilon^2} \left(\overline{S}_{ik} \overline{S}_{jk} - \frac{1}{3} \overline{S}_{kl} \overline{S}_{kl} \delta_{ij} \right) \\
& - 4 C_E C_\mu^2 \frac{K^3}{\epsilon^2} \left(\frac{D_o \overline{S}_{ij}}{Dt} - \frac{1}{3} \frac{D_o \overline{S}_{kk}}{Dt} \delta_{ij} \right),
\end{aligned}\qquad (8.62)$$

which is commonly referred to as the *nonlinear K–ϵ model* [58]. It is among the more commonly used closures of this type. The constants C_D and C_E are found by calibrating against experimental measurements of the normal Reynolds stresses in channel flow. The values $C_D = C_E = 1.68$ are typically used.

In summary, (8.57) and (8.62) contain as special cases most of the quadratic generalizations of (8.1) that have been derived. Such models offer greater flexibility in comparison to (8.1) without the need for significantly greater numerical effort than is involved in applying the standard K–ϵ closure. A nice property of these models is that they can accommodate situations where it is essential to include anisotropic effects. In recent years, cubic and quartic NLEVMs have been developed with a view toward encompassing a greater range of physical phenomena, including low-Reynolds-number effects near boundaries. Despite the many apparent advantages of NLEVMs as a methodology, this approach is nonetheless local in nature and cannot be expected to capture completely the nonlocal physics of turbulent transport. This has encouraged the development of closures to the Reynolds stress equations themselves, since by their very nature, these should yield what is, in effect, nonlocal representations of the Reynolds stress.

8.7   REYNOLDS STRESS EQUATION MODELS

Second moment closures avoid the explicit assumption of a constitutive law such as (8.57) or (8.62) by instead effecting a closure to the Reynolds stress equation. The latter, derived in Section 2.3, consists of

$$\frac{\partial R_{ij}}{\partial t} + \overline{U}_k \frac{\partial R_{ij}}{\partial x_k} = -R_{ik}\frac{\partial \overline{U}_j}{\partial x_k} - R_{jk}\frac{\partial \overline{U}_i}{\partial x_k} - \epsilon_{ij} - \frac{\partial \beta_{ijk}}{\partial x_k} + \Pi_{ij} + \nu\,\nabla^2 R_{ij}, \quad (8.63)$$

where closure depends on modeling the pressure-strain correlation, Π_{ij}, defined in (2.54), the transport correlation β_{ijk} defined in (2.53), and the anisotropic dissipation rate, ϵ_{ij}, defined in (2.51). Among these three correlations, the pressure-strain term has been the subject of the greatest modeling effort. This reflects its critical role in controlling the division of energy between components, as was seen in Section 4.2.5, as well as in the fact that it is somewhat more amenable than the other terms to physical interpretation and analysis.

The dissipation rate term is usually assumed to have the isotropic form

$$\epsilon_{ij} = \tfrac{2}{3}\epsilon\delta_{ij}, \quad (8.64)$$

where ϵ is computed through its own modeled equation, (8.35), derived previously in the context of the K–ϵ closure. Note that the more general assumption (8.32) turns out not to be necessary in this case since its only effect is to introduce a term $\sim \epsilon b_{ij}$, which is usually included in models for Π_{ij}. Next, modeling of the pressure-strain and transport correlations is considered.

8.7.1   Modeling of the Pressure–Strain Correlation

The importance and functioning of the pressure–strain term was demonstrated in Section 4.2.5, where it was seen to be responsible for the redistribution of energy between components. Models for Π_{ij} are usually constructed in the context of homogeneous shear flow, where the transport term in (8.63) does not make a contribution and (8.64) is assumed to hold.

In Section 2.3 the decomposition of Π_{ij} into slow and fast contributions was introduced:

$$\Pi_{ij} = A_{ij} + M_{ijkl}\frac{\partial \overline{U}_k}{\partial x_l}, \quad (8.65)$$

where A_{ij} and M_{ijkl} are defined in (2.58) and (2.59), respectively. Traditionally, Π_{ij} is modeled via separate models for A_{ij} and M_{ijkl} (i.e., the slow and fast parts), although recent efforts tend to treat the term as a whole. The latter point of view is adopted here, because, as a practical matter, there does not appear to be a significant advantage to either methodology since the resulting models share many similar properties. In addition, it is often clear by inspection which terms in a model of Π_{ij} may be attributed to the fast and slow parts of (8.65) if one wishes to do so.

The first step in modeling Π_{ij} is a hypothesis as to which characteristic fields it should depend upon functionally. In view of its role in the creation, maintenance, and destruction of the anisotropy of turbulence, it is natural to insist that it depends on the tensor **b** defined in (8.31). Its dependence on mean velocity gradient is also apparent from (8.65). Thus one is led to the assumption that

$$\Pi_{ij} = F_{ij}\left(\mathbf{b}, \frac{K}{\epsilon}S, \frac{K}{\epsilon}W\right), \tag{8.66}$$

where for convenience the mean velocity gradient is included in the form of the mean rate of strain tensor, S, and the vorticity tensor, W. The factor K/ϵ is used to nondimensionalize the mean velocity derivative tensors. Since (8.65) is linear in $\partial \overline{U}_k / \partial x_l$, the functional F_{ij} is limited to terms that are linear in its second two arguments. Moreover, since the trace of the left-hand side of (8.66) is zero, so too must be that of the right-hand side, and F_{ij} must be chosen accordingly.

If (8.66) is to represent a physical law, its form should be invariant under a rotation of the coordinate system. This means that for an arbitrary rotation tensor Q,

$$QF\left(\mathbf{b}, \frac{K}{\epsilon}S, \frac{K}{\epsilon}W\right)Q^t = F\left(Q\mathbf{b}Q^t, \frac{K}{\epsilon}QSQ^t, \frac{K}{\epsilon}QWQ^t\right) \tag{8.67}$$

(i.e., the form of F_{ij} seen in the rotated coordinates, given on the left-hand side, should be no different than having evaluated F using the tensor forms seen by the rotated observer). The most general form of Π_{ij}, nonlinear in **b**, which satisfies (8.67) is [60,62]

$$\Pi_{ij} = A_0\epsilon b_{ij} + A_1\epsilon\left(b_{ik}b_{kj} - \tfrac{1}{3}\mathrm{II}\delta_{ij}\right) + A_2 K\overline{S}_{ij} + (A_3 b_{kl}\overline{S}_{lk}$$

$$+ A_4 b_{kl}b_{lm}\overline{S}_{mk})Kb_{ij} + (A_5 b_{kl}\overline{S}_{lk} + A_6 b_{kl}b_{lm}\overline{S}_{mk})K\left(b_{ik}b_{kj} - \tfrac{1}{3}\mathrm{II}\delta_{ij}\right)$$

$$+ A_7 K(b_{ik}\overline{S}_{jk} + b_{jk}\overline{S}_{ik} - \tfrac{2}{3}b_{kl}\overline{S}_{lk}\delta_{ij}) + A_8 K(b_{ik}b_{kl}\overline{S}_{jl} + b_{jk}b_{kl}\overline{S}_{il} \tag{8.68}$$

$$- \tfrac{2}{3}b_{kl}b_{lm}\overline{S}_{mk}\delta_{ij}) + A_9 K(b_{ik}\overline{W}_{jk} + b_{jk}\overline{W}_{ik})$$

$$+ A_{10}K(b_{ik}b_{kl}\overline{W}_{jl} + b_{jk}b_{kl}\overline{W}_{il}).$$

Note

$$\mathrm{III} = b_{ik}b_{kl}b_{li}, \tag{8.69}$$

together with (8.59), are two of the three scalar invariants of the tensor **b**. The third invariant is b_{ii}, which is identically zero in the present case. The coefficients $A_0, \ldots, A_{10}$ are most generally functions of II and III. As it is, (8.68) is a formidable expression whose use in closure schemes is not practical. Besides the numerical expense of evaluating so many terms, the nonlinearity in **b** can lead to numerical instability. Furthermore, there is no clearcut means for determining the large number of empirical coefficients. The main value of (8.68) is in providing a backdrop on which further assumptions can be made, leading to more practical models.

Some insight in this direction can be had by considering the case of an initially homogeneous anisotropic turbulent flow—created, for example, by the action of a given mean shear—when the shear is suddenly removed. As time proceeds, the turbulence relaxes back to an isotropic state. This is referred to as the *return-to-isotropy problem*. In this flow the terms in (8.68) depending on $\overline{S}$ and $\overline{W}$ are zero, and the remaining terms may be identified with that part of (8.68) appropriate to the slow term in (8.65). It is important to note that the opportunity provided by Reynolds stress equation models to accommodate the return to isotropy is not shared by the constitutive laws considered in Section 8.6, including (8.57) and (8.62). In fact, the latter have the unphysical property that as soon as the mean shear is removed, R_{ij} is predicted to be isotropic, so the natural relaxation toward isotropy cannot be accounted for in such models. The ARMs discussed in Section 8.8 will be seen to be similarly limited.

To summarize, the model for the slow term contained in (8.68) is

$$\Pi_{ij}^S = A_0 \epsilon b_{ij} + A_1 \epsilon \left(b_{ik} b_{kj} - \tfrac{1}{3} \mathrm{II} \delta_{ij} \right). \tag{8.70}$$

With $A_1 \neq 0$, this is a nonlinear model in the isotropy tensor **b**, whereas if $A_1 = 0$, it is linear. The earliest models of Π_{ij}^S were strictly linear, as in the model proposed by Rotta [45], which has

$$\Pi_{ij}^S = -C_1 \epsilon b_{ij}. \tag{8.71}$$

Here $C_1 > 0$ is referred to as the *Rotta constant*. If (8.71) is used in (8.63), the computed solutions display a power law decay in b_{ij} at the same rate for all components when the mean shear is removed. Comparisons with experiment suggest that within the limitations of this model, $C_1 \approx 3$ is an optimal value. To include the likely possibility that different components of b_{ij} decay at different rates, it is necessary to include the nonlinear term in (8.70).

An opposite situation to the one just considered occurs when turbulence is initially isotropic and then a mean shear is imposed. This is an example of the limiting flow regime where RDT is applicable. At the initial instant in this case, **b** is very small or zero, and it may be assumed that the only contribution to Π_{ij} comes via the fast term in (8.65). Using the identities

$$M_{iikl} = M_{ijkk} = 0 \tag{8.72}$$

and

$$M_{ikkj} = M_{ikjk} = 2R_{ij}, \tag{8.73}$$

which follow from the definition (2.59), it may be proven [11] that

$$\Pi_{ij}^F \sim \tfrac{4}{5} K \overline{S}_{ij} \tag{8.74}$$

in these circumstances. Thus (8.68) is compatible with (8.74) as long as $A_2 = \tfrac{4}{5}$. Equation (8.74) is referred to as the *Crow constraint*.

One of the first complete pressure–strain models including both fast and slow contributions is that due to Launder, Reece, and Rodi (LRR) [31], in which (8.68) is simplified by assuming linearity in **b**. The rationale for this is grounded in the fact that the eigenvalues of **b**, say, b^1, b^2, and b^3, are bounded according to

$$-\tfrac{1}{3} \leq b^i \leq \tfrac{2}{3}, \qquad i = 1, 2, 3. \tag{8.75}$$

The lower limit is derived by noting that the eigenvalues of the real-symmetric matrix R_{ij} must be positive, while the upper limit follows from an application of the Schwarz inequality to R_{ij}. In fact, many experiments show that the largest of the three eigenvalues has magnitude less than 0.25. This implies a similar bound on the magnitude (i.e., norm) of **b**. Thus it is reasonable to imagine that if a Taylor series of (8.68) were made in powers of **b**, it might be legitimate to keep just the first few terms in the expansion. For the LRR model, all nonlinear terms are dropped. In later models, just a few quadratic or cubic terms are kept.

With the assumption of linearity and the need to satisfy (8.72) and (8.73), the following model is produced:

$$\Pi_{ij} = - C_1 \epsilon b_{ij} + \tfrac{4}{5} K \overline{S}_{ij} + A_7 K \left(b_{ik} \overline{S}_{jk} + b_{jk} \overline{S}_{ik} - \tfrac{2}{3} b_{kl} \overline{S}_{lk} \delta_{ij} \right) \\ + A_9 K \left(b_{ik} \overline{W}_{jk} + b_{jk} \overline{W}_{ik} \right), \tag{8.76}$$

where

$$A_7 = \frac{18 C_2 + 12}{11} \tag{8.77}$$

and

$$A_9 = \frac{20 - 14 C_2}{11}. \tag{8.78}$$

Thus the LRR model depends on values of just two constants, C_1 and C_2. These are usually taken to be 3 and 0.4, respectively, based on calibration of the model with experimental results for homogeneous turbulence. A calculation then gives $A_7 = 1.75$ and $A_9 = 1.31$.

A simplification of the LRR model given by

$$\Pi_{ij} = -C_1 \epsilon b_{ij} + \tfrac{4}{5} K \overline{S}_{ij} \tag{8.79}$$

is known as the *isotropization of production* (IP) *model* [17,37]. It is the minimal linear model compatible with (8.71) and (8.74), and it finds considerable use today.

Speziale, Sarkar, and Gatski [61] approached the problem of developing a practical pressure–strain model out of (8.68) by studying the form that this relation should have under equilibrium conditions of homogeneous shear flow with a two-dimensional mean field, and then allowing for some small departures from equilibrium. Here, *equilibrium* is meant in the same sense as it was used in Section 7.8, namely, that b_{ij} and related statistics have converged to asymptotic constant values $(b_{ij})_\infty$ as $t \to \infty$. This analysis shows that at equilibrium, the terms depending on A_8 and A_{10} in (8.68) can be expressed as linear combinations of the remaining terms, so they do not need to appear explicitly. The resulting model, which technically applies only under the restriction of two-dimensional mean fields, is known as the SSG model and has the form

$$\begin{aligned}
\Pi_{ij} = &- (C_1 \epsilon + C_1^* \mathcal{P}) b_{ij} + C_2 \epsilon \left(b_{ik} b_{kj} - \tfrac{1}{3} \mathrm{II} \delta_{ij} \right) \\
&+ \left[C_3 - C_3^* (\mathrm{II})^{1/2} \right] K \overline{S}_{ij} \\
&+ C_4 K \left(b_{ik} \overline{S}_{jk} + b_{jk} \overline{S}_{ik} - \tfrac{2}{3} b_{kl} \overline{S}_{lk} \delta_{ij} \right) \\
&+ C_5 K (b_{ik} \overline{W}_{jk} + b_{jk} \overline{W}_{ik}),
\end{aligned} \tag{8.80}$$

which, it may be noticed, is consistent with the Crow constraint when $C_3 = \tfrac{4}{5}$. Unlike the LRR model, which can be extracted as a special case of this expression, (8.80) keeps the nonlinear part of the slow term (i.e., the term depending on C_2). When $C_1^* = C_3^* = 0$, (8.80) is identical to the form that (8.68) takes for plane mean flows under equilibrium conditions. Unlike the LRR model there is no need to constrain C_4 and C_5 according to (8.72) and (8.73). This is a consequence of relating the nonlinear terms to the linear terms at equilibrium [60].

The terms depending on C_1^* and C_3^* enable (8.80) to take into account some departures from equilibrium. Their appearance is consistent with the fact that the coefficients in (8.80) are allowed to be functions of II and III. Their mathematical forms are designed to bring the respective terms up to a **b** dependence matching that of the other terms. In the case of C_1^*, it is to match the quadratic term with coefficient C_2 (note that $\mathcal{P} b_{ij}$ is quadratic in **b** since $\mathcal{P}$ contains a linear term in **b**). For C_3^* it is to add a linear **b** dependence matching the other part of the fast model.

It is instructive to consider the process by which the SSG model is calibrated (i.e., how its constants are selected). This entails first finding C_1 and C_2 by applying SSG to the return-to-isotropy problem, where all other terms in (8.80) are identically zero. In these circumstances, (8.20) and (8.63) may be manipulated [47] to create coupled dynamical equations for the invariants II and III, namely,

$$\frac{d\text{II}}{d\tau} = -2[(C_1 - 2)\text{II} - C_2\text{III}] \tag{8.81}$$

$$\frac{d\text{III}}{d\tau} = -3\left[(C_1 - 2)\text{III} - C_2\frac{\text{II}^2}{6}\right], \tag{8.82}$$

where τ is the same time coordinate as appeared in (7.57). Examination of (8.81) and (8.82) makes clear that the condition $C_1 > 2$ is necessary to ensure that solutions of the system of equations do not become unbounded. In addition, the long-time solutions of these equations, no matter what the initial state, should be a state of isotropy wherein $b_{ij} = 0$, and so too are II and III. It may be shown that this requires $C_2 \leq 3(C_1 - 2)$. It is also interesting to note that this same condition enforces realizability to the effect that the normal Reynolds stresses remain positive during the decay. Comparing predictions of (8.80) with experimental data suggests that the best agreement for any given value of C_1 occurs when C_2 has its maximum allowable value of $3(C_1 - 2)$. The choice of $C_1 = 3.4$ appears to be the best overall, in which case $C_2 = 4.2$.

The success of this part of the calibration of the SSG model is illustrated in Fig. 8.2, wherein the normalized invariant $\text{II}^{1/2}$ is plotted versus $\text{III}^{1/3}$ during the return to isotropy process. A curved trajectory matching the data is achieved with the SSG model. This is attributable to keeping a quadratic form for the slow term and hence different decay rates for different components of **b**. This is to be contrasted with LRR and other linear models, which, as shown in the figure, predict an unphysical straight-line return to isotropy.

Despite the advantages of the quadratic model evident in Fig. 8.2, in practice these terms often make only a small contribution to the pressure–strain term and can be dropped without significant penalty. In fact, the advantages of the nonlinear form are often not sufficiently significant to warrant the extra expense of including them. For example, Fig. 8.3 demonstrates that despite the situation in Fig. 8.2, when the prediction of the Reynolds stress components is considered, both the LRR and SSG models have the same degree of success in modeling the return to isotropy, although the latter model is slightly more accurate.

The remaining constants in the SSG model are selected by optimizing the prediction of homogeneous shear flow [64] and RDT results [6] for rotating shear flow. This yields $C_1^* = 1.8$, $C_3^* = 1.3$, $C_4 = 1.25$, and $C_5 = 0.4$. These values represent a departure from those in the LRR and IP models and reflect the belief that the performance of SMCs can be enhanced not by keeping additional terms in (8.68), but rather, by taking more care in selecting the model constants. In particular, the idea is to take better advantage of the fact that the coefficients of the tensor quantities in (8.68) are allowed to depend on the invariants of the anisotropy tensor.

As a further step in this direction [19], the coefficients of the terms in (8.68) were calibrated over a wider range of equilibrium flows than had been attempted in calibrating SSG. In this, the parameter

$$\eta_1 \equiv \frac{\overline{S}_{ij}\overline{S}_{ij}}{\eta} \tag{8.83}$$

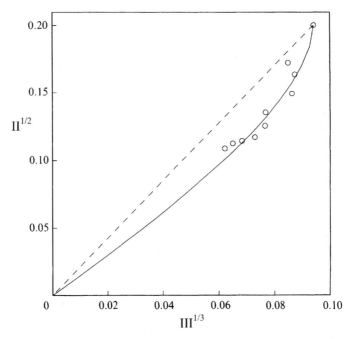

Fig. 8.2 *Prediction of return to isotropy in plane strain. —, SSG model ; — —, LRR model [31]; ○, experiments [9]. (From [61]. Reprinted with the permission of Cambridge University Press.)*

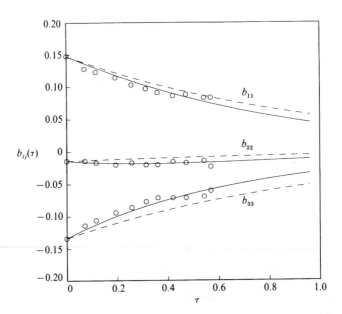

Fig. 8.3 *Normal components of* **b** *during return to isotropy. —, SSG model; — —, LRR model [31]; ○, experiments [9]. (From [61]. Reprinted with the permission of Cambridge University Press.)*

is introduced, where $\eta \equiv \overline{S}_{ij}\overline{S}_{ij} + \overline{W}_{ij}\overline{W}_{ij}$. $\eta_1 = 0$ corresponds to a rotational flow with no mean strain, while the opposite extreme occurs with $\eta_1 = 1$ (i.e., the flow has strain but no mean rotational component). The class of flows for which $\eta_1 < 0.5$ are referred to as *elliptical flows* since rotation dominates strain and the streamlines are elliptical (i.e., reveal the presence of rotational motion).

For any closure based on a particular pressure–strain model, the fundamental properties of solutions to the evolving homogeneous shear flow can be determined as a function of η_1. This entails taking a dynamical systems view of the nonlinear equations in which the dependence on η_1 of the fixed and bifurcation points of the closure scheme are determined. Whatever basic properties are discovered about the closure equations may be judged against what is known to occur for the physical system. This then suggests a strategy that may be used to improve the physical relevance of the parameters in the pressure–strain model. In particular, the idea is to force the model to take on characteristics known to be true about the physical system.

To be specific, it is known from the physical system that as long as η_1 is not exactly zero (i.e., some strain exists), eventually there will be growth in turbulent kinetic energy such as occurs in strain-dominant flows. A dynamical systems analysis of the SSG model, for example, reveals that it conflicts with this property. In fact, the value of η_1 separating asymptotic growth and nongrowth in the SSG model is not at $\eta_1 = 0$. This is partly due to the fact that the SSG model has been calibrated strictly after examining experiments for which $\eta_1 > 0.5$, and the constants appearing in it are such as to represent this class of flows successfully. A technical fix to this problem involves forcing the bifurcation point separating growth and nongrowth in the SSG solution to the point $\eta_1 = 0$, so that flows without finite strain will decay while those with finite strain will eventually grow in turbulent energy.

As a result of this analysis, a new choice for C_5 has been suggested [19], namely,

$$C_5 = \begin{cases} 0.4 & \eta_1 \geq 0.5 \\ 2 - 1.6\left(\dfrac{\eta_1}{1 - \eta_1}\right)^{3/4} & \eta_1 < 0.5. \end{cases} \tag{8.84}$$

In this, for strain-dominated flows $C_5 = 0.4$, the value traditionally used in the SSG model. A continuous change of C_5 to the value of 2 occurs as η_1 decreases to zero.

Another instance where improvements in the SSG calibration have been achieved center on the coefficient $C_3 - C_3^*(\text{II})^{1/2}$ in (8.80). With $C_3 = \frac{4}{5}$, it is designed specifically to satisfy the Crow constraint far from equilibrium when b_{ij} is small, under an imposed shear, yet also applicable to flows near equilibrium. The SSG closure takes $C_3^* = 0.36$ as an optimal value based on predictions in the equilibrium range. A reexamination of the determination of this coefficient over a wider range of conditions has led to better means for adapting it to the RDT limit at one extreme and the equilibrium condition at the other. This makes use of the parameter

$$\omega_1 \equiv \frac{\epsilon}{\sqrt{\eta}K}, \tag{8.85}$$

which is a generalized form of the ratio of time scales previously considered in (8.54). As is evident from the discussion surrounding (8.54), $\omega_1 \to 0$ in the RDT limit. For equilibrium conditions it converges to a value, say, ω_∞. Then an improved determination of C_3 consists of

$$C_3 = \begin{cases} 0.36 & \dfrac{\omega_1}{\omega_\infty} \geq 1 \\ \dfrac{4}{5} - 0.44 \left(\dfrac{\omega_1}{\omega_\infty} \right)^{1/4} & \dfrac{\omega_1}{\omega_\infty} < 1. \end{cases} \tag{8.86}$$

Thus at equilibrium, $C_3 = 0.36$, while in the RDT limit, $C_3 = 0.8$. Equations (8.84) and (8.86) constitute the Girimaji model.

The benefit of (8.84) and (8.86) is illustrated in Fig. 8.4, comparing the prediction of K in elliptic flow as computed by the SSG and Girimaji models. This particular flow has $\eta_1 = 0.26$, which is in the elliptic range where the traditional SSG model, as confirmed by the figure, is known to misjudge the proper growth of kinetic energy. The calculation with the updated model coefficients captures the trend in K as predicted via DNS [7].

A number of models have been developed that commit to using the nonlinear terms contained in (8.68). Among these are the models of Shih and Lumley [48] and Fu et al. [15], which have been built around the idea of satisfying the realizability condition. A difficult aspect of the non-linear terms is that, even if they provide greater flexibility in modeling equilibrium states of turbulence, at the same time they can promote unphysical responses of the system to perturbations that are difficult to control. As a result of these and other problems, nonlinear models tend not to find widespread engineering use today.

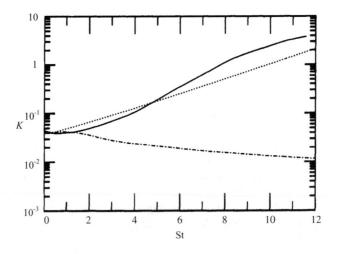

Fig. 8.4 *Prediction of K in elliptic flow with $\eta_1 = 0.26$. —, DNS [7]; $\cdots$, Girimaji model $- \cdot -$, SSG model. (From [19]. Reprinted with the permission of Cambridge University Press.)*

8.7.2 Transport Correlation

The turbulent transport correlation β_{ijk} is the sum of the Reynolds stress transport term, $\overline{u_i u_j u_k}$, and the pressure–velocity correlation, $\overline{pu_i}\delta_{jk}$. The former is symmetric in the sense that its value is unchanged by any reordering of the indices. The pressure–velocity correlation, on the other hand, is not symmetric. This means that β_{ijk} can be thought of as symmetric only if the neglect of $\overline{pu_i}$ can be justified. DNS measurements in a channel flow [36], for example, show that there is some justification for taking β_{ijk} to be symmetric away from boundaries, even if the pressure term is not exactly zero. The foregoing observations help justify the common practice of treating β_{ijk} as if it were exactly symmetric and thus modeling it in a similar fashion. Among the popular approaches adopting this point of view is that due to Hanjalic and Launder [23] which assumes that

$$\beta_{ijk} = -C_s \frac{K}{\epsilon} \left(R_{im} \frac{\partial R_{jk}}{\partial x_m} + R_{jm} \frac{\partial R_{ik}}{\partial x_m} + R_{km} \frac{\partial R_{ij}}{\partial x_m} \right). \tag{8.87}$$

This model is often used with the LRR and SSG closures. An alternative expression is

$$\beta_{ijk} = -C_s \frac{K^2}{\epsilon} \left(\frac{\partial R_{jk}}{\partial x_i} + \frac{\partial R_{ik}}{\partial x_j} + \frac{\partial R_{ij}}{\partial x_k} \right), \tag{8.88}$$

which follows from (8.87) after invoking an isotropy assumption. In both these models it is usually assumed that $C_s = 0.11$. Another approach, often used with the IP model, is the nonsymmetric model due to Daly and Harlow [12], namely,

$$\beta_{ijk} = -C_s \frac{K}{\epsilon} R_{km} \frac{\partial R_{ij}}{\partial x_m}, \tag{8.89}$$

where $C_s = 0.22$.

In view of the overall intractability of the correlations in β_{ijk}, there has been little progress to date in developing alternative forms of the transport correlation which better agree with DNS or experimental results. However, it is also the case that it is not generally perceived that finding better transport models should be a high priority compared to more pressing issues, such as better accommodating the pressure–strain term.

An interesting aspect of the three transport models given here is that if they are each made the basis for a K equation model [e.g., by contracting indices and substituting them into (8.63)], the K transport models they yield differ from the expression (8.23) incorporated in (8.24).

8.7.3 Complete Second-Moment Closure

Second-moment closures require the solution of the RANS and continuity equations (2.10) and (2.11), respectively, for $\overline{U}_i$ and $\overline{P}$, the dissipation equation (8.35) for ϵ and a closed form of (8.63) for R_{ij}. To give concrete expression to the modeling used in the latter, the closed Reynolds stress equation that is shared by both the LRR and SSG models is given in detail. In particular, after substituting (8.64), (8.80), and (8.87) into (8.63), it is found that

$$
\frac{\partial R_{ij}}{\partial t} + \overline{U}_k \frac{\partial R_{ij}}{\partial x_k} = -R_{ik}\frac{\partial \overline{U}_j}{\partial x_k} - R_{jk}\frac{\partial \overline{U}_i}{\partial x_k} - \frac{2}{3}\epsilon\delta_{ij}
$$

$$
+ \frac{\partial}{\partial x_k}\left[C_s \frac{K}{\epsilon}\left(R_{im}\frac{\partial R_{jk}}{\partial x_m} + R_{jm}\frac{\partial R_{ik}}{\partial x_m} + R_{km}\frac{\partial R_{ij}}{\partial x_m} \right) \right]
$$

$$
- (C_1\epsilon + C_1^*\mathcal{P})b_{ij} + C_2\epsilon\left(b_{ik}b_{kj} - \frac{1}{3}\mathrm{II}\delta_{ij} \right) + \left[\frac{4}{5} - C_3^*(\mathrm{II})^{1/2} \right] K\overline{S}_{ij} \qquad (8.90)
$$

$$
+ C_4 K\left(b_{ik}\overline{S}_{jk} + b_{jk}\overline{S}_{ik} - \frac{2}{3}b_{kl}\overline{S}_{lk}\delta_{ij} \right)
$$

$$
+ C_5 K(b_{ik}\overline{W}_{jk} + b_{jk}\overline{W}_{ik}) + \nu\,\nabla^2 R_{ij}.
$$

Despite the restriction to two-dimensional mean flows in the development and calibration of (8.90), SMC of this type has been and continues to be applied to a wide range of three-dimensional flows. For the most complex situations, this class of models involves the solution of as many as 11 coupled nonlinear partial differential equations. Such systems can be expensive to solve, even in the present era of fast computers, and, moreover, can be difficult to solve numerically. For example, the Reynolds stresses appear in differentiated form in the mean velocity equation, and depending on how they are predicted, there is a potentiality for numerical instability. In the case of (8.1), this is generally not a problem since the structure of (8.2) in this case is not unlike that of the Navier–Stokes equation and can be solved similarly. These limitations have been a motivating force in attempts to develop methods that have the same perceived benefits as SMCs insofar as modeling the physics of turbulence is concerned, but without the associated cost. These are the ARMs considered in the next section.

8.8   ALGEBRAIC REYNOLDS STRESS MODELS

The goal of algebraic Reynolds stress models is to harness much of the generality of SMCs, such as their capability for predicting flow anisotropy, in a more computationally affordable context. What is desired is an algebraic relationship between the Reynolds stresses and mean velocity gradients, in effect a constitutive law, that improves upon the ideas discussed in Section 8.6. For the most part, ARMs as they are most often encountered represent a line of reasoning that began with the work of Rodi [44] and has been carried through several generations of development [16,18,40,63]. The basis for ARMs is the hypothesis that in a small local flow region convecting with the mean velocity, the anisotropy of turbulence, as reflected in the anisotropy tensor b_{ij} defined in (8.31), will often be mostly a function of the local interaction between turbulence production, dissipation, and pressure–strain terms. In other words, it is imagined that an equilibrium condition exists wherein spatial and time gradients of the Reynolds stress are of secondary importance to other physical processes in setting the Reynolds stress. Such a viewpoint is met exactly in the equilibrium condition of a homogeneous shear flow after sufficient time has developed. In fact, as was noted previously, b_{ij} attains an asymptotic form under these conditions. The goal of ARMs, then, is to use the equilibrium concept to develop a constitutive model of the Reynolds stress.

ARMs are derived by first deciding on a form of the pressure–strain correlation. Here, following [18], the class of quasi-linear models of the type (8.80) are considered with the restriction that $C_2 = C_3^* = 0$. The term with coefficient C_2 is omitted since its influence is relatively minor, as noted previously. The rational for discarding C_3^* is that in equilibrium its associated term becomes absorbed in C_3 and need not be considered separately.

When equilibrium is achieved,

$$\frac{db_{ij}}{dt} = 0, \tag{8.91}$$

from which it follows from (8.31) that

$$K\frac{dR_{ij}}{dt} - R_{ij}\frac{dK}{dt} = 0. \tag{8.92}$$

An expression for dK/dt is given in (8.24), while dR_{ij}/dt is given from (8.90). It is at this point that the assumed pressure-strain model enters the analysis. With these expressions, (8.92) becomes, after dropping the transport and diffusion terms, and using (8.20):

$$b_{ij}\left[C_1 - 2 - 2(C_1^* + 2)b_{mn}\frac{K}{\epsilon}\overline{S}_{mn}\right]$$
$$= \left(C_3 - \frac{4}{3}\right)\frac{K}{\epsilon}\overline{S}_{ij} + (C_4 - 2)\frac{K}{\epsilon}\left(b_{ik}\overline{S}_{jk} + b_{jk}\overline{S}_{ik} - \frac{2}{3}b_{mn}\overline{S}_{mn}\delta_{ij}\right) \tag{8.93}$$
$$+ (C_5 - 2)\frac{K}{\epsilon}(b_{ik}\overline{W}_{kj} + b_{jk}\overline{W}_{ki}).$$

This is a nonlinear system of equations out of which a solution for the tensor b_{ij} is to be found, and from this, R_{ij} can be computed from (8.31). If the restriction to two-dimensional mean velocity fields is imposed, a number of simplifying identities are available which result in the opportunity to solve (8.93) exactly.

It is convenient in solving (8.93) first to simplify the notation through the definitions

$$L_1 \equiv \tfrac{1}{2}C_1 - 1, \qquad L_1^* \equiv C_1^* + 2, \qquad L_2 \equiv \tfrac{1}{2}C_3 - \tfrac{2}{3},$$
$$L_3 \equiv C_4 - 1, \qquad L_4 \equiv \tfrac{1}{2}C_5 - 1,$$

in which case (8.93) becomes

$$b_{ij}\left(L_1 - L_1^* b_{mn} \overline{S}_{mn}^*\right) = L_2 \overline{S}_{ij}^* + L_3 \left(b_{ik}\overline{S}_{jk}^* + b_{jk}\overline{S}_{ik}^* - \frac{2}{3}b_{mn}\overline{S}_{mn}^*\delta_{ij}\right)$$

$$+ L_4 \left(b_{ik}\overline{W}_{kj}^* + b_{jk}\overline{W}_{ki}^*\right). \tag{8.94}$$

Here $\overline{S}_{ij}^* \equiv (K/\epsilon)\overline{S}_{ij}$ and $\overline{W}_{ij}^* \equiv (K/\epsilon)\overline{W}_{ij}$ are dimensionless quantities. The desired solution for b_{ij} will depend on the tensors $\overline{S}_{ij}^*$ and $\overline{W}_{ij}^*$. All such relations between tensor quantities must be consistent with the transformation properties of tensors between coordinate systems. In fact, it can be shown formally [16] that for a general three-dimensional mean flow, b_{ij} is limited to being a linear combination of at most 10 particular tensorial forms built up out of $\overline{S}_{ij}^*$ and $\overline{W}_{ij}^*$. These are referred to as the *integrity basis* [67]. If all such terms are included, the resulting model will be too cumbersome to be practical. However, if it is assumed that the mean field is two-dimensional, then just three of the integrity bases are linearly independent of one another, and they can be combined to create the model

$$b_{ij} = G_1 \overline{S}_{ij}^* + G_2(\overline{S}_{ik}^*\overline{W}_{kj}^* - \overline{W}_{ik}^*\overline{S}_{kj}^*) + G_3 \left(\overline{S}_{ik}^*\overline{S}_{kj}^* - \frac{1}{3}\overline{S}_{mn}^*\overline{S}_{mn}^*\delta_{ij}\right), \tag{8.95}$$

where the coefficients G_1, G_2, and G_3 are functions of the quantities L_1, L_1^*, L_3, L_4, and L_5 as well as the scalar invariants of the tensors $\overline{S}_{ij}^*$ and $\overline{W}_{ij}^*$. In the present case the latter are $\eta_1^* \equiv \overline{S}_{ij}^*\overline{S}_{ij}^*$ and $\eta_2^* \equiv \overline{W}_{ij}^*\overline{W}_{ij}^*$.

Substitution of (8.95) into (8.94) and grouping together terms having the same tensor factors yields a coupled system of equations for the three unknowns G_1, G_2, and G_3. A further calculation shows that G_2 and G_3 can be given in terms of G_1 by the relations

$$G_2 = \frac{-L_4 G_1}{L_1 - \eta_1^* L_1^* G_1} \tag{8.96}$$

and

$$G_3 = \frac{2L_3 G_1}{L_1 - \eta_1^* L_1^* G_1}. \tag{8.97}$$

Finally, it may be shown that a cubic equation is derived for G_1 in the form

$$\left(\eta_1^* L_1^*\right)^2 G_1^3 - \left(2\eta_1^* L_1 L_1^*\right) G_1^2$$

$$+ \left(L_1^2 + \eta_1^* L_1^* L_2 - \frac{2}{3}\eta_1^* L_3^2 + 2\eta_2^* L_4^2\right) G_1 - L_1 L_2 = 0. \tag{8.98}$$

Fortunately, cubic equations are exactly solvable, so that an exact solution to (8.98) is achievable. Since cubic equations have three solutions, some care must be taken to make sure that only the "physically correct" root is kept. Depending on circumstances, the solutions can consist of either a single real root for G_1 with two complex roots, or else, three real roots. The first possibility makes identifying the physical root simple. The second case makes for a nontrivial consideration of the mathematical and physical properties of the roots. In essence, the physical root is identified by insisting on continuity of the roots as a change in parameters causes a change from one to three real roots. Moreover, in some cases the selection is based on requiring that the predicted production term in the kinetic energy equation, which depends directly on G_1, not always be negative. How these conditions are implemented is beyond the scope of this treatment, but the final result is that

$$
G_1 = \begin{cases}
\dfrac{L_1 L_2}{L_1^2 + 2\eta_2 L_4^2} & \text{if} \quad \eta_1^* = 0; \\[3ex]
\dfrac{L_1 L_2}{L_1^2 - \frac{2}{3}\eta_1^* L_3^2 + 2\eta_2^* L_4^2} & \text{if} \quad L_1^* = 0; \\[3ex]
-\dfrac{p}{3} + \sqrt[3]{-\dfrac{b}{2} + \sqrt{D}} + \sqrt[3]{-\dfrac{b}{2} - \sqrt{D}} & \text{if} \quad D > 0; \\[3ex]
-\dfrac{p}{3} + 2\sqrt{\dfrac{-a}{3}} \cos\dfrac{\theta}{3} & \text{if} \quad D < 0, b < 0; \\[3ex]
-\dfrac{p}{3} + 2\sqrt{\dfrac{-a}{3}} \cos\left(\dfrac{\theta}{3} + \dfrac{2\pi}{3}\right) & \text{if} \quad D < 0, b > 0,
\end{cases}
\tag{8.99}
$$

where $p \equiv -2L_1/\eta_1^* L_1^*$, $r \equiv -2L_1 L_2/(\eta_1^* L_1^*)^2$, $q \equiv [L_1^2 + \eta_1^*(L_1^* L_2 - 2/3L_3^2) + 2\eta_2^* L_4^2]/(\eta_1^* L_1^*)^2$, $a \equiv q - p^2/3$, $b \equiv (2p^3 - 9pq + 27r)/27$, $D \equiv b^2/4 + a^3/27$, and $\cos\theta \equiv -b/2/\sqrt{-a^3/27}$. Note that $D < 0 \Rightarrow a < 0$. In view of the fact that (8.95) incorporating (8.96), (8.97), and (8.99) is an explicit solution of (8.94), this model is referred to as an *explicit algebraic Reynolds stress model* (EARSM).

Although ARMs are generally derived under equilibrium assumptions, in practice they are applied to flows of all types, including those that are far from equilibrium. It has been speculated that this can lead to the appearance of singularities in the computed solutions, particularly for ARMs that are based on approximate solutions to the equilibrium system (8.94) [16]. Evidently, the approach culminating in the exact solution in (8.95) is not subject to the same potential numerical instabilities. This may represent a considerable advantage, apart from any gains in accuracy.

The fact that (8.95) is limited to quadratic terms in the rate of strain and rotation tensors is a result of the limitation of the formal analysis to two-dimensional mean flows. To formally include three-dimensional mean fields by this approach requires many more terms of great complexity. A simpler approach is to add cubic and quartic terms as is done for NLEVMs, (e.g., [1]). At this point, however, for all practical purposes the ARM methodology is equivalent to that of NLEVM.

8.9 NEAR-WALL MODELS

For the most part, the modeling discussed in previous sections is concerned with high-Reynolds-number conditions and so cannot be expected to apply next to boundaries where the turbulence magnitude goes to zero and the effective turbulence Reynolds number is small. Modifications have to be made to closure schemes to accommodate wall conditions, much the same way that empirical formulas for the mixing length had to have different forms at different distances from boundaries. Unfortunately, it is difficult to model the subtle trends of the many terms appearing in the K, ϵ, and R_{ij} equations produced by the presence of solid boundaries. Many of the near-wall models that have been proposed are derived using empirical results and thus are relevant to only a narrow range of flows. There is no established best model for near-wall flows, and there continues to be considerable interest in developing new and better models.

If the wall region is treated by solving suitably modified RANS equations, the natural, physical boundary conditions for the unknown mean fields can be expected to apply. An alternative strategy, which largely predates current efforts at deriving near-wall closures, is the practice of applying boundary conditions outside the immediate wall vicinity so as to avoid solving closure equations next to the wall. This method depends on the use of *wall functions*, which are specially formulated relations allowing one to estimate the mean velocity and Reynolds stresses close to a wall, most often within the log-law region, under a range of conditions. Wall functions serve as boundary conditions for numerical solutions to the closure equations at the first mesh point off the wall, which is purposefully sited to be where the wall functions are legitimate. Under some circumstances, most notably attached boundary layers under favorable pressure gradients, it is possible to devise useful wall functions based on theoretical results about flow under these conditions (e.g., by using the log law). In more general flow conditions, physically appropriate wall functions are difficult or impossible to devise, so the present trend is away from such heavily empirical formulations and toward the development of all-purpose near-wall RANS models. The latter will be considered here, after first presenting some of the typical ideas that go into the making of wall functions.

8.9.1 Wall Functions

There are a number of different means for numerically implementing boundary conditions in the form of wall functions. For a simple flat plate boundary layer, assuming that the first grid point is placed in the log region, a wall function may give the mean velocity by using the law of the wall, (4.43). A technique for getting wall function boundary conditions to the K, ϵ, and other equations is to invoke the constant-stress hypothesis as it was used in calibrating the K–ϵ closure in Section 8.4. For example, a relation for ϵ is readily obtainable from (8.38) and (8.40) in the form

$$\epsilon_2 = \frac{C_\mu^{3/4} K_2^{3/2}}{\kappa y_2}, \tag{8.100}$$

where the subscript "2" denotes the first mesh point off the wall. The value of K_2 is whatever exists on the mesh at this time and location. Since it appears in the boundary condition for ϵ, an independent K boundary condition is required. A common approach is to use the natural boundary conditions $K = 0$ or $\partial K / \partial y = 0$ at the surface while modeling the turbulence production term near the wall using the constant-stress region assumptions.

The chief difficulties of the wall function approach occur in complex situations where flow separation and other complicating factors are present. In such circumstances the simplifying assumptions upon which the wall functions are based are not applicable, so it can be considered at best fortuitous if closure predictions yield acceptable results. More often than not, results will be seriously in error. To have any hope of accurately representing general flows through wall function methods, the wall functions must be designed to cover a wide range of flow conditions. For example, they should take into account such information as the local mean pressure gradient. Relations of this type that have been developed are highly empirical and fraught with problems when applied to practical flows.

Apart from considerations of accuracy, wall functions provide a means for avoiding the potentially large numerical expense of resolving the fine-grained variations in turbulence quantities near boundaries. In and of itself this has been a significant reason for their past popularity in RANS modeling. It is also an incentive for using wall functions as boundary conditions for large eddy simulations.

8.9.2 Near-Wall K–ϵ Models

Some insight into how the high-Reynolds-number K–ϵ closure ought to be modified near boundaries comes from examining the properties of the exact governing equations for $\overline{U}_i$, K, and ϵ in the region closest to a solid surface. In the case of the momentum equation, (8.2), the main concern is modeling the correct behavior of the Reynolds shear stress, $\overline{uv}$, near the wall. Considering unidirectional flow where (8.19) and (8.4) hold, it falls upon the choice of ν_t to guarantee that $\overline{uv}$ is well described. Some control of how $\overline{uv}$ behaves is achieved by introducing a function f_μ (also referred to as a wall function) into (8.19) so that it becomes

$$\nu_t = C_\mu f_\mu \frac{K^2}{\epsilon}, \tag{8.101}$$

where f_μ depends on y, K, or other variables. The primary purpose of f_μ is to bring (8.101) into line with what is known variously as the *wall blocking effect* or *viscous damping*, which is to say that transport normal to the wall, as in $\overline{uv}$, is inhibited by the boundary presence, so ν_t must be decreased accordingly. This idea was already present in the previous analysis in (6.22), which showed that the eddy viscosity behaves like $\mathcal{T}_{22}\overline{v^2}$. The factor $\overline{v^2}$, much more so than $\overline{u^2}$ or $\overline{w^2}$, reflects the damping effect of the wall. In practical terms this means that one of the factors, K, in (8.101) must be altered to more closely behave like $\overline{v^2}$. This is the role of f_μ. Note that in this analysis K/ϵ takes on the role of the time scale, $\mathcal{T}_{22}$. The general practice of using multiplicative functions, such as f_μ in (8.101), to force wall modification of high-Reynolds-number terms is referred to as viscous damping. This may be contrasted with more elaborate near-wall models, usually developed in the context of NLEVM, which incorporate specially designed near-wall terms that do not appear in the high-Reynolds-number equations.

In recognition of the importance of $\overline{v^2}$ in accounting for near-wall transport, there has been some interest in developing models that include a separate equation for $\overline{v^2}$. The hope is that this is simpler than trying to force a very different behavior from the K distribution. This point is returned to below.

At a wall $y = 0$, the K equation, (8.20), reduces to

$$\epsilon(0) = \nu \frac{\partial^2 K}{\partial y^2}(0), \tag{8.102}$$

a relation that also follows directly from the definitions of K and ϵ. This serves as a boundary condition for ϵ. It is not hard to show using Taylor series expansions of K and ϵ that (8.102) is equivalent to the relation

$$\lim_{y \to 0} \frac{\nu K}{\epsilon y^2} = \frac{1}{2}. \tag{8.103}$$

In practice, if a near-wall K–ϵ closure is to be considered asymptotically consistent with the physics of the near-wall region, the solutions it produces for K and ϵ should satisfy (8.103). For example, this means that the modeling should not create conditions where $\partial K(0)/\partial y \neq 0$. There is substantial evidence that satisfying asymptotic consistency in this and other ways has a beneficial effect on the accuracy of predictions. Some illustrations of this will be apparent in calculations shown in Chapter 9.

A Taylor series expansion of each of the terms in (8.20) in the vicinity of the surface reveals that the nonzero terms of $O(y)$ consist of the two terms in (8.102) joined by the pressure diffusion term. The contribution from the latter is not large, at least in a turbulent channel flow, as shown in Fig. 4.10. Consequently, at a minimum, near-wall models for the K equation should result in a balance of dissipation and viscous diffusion near the surface, with (8.102) forming a boundary condition for the ϵ equation. In fact, the usual modeled high-Reynolds-number K equation (8.24) satisfies (8.102), so there has not been a strong motivation for developing near-wall models for the production, transport, and pressure diffusion terms appearing in this equation.

It should also be noted that by virtue of the identity

$$\frac{\partial^2 K}{\partial y^2} = 2\sqrt{K}\frac{\partial^2 \sqrt{K}}{\partial y^2} + 2\left(\frac{\partial \sqrt{K}}{\partial y}\right)^2, \tag{8.104}$$

an alternative boundary condition for ϵ at $y = 0$ is

$$\epsilon(0) = 2v \left(\frac{\partial \sqrt{K}}{\partial y}(0) \right)^2. \tag{8.105}$$

In some instances this is used in place of (8.102). Some of the original near-wall versions of the K–ϵ closure do not use either (8.102) or (8.105) as boundary condition, preferring instead to use the pseudocondition $\epsilon = 0$. For such models the variable ϵ appearing in the ϵ equation needs to be reinterpreted as being a reduced dissipation [e.g., as in $\epsilon - 2v(\partial \sqrt{K}/\partial y)^2$]. If this is done, it is also necessary to add a term to the K equation to compensate for the fact that ϵ appearing in it will not be reproducing the physically correct amount of dissipation near the surface. For example, in the original Jones and Launder version [27] of the K–ϵ closure, the term $-2v(\partial \sqrt{K}/\partial y)^2$ is added to the K equation. Clearly, if this is done, the exact dissipation rate will be recovered at the surface.

In the case of the ϵ equation at solid surfaces, (8.25) reduces to

$$\Pi_\epsilon(0) + D_\epsilon(0) = \Upsilon_\epsilon(0), \tag{8.106}$$

with $D_\epsilon(0)$ and $\Upsilon_\epsilon(0)$ the same order of magnitude and much larger than $\Pi_\epsilon(0)$. This property was illustrated previously in the dissipation equation budget shown in Fig. 4.12. As it stands, (8.106) is not satisfied by the high-Reynolds-number model (8.35). In fact, Π_ϵ has heretofore not been modeled explicitly, since it was grouped with T_ϵ in (8.34) and made no contribution at the wall. Moreover, the high-Reynolds-number model for Υ_ϵ in (8.26) is unbounded at solid surfaces and clearly needs significant modification if it is to apply at walls. Low-Reynolds-number models for the ϵ equation are designed to overcome these and other difficulties. The traditional route is to incorporate several wall functions with specific goals in mind. Thus, the low-Reynolds-number ϵ equation is assumed to have the basic form

$$\frac{\partial \epsilon}{\partial t} + \overline{U}_j \frac{\partial \epsilon}{\partial x_j} = C_{\epsilon_1} f_1 \frac{\epsilon}{K} \mathcal{P}_K - C_{\epsilon_2} f_2 \frac{\epsilon \tilde{\epsilon}}{K} + \frac{\partial}{\partial x_i} \left[\left(v + \frac{v_t}{\sigma_\epsilon} \right) \frac{\partial \epsilon}{\partial x_i} \right] + F, \tag{8.107}$$

where f_1 and f_2 are wall functions placed in the production and dissipation terms, respectively, to help make them conform to the known wall behavior. F generally constitutes a wall model for Π_ϵ and the quantity $\tilde{\epsilon}$ in (8.107) may be ϵ in some models, whereas in others it is taken to be $\epsilon - 2v(\partial \sqrt{K}/\partial y)^2$, or equivalent, so that it goes to zero at the boundary. The latter choice is usually made to prevent the dissipation term in (8.107) from being unbounded at $y = 0$ (i.e., the ratio $\epsilon \tilde{\epsilon}/K$ is finite as $y \to 0$, while $\epsilon \epsilon/K$ is not). It is not usually a priority to make sure that wall functions are chosen so that (8.106) is satisfied exactly. Thus, approximate agreement is generally all that is usually achieved for this condition.

The wall functions f_μ, f_1, f_2, and F are typically derived through analysis of experimental and numerical simulation results. Different low-Reynolds-number K–ϵ closures are distinguished mostly by the different choices made for these functions.

Ten particular sets of choices have been published in tabular form [46]. By way of illustration, the classic Jones and Launder model has

$$f_\mu = e^{-2.5/(1+R_t/50)} \tag{8.108}$$

$$f_1 = 1, \qquad f_2 = 1 - 0.3e^{-R_t^2} \tag{8.109}$$

and

$$F = 2\nu\nu_t \left(\frac{d^2\overline{U}}{dy^2}\right)^2, \tag{8.110}$$

where $R_t = K^2/\nu\epsilon$. A more recent model [46] shows the influence of DNS data in providing a clearer picture of the near-wall region. Here, some of the wall functions have gotten very elaborate:

$$f_\mu = \left(1 + 3R_t^{-3/4}\right)\left(1 + 80\ e^{-R_\epsilon}\right)\left(1 - e^{-R_\epsilon/43 - R_\epsilon^2/330}\right)^2, \tag{8.111}$$

$$f_1 = 1 - \frac{2.25}{C_{\epsilon 1}}e^{-(R_t/40)^2}, \qquad f_2 = 1 + \frac{0.57}{C_{\epsilon 2}}e^{-(R_t/40)^2}, \tag{8.112}$$

and

$$F = e^{-(R_t/40)^2}0.5\frac{(\epsilon - 2\nu K/y^2)^2}{K} \tag{8.113}$$

where $R_\epsilon = (\nu\epsilon)^{1/4}y/\nu$. Unlike the Jones and Launder model, the ϵ equation boundary condition in this model is taken to be (8.105).

One important difference between the two models illustrated here is that (8.111) through (8.113) depend explicitly on the wall-normal coordinate, y. This gives greater control of how the wall model behaves than relying purely on K, ϵ, and their combinations, since the latter are not known a priori; in fact, they are determined only during the solution process. Using y allows the development of expressions that satisfy asymptotic consistency of the model terms with the known behavior of the exact expressions near boundaries. This is believed to have a significant beneficial effect in the performance of near-wall models, as will be seen in Chapter 9. It must also be cautioned that the use of y in this context prevents models from achieving true generality since they will not transform properly from one observer to the next. It is also the case that while (8.111) through (8.113) is a near-wall model, and the near-wall region is one where the turbulence Reynolds number is low, it is not a low-Reynolds-number model in the sense that it can successfully model such flows as the final period of isotropic decay. In essence, the price for superior performance in one class of flows can, in fact, inhibit performance in other classes of flows.

Another approach to wall region modeling, which has sometimes been implemented as an adjunct to the K–ϵ closure, is the elliptic relaxation method [13]. This scheme includes a model equation for $\overline{v^2}$ as part of computing the eddy viscosity in the form

$$\nu_t = C_\mu \overline{v^2} T_t, \tag{8.114}$$

where T_t is a mixing time scale. The latter is assumed to be given by $T_t = \max(K/\epsilon, 6\sqrt{\nu/\epsilon})$, which is based on the idea that the first of these expressions is a reasonable characterization of the mixing time away from boundaries. In near-wall regions this must be replaced by the second expression, which is proportional to the Kolmogorov scale, in order to prevent the unphysical possibility that the mixing time is smaller than the Kolmogorov scale.

Empirical observation shows that the near-wall behavior of ν_t is mostly determined by that of $\overline{v^2}$, and in recognition of this, the elliptic relaxation model aims to confront the behavior of $\overline{v^2}$ directly rather than through damping factors or other models that force the eddy viscosity to have the physical properties expressed by (8.114). The physical argument behind elliptic relaxation notes that whereas viscosity, through the no-slip condition, forces the normal Reynolds stresses to zero at solid boundaries, the wall-normal stress $\overline{v^2}$ is forced to zero from farther out in the flow due to the blocking effect. In this, v is conveniently thought of as being induced by a collection of vortices in the wall vicinity which are paired with image vortices so that their net contribution to v is zero at the surface. Loosely speaking, the blockage effect is caused by the action of the pressure field, which itself satisfies a Poisson equation. This motivates the idea of setting up an elliptic equation for a forcing term in a modeled form of the $\overline{v^2}$ equation.

The implementation of these ideas lacks formal derivation, and several modifications of the original model have been proposed as a result of accumulated experience with its performance. In its initial form, the model equation for $\overline{v^2}$ is hypothesized to be of the form

$$\frac{\partial \overline{v^2}}{\partial t} + \overline{U}_j \frac{\partial \overline{v^2}}{\partial x_j} = K f_{22} - \overline{v^2} \frac{\epsilon}{K} + \frac{\partial}{\partial x_i} \left[\left(\nu + \frac{\nu_t}{\sigma_\epsilon} \right) \frac{\partial \overline{v^2}}{\partial x_i} \right], \tag{8.115}$$

where the function f_{22} is determined as the solution of the Poisson equation

$$L^2 \nabla^2 f_{22} - f_{22} = \frac{1 - C_1}{T_t} \left(\frac{2}{3} - \frac{\overline{v^2}}{K} \right) - C_2 \frac{\mathcal{P}}{K}. \tag{8.116}$$

Here, $L = C_L \max[(K^{3/2}/\epsilon), C_\eta (\nu^3/\epsilon)^{1/4}]$, $C_1 = 1.4$, $C_2 = 0.3$, $C_L = 0.3$, and $C_\eta = 70.0$. The condition $\overline{v^2} \to \epsilon(0) f_{22}(0)(x_j n_j)^4/20\nu^2$ is satisfied at the wall, where **n** is a unit normal at the surface. Moreover, in this approach the boundary conditions $K(0) = 0$ and $n_i \partial K/\partial x_i = 0$ are satisfied at the wall instead of boundary conditions for K and ϵ.

A favorable property of this and other forms of the elliptic relaxation scheme is that they tend to enhance the stability of models in the near-wall region. On the other hand, the additional numerical expense of solving the extra relaxation equations can be significant, particularly in general, three-dimensional mean flows.

8.9.3   Near-Wall Reynolds Stress Equation Models

Although second-moment closures have a better opportunity than lower-order methods to account for flow anisotropies, the anisotropies found in turbulent wall flow are as much a challenge for these methods as they are for less sophisticated schemes. In particular, the individual tensor components of the terms in the R_{ij} equation tend to have different behaviors near the wall. Devising tensorially correct models for each of these effects is a daunting challenge and one that has yet to be fully realized. Although there are some common themes to how the different high-Reynolds-number SMC have been specialized to the wall region flow, there are often as not many significant differences. It is not practical to cover in detail the various approaches that have been taken, so instead, this discussion concentrates on the general outline of the problem and includes some specifics for just one of the closure schemes.

Similar to the analysis of the K and ϵ equations, it is helpful to consider which terms in the Reynolds stress equation are most significant in the near-wall region. Via Taylor series expansions, it follows that through terms of $O(y)$ near the wall, (8.63) reduces to

$$\epsilon_{ij} = D_{ij}^p + \Pi_{ij} + \nu\, \nabla^2 R_{ij}, \tag{8.117}$$

where

$$D_{ij}^p \equiv -\frac{1}{\rho}\frac{\partial \overline{pu_i}}{\partial x_k}\delta_{jk} - \frac{1}{\rho}\frac{\partial \overline{pu_j}}{\partial x_k}\delta_{ik} \tag{8.118}$$

is the pressure diffusion term. Formerly, this term was included as part of

$$-\frac{\partial \beta_{ijk}}{\partial x_k} = -\frac{\partial \overline{u_i u_j u_k}}{\partial x_k} + D_{ij}^p, \tag{8.119}$$

which is now divided for convenience. In other words, only the part of β_{ijk} originating in the pressure contributes to the balance in (8.117).

It may be shown that the $(1, 1)$, $(3, 3)$, and $(1, 3)$ components in (8.117) are $O(1)$ near the surface, the $(2, 2)$ component is $O(y^2)$, and the remainder are $O(y)$. These are trends that one hopes to capture in turbulence models, although it often proves difficult to achieve in practice. Another set of conditions is given by the asymptotic relations

$$\frac{\epsilon}{K} = \frac{\epsilon_{11}}{R_{11}} = \frac{\epsilon_{33}}{R_{33}} = \frac{\epsilon_{13}}{R_{13}} = \frac{1}{2}\frac{\epsilon_{12}}{R_{12}} = \frac{1}{2}\frac{\epsilon_{23}}{R_{23}} = \frac{1}{4}\frac{\epsilon_{22}}{R_{22}}, \tag{8.120}$$

which express the anisotropy just adjacent to the wall surface. These are readily derived from Taylor series expansions of the numerator and denominators [32]. Closure models for ϵ_{ij} aim to satisfy these conditions.

A convenient way of organizing the dissipation rate model so as to accommodate anisotropy is first to rewrite it in terms of its deviatoric part, ϵ_{ij}^{w}, so that

$$\epsilon_{ij} = \frac{2}{3}\epsilon\delta_{ij} + \epsilon_{ij}^{w}. \tag{8.121}$$

The goal, then, is to model ϵ_{ij}^{w} to be at least partially consistent with (8.120) besides the requirement of having zero trace. Models that pursue this approach generally make use of the wall-normal coordinate, y, and so are not suitable for arbitrary geometries. The condition on ϵ_{22} in (8.120) is usually the hardest to satisfy, and often is not met.

It is clear from (8.121) that it is also desirable to capture the near-wall behavior of ϵ in the context of SMC. A suitable near-wall model of the ϵ equation is therefore an integral part of second moment closures. In the best of circumstances, the near-wall form of the ϵ equation is coordinated with that of the second moment closure with which it is associated. Although this has not generally been the practice, one common step toward bringing the R_{ij} and ϵ equation models into closer alignment is to replace the standard transport model in (8.34) with the anisotropic formula

$$T_\epsilon = \frac{\partial}{\partial x_j}\left(C_{\epsilon_d}\frac{K}{\epsilon}R_{ij}\frac{\partial \epsilon}{\partial x_i}\right). \tag{8.122}$$

Moreover, Π_ϵ may be modeled through a function F as was done previously for the K–ϵ closure. A number of special forms of F have been developed specifically for second-moment closures.

How best to model pressure effects near boundaries in SMC has been a particularly contentious area, with many different methodologies proposed. According to the formal analysis of the pressure–strain correlation discussed in Section 2.3, the general decomposition

$$\Pi_{ij} = \Pi_{ij}^{S} + \Pi_{ij}^{F} + \Pi_{ij}^{W} \tag{8.123}$$

into slow, fast, and wall terms, respectively, can be made. Methods can be classified according to which of these terms are specifically modified for near-wall conditions. Moreover, special modeling of D_{ij}^{p} is also sometimes included together with that of Π_{ij}. Some models (e.g., [28,29,42,49,50,54,55]) choose to add an expression for Π_{ij}^{W} while the fast and slow terms are left unchanged and D_{ij}^{p} is not modeled. In others, Π_{ij}^{W} is omitted and just a special model for D_{ij}^{p} is used, reminiscent of the K equation modeling [35]. A third strategy is to model both Π_{ij}^{W} and D_{ij}^{p} [53]. Yet another common approach is not to model either D_{ij}^{p}, or Π_{ij}^{W}, but instead, to develop wall variants of the fast and slow terms [23,30,33,34].

Similar to the case of the K equation, the transport correlation D_{ij}^t is not directly involved in the physical balance at the wall since it is of higher order in y. The different components of D_{ij}^t have behaviors running from $O(y^3)$ to $O(y^5)$. Capturing such effects does not appear to be a priority, and thus the tendency has been not to develop special near-wall forms for this term.

To give one representative example of how SMCs are modified for near-wall flow, we describe such a modification of the high-Reynolds-number SSG model given in Section 8.7. In this approach [53,62], near-wall models of Π_{ij}^W and D_{ij}^p are incorporated in the form

$$\Pi_{ij}^W = f_{w_1}[(2C_1\epsilon + C_1^*\mathcal{P})b_{ij} + 4C_1'\epsilon(b_{ik}b_{kj} - b_{kl}b_{kl}\delta_{ij}/3)$$
$$+ \alpha^*(\mathcal{P}_{ij} - 2\mathcal{P}\delta_{ij}/3) + 2\gamma^* K\overline{S}_{ij}] \tag{8.124}$$

and

$$D_{ij}^p = -\frac{1}{3}\left[\nu \nabla^2(\overline{u_i u_k})n_k n_j + \nu \nabla^2(\overline{u_j u_k})n_k n_i\right] + \frac{1}{3}\delta_{ij}\nu \nabla^2(\overline{u_k u_l})n_k n_l. \tag{8.125}$$

Moreover, the deviatoric part of the dissipation tensor is modeled via

$$\epsilon_{ij}^w = -\frac{2}{3}f_{w_1}\epsilon\delta_{ij} + f_{w_1}\frac{\epsilon}{K}\frac{R_{ij} + R_{ik}n_k n_j + R_{jk}n_k n_i + R_{kl}n_k n_l\delta_{ij}}{1 + \frac{3}{2}R_{kl}n_k n_l/K}. \tag{8.126}$$

In these relations

$$f_{w_1} = e^{-(R_t/200)^2} \tag{8.127}$$

is a damping function, $C_1^* = 1.8, \alpha^* = -0.029, \gamma^* = 0.065$, and $C' = -1.05$. Moreover, (8.122) is used for dissipation transport, (8.113) is used as the wall function, F, in the ϵ equation, $C_{\epsilon_1} = 1.50, C_{\epsilon_2} = 1.83$, and $C_{\epsilon_d} = 0.12$. The boundary condition on the ϵ equation for this model is (8.105). In all other aspects the model has the same constants as the standard SSG model. A desirable feature of this model is its avoidance of direct dependence on y, so that it may be applied to general flows without further consideration. It should be emphasized that there are many competing near-wall SMCs which are under active development at the present time, far too many to begin to account for the range of modeling assumptions employed in their derivation. The predictions of several flows via these methods are included in Chapter 9, although their details are not described in this book.

8.10   VORTICITY TRANSPORT CLOSURE

The potential advantages of a vorticity transport analysis of turbulent flow were demonstrated in Section 6.2. Perhaps its most significant property was a capability for explaining the complex anisotropic behavior of the vorticity fluxes in the near-wall region. Since reproducing such anisotropies is a particularly challenging area for traditional Reynolds stress models, it is of some interest to consider how vorticity transport modeling could be incorporated into a useful closure scheme.

At first sight, the previously derived vorticity flux formula (6.38) appears to depend on a number of complicated correlations that would inhibit its development as a practical model. However, it can be shown that despite appearances to the contrary, with the help of several identities deriving from the continuity equation and the vorticity definition, (6.38) can be rewritten in an equivalent form involving just the normal Reynolds stresses [5,20]. In this form the transport law may have application in turbulence modeling (e.g., [22]).

The essential steps in this development can be illustrated most simply in the case of fully developed channel flow. Thus, using the identities $\overline{u(\partial w/\partial z)} = -\overline{w(\partial u/\partial z)}$ and $\overline{v(\partial w/\partial z)} = -\overline{w(\partial v/\partial z)}$, and substituting the definitions $\omega_1 = \partial w/\partial y - \partial v/\partial z$ and $\omega_2 = \partial u/\partial z - \partial w/\partial x$ into (6.40) and (6.42) yields the following coupled equations for $\overline{w(\partial u/\partial z)}$ and $\overline{w(\partial v/\partial z)}$:

$$\frac{1}{2}\frac{d\overline{w^2}}{dy} - \overline{w\frac{\partial v}{\partial z}} = Q_3 \overline{w\frac{\partial u}{\partial z}}\Omega \tag{8.128}$$

$$\overline{w\frac{\partial u}{\partial z}} = Q_4 \overline{w\frac{\partial v}{\partial z}}\Omega, \tag{8.129}$$

where the fact that $\overline{w\partial w/\partial x} = 0$ has been used, $\Omega = -d\overline{U}/dy$ and $Q_2 \equiv \mathcal{Q}_{233}$, $Q_3 \equiv \mathcal{Q}_{313}$, and $Q_4 \equiv \mathcal{Q}_{323}$. Solving these equations gives

$$\overline{w\frac{\partial u}{\partial z}} = \frac{\frac{1}{2}Q_4\Omega\left(d\overline{w^2}/dy\right)}{1 + Q_3Q_4\overline{\Omega}^2} \tag{8.130}$$

and

$$\overline{w\frac{\partial v}{\partial z}} = \frac{\frac{1}{2}\left(d\overline{w^2}/dy\right)}{1 + Q_3Q_4\overline{\Omega}^2}. \tag{8.131}$$

Substituting these results into (6.30) then yields

$$0 = -\frac{1}{\rho}\frac{\partial \overline{P}}{\partial x} + (\nu + \mathcal{T}_{22}\overline{v^2})\frac{d^2\overline{U}}{dy^2} + \frac{1}{2}\frac{d\overline{U}}{dy}\frac{d\overline{w^2}}{dy}\frac{Q_2 + Q_4}{1 + Q_3Q_4\left(d\overline{U}/dy\right)^2}, \tag{8.132}$$

which is the vorticity transport closure to the x-momentum equation.

Comparing (8.132) with (4.1) makes evident that there is an implied model for the Reynolds shear stress involving (8.130) and (8.131). In fact, a calculation reveals that

$$\overline{uv} = -\mathcal{T}_{22}\overline{v^2}\frac{d\overline{U}}{dy} + \int_0^y \frac{d\overline{U}}{dy}\left(\frac{d(\mathcal{T}_{22}\overline{v^2})}{dy} - \frac{\frac{1}{2}(Q_2 + Q_4)}{1 + Q_3 Q_4 \left(d\overline{U}/dy\right)^2}\frac{d\overline{w^2}}{dy}\right)dy, \quad (8.133)$$

where the integral expression evidently is related to the various effects that are represented by the nongradient correlations in (6.22). Equation (8.133) underscores the expectation that the Reynolds stress at a point has a nonlocal dependence on the statistical properties of the flow.

The vorticity transport theory shows that besides $\overline{v^2}$, which was previously shown to have a natural association with wall-normal transport near boundaries, the correlation $\overline{w^2}$ also plays an important role in the physics. Its appearance in (8.133) can be traced to the vortex stretching term and it is therefore reasonable to interpret the second part of the integral in (8.133) as accounting for this effect on momentum transport.

If a closure is to be built around (8.132), then it needs a means for predicting the normal Reynolds stresses. There are at least two ways to proceed: $K - \epsilon$ equations can be solved and the individual normal stresses determined from K using algebraic formulas [20], or else closed equations for $\overline{v^2}$ and $\overline{w^2}$ as given by a SMC can be solved in conjunction with (8.132) to form a complete closed scheme. The unknown time scales would have to be determined through calibration of the model.

The vorticity transport closure scheme can be generalized, in principle, to encompass arbitrary three-dimensional mean flows [5]. A system of equations can be derived expressing the full set of vorticity fluxes in terms of the mean vorticity components, the normal Reynolds stresses and the time scales. No three-dimensional applications have yet been attempted, but some investigation has been done for the case of two-dimensional mean flows in which a very substantial simplification of the closure model is possible [20].

8.11   SUMMARY

This chapter has tried to put into perspective the major themes in turbulence closure development. In the modern era, the $K-\epsilon$ closure is most often encountered in practical applications, although in many specialized applications low-order methods continue to be used. The front lines in closure development are mainly at the level of SMC and ARM and in the treatment of boundaries. Even though there is a sense that closure models will never achieve the kinds of accuracy that engineers would prefer, nonetheless they are not likely to be replaced any time soon by alternative schemes. This is true largely because of their relative cost compared to alternative methodologies, primarily LES. It is also the case, although it is not discussed at length in this book, that RANS modeling has as yet a unique capability in comparison to LES or DNS in economically modeling flows containing the widest range of physics (e.g., compressibility, combustion, and other phenomena).

REFERENCES

1. Apsley, A. D. and Leschziner, M. A. (1997) "A new low-Reynolds-number non-linear two-equation turbulence model for complex flows," *Int. J. Heat Fluid Flow* **19**, 209–222.

2. Baldwin, B. S. and Barth, T. J. (1990) "A one-equation turbulence transport model for high Reynolds number wall-bounded flows," *NASA TM-102847*.

3. Baldwin, B. S. and Lomax, H. (1978) "Thin-layer approximation and algebraic model for separated turbulent flows," *AIAA Pap. 78-257*.

4. Batchelor, G. K. (1967) *An Introduction to Fluid Dynamics*, Cambridge University Press, Cambridge.

5. Bernard, P. S. (1990) "Turbulent vorticity transport in three dimensions," *Theor. Comput. Fluid Dyn.* **2**, 165–183.

6. Bertoglio, J. P. (1982) "Homogeneous turbulent field within a rotating frame," *AIAA J.* **20**, 1175–1181.

7. Blaisdell, G. A. and Shariff, K. (1996) "Simulation and modeling of elliptic streamline flow," *Proc. Summer Program, Center for Turbulence Research*.

8. Cebeci, T. and Smith, A. M. O. (1974) *Analysis of Turbulent Boundary Layers*, Academic Press, New York.

9. Choi, K. S. and Lumley, J. L. (1984) "Return to isotropy of homogeneous turbulence revisited," in *Turbulence and Chaotic Phenomena in Fluids* (T. Tatsumi, Ed.) North-Holland, New York, pp. 267–272.

10. Craft, T. J., Launder, B. E. and Suga, K. (1995) "A non-linear eddy viscosity model including sensitivity to stress anisotropy," *Proc. Turbulent Shear Flows 10*, Vol. 3, pp. 23.19–23.24.

11. Crow, S. C. (1968) "Viscoelastic properties of fine-grained incompressible turbulence," *J. Fluid Mech.* **33**, 1–20.

12. Daly, B. J. and Harlow, F. H. (1970) "Transport equations in turbulence," *Phys. Fluids* **13**, 2634–2649.

13. Durbin, P. A. (1991) "Near wall turbulence models without damping functions," *Theor. Comput. Fluid Dyn.* **3**, 1–13.

14. Durbin, P. A. and Speziale, C. G. (1994) "Realizability of second-moment closure via stochastic analysis," *J. Fluid Mech.* **280**, 395–407.

15. Fu, S., Launder, B. E. and Tselepidakis, D. P. (1987) "Accommodating the effects of high strain rates in modeling the pressure–strain correlation," *UMIST Tech. Rep. TFD/87/5*.

16. Gatski, T. B. and Speziale, C. G. (1993) "On explicit algebraic stress models for complex turbulent flows," *J. Fluid Mech.* **254**, 59–78.

17. Gibson, M. M. and Launder, B. E. (1978) "Ground effects on pressure fluctuations in the atmospheric boundary layer," *J. Fluid Mech.* **86**, 491–511.

18. Girimaji, S. S. (1996) "Fully explicit and self-consistent algebraic Reynolds stress model," *Theor. Comput. Fluid Dyn.* **8**, 387–402.

19. Girimaji, S. S. (2000) "Pressure–strain correlation modeling of complex turbulent flows," *J. Fluid Mech.* **422**, 91–123.

20. Gorski, J. J. and Bernard, P. S. (1995) "Vorticity transport analysis of turbulent flows", *ASME J. Fluids Eng.* **117**, 410–416.

21. Gurtin, M. E. (1981) *An Introduction to Continuum Mechanics*, Academic Press, New York.

22. Hanjalić, K., Jakirlić, S. and Ristorcelli, J. R. (1996) "Alternative approach to modeling the dissipation equation," in *Advances in Turbulence VI* (S. Gavrilakis, L. Machiels, and P. A. Monkewitz, Eds.), Kluwer Academic, Dordrecht, The Netherlands, pp. 27–30.

23. Hanjalić, K. and Launder, B. E. (1976) "Contribution towards a Reynolds-stress closure for low-Reynolds-number turbulence," *J. Fluid Mech.* **74**, 593–610.

24. Hunt, J. C. R. and Carruthers, D. J. (1990) "Rapid distortion theory and the 'problems' of turbulence," *J. Fluid Mech.* **212**, 497–532.

25. Johnson, D. A. and King, L. S. (1984) "A new turbulence closure model for boundary layer flows with strong adverse pressure gradients and separation," *AIAA Pap. 84-0175*.

26. Johnson, D. A. and King, L. S. (1985) "A mathematically simple turbulence closure model for attached and separated turbulent boundary layers," *AIAA J.* **23**, 1684–1692.

27. Jones, W. P. and Launder, B. E. (1972) "The prediction of laminarization with a two-equation model of turbulence," *Int. J. Heat Mass Transfer* **15**, 301–314.

28. Kebede, W., Launder B. E., and Younis, B. A. (1985) "Large amplitude periodic pipe flow: a second moment closure study," *Proc. Turbulent Shear Flows 5*, Pap. 16, Ithaca, N.Y.

29. Lai, Y. G. and So, R. M. C. (1990) "On near-wall turbulent flow modeling," *J. Fluid Mech.* **221**, 641–673.

30. Launder, B. E. and Li, S.-P. (1994) "On the elimination of wall-topography parameters from second-moment closure," *Phys. Fluids* **6**, 999–1006.

31. Launder, B. E., Reece, G. J. and Rodi, W. (1975) "Progress in the development of a Reynolds stress turbulence closure," *J. Fluid Mech.* **68**, 537–566.

32. Launder, B. E. and Reynolds, W. C. (1983) "Asymptotic near-wall stress dissipation rates in a turbulent flow," *Phys. Fluids* **26**, 1157–1158.

33. Launder, B. E. and Shima, N. (1989) "Second-moment closure for the near-wall sublayer: development and application," *AIAA J.* **27**, 1319–1325.

34. Launder, B. E. and Tselepidakis, D. P. (1988) "Contribution to the second-moment modeling of sub-layer turbulent transport," *Zaric Memorial International Seminar on Near-Wall Turbulence,* Yugoslavia, May 16–20.

35. Launder, B. E. and Tselepidakis, D. P. (1993) "Progress and paradoxes in modeling near-wall turbulence," *Proc. Turbulent Shear Flows 8*, pp. 81–96.

36. Mansour, N. N., Kim, J. and Moin, P. (1988) "Reynolds-stress and dissipation-rate budgets in a turbulent channel flow," *J. Fluid Mech.* **194**, 15–44.

37. Naot, D., Shavit, A. and Wolfshtein, M. (1973) "Two-point correlation model and the redistribution of Reynolds stress," *Phys. Fluids* **16**, 738–740.

38. Norris, L. H. and Reynolds, W. C. (1975) "Turbulent channel flow with a moving wavy boundary," *Stanford Univ. Rep. FM-10*.

39. Pedlosky, J. (1979) *Geophysical Fluid Dynamics*, Springer-Verlag, New York.

40. Pope, S. B. (1975) "A more general effective-viscosity hypothesis," *J. Fluid Mech.* **72**, 331–340.

41. Prandtl, L. (1925) "Über die ausgebildete Turbulenz," *Z. Angew. Math. Mech.* **5**, 136–139.

42. Prud'homme, M. and Elghobashi, S. (1983) "Prediction of wall-bounded turbulent flows with an improved version of a Reynolds stress model," *Turbulent Shear Flows 4*, Pap. 1.2, Karlsruhe, Germany.

43. Reynolds, W. C. (1989) "Effects of rotation on homogeneous turbulence," In *Proc. Australasian Conference on Fluid Mechanics.*

44. Rodi, W. (1976) "A new algebraic relation for calculating Reynolds stress," *Z. Angew. Math. Mech.* **56**, 331–340.

45. Rotta, J. C. (1951) "Statistische Theorie nichthomogener Turbulenz," *Z. Phys.* **129**, 547–572.

46. Sarkar, A. and So, R. M. C. (1997) "A critical evaluation of near-wall two-equation models against direct numerical simulation data," *Int. J. Heat Fluid Flow* **18**, 197–208.

47. Sarkar, S. and Speziale, C. G. (1990) "A simple nonlinear model for the return to isotropy in turbulence," *Phys. Fluids A* **2**, 84–93.

48. Shih, T.-H. and Lumley, J. L. (1993) "Critical comparison of second-order closures with direct numerical simulation of homogeneous turbulence," *AIAA J.* **31**, 663–670.

49. Shih, T.-H. and Mansour, N. N. (1990) "Modeling of near-wall turbulence," in *Engineering Turbulence Modeling and Experiments* (W. Rodi and E. N. Ganic, Eds.), Elsevier Science, Amsterdam, The Netherlands, pp. 13–19.

50. Shima, N. (1988) "A Reynolds stress model for near-wall and low-Reynolds-number regions," *ASME J. Fluids Eng.* **110**, 38–44.

51. Smagorinsky, J. (1963) "General circulation experiments with the primitive equations," *Mon. Weather Rev.* **91**, 99–165.

52. Smith, G. F. (1971) "On isotropic functions of symmetric tensors, skew-symmetric tensors and vectors," *Int. J. Eng. Sci.* **9**, 899–916.

53. So, R. M. C., Aksoy, H., Sommer, T. P., and Yuan, S. P. (1994) "Development of a near-wall Reynolds-stress closure based on the SSG model for the pressure strain," *NASA Contractor Rep. 4618.*

54. So, R. M. C., Lai, Y. G. Zhang, H. S. and Hwang, B. C. (1991) "Second-order near-wall turbulence closures: a review," *AIAA J.* **29**, 1819–1835.

55. So, R. M. C. and Yoo, G. J. (1987) "Low-Reynolds-number modeling of turbulent flows with and without wall transpiration," *AIAA J.* **25**, 1556–1564.

56. Spalart, P. R. and Allmaras, S. R. (1992) "A one-equation turbulence model for aerodynamic flows," *AIAA Pap. 92-439.*

57. Speziale, C. G. (1985) "Modeling the pressure gradient-velocity correlation of turbulence," *Phys. Fluids* **28**, 69–71.

58. Speziale, C. G. (1987) "On nonlinear $K–l$ and $K–\epsilon$ models of turbulence," *J. Fluid Mech.* **178**, 459–475.

59. Speziale, C. G. (1989) "Turbulence modeling in non-inertial frames of reference," *Theor. Comput. Fluid Dyn.* **1**, 3–19.

60. Speziale, C. G. (1991) "Analytical method for the development of Reynolds-stress closures in turbulence," *Annu. Rev. Fluid Mech.* **23**, 107–157.

61. Speziale, C. G., Sarkar, S. and Gatski, T. B. (1991) "Modeling the pressure–strain correlation of turbulence: an invariant dynamical systems approach," *J. Fluid Mech.* **227**, 245–272.

62. Speziale, C. G. and So, R. M. C. (1998) "Turbulence modeling and simulation," in *The Handbook of Fluid Dynamics* (R. W. Johnson, Ed.), CRC Press, Boca Raton, Fla., Chap. 14.

63. Taulbee, D. B. (1992) "An improved algebraic Reynolds stress model and corresponding nonlinear stress model," *Phys. Fluids A* **4**, 2555–2561.

64. Tavoularis, S. and Karnik, U. (1989) "Further experiments on the evolution of turbulent stresses and scales in uniformly sheared turbulence," *J. Fluid Mech.* **204**, 457–478.

65. Wilcox, D. (1993) *Turbulence Modeling for CFD*, DCW Industries and Griffin Printing, Glendale, Calif.

66. Yoshizawa, A. (1984) "Statistical analysis of the deviation of the Reynolds stress from its eddy-viscosity representation," *Phys. Fluids* **27**, 1377–1387.

67. Zemach, C. (1998) "Mathematics of fluid mechanics," in *The Handbook of Fluid Dynamics* (R. W. Johnson, Ed.), CRC Press, Boca Raton, Fla., App. A.

9

Applications of Turbulence Modeling

Part of the art of turbulence modeling is knowing which model, if any, has the best likelihood of yielding useful information about a flow of interest. Since the spectrum of possible flows is vast and there are a seemingly infinite number of models (just a few of which were given in detail in Chapter 8), the determination of a "best" closure scheme in any instance appears to be a formidable undertaking. Further complicating the story is the fact that different closures can have different degrees of success in accommodating different aspects of a particular flow. For example, the closure scheme that best predicts a reattachment point in a separated flow may not necessarily be the one that best predicts the skin friction coefficient, or mean velocity profile, or pressure distribution. It is also true that closure predictions vary with the choice of constants appearing in the models, and if one is willing to make such adjustments, parameter selection becomes part of the decision-making process.

In reality, the process of closure selection is not as complicated as it may appear since similar closures tend to behave similarly in many circumstances. Furthermore, much is known from empirical evidence accumulated over many years how different classes of closures behave in different types of flows. Moreover, a practical constraint on the range of models that one can consider is the fact that commercial CFD codes tend to include just a small selection of the most popular models, such as the $K-\epsilon$ closure.

The object of this chapter is to reveal, by way of example, what can be expected in the performance of some of the most commonly encountered closure schemes. The priority is on showing the strengths of closures so as to give an idea of what is achievable in practice and where they might be useful. A number of the following sample calculations are for flows used in model calibration. Although these often show closures at their very best, they are nonetheless instructive for demonstrating where the fundamental limitations of closures may lie. For example, the calibration process often reveals fundamental reasons why a particular category of closures is incapable of modeling the physical phenomena appearing in a particular flow. Although this discussion will not highlight the less successful side of closure modeling, the reader is cautioned that it is good practice when using RANS modeling to view predicted results with a healthy degree of skepticism.

In the chapter we first consider flows in a turbulent channel and zero-pressure-gradient boundary layer. These are among the most basic of the calibration flows in view of their relative simplicity and the extensive documentation of their properties from physical experiments and DNS. A perspective will be given on what is meant by "success" in a closure prediction. This is helpful in forming an idea as to the performance to be expected in predicting more complex flows. For example, if a closure has obvious errors in resolving near-wall anisotropy in a channel flow, it is not likely to capture similar effects in more complex situations. On the other hand, it is reasonable to expect a model to achieve performance similar to that in a channel flow when applied to related flows that do not have additional complicating physics.

The assessment of closure schemes in more general situations than a channel or attached boundary layer can be done somewhat systematically by concentrating on how closures deal with the presence of certain specific complicating phenomena which either in isolation or in combination appear in most engineering flows. Thus, after the channel flow discussion, consideration is given to flows with separation, rotation, secondary motions, curvature, and stagnation points. As in the simpler situations, one expects that a model which is successful in capturing a particular physical property (e.g., curvature) should also do as well in other flows containing curvature. Finally, some results on the prediction of flow past a prolate spheroid are given as an example of a complex flow where models must account for several difficult factors simultaneously.

9.1 CHANNEL AND ZERO-PRESSURE-GRADIENT BOUNDARY LAYER FLOW

The channel and zero-pressure-gradient boundary layer flows have the property that accurate prediction of the mean velocity $\overline{U}$ is tantamount to predicting the Reynolds shear stress $\overline{uv}$ accurately, since the two quantities are closely related to each other. For example, in a channel flow an exact correspondence between these quantities was given in (4.9). In evaluating the predictions of these flows, attention can be paid to either $\overline{U}$ or $\overline{uv}$. In contrast, it is a separate question entirely whether K, ϵ, and possibly the normal Reynolds stresses are well predicted.

Simple mixing-length models tend to be well tuned to canonical flows such as a channel or attached boundary layer and can well reproduce the mean velocity profiles in such cases. Of course, such schemes do not predict the other turbulence quantities. If K and the individual Reynolds stresses must be predicted, more elaborate models must be used. The earliest schemes for predicting K, such as the K–ϵ closure, achieve only modest success (e.g., the near-wall peak in K cannot be accounted for). Over the years the quality of predictions has improved greatly, largely as a result of developing better near-wall models with the help of DNS evaluations of the energy and dissipation rate budgets. These reveal important information about near-wall physics [43] which can be used to pinpoint weaknesses in specific closure models and suggest where and possibly how they need to be improved.

Figures 9.1 through 9.4 illustrate the predictions of K and ϵ (scaled by U_τ and ν) in a channel flow at two different Reynolds numbers resulting from two variants of the $K-\epsilon$ closure. One is the traditional Jones and Launder (JL) model [24], and the other is the recent SSA model [47] contained in (8.111) through (8.113). Evidently, the much greater attention to near-wall behavior in the SSA model results in a scheme that can virtually match the predicted K and ϵ fields, whereas the older model seriously misses the peak values. Note as well that by its use of the artificial $\epsilon = 0$ boundary condition, the JL model cannot be expected to capture the physical trend in ϵ near the wall. Similar results occur for boundary layer flow as shown in Figs. 9.5 and 9.6, where the K field in boundary layer flow at two Reynolds numbers are compared for the JL and SSA variants of the $K-\epsilon$ closure.

Until the development of near-wall variants of second moment closures and non-linear eddy viscosity models, predictions of the individual components of the normal Reynolds stresses (i.e., the anisotropy of the turbulence) were confined essentially to the region away from the wall. For example, the standard high-Reynolds-number form of the SSG model [55] can capture the anisotropy outside the wall region of a channel flow but cannot provide a credible prediction of the near-wall anisotropy unless it is given a near-wall modification. The explicit ARM [19] presented in Section 8.8 also captures accurately the normal stresses away from the wall, as shown in

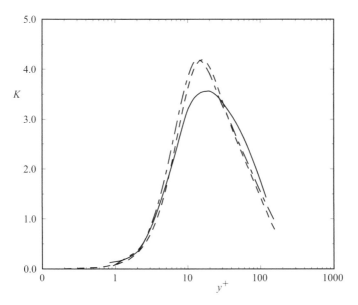

Fig. 9.1 *$K-\epsilon$ closure predictions of K in channel flow at $R_\tau = 180$. —, JL model [24]; — · —, SSA model [47]; — —, DNS [29]. (Reprinted from [47], Copyright (1997) with permission from Elsevier Science.)*

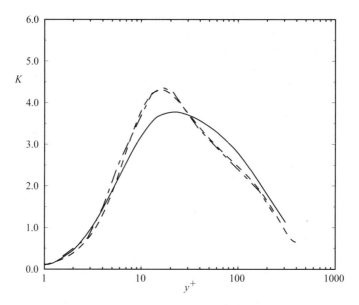

Fig. 9.2   *K–ϵ closure predictions of K in channel flow at $R_\tau = 395$. —, JL model [24]; — · —, SSA model [47]; – –, DNS [45]. (Reprinted from [47], Copyright (1997) with permission from Elsevier Science.)*

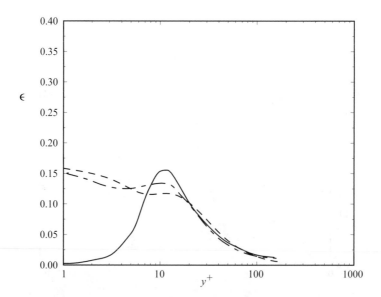

Fig. 9.3   *K–ϵ closure predictions of ϵ in channel flow at $R_\tau = 180$. —, JL model [24]; — · —, SSA model [47]; – –, DNS [29]. (Reprinted from [47], Copyright (1997) with permission from Elsevier Science.)*

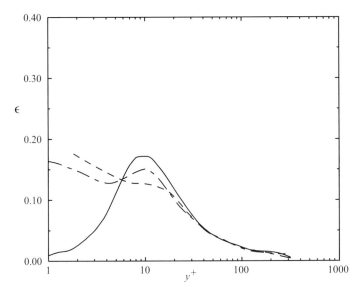

Fig. 9.4   *K – ε closure predictions of ε in channel flow at $R_\tau = 395$. —, JL model [24]; — · —, SSA model [47]; — —, DNS [45]. (Reprinted from [47], Copyright (1997) with permission from Elsevier Science.)*

Fig. 9.7 for boundary layer flow. In this case, without special wall modification, the prediction of the near-wall anisotropy includes a significant underprediction of the wall-normal Reynolds stress and overprediction of the streamwise Reynolds stress.

With an appropriate investment in modeling near-wall trends, accurate predictions of the Reynolds stresses over the entire flow domain are possible with advanced closure schemes. This is illustrated in Fig. 9.8, showing the three normal stresses versus DNS data for a special near-wall variant of a NLEVM [13]. This Reynolds stress model includes quartic terms besides the quadratic terms in (8.57) and the cubic terms in (8.58). Moreover, the model coefficients are specially calibrated for near-wall flow and a separate modeled equation for the anisotropy tensor invariant, II, defined in (8.59), is solved. Predictions of an equivalent caliber have been obtained with an SMC [10] combined with the elliptic relaxation treatment of the wall region [63]. These kinds of calculations make clear that with sufficient care for detail and the availability of DNS results, virtually flawless reproductions of such complicated features of turbulent flow as the anisotropy of the wall region are possible.

An important sidelight to these calculations is the issue of how wide a range of Reynolds numbers they are able to accommodate without need for recalibration of the model parameters. Ideally, one would hope that such modifications are unnecessary; often, the question has not been investigated, and it is difficult to know one way or the other whether the correct Reynolds number sensitivity exists. The Reynolds number dependency of a near-wall version of the SSG model has been examined extensively against measurements in channel, pipe, and boundary layer flows [49]. This suggests that the proper trends are achievable with the help of a near-wall model which is

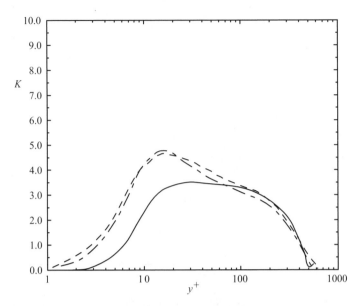

Fig. 9.5 *K–ε closure predictions of K in boundary layer flow at $R_\theta = 1410$. —, JL model [24]; — · —, SSA model [47]; — —, DNS [51]. (Reprinted from [47], Copyright (1997) with permission from Elsevier Science.)*

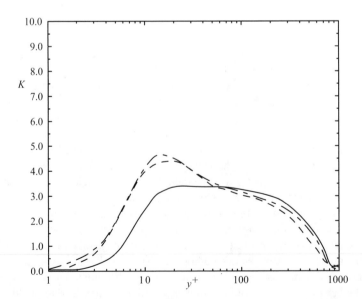

Fig. 9.6 *K–ε closure predictions of K in boundary layer flow at $R_\theta = 2420$. —, JL model [24]; — · —, SSA model [47]; — —, measured [27]. (Reprinted from [47], Copyright (1997) with permission from Elsevier Science.)*

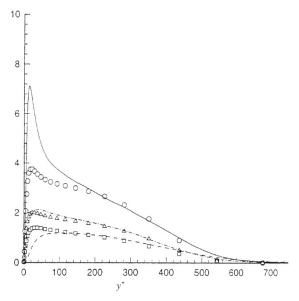

Fig. 9.7 *ARM [19] prediction of normal Reynolds stresses in boundary layer flow (scaled by U_τ). —, $\overline{u^2}$; — · —, $\overline{v^2}$; — —, $\overline{w^2}$; symbols are from DNS [52]. (From [19] copyright © Springer-Verlag.)*

appropriately calibrated and designed to enforce the correct asymptotic conditions [e.g., the near-wall relations given in (8.120)].

9.2 FLOW SEPARATION

Separation of boundary layers is a prevalent aspect of engineering flows, and capturing the size and nature of separated regions has been a central focus of RANS modeling since the inception of such techniques. For example, knowing the exact size and position of separation zones is essential to accurate drag and lift predictions for the flow past bluff bodies or around airfoils during stall conditions. Indeed, it is the extent of the separated region that is most important in determining the pressure field and hence the forces exerted on bodies.

Separated flows can occur with fixed separation points, as in a backstep flow where fluid traveling through a duct encounters a suddenly enlarged region, or the separation point can be determined as the consequence of a delicate force balance in which low momentum fluid next to a body separates upon encountering a sufficiently large unfavorable pressure gradient. The latter situation is encountered in diffusers, for example, when the flow separates from the expanding walls. In this section we illustrate the capabilities of closures in predicting separated flow in several standard applications.

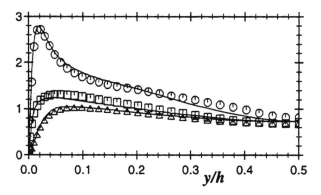

Fig. 9.8 *NLEVM prediction of normal Reynolds stresses (scaled by U_τ) in channel flow at $R_e = 13{,}750$. Upper to lower curves represent $\overline{u^2}$, $\overline{w^2}$ and $\overline{v^2}$, respectively. Symbols are from DNS [45]. (From [13].)*

9.2.1 Backward-Facing Step

The geometry of a backward-facing step flow is shown in Fig. 9.9 together with mean streamlines from a calculation using the nonlinear K–ϵ closure. In this flow, the boundary layer on the lower surface separates at the sharp corner and reattaches downstream, after which point the boundary layer re-forms, eventually returning to a canonical state a significant distance downstream. It should be noted that the streamlines in Fig. 9.9 are meant to be interpreted as the end result of time-averaging the flow field. In actual physical experiments the recirculation region contains large vortices which grow and shed off the corner (e.g., as in the mixing layer), so the reattachment point will dynamically move up and downstream of the average location. Clearly, the mean flow indicated in Fig. 9.9 is quite distinct from the physics of the periodically shedding and growing vortices.

Closure schemes that are adept at solving the nonsteady RANS equations may predict periodic vortex shedding in the back step flow. Such solutions are sometimes referred to as unsteady RANS (URANS) or very large eddy simulations (VLESs) and may be viewed conceptually as solving for the mean field determined by ensemble averaging over many periods of the vortex shedding. There are a number of fundamental problems with this approach [53] (e.g., periodic shedding of vortices in turbulent conditions has much variability in phase and amplitude). Moreover, in two-dimensional mean fields it is not certain that variability in the third direction can be ignored. Nonetheless, applications show that the approach is not without some advantages. For problems such as the backward-facing step, time averaging of a URANS solution over a period should yield results qualitatively similar to Fig. 9.9. The present discussion is limited to studies that directly seek the time-averaged steady solution, since this case has received greater systematic attention, both computationally and experimentally, than has the nonsteady regime.

Historically, predictions of backstep flow have been plagued by confusion between numerical and modeling effects (i.e., often, properties of the numerical calculation

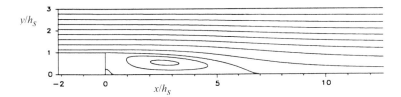

Fig. 9.9 *Geometry of backward-facing step flow. Mean streamlines predicted by the nonlinear $K-\epsilon$ closure [54]. (From [59]. Copyright © 1992 by the American Institute of Aeronautics and Astronautics, Inc. Reprinted with permission.)*

have been misidentified as properties of the turbulence model) [58]. For example, only after a numerical solution of the RANS equations is known to be grid independent can it be certain that the computed results reflect the modeling decisions. Recent studies using faster computers and finer meshes have been more successful in establishing grid independence. As mentioned previously, numerical issues are particularly acute for SMCs since, among other reasons, they require numerical differentiation of the Reynolds stresses in the mean momentum equation—a potentially significant source of error. These kinds of numerical issues have been observed to cause serious problems in backstep flow calculations [1] and may very well be magnified in the treatment of more complex flows.

Information about backstep flow comes from both experiments [15,26,28] and, more recently, DNS calculations [36]. These are sometimes characterized by a Reynolds number based on centerline mean velocity of the upstream channel flow and the height of the outlet channel, say $\mathrm{Re}_1 \equiv U_c h/\nu$, and sometimes are based on the bulk velocity U_m and the step height h_s, say $\mathrm{Re}_2 \equiv U_m h_s/\nu$. The backstep flow at a relatively low $\mathrm{Re}_2 = 5100$ has been measured [26] and studied via DNS [36]. High Reynolds numbers have been studied experimentally (e.g., $\mathrm{Re}_1 = 35,000$ in [15] and $\mathrm{Re}_1 = 1.32 \times 10^5$ in [28]). The DNS supplies comprehensive data about average properties of the low-Reynolds-number case, while experiments provide just partial information about the flow at the two larger Reynolds numbers. For the largest Reynolds number the average location of the reattachment point is known as well as mean velocities, Reynolds shear stress, and the streamwise normal Reynolds stress at some streamwise locations. Apart from the mean velocity, the experimental data tend not to be close to the boundary.

The average reattachment point for the $\mathrm{Re}_1 = 1.32 \times 10^5$ experiment occurs at $7.1 h_s$. A calculation with the standard $K-\epsilon$ closure and wall function boundary conditions predicts reattachment in the range $6.1 h_s$ through $6.25 h_s$. This error, which is approximately 12%, is considered to be fairly significant. The same case treated with the nonlinear $K-\epsilon$ closure [54], and shown in Fig. 9.9, improves the prediction of the reattachment point to $6.9 h_s$, which is within $\approx 3\%$ of the measured value and well within the experimental uncertainty. The solutions for $\overline{U}$, u_{rms} and $\overline{uv}$ from the standard form of the $K-\epsilon$ closure are shown in Figs. 9.10, 9.11, and 9.12, respectively. The predictions of the mean velocity are seen to be excellent. The errors in the Reynolds stresses are more significant, but the trends are reasonably well captured. It

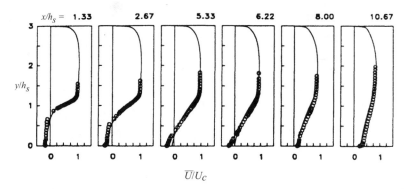

Fig. 9.10   *Predictions of $\overline{U}$ in backward-facing step flow at* $Re_1 = 1.32 \times 10^5$. —, *K–ϵ closure [59]; o, experiments [28]. (From [59]. Copyright © 1992 by the American Institute of Aeronautics and Astronautics, Inc. Reprinted with permission.)*

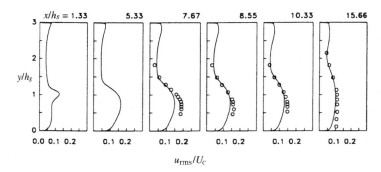

Fig. 9.11   *Predictions of u_{rms} in backward-facing step flow at* $Re_1 = 1.32 \times 10^5$. —, *K–ϵ closure [59]; o, experiments [28]. (From [59]. Copyright © 1992 by the American Institute of Aeronautics and Astronautics, Inc. Reprinted with permission.)*

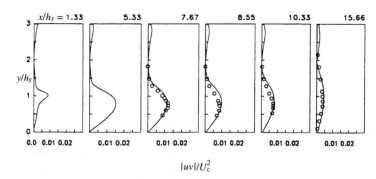

Fig. 9.12   *Predictions of $\overline{uv}$ in backward-facing step flow at* $Re_1 = 1.32 \times 10^5$. —, *K–ϵ closure [59]; o, experiments [28]. (From [59]. Copyright © 1992 by the American Institute of Aeronautics and Astronautics, Inc. Reprinted with permission.)*

so happens that despite the better prediction of the reattachment point by the nonlinear K–ϵ model, it is actually slightly less adept at predicting $\overline{U}$ than the standard K–ϵ model. The opposite is true with regard to u_{rms}, although the differences between predictions are not large.

It is instructive to consider why the nonlinear K–ϵ closure has better success at predicting the reattachment point than the linear EVM. A likely explanation may be found in the superior capability of the former in predicting normal Reynolds stress anisotropy. To see this, consider the spanwise mean vorticity equation in the plane backstep flow, which has the form

$$\overline{U}\frac{\partial \overline{\Omega}}{\partial x} + \overline{V}\frac{\partial \overline{\Omega}}{\partial y} = \frac{\partial^2}{\partial x\, \partial y}\left(\overline{u^2} - \overline{v^2}\right) - \frac{\partial^2 \overline{uv}}{\partial x^2} + \frac{\partial^2 \overline{uv}}{\partial y^2}, \qquad (9.1)$$

where $\overline{\Omega} \equiv \overline{\Omega}_3$. Equation (9.1) emphasizes the fact that both $\overline{uv}$ and the difference $(\overline{u^2} - \overline{v^2})$ have the potential to strongly affect the mean flow prediction. In a comparison of an EVM and a NLEVM, the latter are expected to give better predictions of the normal Reynolds stress anisotropy, even if the accuracy of the $\overline{uv}$ correlation is comparable. For example, in a plane channel flow, an EVM such as the K–ϵ closure predicts that $\overline{u^2} = \overline{v^2}$, whereas this is not the case for a NLEVM. Thus NLEVMs may better predict reattachment because their modeling of the right-hand side of (9.1) is more realistic, leading to better $\overline{\Omega}$ predictions.

Another interesting comparison with data may be found in the prediction of the $Re_2 = 5100$ backstep flow using a near-wall formulation of a SMC [8,12]. For these calculations the natural boundary conditions are applied instead of wall functions. Plots of $\overline{U}$, the normal Reynolds stresses, $\overline{u^2}$ and $\overline{v^2}$, as well as $\overline{uv}$ are shown in Figs. 9.13, 9.14, and 9.15, respectively. The mean velocity prediction is seen to be virtually flawless. Some small errors and slightly unphysical behaviors appear in the Reynolds stresses. At the higher Reynolds number of 35,000, predictions with the same model show some degradation, but on the whole the modeling appears to be capable of capturing the essential properties of the backstep flow.

A particularly sensitive indicator of the quality of the backstep flow modeling is the distribution of the friction coefficient $C_f \equiv 2\tau_w/\rho U_m^2$ along the wall downstream of the step. In fact, despite reasonable predictions of other mean statistics, such as $\overline{U}$, it is possible for C_f to be seriously in error. Figure 9.16 shows a comparison of C_f predicted by a near-wall NLEVM [47] with measured values for the $Re_2 = 5100$ backstep flow. The particular model used here performed the best among a group of models, including both NLEVM and SMC [50]. Many of the latter give relatively poor predictions of C_f even if other aspects of the flow are well predicted. The favorable performance of the NLEVM depicted in the figure is attributable largely to the effort that has been spent in forcing consistency with asymptotic expansions of the flow variables near the boundary. In Fig. 9.16 the most significant error is in the region from the step to the minimum in C_f. The reattachment point of the separated flow occurs at $C_f = 0$ (i.e., $\tau_w = 0$), which is at $x/h_s \approx 6$. The model calculation gives reattachment at $x = 6.5$.

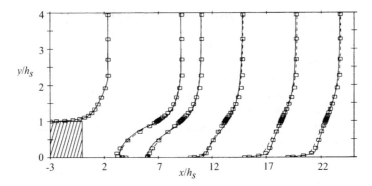

Fig. 9.13 *Predictions of $\overline{U}$ in backward-facing step flow at* $Re_2 = 5100$. —, *SMC closure [8];* □, *DNS [36]. (Reprinted from [8], Copyright (1998) with permission from Elsevier Science.)*

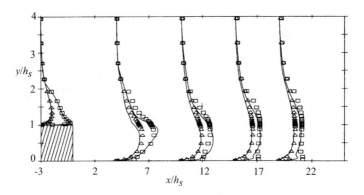

Fig. 9.14 *Predictions of normal Reynolds stresses in backward-facing step flow at* $Re_2 = 5100$. —, *SMC [8];* □, u_{rms} *and* △, v_{rms} *from a DNS [36]. (Reprinted from [8], Copyright (1998) with permission from Elsevier Science.)*

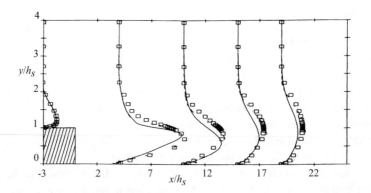

Fig. 9.15 *Predictions of $\overline{uv}$ in backward-facing step flow at* $Re_2 = 5100$. —, *SMC [8];* □, *DNS [36]. (Reprinted from [8], Copyright 1998 with permission from Elsevier Science.)*

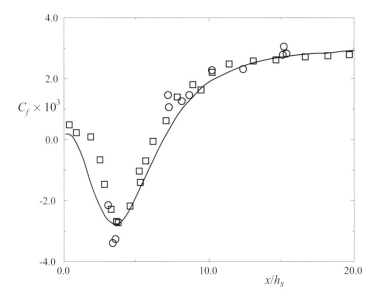

Fig. 9.16 C_f *downstream of the step on the lower wall.* —, *near-wall NLEVM [47]; o, experiments [26]; □, DNS [36]. (Reprinted from [50], Copyright 1998 with permission from Elsevier Science.)*

9.2.2 Hill Flow

The flow past a two-dimensional hill of height h_h located on one wall of a channel flow [2] has slightly greater complexity than the backstep flow. Specifically, the location of the separation point has some dependency on conditions computed on the windward side of the hill, so it is not definitely positioned. The standard K–ϵ model with wall function boundary conditions drastically underestimates the size of the separation zone behind the hill. This is generally understood to be the result of overpredicting turbulent energy in the upstream boundary layer, leading to a greater diffusion of momentum toward the wall and thus a delay in separation. Some improvement can be had by replacing the wall function boundary conditions with a low Reynolds number form of the K–ϵ closure. An even better prediction of the reattachment point, one that is slightly smaller than the experimental value can be achieved by the IP second moment closure with wall function boundary conditions, illustrated in Fig. 9.17. Apart from the near-wall behavior, which is neither predicted nor given by the experiment, the computed solution well captures the K distribution (scaled by U_c^2) at the top of the hill, a result that is shown in Fig. 9.18. It may be noticed that the model prediction of K rises to a peak very close to the surface which is determined by the wall function.

9.2.3 Diffuser Flow

The flow in diffusers is accompanied by an adverse pressure gradient associated with the deceleration of the bulk velocity caused by the widening of the passage. In many

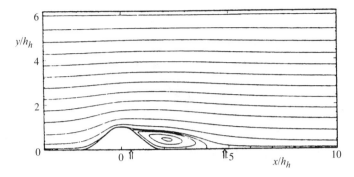

Fig. 9.17 *Mean streamlines for flow past a hill as computed using the IP model. Arrows indicate the separation and reattachment points in an experiment [2]. (From [35].)*

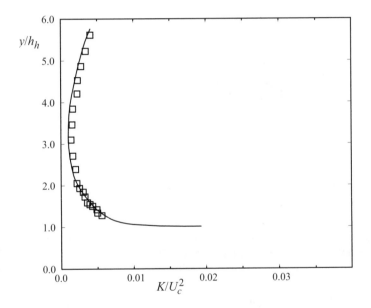

Fig. 9.18 *Prediction of K at the hill summit. —, IP model; □, experiment [2]. (From [35].)*

instances the slow-moving fluid adjacent to the wall is vulnerable to separation. It is a significant challenge to closure schemes to capture accurately the delicate force balance at play in these circumstances so that the size, location, and properties of the separated regions can be well predicted.

If interest is confined to predicting just the mean velocity, (i.e., higher order statistics such as K are not required), then one may hope that an EVM would be sufficient to make the prediction. It turns out, however, that simple EVM such as (8.14) [5] are unsuccessful in this instance, though credible predictions of $\overline{U}$ are possible with the extra empiricism built into the Johnson–King EVM [22]. For the diffuser with

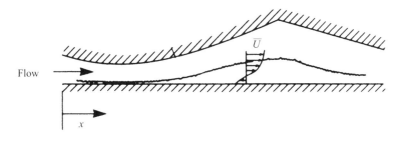

Fig. 9.19   *Diffuser flow field showing separation on the lower wall. (From [22].)*

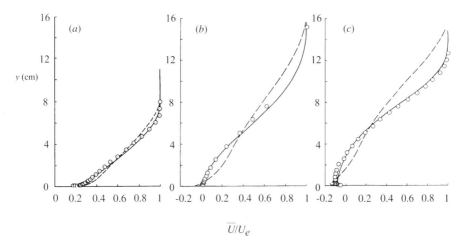

$$\overline{U}/U_e$$

Fig. 9.20 $\overline{U}$ *at three x locations in the diffuser flow shown in Fig. 9.19: (a) x = 3.01 m; (b) x = 3.42 m; (c) x = 3.972 m. U_e is the velocity at the boundary layer edge. − −, mixing-length model [5]; —, Johnson–King model; ○, experiments [48]. (From [22].)*

geometry shown in Fig. 9.19, this is illustrated in Fig. 9.20. The configuration on the upper wall leads to a significant adverse pressure gradient and separation off the lower wall just beyond $x = 3.42$ m. Predicted [22] and experimentally measured [48] mean velocity profiles upstream, within, and downstream of the separation zone are seen to be well matched, including the reverse flow in the separated region at $x = 3.972$ m.

The capabilities of higher-order closures in diffuser flows are illustrated for the plane asymmetric diffuser [4,46] shown in Fig. 9.21, in which flow separation occurs in the upper corner, where the duct returns to a constant area. In this case the Reynolds number is 21,200, based on the upstream centerline velocity and the inlet channel height. Compared in Fig. 9.22 are predictions of mean velocity and Reynolds stresses from a linear version of the K–ϵ closure [40] containing Reynolds-number-dependent coefficients, the nonlinear K–ϵ closure [54], and two relatively recent NLEVMs [4,39] that include cubic terms. The latter models differ from each other according

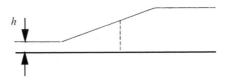

Fig. 9.21   *Asymmetric plane diffuser. Vertical line indicates where data in Fig. 9.22 are taken. (Reprinted from [4], Copyright (1997) with permission from Elsevier Science.)*

to how they treat the wall region. In one case [39], viscous damping functions such as those discussed in Section 8.9.2 are used. In the other low-Reynolds-number near-wall models are used whose coefficients depend on the shear and rotation invariants. The results in Fig. 9.22 are at a station approximately halfway through the diffuser. It is seen that the additional empiricism in [4] affects the predictions significantly by producing accuracy superior to that of the other closures. This model successfully predicts the asymmetry of the mean velocity and the Reynolds stresses and is the only one of the models tested that predicts separation at the upper corner, although judging by the predicted mean velocity profile, this is not as extensive as that seen in the experiments. On the whole, it is evident from Fig. 9.22 that the quantitative accuracy of the predictions has slipped significantly below that achieved in the calibration flows considered in Section 9.1.

9.3   ROTATION

The presence of system rotation at angular velocity $\mathbf{\Omega}^o$ is manifested in the equation for the velocity fluctuation (8.46) by the appearance of a Coriolis force term, $-2(\mathbf{\Omega}^o \times \mathbf{u}^*)_i$, where, as before, $(\ )^*$ denotes quantities as viewed by the rotating observer. The effect of this in the derived equation for $R_{ij}^* = \overline{u_i^* u_j^*}$, [in contrast to (8.63)], is to create a Coriolis term of the form

$$- 2\epsilon_{imn}\Omega_m^o R_{jn}^* - 2\epsilon_{jmn}\Omega_m^o R_{in}^*. \tag{9.2}$$

In the case of the rotating channel flow shown in Fig. 9.23, where the average motion is unidirectional (i.e., the mean streamwise velocity remains independent of x position), the influence of (9.2) on the dynamics is relatively easy to assess. Thus, with a rotation vector $\mathbf{\Omega}^o = (0, 0, \Omega^o)$, it follows from (9.2) that the $\overline{u^* v^*}$ equation contains the additional term

$$- 2\Omega^o (\overline{u^{*2}} - \overline{v^{*2}}). \tag{9.3}$$

Since, nominally, $\overline{u^{*2}} > \overline{v^{*2}}$ in channel flow, (9.3) is negative, in which case it acts to add to the Reynolds shear stress on the lower wall of the channel (where $\overline{u^* v^*} < 0$) and decrease it on the upper wall. Since enhanced Reynolds shear stress may be associated with heightened turbulence activity, the lower channel wall in this problem

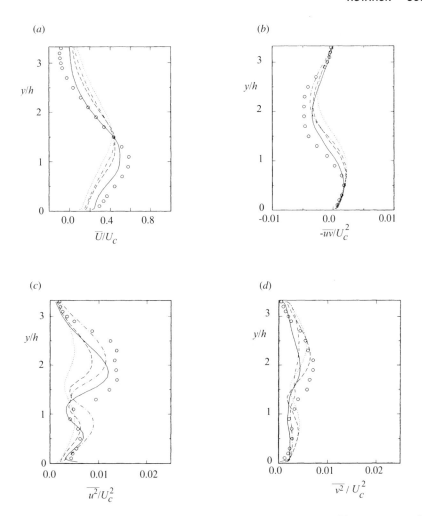

Fig. 9.22 *Predicted and measured turbulence statistics in plane diffuser flow: (a)* $\overline{U}/U_c$; *(b)* $-\overline{uv}/U_c^2$; *(c)* $\overline{u^2}/U_c^2$; *(d)* $\overline{v^2}/U_c^2$. ···, *K–ε closure [40];* – · –, *nonlinear K–ε closure [54];* – –, *NLEVM [39];* —, *NLEVM [4];* ○, *measurements [46]. (Reprinted from [4], Copyright (1997) with permission from Elsevier Science.)*

is referred to as the unstable side, while the upper wall is the stable side. Experiments show that when the rotation number,

$$R_o \equiv \frac{\Omega_o h}{U_m}, \tag{9.4}$$

where h is the channel width and U_m is the average mass flow velocity, is large enough, the flow on the upper wall relaminarizes. The lower and upper sides of the channel are also referred to as the "pressure" and "suction" sides, respectively, in

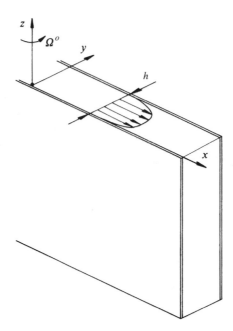

Fig. 9.23 *Geometry of rotating channel flow. (From [34]. Reprinted with the permission of Cambridge University Press.)*

analogy to similar phenomena seen in turbine and rotorcraft flows, in which the front side of the blade has high pressure and the rear side has low pressure.

To see why the Coriolis force has the effect it has on the Reynolds shear stress, first consider the part of (9.3) depending on $\overline{u^{*2}}$. This term has arisen from multiplying the Coriolis term in the v^* equation, namely $-2\Omega^o u^*$, by u^*. The expression $-2\Omega^o u^*$ represents a force deflecting $u^* > 0$ motions toward the lower channel wall and motions $u^* < 0$ toward the upper channel wall. In other words, $u^* > 0$ motions are accompanied by a move negative v^*, while $u^* < 0$ motions make v^* move positive. This is equivalent to a heightening of the potency of Q2 and Q4 events (i.e., events with $u^* < 0, v^* > 0$ and $u^* > 0, v^* < 0$), which were seen in Section 4.5.1 to underlie the $\overline{u^*v^*}$ correlation near the lower channel wall. Thus it is not surprising that the Reynolds shear stress is amplified on this side of the channel and that the opposite effect happens on the upper channel wall. This argument explains the influence of the term $-2\Omega^o \overline{u^{*2}}$ in (9.3).

A similar argument can be used to explain why the term $2\Omega^o \overline{v^{*2}}$ in (9.3) acts to diminish the Reynolds shear stress. In this case the Coriolis force in the u^* equation, namely $2\Omega^o v^*$, deflects motion in the $\pm y$ directions toward the $\pm x$ directions, respectively, which is, in effect, to reduce the potency of Q2 and Q4 events and with it the Reynolds stress. The cumulative effect of the two processes represented in (9.3) is to enhance Reynolds stress on the lower wall while decreasing it on the upper wall, presumably because the v^* fluctuations are much smaller than the u^* fluctuations, so the corresponding Coriolis effect is smaller.

Another interesting consequence of (9.2) is that there is an extra production term

$$4\Omega^o \, \overline{u^* v^*} \tag{9.5}$$

in the $\overline{u^{*2}}$ equation and an equal and opposite term in the $\overline{v^{*2}}$ equation. In view of the sign of $\overline{u^* v^*}$, (9.5) means that rotation diminishes $\overline{u^{*2}}$ and enhances $\overline{v^{*2}}$ on the lower channel wall, and vice versa on the upper wall. For a high enough R_o value, it can happen that $\overline{u^{*2}} < \overline{v^{*2}}$, in which case, according to (9.3), there is no new effect of rotation on enhancing or diminishing Reynolds stress. In fact, experiments show that the wall shear stress on the lower wall achieves a maximum when $R_o \approx 0.08$ and decreases slowly as R_o increases beyond this point [31].

Since (9.5) appears with equal and opposite magnitude in the normal Reynolds stress equations, there is clearly no net effect of rotation in the turbulent kinetic energy equation. Moreover, rotation effects do not appear in the mean momentum equation, except indirectly through the action of the Reynolds shear stress. Consequently, there is no mechanism by which a model such as the $K-\epsilon$ closure, which depends on (8.1), is able to account for the effects of rotation in a channel flow. Clearly, the minimum requirement necessary to do so is to have the capacity to model the anisotropy of the Reynolds stress tensor in unidirectional mean flows. Only more advanced closures such as SMC, NLEVM, or ARM are potentially useful for treating rotating channel flow.

Numerical calculations show that at relatively mild rotation rates more or less accurate predictions of mean velocity are achieved by the SSG model [55] and other SMCs, such as the Girimaji model [20], using wall function boundary conditions. For example, the mean velocity predicted by the latter model is compared to experiment in Fig. 9.24 for flow with $R_o = 0.069$ and $R_e = 11,500$. The model is successful in

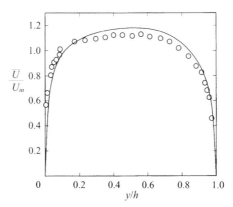

Fig. 9.24 $\overline{U}/U_m$ for rotating channel flow at $R_e = 11,500$, $R_o = 0.069$. —, Girimaji model [20]; ○, experiments [23]. (From [20]. Reprinted with the permission of Cambridge University Press.)

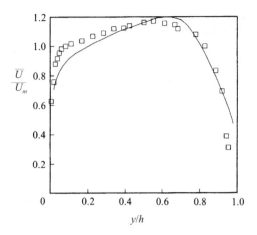

Fig. 9.25 $\overline{U}/U_m$ *for rotating channel flow at* $R_e = 11{,}500$, $R_o = 0.21$. —, *SSG model [55]; □,*
experiments [23]. (From [16]. Reprinted with the permission of Cambridge University Press.)

capturing the distinctive aspects of the stable and unstable sides of the channel. For
larger rotation rates, the qualitative behavior of $\overline{U}$ can still be captured, although
there is a sizable degradation in quantitative accuracy, as illustrated in Fig. 9.25 for a
rotating channel flow with the same Reynolds number as before, but now $R_o = 0.21$.

Some indication of how well the normal Reynolds stresses can be predicted via
SMC is shown in Fig. 9.26 for a low-Reynolds-number flow which has been simulated
using a DNS [63]. Here, $R_\tau = 194$ and $R_o = 0.05$. The SSG model is used with
a modified version of the elliptic relaxation model near the wall and an additional
production term added to the ϵ equation [63]. The excellent quality of the results in
Fig. 9.26 depends strongly on the attention paid to the near-wall modeling.

9.4 DUCT FLOWS

Another context where the success of closure models depends on making a credible
representation of turbulence anisotropy is the flow in a rectangular duct (i.e., channel
of finite width). Here, anisotropy of the normal Reynolds stresses, which is a fixture
of near-wall flow, causes the creation of secondary flows (i.e., mean flows in the
plane perpendicular to that of the bulk motion). The relationship between the two can
be seen clearly in the context of the steady equation for mean streamwise vorticity
assuming unidirectional flow, which may be expressed in the form

$$\overline{V}\frac{\partial \overline{\Omega}_1}{\partial y} + \overline{W}\frac{\partial \overline{\Omega}_1}{\partial z} = \overline{\Omega}_2 \frac{\partial \overline{U}}{\partial y} + \overline{\Omega}_3 \frac{\partial \overline{U}}{\partial z}$$

$$+ \nu \left(\frac{\partial^2 \overline{\Omega}_1}{\partial y^2} + \frac{\partial^2 \overline{\Omega}_1}{\partial z^2} \right) + \frac{\partial^2}{\partial y \partial z} \left(\overline{v^2} - \overline{w^2} \right) + \frac{\partial^2 \overline{vw}}{\partial y^2} - \frac{\partial^2 \overline{vw}}{\partial z^2}. \tag{9.6}$$

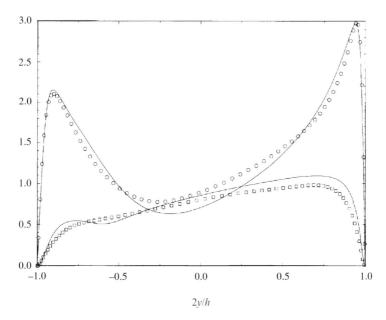

Fig. 9.26 *Turbulence statistics in rotating channel flow predicted by SMC [63] for $R_\tau = 194$, $R_o = 0.05$. Symbols denote DNS [3]:* o, $\overline{u^2}$; □, $\overline{v^2}$. *(Reprinted from [63], Copyright (1996) with permission from Elsevier Science.)*

Here the coordinates adhere to the previous notation (i.e., x is the streamwise and y and z the cross-stream coordinates). Equation (9.6) reveals circumstances under which it can be expected that flow with axial mean vorticity $\overline{\Omega}_1$, which is synonymous with secondary flows, is possible. First note that in a channel flow with homogeneity in the x and z directions, the right-hand side of (9.6) is identically zero and thus $\overline{\Omega}_1$ will always be zero. By adding walls so as to create a duct, homogeneity in the z direction is lost, and even if $\overline{\Omega}_1 = 0$ initially, the term in (9.6) depending on $\overline{v^2} - \overline{w^2}$ (a correlation that is nonzero even for channel flow), will then be nonzero and lead to $\overline{\Omega}_1 \neq 0$ and the presence of secondary mean motions. To the extent that the correlation $\overline{vw}$ (which vanishes in channel flow) develops, this will also affect the presence of secondary flows.

If a closure scheme is to capture secondary motions, it must be able to compute the development of $\overline{V}$ and $\overline{W}$ as they are affected by differences in $\overline{v^2}$ and $\overline{w^2}$, and $\overline{vw}$ if it is nonzero. If $\overline{V} = \overline{W} = 0$ initially, and (8.1) is used to determine the Reynolds stress so that both $\overline{v^2} = \overline{w^2}$ and $\overline{vw} = 0$, no secondary flow can be predicted. In contrast, higher-order models, such as the nonlinear K–ϵ closure, even with a zero initial state, will still generate secondary flows since such models generally predict a difference in $\overline{v^2}$ and $\overline{w^2}$ whenever $\partial \overline{U}/\partial y$ is not zero.

Predictions of secondary motions in duct flows were achieved by some of the first SMCs [14] and later by ARM and NLEVM methods. An example of the attainable

(a)   (b)

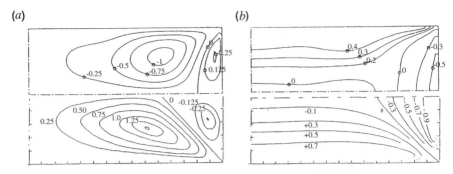

Fig. 9.27 *End-on view of rectangular duct flow: (a) mean streamlines; (b) $\overline{v^2} - \overline{w^2}$. Upper half, nonlinear K–ϵ closure [54]; lower half, experiment [21]. (Reprinted from [56], Copyright (1993) with permission from Elsevier Science.)*

(a)   (b)

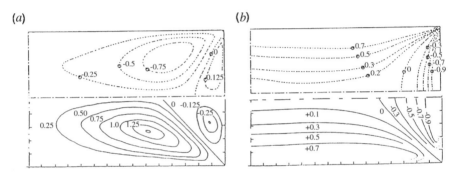

Fig. 9.28 *End-on view of rectangular duct flow: (a) mean streamlines; (b) $\overline{v^2} - \overline{w^2}$. Upper half, LRR closure [32]; lower half, experiment [21]. (Reprinted from [56], Copyright (1993) with permission from Elsevier Science.)*

accuracy is shown in Figs. 9.27 and 9.28, where predictions of the nonlinear K–ϵ closure [54] and the LRR form of a SMC [32] are compared to experimental measurements [21] of flow in a 3 × 1 rectangular duct [56]. Contours of secondary flow streamlines and the normal Reynolds stress difference, $\overline{v^2} - \overline{w^2}$, are given. The qualitative similarity of the predictions with experiments is reasonably good. Some quantitative differences are evident, with neither model being clearly superior with regard to both statistics.

9.5 CURVATURE

Fully developed flow in a straight channel (i.e., without curvature) is unidirectional and may be described via the mean velocity $\overline{U}(y)$ with a single shearing gradient, $d\overline{U}/dy$, as noted previously. If, on the other hand, the channel has streamwise curvature (i.e., changes direction as x increases), then $\overline{V} \neq 0$ and both $\overline{U}$ and $\overline{V}$ depend on both x and y. For this more general situation, the gradient $\partial \overline{U}/\partial x$ is nonzero, as are the derivatives of $\overline{V}$. The presence of such additional shearing and strains strongly affects the Reynolds stresses and must be accommodated by models if flow curvature is to be simulated accurately. To get a clear sense of what this entails, it is helpful to trace the effect of curvature through the Reynolds stress equation for a curved channel flow. For the straight channel it was shown in Section 4.2.5 that among the normal stresses, only $\overline{u^2}$ is produced directly by the mean shear. The cross-stream normal stresses $\overline{v^2}$ and $\overline{w^2}$ develop from the pressure–strain correlation. In addition, production of $\overline{uv}$ depends on the single term $-\overline{v^2}d\overline{U}/dy$, as shown in (4.31).

With curvature, the production term in the $\overline{uv}$ equation is now

$$-\overline{v^2}\frac{\partial \overline{U}}{\partial y} - \overline{u^2}\frac{\partial \overline{V}}{\partial x}. \tag{9.7}$$

In view of the relative sizes of $\overline{u^2}$ and $\overline{v^2}$, it may be imagined that the second term in (9.7) has the potentiality to exert a large influence on $\overline{uv}$. For convex flows, as in the case of flow over the crest of a hill, $\partial \overline{V}/\partial x < 0$, so the second term in (9.7) counteracts the production implied by the first term wherever $\overline{uv} < 0$. In a channel this would be the lower channel wall. Conversely, for concave flow, as happens in and out of a valley, for example, $\partial \overline{V}/\partial x > 0$, so (9.7) suggests that $\overline{uv}$ will be intensified on the lower wall. Looking at the channel as a whole, it is not hard to see that if it curves downward, $|\overline{uv}|$ decreases on the lower surface and rises on the upper surface and the opposite happens if the channel curves upward.

Another important aspect of the physics of curvature is felt in the production term in the normal Reynolds stress equation. Of most interest is $\overline{v^2}$, which, unlike the straight-channel-flow case, has nonzero components in the production term of the form

$$-\overline{uv}\frac{\partial \overline{V}}{\partial x} - \overline{v^2}\frac{\partial \overline{V}}{\partial y}. \tag{9.8}$$

The first term is likely to be the more significant because it contains the larger $\overline{V}$ gradient. Near the lower wall of a convex channel, (9.8) will represent destruction of $\overline{v^2}$. According to (9.7), this then accelerates the reduction in $\overline{uv}$ production. Conversely, for concave flow, there is heightened production of the Reynolds shear stress. These mechanisms help explain the great sensitivity of curved flows to anisotropy of the Reynolds stress and the fact that it is a challenge to model them with closure schemes.

The advantages of higher-order closures in comparison to the K–ϵ closure becomes apparent in computing flows where curvature is a significant factor. One classic example is the flow in the annular diffuser shown in Fig. 9.29. Here the flow undergoes two changes in direction. Near the inner wall the fluid experiences concave flow followed by convex flow, whereas the opposite occurs for flow near the outer wall. The mechanism discussed above should cause $\overline{uv}$ first to increase and then decrease in magnitude on the inner wall, while the opposite occurs on the outer wall.

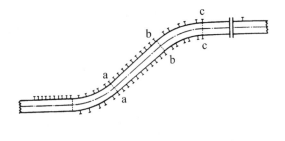

Fig. 9.29 *Geometry of annular diffuser flow. The dashed line is the axis of revolution. (From [31]. Copyright © Springer-Verlag.)*

The modeling of curvature effects is not well suited to EVMs such as the K–ϵ closure, since such methods cannot account for the subtle mechanisms associated with (9.7) and (9.8). On the other hand, SMCs fare much better in this situation, as illustrated in Fig. 9.30, where experimental measurements [57] of the mean velocity and Reynolds stresses in the annular diffuser are compared to predictions [31] of the K–ϵ closure and a SMC [25] which is closely related to the LRR model [32]. The first results are for Reynolds shear stress across the diffuser at a station immediately after the first turn. The absiccas of the plots are the normalized distance from the inner wall of the diffuser to the outer wall. The experiment shows that in agreement with the earlier discussion, $|\overline{uv}|$ increases on the lower, concave wall while it diminishes significantly on the upper, convex wall. This effect is underpredicted by the K–ϵ closure, but reasonably well accommodated by the SMC. At the second station, which is just prior to the final curve, some recovery of $|\overline{uv}|$ toward its symmetric state has taken place, particularly on the outer wall. Once again the K–ϵ closure fails to accommodate the lingering asymmetry between sides adequately, although the SMC has done a better job of this. The final figure shows that the mean velocity is asymmetric at the exit of the diffuser as a legacy of the different amounts of accumulated shearing on each wall. This effect is reproduced by the SMC and not captured successfully by the K–ϵ closure.

9.6 STAGNATION POINT FLOW

Numerous engineering flows contain flow impingement against a solid surface accompanied by a stagnation point: for example, the leading edge of an airfoil in flight, turbine blades, tube banks in boilers, and ramjet combustors. One common example that has served as an important test problem in the development of closure models is the case of a round jet impinging against a flat wall, as illustrated in Fig. 9.31. Several aspects of this flow are challenging to accommodate in turbulence models, including the modification of the turbulence in the incoming jet by curvature and the behavior of

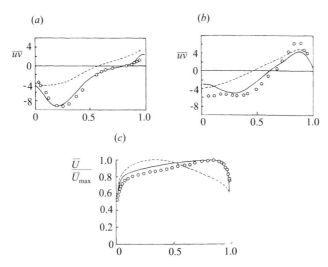

Fig. 9.30 *Statistics of annular diffuser flow: (a) $\overline{uv}$ on line a–a; (b) $\overline{uv}$ on line b–b; (c) $\overline{U}$ on line c–c.
—, SMC [25]; − −, K−ε closure; ○, experiments [57]. (From [31]. Copyright © Springer-Verlag.)*

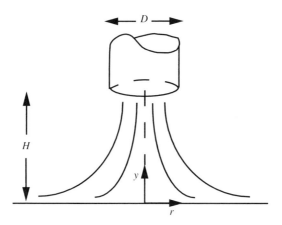

Fig. 9.31 *Geometry of stagnation flow in a round jet impinging on a flat surface.*

the turbulent field near the center of the jet as the fluid is forced to decelerate toward the stagnation point.

Of particular interest here is the responsiveness of turbulence models to the flow in the vicinity of the stagnation point. The mean velocity in this region is approximately of the form of an irrotational straining motion with components $\overline{U}(r)$ in the radial direction and $\overline{V}(y)$ in the axial direction. With increasing radial distance from the stagnation point, shearing effects grow in importance and the flow acquires more of the characteristics of a typical boundary layer. Close to the stagnation point, shearing can be largely neglected, and it is reasonable to imagine that the flow consists of the pure straining field $(\overline{U}(r), \overline{V}(y))$. In fact, the mean velocity field near the stagnation point is determined primarily by the balance of the pressure and inertia forces, with a relatively minor role for the Reynolds stresses. Attention may then be focused on how well turbulence models are able to reproduce the Reynolds stresses in the impingement region. These are of great consequence for problems where temperature fields and cooling rates are desired, since heat transfer is strongly tied to the Reynolds stress levels. Moreover, any turbulence generated at the stagnation point will be convected downstream to interact with the developing boundary layers.

The fact that the impinging flow is strain dominated, as against the shear-dominated mean fields that were encountered previously, has some significant consequences for the balance in the K equation. In particular, the production term in this case becomes

$$-\overline{u^2}\frac{\partial \overline{U}}{\partial r} - \overline{v^2}\frac{\partial \overline{V}}{\partial y} = -(\overline{u^2} - \overline{v^2})\frac{\partial \overline{U}}{\partial r} \tag{9.9}$$

since continuity implies that

$$\frac{\partial \overline{U}}{\partial r} + \frac{\partial \overline{V}}{\partial y} = 0. \tag{9.10}$$

Note that by the coordinate definitions, $\overline{v^2}$ is the normal stress in the flow direction and will tend to be larger than $\overline{u^2}$, which is for the transverse direction. Moreover, $\partial \overline{U}/\partial r > 0$, so (9.9) acts like a positive production term. For an EVM the production term is modeled via an expression of the form

$$-R_{ij}\frac{\partial \overline{U}_i}{\partial x_j} = \nu_t \frac{\partial \overline{U}_i}{\partial x_j}\left(\frac{\partial \overline{U}_i}{\partial x_j} + \frac{\partial \overline{U}_j}{\partial x_i}\right) = 4\nu_t\left(\frac{\partial \overline{U}}{\partial r}\right)^2. \tag{9.11}$$

To the extent that (9.11) disagrees with (9.9) [i.e., $4\nu_t \, \partial \overline{U}/\partial r$ differs from $(\overline{v^2} - \overline{u^2})$], the production of K near the stagnation point will be in error. In fact, it is usually the case that the standard K–ϵ closure significantly overpredicts production, with significant consequences for the downstream evolution of the impinging flow. For example, for flows striking a bluff body, the extra turbulence energy produced near the stagnation point by an EVM will move into the wake, causing excessive dissipation of vortices that have shed off the body. One obvious advantage that SMCs have in this flow is that they do not model production terms of the form (9.9). To capitalize on this advantage, a SMC must compute the difference $\overline{u^2} - \overline{v^2}$ accurately. It is less than assured that this can be done in practice.

Figure 9.32 compares experimental measurements of v_{rms}/U_m, where U_m is the average mass flow velocity of the jet exiting the orifice at $R_e \equiv U_m d/\nu = 23{,}000$ with predictions based on a low Reynolds number K–ϵ formulation [33], and several variants of a SMC. The latter uses the K–ϵ formulation within the viscous sublayer and incorporates wall reflection terms to modify the pressure-strain term in the near-

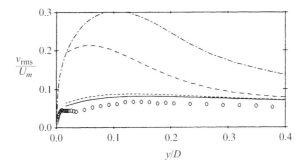

Fig. 9.32 v_{rms}/U_m *on the stagnation line* $r = 0$ *for the impinging jet with* $R_e = 23,000$ *and* $H/D = 2$. o, *experiments [7];* — · —, K–ϵ *closure [33];* — —, *SMC [18];* —, *SMC [11];* - - -, *SMC [10]. (Reprinted from [9], Copyright (1993) with permission from Elsevier Science.)*

wall region. The SMCs vary from a traditional form [18], to recent elaborations resulting from calibrating against the impinging flow data. According to Fig. 9.32, the enhanced modeling pays off in dramatically reducing the unacceptably large overprediction of turbulent energy by the K–ϵ closure, a trend that it may be noted is not reduced significantly by the unadorned SMC.

Despite very different experiences with regard to modeling the Reynolds stresses, the models under consideration do not give dramatically different predictions of the mean velocity amplitude at fixed radii from the jet axis, as shown in Fig. 9.33 for the distance $r/D = 1$. In all cases the agreement with experiment is respectable, although the specially calibrated SMC gives the best overall results.

9.7 PROLATE SPHEROID FLOW

The flow past a prolate spheroid[1] at angle of attack, as shown in Fig. 9.34, has the kind of multiple complex features that are often encountered in engineering design studies. Besides full three-dimensionality of the mean field, including extra shear and strains, there are curved boundary layers, separation, impingement, transitional regions, and wake flow. Drag and lift forces are important to predict, as well as velocity profiles and the locations of separated regions. Extensive experimental measurements of the flow past a 6:1 prolate spheroid (i.e., where the length is six times the width) have been reported [6,30,44,62]. These are a rich source of information about the flow physics. Attempts to model the prolate–spheroid flow run the gamut from simple EVM, including the Baldwin–Lomax and Johnson–King models [17,61], to the K–ϵ closure with wall functions [60], a near-wall formulation of the K–ϵ closure [42], and to anisotropy resolving methods, including NLEVM and SMC [42]. A strong argument can be made that it is only with the anisotropy predicting methods that some

[1]The three-dimensional body of revolution formed by an ellipse.

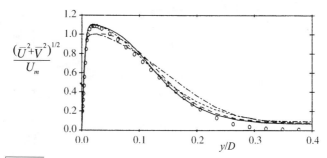

Fig. 9.33 $\sqrt{\overline{U^2} + \overline{V^2}}/U_m$ *on the line,* $r = D$ *for the impinging jet with* $R_e = 23{,}000$ *and* $H/D = 2$. ∘, *experiments [7];* — · —, K–ϵ *closure [33];* — —, *SMC [18];* —, *SMC [11];* - - -, *SMC [10]. (Reprinted from [9], Copyright (1993) with permission from Elsevier Science.)*

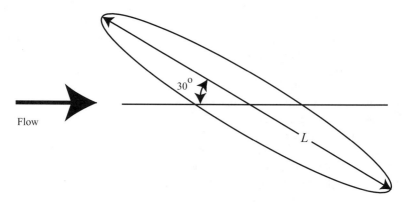

Fig. 9.34 *Geometry of flow past a 6:1 prolate spheroid.*

of the important physical features of the prolate spheroid flow succumb to modeling, although there is still considerable room to improve accuracy of the predictions.

The flow to be considered here is at $R_e = 6.5 \times 10^6$ [30,44] based on body length L and freestream velocity, and has 30° angle of attack. Despite how large R_e is, the flow in a significant region centered on the impingement line along the lower (windward) surface is laminar and/or transitional. The flow on the back (leeward) side is more definitely turbulent, as is the flow in the wake. This poses a problem for RANS calculations, since they are predicated on the assumption that the underlying flow is definitely turbulent (i.e., they are usually not designed with a capability for accommodating laminar and transitional flows). This places a severe test on the ability of the closure not to seriously distort calculation of the turbulent part of the flow due to errors in predicting the nonturbulent part.

The prolate spheroid flow at an angle of attack is another example of one where large-scale periodic features may be present. As in the case of the backstep flow, for example, there is likely to be shedding of vortices from the rear of the prolate spheroid, in this instance alternating from one side to the other. Such physical attributes may possibly be resolved by a nonsteady RANS calculation, although the prior analysis in this area has been exclusively to predict the averaged steady-state flow (i.e., as if the flow statistics are obtained by long-time averaging).

The particular solutions presented here are from a recent comparison of the capabilities of a low-Reynolds-number variant of the $K-\epsilon$ closure [41], a quadratic NLEVM [42], and a SMC [18]. In the latter case the near-wall $K-\epsilon$ closure is used in the near-wall region. Since turbulence models produce flow fields that are "turbulent" whenever they are applied, they will not normally give a prediction of the regions in a flow that are laminar or transitional, even if a flow happens to contain such regions. For example, models that predict turbulent energy are not necessarily likely to predict that $K = 0$ in laminar regions. Consequently, it is to be expected that this will affect the accuracy of RANS solutions to prolate spheroid flow. An exception to these observations occurs in the case of the NLEVM, which, it turns out, does display some sensitivity to the flow state. This is illustrated in Fig. 9.35, giving the magnitude of the skin friction coefficient, C_f, at the streamwise station $x/L = 0.139$, which is near the front of the prolate spheroid. The horizontal axis denotes the azimuthal direction in degrees running from 0° at the windward (bottom) surface symmetry line to 180° at the leeward (top) surface symmetry line. The average flow is symmetric about these lines.

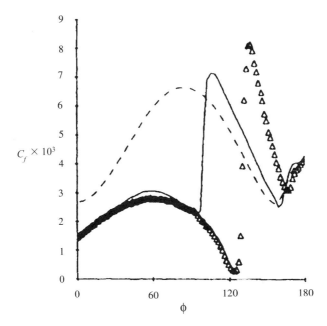

Fig. 9.35 *Skin friction coefficient magnitude at $x/L = 0.139$ computed using NLEVM [42]. —, 0.2% free-stream turbulence; − −, 3% free-stream turbulence level. Experimental result [30,44] indicated by △. (From [42] by permission of The Royal Aeronautical Society.)*

The result of calculations in which the free-stream turbulence level was set at 0.2%, matching the experimental value, and 3%, which is substantially higher, are shown in the figure. For $0° \leq \theta \leq 120°$ the experimental flow is laminar, and transition is indicated by the large increase in C_f just beyond $\theta = 120°$. The model prediction for the case with a small background turbulence level follows the laminar result until a sharp transition occurs somewhat prematurely. For the model case with larger free-stream turbulence, there is no transition and C_f is turbulentlike and overpredicted everywhere. The favorable performance of the NLEVM in this instance, at least when compared to the other schemes, may be attributed to how it handles turbulence production in the neighborhood of the impingement point. In particular, this model shows considerable restraint, consistent with the true physics, in not overproducing turbulence so that "transition" is not provoked too early. Eventually, the accumulating effect of turbulence diffusion to the wall region from the free stream and the turbulence generated by shear does lead to transition. Clearly, this is a delicate process for any turbulence model to get correct.

On the leeward side of the prolate spheroid, Fig. 9.35 shows that the model predictions become very close to each other despite the very different experiences on the windward side. In fact, at later positions farther along the prolate spheroid, the model solutions are essentially independent of free-stream turbulence levels. The conclusion may be drawn that at least for this particular flow, accurate treatment of the laminar and transitioning flow on the windward side is not necessarily critical to the overall performance of the closure models.

Figure 9.36 shows the transverse velocity vectors on a cut through the flow field at $x/L = 0.917$, as predicted by the models and experiment. The separation off the side leading to a large vortex is clear. Reattachment occurs before the leeward symmetry line. For the experiment there is evidence of a secondary vortex that is not found in the model calculations. Note that the recirculation zones predicted by the NLEVM and SMC are more nearly similar to the experiment than the EVM. In all cases the strength of the physical vortex appears to be greater than that of the model predictions.

A view of the overall flow topology is provided by the surface streaklines shown in Fig. 9.37, determined by experiment and the three turbulence models. These are created by projecting half the surface of the prolate spheroid onto a plane surface. The lower axis is along the windward symmetry line from which the flow curls around the body toward a separation line enclosing an increasing part of the leeward side of the body. The separation line forms the outer edge of a streamwise vortex formed by the separating flow. The spiraling fluid reattaches to the body at the line at approximately 150° bounding the separated zone. The secondary flow found in the experimental result is indicated by a third separation line at 120°. The three models capture a qualitative sense of the flow in many but not all aspects. For example, besides the absence of secondary motion, the extent of the separated region is exaggerated on the leeward side near the end.

A quantitative sense of how well the closures do is provided by Fig. 9.38, showing the friction coefficient magnitude and direction at $x/L = 0.565$, and Fig. 9.39, showing the pressure coefficient C_p at $x/L = 0.835$. In the former case it is clear that the laminar behavior on the windward side is entirely missed, as discussed previously.

(a)

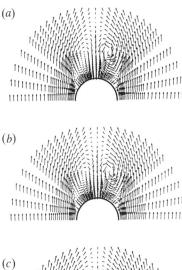

(b)

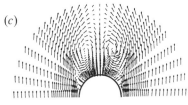

(c)

Fig. 9.36   *Predicted transverse velocity field at $x/L =$ 0.917 (on the left side) compared with experiment [30, 44] (on the right side): (a) EVM [41]; (b) SMC [18]; (c) NLEVM [42]. (From [42] by permission of The Royal Aeronautical Society.)*

More hopefully, some of the complexity of the separated turbulent flow on the back is captured successfully. Generally, the EVM is least capable among the models. The same conclusion follows for the pressure coefficient with regard to how well the models capture the extent of the plateau beginning at $100°$. There is substantially less agreement for the EVM, which also predicts too much pressure recovery.

9.8 SUMMARY

This chapter has surveyed the capabilities of RANS models in predicting turbulent flows. The conclusion may be reached that with proper attention to the physics, many flows, including those of practical importance, can be reasonably well modeled through the RANS approach. The accuracy of the representation depends strongly on the complexity of the flow. Moreover, some degree of empiricism appears to be necessary to get the most favorable results. This raises a question as to whether or not RANS modeling can deliver true predictability in flows for which nothing is known a priori. It is hard to answer this question for any given flow, but it may be deduced from the examples given above that the best hope of success depends on having some sense of which physical phenomena are present and important and then using a RANS model that has known capabilities in this regard in other contexts.

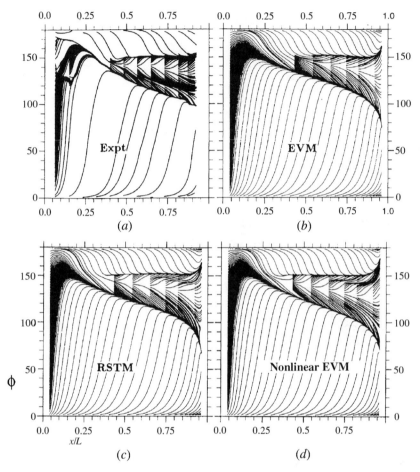

Fig. 9.37 *Surface streaklines on prolate spheroid at a 30° angle of attack: (a) experiment [30,44]; (b) EVM [41]; (c) SMC [18]; (d) NLEVM [42]. S is the fractional arc length along the surface. (From [42] by permission of The Royal Aeronautical Society.)*

It has been made clear that despite their favorable numerical advantages, relatively simple models that predict only mean velocity ought to find application mostly to simple flows. Even the flexibility provided by the $K-\epsilon$ closure is insufficient for many applications where anisotropy plays an important role in the physics. In fact, the complexity of many flows (e.g., curvature, secondary flows, impingement, and rotation) is strongly tied to the dynamical effect of anisotropy, and there would appear to be little alternative in such cases except to utilize the more sophisticated closures, which model anisotropy [37,38]. It is also clear that the mere fact of accommodating anisotropy is not sufficient to expect acceptable performance from a closure. Many of the best predictions have followed from closures that extensively model the particulars of the near-wall region where the effective Reynolds number is low. Demanding

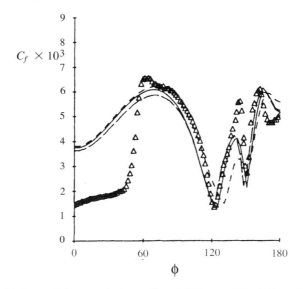

Fig. 9.38 *Skin friction coefficient magnitude at $x/L = 0.565$. $- -$, EVM [41]; —, SMC [18]; ——, NLEVM [42]; $\triangle$, experiments [30,44]. (From [42] by permission of The Royal Aeronautical Society.)*

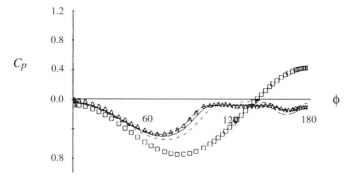

Fig. 9.39 *Pressure coefficient at $x/L = 0.835$. $- -$, EVM [41]; —, SMC [18]; ——, NLEVM [42]; $\triangle$, experiments [30,44]; $\square$, inviscid solution. (From [42] by permission of The Royal Aeronautical Society.)*

consistency with asymptotic relations at the surface appears to be helpful, as well as calibration against DNS or experimental data for similar flow fields.

In comparison to more computationally intensive approaches, the cost-effectiveness of RANS modeling and the large body of achievement in many flows of engineering relevance assure that classical closure modeling will continue to be a dominant tool into the future. On the other hand, its inability to break free of a significant need for empiricism has encouraged the development of DNS and LES techniques, which have less or no dependence on modeling. The rise of LES has been particularly fast in recent years, and this methodology is described in Chapter 10.

REFERENCES

1. Aksoy, H. and So, R. M. C. (1995) "Effects of numerical methods and near-wall Reynolds stress modeling on the prediction of backstep flow," in *Separated and Complex Flows* (M. V. Otugen et al., Eds.), ASME International, pp. 265–272.

2. Almeida, G. B., Durao, D. F. G. and Heitor, M. V. (1992) "Wake flow behind two-dimensional model hills," *Exp. Thermal Fluid Sci.* **7**, 87–101.

3. Andersson, H. I. and Kristoffersen, R. (1993) "Reynolds stress budgets in rotating channel flow," *Proc. 9th Symposium on Turbulent Shear Flows*, Kyoto, Japan.

4. Apsley, A. D. and Leschziner, M. A. (1997) "A new low-Re non-linear two-equation turbulence model for complex flows," *Int. J. Heat Fluid Flow* **19**, 209–222.

5. Cebeci, T. and Smith, A. M. O. (1974) *Analysis of Turbulent Boundary Layers*, Academic Press, New York.

6. Chesnakas, C. J. and Simpson, R. L. (1997) "Detailed investigation of the three-dimensional separation about a 6:1 prolate spheroid," *AIAA J.* **35**, 990–999.

7. Cooper, D., Jackson, D. C., Launder, B. E. and Liao, G. X. (1993) "Impinging jet studies for turbulence model assessment. I. Flow-field experiments," *Int. J. Heat Mass Transfer* **36**, 2675–2684.

8. Craft, T. J. (1998) "Developments in a low-Reynolds-number second-moment closure and its application to separating and reattaching flows," *Int. J. Heat Fluid Flow* **19**, 541–548.

9. Craft, T. J., Graham, L. J. W. and Launder, B. E. (1993) "Impinging jet studies for turbulence model assessment. II. An examination of the performance of four turbulence models," *Int. J. Heat Mass Transfer* **36**, 2685–2697.

10. Craft, T. J. and Launder, B. E. (1991) "Computation of impinging flows using second moment closures," *Proc. Turbulent Shear Flows 8*, Munich, Germany, pp. 5–8.

11. Craft, T. J. and Launder, B. E. (1992) "New wall-reflection model applied to the turbulent impinging jet," *AIAA J.* **30**, 2970–2972.

12. Craft, T. J. and Launder, B. E. (1996) "A Reynolds stress closure designed for complex geometries," *Int. J. Heat Fluid Flow* **17**, 245–254.

13. Craft, T. J., Launder, B. E. and Suga, K. (1995) "A non-linear eddy viscosity model including sensitivity to stress anisotropy," *Proc. Turbulent Shear Flows 10*, Vol. 3, pp. 23-19 to 23-24.

14. Demuren, A. O. and Rodi, W. (1984) "Calculation of turbulence-driven secondary motion in non-circular ducts," *J. Fluid Mech.* **140**, 189–222.

15. Driver, D. M and Seegmiller, H. L. (1983) "Features of a reattaching turbulent shear layer in divergent channel flow," *AIAA J.* **23** 163–171.

16. Gatski, T. B. and Speziale, C. G. (1993) "On explicit algebraic stress models for complex turbulent flows," *J. Fluid Mech.* **254**, 59–78.

17. Gee, K., Cummings, R. M. and Schiff, L. B. (1992) "Turbulence model effects on separated flow about a prolate spheroid," *AIAA J.* **30**, 655–664.

18. Gibson, M. M. and Launder, B. E. (1978) "Ground effects on pressure fluctuations in the atmospheric boundary layer," *J. Fluid Mech.* **86**, 491–511.

19. Girimaji, S. S. (1996) "Fully explicit and self-consistent algebraic Reynolds stress model," *Theor. Comput. Fluid Dyn.* **8**, 387–402.

20. Girimaji, S. S. (2000) "Pressure–strain correlation modeling of complex turbulent flows," *J. Fluid Mech.* **422**, 91–123.

21. Hoagland, L. C. (1960) Ph.D. dissertation, MIT.

22. Johnson, D. A. and King, L. S. (1985) "A mathematically simple turbulence closure model for attached and separated turbulent boundary layers," *AIAA J.* **23**, 1684–1692.

23. Johnston, J. P., Halleen, R. M. and Lezius, D. K. (1972) "Effects of spanwise rotation on structure of 2 dimensional fully developed turbulent channel flow," *J. Fluid Mech.* **56**, 533–557.

24. Jones, W. P. and Launder, B. E. (1972) "The prediction of laminarization with a two-equation model of turbulence," *Int. J. Heat Mass Transfer* **15**, 301–314.

25. Jones, W. P. and Manners, A. (1989) "The calculation of the flow through a two-dimensional faired diffuser," in *Proc. Turbulent Shear Flows 6* (J. C. Andre et al., Eds.), Springer-Verlag, New York, pp. 18–31.

26. Jovic, S. and Driver, D. (1994) "Backward-facing step measurements at low Reynolds number," *NASA TM-108870.*

27. Karlsson, R. I. and Johansson, T. G. (1988) "LDV measurements of higher order moments of velocity fluctuations in a turbulent boundary layer," in *Laser Anemometry in Fluid Mechanics* (D. F. G. Durao et al., Eds.), Ladoan-Instituto Superior Tecnico, Portugal, pp. 273–289.

28. Kim, J., Kline, S. J. and Johnston, J. P. (1980) "Investigation of a reattaching turbulent shear layer: flow over a backward-facing step," *J. Fluids Eng.* **102**, 302–308.

29. Kim, J., Moin, P. and Moser, R. (1987) "Turbulence statistics in fully developed channel flow at low Reynolds number," *J. Fluid Mech.* **177**, 133–166.

30. Kreplin, H. P., Vollmers, H. and Meier, H. U. (1985) "Wall shear stress measurements on an inclined prolate spheroid in the DFVLR 3 m × 3 m low speed wind tunnel, Gottingen. *DFVLR Rep. IB 222-84 A 33.*

31. Launder, B. E. (1990) "Phenomenological modeling: present . . . and future? *Proc. Whither Turbulence Workshop*, Lecture Notes in Physics (J. L. Lumley, Ed.), Springer-Verlag, Berlin, pp. 439–485.

32. Launder, B. E., Reece, G. J. and Rodi, W. (1975) "Progress in the development of a Reynolds stress turbulence closure," *J. Fluid Mech.* **68**, 537–566.

33. Launder, B. E. and Sharma, B. I. (1974) "Application of the energy dissipation model of turbulence to the calculation of flow near a spinning disc," *Lett. Heat Mass Transfer* **1**, 131–138.

34. Launder, B. E., Tselepidakis, D. P. and Younis, B. A. (1987) "A 2nd moment closure study of rotating channel flow," *J. Fluid Mech.* **183**, 63–75.

35. Laurence, D. (1997) "Applications of Reynolds averaged Navier–Stokes equations to engineering problems," *Von Kármán Inst. Lect. Ser. 1997-03.*

36. Le, H., Moin, P. and Kim, J. (1997) "Direct numerical simulation of turbulent flow over a backward-facing step," *J. Fluid Mech.* **330**, 349–374.

37. Leschziner, M. A. (2000) "The computation of turbulent engineering flows with turbulence-transport closures," in *Advanced Turbulent Flow Computations* (R. Peyret and E. Krause, Eds.), Springer-Verlag, New York, pp. 209–278.

38. Leschziner, M. A. (2000) "Turbulence modeling for separated flows with anisotropy-resolving closures," *Philos. Trans. R. Soc. Ser. A* **358**, 3247–3277.

39. Lien, F. S., Chen, W. L. and Leschziner, M. A. (1996) "Low-Reynolds-number eddy-viscosity modeling based on nonlinear stress-strain/vorticity relations," *Proc. 3rd Symposium on Engineering Turbulence Modeling and Measurements*, Crete.

40. Lien, F. S. and Leschziner, M. A. (1993) "A pressure–velocity solution strategy for compressible flow and its application to shock/boundary-layer interaction using second-moment turbulence closure," *J. Fluids Eng.* **115**, 717–725.

41. Lien, F. S. and Leschziner, M. A. (1993) "Computational modeling of 3D turbulent flow in S-diffuser and transition ducts," *Engineering Turbulence Modeling and Measurements 2*, Elsevier, Amsterdam, pp. 217–228.

42. Lien, F. S. and Leschziner, M. A. (1997) "Computational modeling of separated flow around a streamlined body at high incidence," *Aeronaut. J.* **101**, 269–275.

43. Mansour, N. N., Kim, J. and Moin, P. (1988) "Reynolds-stress and dissipation-rate budgets in a turbulent channel flow," *J. Fluid Mech.* **194**, 15–44.

44. Meier, H. U., Kreplin, H. P., Landhauser, A. and Baumgarten, D. (1984) "Mean velocity distributions in three-dimensional boundary layers developing on a 1:6 prolate spheroid with artificial transition ($\alpha = 10°$, $U_\infty = 55\mathrm{ms}^{-1}$, cross sections $x_0/2a = 0.48, 0.56, 0.64$ and 0.73). *DFVLR Rep. IB 222 A 11.*

45. Moser, R. D., Kim, J. and Mansour, N. N. (1999) "DNS of turbulent channel flow up to $R_\tau = 590$," *Phys. Fluids* **11**, 943–945.

46. Obi, S., Aoiki, K. and Masuda, S. (1993) "Experimental and computational study of turbulent separated flow in an asymmetric diffuser," *Proc. 9th Symposium on Turbulent Shear Flows*, Kyoto, Vol. 3, p. 305.

47. Sarkar, A. and So, R. M. C. (1997) "A critical evaluation of near-wall two-equation models against direct numerical simulation data," *Int. J. Heat Fluid Flow* **18**, 197–208.

48. Simpson, R. L., Chew, Y. T. and Shivaprasad, B. G. (1981) "The structure of a separating turbulent boundary layer. I. Mean flow and Reynolds stresses," *J. Fluid Mech.* **113**, 23–51.

49. So, R. M. C., Aksoy, H. and Yuan, S. P. (1996) "Modeling Reynolds-number effects in wall-bounded turbulent flows," *ASME J. Fluids Eng.* **118**, 260–267.

50. So, R. M. C. and Yuan, S. P. (1998) "Near-wall two-equation and Reynolds-stress modeling of backstep flow," *Int. J. Eng. Sci.* **36**, 283–298.

51. Spalart, P. R. (1988) "Direct simulation of a turbulent boundary layer up to $\mathrm{Re}_\theta = 1410$," *J. Fluid Mech.* **187**, 61–98.

52. Spalart, P. R. (1986) "Numerical study of sink flow boundary layers," *J. Fluid Mech.* **172**, 307–326.

53. Spalart, P. R. (2000) "Strategies for turbulence modeling and simulations," *Int. J. Heat Fluid Flow* **21** 252–263.

54. Speziale, C. G. (1987) "On nonlinear $K–l$ and $K–\epsilon$ models of turbulence," *J. Fluid Mech.* **178**, 459–475.

55. Speziale, C. G., Sarkar, S. and Gatski, T. B. (1991) "Modeling the pressure–strain correlation of turbulence: an invariant dynamical systems approach," *J. Fluid Mech.* **227**, 245–272.

56. Speziale, C. G., So, R. M. C. and Younis, B. A. (1993) "On the prediction of turbulent secondary flows," in *Near Wall Turbulent Flows* (R. M. C. So et al., Eds.), Elsevier Science, Amsterdam, pp. 105–114.

57. Stevens, S. J. and Fry, P. (1973) "Measurements of the boundary-layer growth in annular diffusers," *J. Aircr.* **10**, 73–80.

58. Thangam, S. and Hur, N. (1991) "A highly resolved numerical study of turbulent separated flow past a backward-facing step," *Int. J. Eng. Sci.* **29**, 607–615.

59. Thangam, S. and Speziale, C. G. (1992) "Turbulent flow past a backward-facing step: a critical evaluation of two-equation models," *AIAA J.* **30**, 1314–1320.

60. Tsai, C.-Y. and Whitney, A. K. (1999) "Numerical study of three-dimensional flow separation for a 6:1 ellipsoid," *AIAA Pap. 99-0172*.

61. Vatsa, V. N., Thomas, J. L. and Wedan, B. W. (1989) "Navier–Stokes computations of prolate spheroids at angle of attack," *J. Aircr.* **26**, 986–993.

62. Wetzel, T. G., Simpson, R. L. and Chesnakas, C. J. (1998) "Measurement of three-dimensional crossflow separation," *AIAA J.* **36**, 557–564.

63. Wizman, V., Laurence, D., Kanniche, M., Durbin, P. and Demuren, A. (1996) "Modeling near-wall effects in second-moment closures by elliptic relaxation," *Int. J. Heat Fluid Flow* **17**, 255–266.

10

Large Eddy Simulations

Turbulence modeling as it is formulated in Chapters 8 and 9 has the onerous burden of deciphering the relationship between transport correlations, such as the Reynolds stresses, and the complex flow physics that might cause them. If there is any lesson to be learned from this survey of RANS models, it is that this is an exceedingly difficult task to accomplish with reliable accuracy. At the same time, the alternative, of not a priori averaging the Navier–Stokes equation, so that solutions to flow problems are in the form of DNS, is not practical for most turbulent engineering applications, owing to the wide range of scales in need of resolution and the limits of computational speed. In consequence of this state of affairs, the last decade has seen the rise of large eddy simulations (LES), which occupy a middle ground between RANS and DNS. In this, the idea is to compute the dynamics of as much of the large energy containing scales of motion as is economically feasible while modeling only the effects of small, unresolved phenomena on the larger, resolved scales. By *resolved* is meant that part of the velocity field, **U**, which is computed, versus the unresolved part, which is not. Formally, the resolved field is that part of **U** remaining after application of a filtering operation which ostensibly removes (i.e., averages away) the small scales. Note that unlike traditional averaging, the fact that a distinction is made between resolved and unresolved scales means that there is an implied length scale upon which any particular LES calculation must depend.

In the traditional approach to LES the filtering operation is applied to the Navier–Stokes equation to obtain a governing relation for filtered velocity. Very much in the same way that averaging the Navier–Stokes equation creates a need for modeling the Reynolds stresses, the end result of filtering the Navier–Stokes equation is to create the need to model *subgrid-scale stresses*. How the latter goal is achieved, resulting in a closed system of equations, distinguishes one LES method from another.

Most of the following discussion considers aspects of traditional LES, which normally depends on the use of a grid-based numerical scheme. It is also possible to formulate LES using grid-free vortex methods in which computational elements are taken to be convecting vortices. A description of one such gridfree LES is presented in Section 10.6.

For all LES methods the choice of scale demarcating the boundary between resolved and unresolved motions establishes both the size of the numerical burden as well as the degree to which reliance is placed on modeling. Usually, the models themselves are meant only to be applied for scales below a certain point (e.g., at the size of the inertial range). It very well may be the case that this length is so small as to demand an impractical level of computational resources (e.g., a grid that is too large to be useful). The end result is then a trade-off between modeling accuracy and the size and speed of computations. As computers continue to get faster, however, the scale of LES calculations is increasing with a natural shift in the direction of less modeling and more simulation. This eases the problems associated with physical modeling and increases the range of flows to which LES can be applied successfully. Even though LES is at present feasible only for a limited range of Reynolds numbers, and the skill of subgrid modeling is not as proficient as one would want, nevertheless, it is widely believed that the development problems of LES are less than those faced by alternative approaches such as RANS modeling, so the pursuit of LES methods has become a major aspect of modern-day turbulent flow research.

10.1 FILTERS

The traditional approach to LES requires the solution of closed equations for the filtered velocity field on a numerical mesh. Implementations of LES vary according to which filter is used and the specifics of the subgrid-scale model. A filter is a function of the form $G(\mathbf{x}, \mathbf{y})$ which is used to produce the filtered velocity field, as in

$$\langle \mathbf{U} \rangle (\mathbf{x}, t) \equiv \int_{\Re^3} G(\mathbf{x}, \mathbf{y}) \mathbf{U}(\mathbf{y}, t) \, d\mathbf{y}, \tag{10.1}$$

and is normalized by the condition that

$$\int_{\Re^3} G(\mathbf{x}, \mathbf{y}) \, d\mathbf{y} = 1, \tag{10.2}$$

which guarantees that filtering of a constant field returns the same constant. The unresolved part of the velocity field is denoted as $\mathbf{u}'$ and is defined according to

$$\mathbf{u}' \equiv \mathbf{U} - \langle \mathbf{U} \rangle. \tag{10.3}$$

Unlike averaging, it is not necessarily true that $\langle \mathbf{u}' \rangle = 0$, although for some filters this condition is satisfied. Whether or not it occurs depends, according to (10.3), on having $\langle \langle \mathbf{U} \rangle \rangle = \langle \mathbf{U} \rangle$, a property that does not necessarily follow from the definition (10.1).

For flows that are homogeneous, it is natural to select a filter function having the homogeneous form

$$G(\mathbf{x}, \mathbf{y}) = G^*(\mathbf{x} - \mathbf{y}), \tag{10.4}$$

in which case (10.1) gives

$$\langle \mathbf{U} \rangle (\mathbf{x}, t) = \int_{\Re^3} G^*(\mathbf{x} - \mathbf{y}) \mathbf{U}(\mathbf{y}, t) \, d\mathbf{y}. \tag{10.5}$$

Assuming further that $G^*(\mathbf{r})$ has bounded support, it may be shown that the filtering operation, just like the averaging operation, has the helpful property of commuting with differentiation [24]:

$$\left\langle \frac{\partial U_i}{\partial x_j} \right\rangle = \frac{\partial \langle U_i \rangle}{\partial x_j}. \tag{10.6}$$

This property is particularly helpful in simplifying the governing equations for filtered velocity, as will be seen below.

For nonhomogeneous flows, which includes most flows, the filter needs to reflect local changes in the flow scale. For example, as a boundary is approached, one expects filtering to be performed over a smaller region. Filters in the form (10.4) are incompatible with this condition, and an inhomogeneous filter has to be used instead. It is usual practice in such cases to assume the validity of (10.6) with the justification that whatever errors are produced from its not being exactly true are typically of the same order of magnitude as the errors associated with solving the governing equations numerically. However, if a higher-order numerical scheme is used to solve the LES equations, the errors generated by (10.6) may have to be given analytic form and evaluated as part of the numerical scheme [15].

It is also common practice to consider filters that can be written as products of one-dimensional filters, say $H(r)$, so that, for example,

$$G^*(\mathbf{x} - \mathbf{y}) = H(x_1 - y_1) H(x_2 - y_2) H(x_3 - y_3). \tag{10.7}$$

Here it is assumed that the integral of H over the real line is unity so that (10.2) is satisfied. It is also frequently the case that different one-dimensional filters are used in different directions, the goal being to take advantage of various flow symmetries. Moreover, the filters may be homogeneous in some directions and inhomogeneous in others. For example, in a channel flow, the streamwise and spanwise directions are homogeneous and the wall-normal direction is not, so filters reflecting this property can be applied. In fact, the inhomogeneous direction in this case is often not given a filter, with reliance instead placed on the inevitable smoothing effect of replacing derivative expressions in this direction with their finite difference counterparts.

There are three particular one-dimensional filters that are commonly used in LES. The first is the *Gaussian filter*,

$$H(r) = \sqrt{\frac{6}{\pi h^2}} \, e^{-6r^2/h^2}, \tag{10.8}$$

in which h is a length scale reflecting the boundary between resolved and unresolved parts of the turbulent field. The *tophat filter* is defined by

$$H(r) = \begin{cases} 1/h & \text{if } |r| \leq h/2 \\ 0 & \text{otherwise,} \end{cases} \qquad (10.9)$$

while the *sharp Fourier cutoff filter* is given by

$$H(r) = \frac{\sin(\pi r/h)}{\pi r}. \qquad (10.10)$$

The plot of these functions in Fig. 10.1 shows that they have quite different properties.

The tophat filter is tantamount to equating filtering with local averaging (i.e., averaging over a moving window). The Gaussian filter is a similar idea except that the Gaussian kernel places greater weight on the local field. Equation (10.10) is called the sharp Fourier cutoff filter because it is the inverse Fourier transform of what amounts to a tophat function in Fourier space (i.e., a filter that cuts off contributions from all wavenumbers with magnitudes greater than a given value). To see this, define the Fourier transform of H by

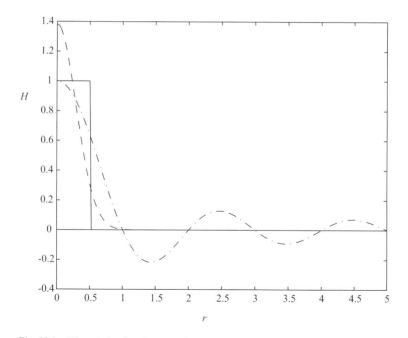

Fig. 10.1 *Filters (when $h = 1$).* —, *tophat;* – –, *Gaussian;* – · –, *sharp Fourier cutoff.*

$$\widehat{H}(k) \equiv \int_{\Re} e^{irk} H(r)\, dr, \tag{10.11}$$

which has the corresponding inverse transform

$$H(r) = \frac{1}{2\pi} \int_{\Re} e^{-irk} \widehat{H}(k)\, dk. \tag{10.12}$$

It can be shown that H given by (10.10) forms a Fourier transform pair with

$$\widehat{H}(k) = \begin{cases} 1 & \text{if } |k| \leq \pi/h \\ 0 & \text{otherwise.} \end{cases} \tag{10.13}$$

The significance of (10.13) is made particularly clear in the case of turbulence in a periodic box, such as was considered in Section 3.6.2. In this case, $\mathbf{U} = \mathbf{u}$ is given by (3.84), and it is straightforward to show that a Fourier transform of (10.5) incorporating (10.7) gives

$$\widehat{\langle U_i \rangle}(\mathbf{k}, t) = \widehat{H}(k_1) \widehat{H}(k_2) \widehat{H}(k_3) \widehat{U}_i(\mathbf{k}, t). \tag{10.14}$$

After applying (10.13), (10.14) yields

$$\langle U_i \rangle(\mathbf{x}, t) = \sum_{|k_1| < \pi/h} \sum_{|k_2| < \pi/h} \sum_{|k_3| < \pi/h} \widehat{U}_i(\mathbf{k}, t) e^{i\mathbf{k}\cdot\mathbf{x}}. \tag{10.15}$$

Thus the sharp Fourier cutoff filter is equivalent to the act of truncating a Fourier expansion of the velocity field.

The sharp Fourier cutoff filter best satisfies our intuitive notion of what a filter should do, in the sense that it exactly eliminates all modes with wavenumbers larger than π/h, so that resolved and unresolved scales have a precise meaning. It also has the property that $\langle u_i' \rangle = 0$. On the other hand, the Fourier representation of spatially localized structures is not efficient since many modes will be excited. The other filters, such as (10.8) and (10.9), behave much differently: They resolve localized events better, but they do not entirely free $\langle U_i \rangle$ from at least partially representing the effect of small scales.

10.2   FILTERED EQUATIONS AND THEIR SOLUTIONS

Applying a filtering operation to the Navier–Stokes equation and taking note of (10.6) yields a relation governing the filtered velocity field:

$$\frac{\partial \langle U_i \rangle}{\partial t} + \langle U_j \rangle \frac{\partial \langle U_i \rangle}{\partial x_j} = -\frac{1}{\rho} \frac{\partial \langle P \rangle}{\partial x_i} + \nu \, \nabla^2 \langle U_i \rangle - \frac{\partial \tau_{ij}}{\partial x_j}, \tag{10.16}$$

where

$$\tau_{ij} \equiv \langle U_i U_j \rangle - \langle U_i \rangle \langle U_j \rangle \tag{10.17}$$

is referred to as the *subgrid-scale stress tensor*. Similarly, the continuity equation, (2.6), becomes

$$\frac{\partial \langle U_i \rangle}{\partial x_i} = 0 \tag{10.18}$$

(i.e., the filtered velocity field is incompressible). Note that here and henceforth the commutativity of filter and differentiation is assumed. It may be noted that the equations produced by filtering are formally identical to the equations produced by averaging, specifically (2.10) and (2.11), although there is a large conceptual difference in how they are to be interpreted. In particular, $\langle U_i \rangle$ is random, whereas $\overline{U}_i$ is deterministic. Furthermore, while τ_{ij} formally represents the influence that the unresolved part of the flow field has on the resolved part, there is no obvious way that a physical interpretation can be assigned to it the way that one can be for the Reynolds stress tensor (i.e., the latter clearly represents the flux of turbulent momentum). On the other hand, it is expected that τ_{ij} will make a substantially smaller contribution to the dynamics of $\langle \mathbf{U} \rangle$ than R_{ij} does to $\overline{\mathbf{U}}$. If this is true, it might be imagined that errors in modeling τ_{ij} will have less impact on predictions than will errors in modeling R_{ij}. This is a premise that motivates much of the interest in LES.

Although a reduced reliance on modeling is a fundamental advantage of LES over RANS, it is achieved only at the price of a great increase in computational expense compared to solving deterministic systems of equations. Moreover, numerical solutions of the LES equations give the filtered velocity only (i.e., they give no information about the unresolved field). Consequently, LES cannot provide unambiguous estimates of $\overline{U}_i$ and other averaged quantities such as K. For example, it follows from (10.3) that

$$\overline{\mathbf{U}} = \overline{\langle \mathbf{U} \rangle} + \overline{\mathbf{u}'}. \tag{10.19}$$

Although $\overline{\langle \mathbf{U} \rangle}$ is readily available by averaging the LES solution, $\overline{\mathbf{u}'}$ is not, and it cannot be said with certainty that $\overline{\mathbf{u}'} = 0$ for any particular filter. Thus, in a formal sense $\overline{\mathbf{U}}$ is unknowable from LES.

If it can be argued that the contribution to $\overline{\mathbf{U}}$ of the small scales (as manifested in $\overline{\mathbf{u}'}$) is small (i.e., $\overline{\mathbf{u}'} \approx 0$), it follows that

$$\overline{\mathbf{U}} \approx \overline{\langle \mathbf{U} \rangle}, \tag{10.20}$$

so that a prediction of $\overline{\mathbf{U}}$ is possible. As a matter of course, LES computations assume that this is a legitimate step. If it were not true, the usefulness of LES would be much restricted. It is also interesting to note that any observed differences between $\overline{\mathbf{U}}$ and $\overline{\langle \mathbf{U} \rangle}$ may possibly be due to nonnegligible contributions from $\overline{\mathbf{u}'}$. This is in addition to any errors deriving from subgrid modeling or numerical errors.

The properties of filtering have an even bigger impact on the possibility of obtaining estimates of quantities such as the turbulent kinetic energy, $K = \overline{u_i^2}/2$. Thus, note the identity

$$u_i = [\langle U_i \rangle - \overline{\langle U_i \rangle}] + [u_i' - \overline{u_i'}], \tag{10.21}$$

which follows from (10.3) and (10.19) and the definition of u_i. Equation (10.21) says that the traditional velocity fluctuation is the sum of the velocity fluctuation of the resolved (i.e., filtered field), say $u_i^r \equiv \langle U_i \rangle - \overline{\langle U_i \rangle}$, plus the velocity fluctuation of the unresolved scales, $u_i^u \equiv u_i' - \overline{u_i'}$. Thus

$$u_i = u_i^r + u_i^u \tag{10.22}$$

and

$$K = \tfrac{1}{2}\overline{(u_i^r)^2} + \overline{u_i^r u_i^u} + \tfrac{1}{2}\overline{(u_i^u)^2}. \tag{10.23}$$

This shows that K is the sum of the energy of the resolved and unresolved scales given by the first and third terms on the right-hand side, respectively, plus a correlation term between the scales. The last two terms in (10.23) cannot be computed, although as far as the first of these is concerned, it is consistent with the meaning of (10.22) to assume that $\overline{u_i^r u_i^u}$ is insignificant. The importance of $\tfrac{1}{2}\overline{(u_i^u)^2}$ depends on how much energy resides in the subgrid scales, and it cannot be assumed that this is always negligible (e.g., it depends on the resolution attained by the particular grid used in the LES). In the absence of a better strategy,[1] however, one has to accept the approximation

$$K \approx \tfrac{1}{2}\overline{(u_i^r)^2}, \tag{10.24}$$

even if there is nonnegligible energy contained in the subgrid scales. It is thus seen that despite the fact that there is a significant distinction between U_i and $\langle U_i \rangle$, practical necessity demands that LES be regarded as equivalent to a DNS when extracting information about the statistical properties of the velocity field.

10.3 NUMERICAL CONSIDERATIONS

A basic requirement of DNS and RANS calculations—if they are to be considered legitimate—is that they be independent of the discretization used in the numerical scheme. In the case of DNS this means that all subgrid motions are inconsequential. In a LES, on the other hand, it is not usual practice to demand grid independence of the resolved field because of the numerical expense necessary to achieve it. In fact, for grid independence the numerical mesh spacing must be substantially finer than the scale of the filtering operation. It is rare that such an excess of mesh points is available

---

[1]An argument may be made that $\tau_{ii} \approx \overline{(u_i^u)^2}$, which would thus supply a means for estimating subgrid energy. However, it is usually the case that only the deviatoric part of τ_{ij} is modeled in a LES, so that an estimate of τ_{ii} commonly is not available. The models used for the deviatoric stress (e.g., τ_{12}), however, do allow for reasonably complete estimates to be made of the Reynolds shear stress.

to guarantee the numerical accuracy of the resolved scales. Typical calculations use all available mesh points to represent the resolved field. In fact, it is common practice to use parameters reflecting properties of the underlying grid in the role of the scales appearing in models for the subgrid stress, τ_{ij}. In such circumstances, the resolved field is explicitly grid dependent, and the best one can hope for is that the subgrid model has the correct grid dependency.

Since $\langle U \rangle$ is normally grid dependent by default, it is possible that some grid dependency may survive the averaging process designed to estimate quantities such as $\overline{U}_i$ and K. Indeed, (10.19) and (10.23) show specifically how the grid may affect mean flow predictions (i.e., whenever $\overline{u'_i}$, $\overline{u^r_i u^u_i}$, and $\overline{(u^u_i)^2}$ are grid dependent). Again, one hopes that such errors are small.

The question of grid resolution is felt strongly in such locales as the region close to boundaries or in free shear layers, where coherent vortices, such as discussed in Chapters 4 and 5, dominate the physics. One expects that such objects need to be included in the LES as part of the resolved field since they contain much of the energy of the local flow. It is often inconvenient or difficult to guarantee that an appropriate mesh is available to provide the desired resolution. For example, for high-Reynolds-number flow next to a wall, the scale of the coherent vortices may be substantially smaller than that of the energy-containing eddies farther away from the wall and the mesh needs to be designed accordingly. If an acceptable mesh cannot be created (e.g., one that is fine enough to resolve near-wall vortices), reliance must be placed on the subgrid model to accommodate the special physics. Typically, this means that parameters in the subgrid model are changed to reflect the reduced scale of the local physics.

The need to fully resolve near-wall structures puts LES in a situation similar to that of DNS and is impractical when the Reynolds number gets too high. For example, if the typical spanwise spacing of low-speed streaks in a boundary layer is 100 wall units, a mesh with $\Delta z^+ \approx 20$ is needed to resolve such structures, as is known from DNS calculations [21]. This requirement is well met for a typical DNS at a low Reynolds number (e.g., $R_\tau = 180$). For higher Reynolds numbers (e.g., $R_\tau = 2000$), if the same mesh size as was adequate at $R_\tau = 180$ is kept, in this case $\Delta z^+ \approx 145$ and there is little that can be done in the way of refining the wall mesh and maintaining affordability [34]. Computations confirm the absence of the expected vortical structures when a coarse mesh is used, so the burden of supplying the missing physics inevitably falls on the subgrid model. In essence, the fundamental problem for LES is finding a means for capturing the effect of the physics of the coherent vortices without computing them in detail. Reducing the scales that appear as parameters in the subgrid-scale models is one useful step, although finding other, more physically meaningful strategies toward this end continues to be a priority.

Another consequence of the limited accuracy of subgrid stress models, which is exacerbated by grid dependency shows up in the presence of unwanted numerical diffusion. For example, in a turbulent flow there are regions where motions at the resolved scale evolve into motions at the unresolved scale. Under ideal circumstances the subgrid model will provide for the correct energy exchange to the small, unresolved scales, so that the resolved field is calculated correctly. However, if the subgrid model miscalculates the exchange, the numerical solution may attempt to resolve finer scales than it is able to. This will typically lead to smoothing of the computed solution by numerical viscosity, and the integrity of the LES is jeopardized. In some cases, the numerical diffusion errors can exceed the subgrid stresses in magnitude, thereby casting a cloud over the legitimacy of the LES simulation [13,22,46].

It should be pointed out that it is sometimes claimed (e.g., [5,20]) that numerical diffusion can be used in place of an explicit subgrid stress model. This requires the use of computational methods whose local numerical diffusion is representative of the influence of unresolved scales. In this point of view, the inevitable appearance of numerical diffusion in a high-Reynolds-number grid-based calculation takes ᴜ ᴇr the job of subgrid transport and energy dissipation. To be useful, such schemes shoᵤ ld remain stable in the face of underresolution. They also tend to use wall function boundary conditions to avoid a large investment in resolving the wall region. This type of analysis is clearly simpler to implement than traditional LES incorporating a subgrid model, although it is not considered to be competitive with traditional LES methods from the standpoint of accuracy. In fact, the possibility of success of this approach to some degree conflicts with the existence of necessary grid requirements for the success of DNS computations. Among the inaccuracies noted with this technique is the effect of coarse resolution in preventing simulation of desired flow regimes (e.g., by introducing an inappropriate amount of free-stream turbulence) [32].

Despite the desirability of eliminating numerical errors in LES calculations, it proves to be a difficult goal to accomplish. For example, mesh refinement is often not feasible since it brings the cost of the calculation in line with that of DNS and thus undermines the rationale for LES. Another possibility is to use higher-order numerical schemes that are less subject to numerical diffusion. Such algorithms, however, are typically much more expensive to run than traditional schemes [23]. A final option is to develop better subgrid models that more accurately calculate energy transfer. Some progress in this direction has been achieved with the use of dynamic models that attempt to be sensitive to local flow conditions. Further improvements in subgrid models can be expected in the future.

Assessment of the performance of LES models can be done in an a priori sense in which DNS solutions are used to evaluate both τ_{ij} and its model to see if a reasonable correlation exists between the two. In practice, it is found that subgrid models are more often than not poorly correlated at the local (i.e., pointwise) level. Moreover, it is found that a priori tests are not a strong prognosticator of the eventual success of LES schemes when they are used in predicting actual flows. A better measure of likely performance comes from comparing the behavior of τ_{ij} and its model over finite regions. High correlation is more achievable in this case and helps to explain some of the notable predictive successes of LES calculations.

10.4   SUBGRID-SCALE MODELS

It was mentioned previously that it is difficult to associate τ_{ij} directly with a physical process involving fluid motion because it is based on filtering rather than averaging. It is thus not surprising that τ_{ij} tends to be modeled formally without a detailed physical picture in mind. In fact, for wont of a reason to the contrary, subgrid models often expropriate the diffusive form (8.1), substituting filtered velocities for average velocities, as in

$$\tau_{ij} - \tfrac{1}{3}\tau_{kk}\delta_{ij} = -2\nu_T \langle S \rangle_{ij}, \tag{10.25}$$

where

$$\langle S \rangle_{ij} = \frac{1}{2}\left(\frac{\partial \langle U_i \rangle}{\partial x_j} + \frac{\partial \langle U_j \rangle}{\partial x_i}\right) \tag{10.26}$$

is the filtered rate of strain tensor and ν_T is a subgrid scale eddy viscosity which can be modeled in various ways. As in the case of RANS modeling, an intrinsic advantage of (10.25) is that it assures the well posedness of the associated numerical problem.

When (10.25) is substituted into (10.16), it may be noticed that τ_{kk} can be combined with the pressure, in which case it does not have to be evaluated separately. In fact, since just the deviatoric part of τ_{ij} is modeled in (10.25), its contraction shows that this relation gives no information about τ_{kk}. The comparable situation was faced in the RANS context, where (8.1) gave no information about K. To get this information it was necessary to develop a separate model of the K equation. This is a strategy that has also been pursued in the context of LES, as will be shown below.

Another important observation about (10.25) is that it has many of the same conceptual problems as (8.1). For example, (10.25) implies that ν_T should serve as the proportionality constant relating the left- and right-hand sides for six different components. Looked at another way, it means that the matrices on either side have the same principal axes. This implicit assumption is not supported in a priori tests showing only a weak correlation between the principal axes of each tensor.

Since (10.25) appears in (10.16) in the guise of a diffusion term, it is generally true that the condition $\nu_T > 0$ is necessary for numerical stability. It turns out, however, that this condition is fundamentally in conflict with the naturally occurring energy exchange between resolved and unresolved scales, an exchange that sometimes locally requires $\nu_T < 0$ if (10.25) is to be physically correct. To see why negative ν_T occurs, consider the equation for resolved kinetic energy, $\langle K \rangle \equiv \tfrac{1}{2}\langle U_i \rangle^2$, which is formally equivalent to equation (2.28) for $\overline{K} = \tfrac{1}{2}\overline{U}_i^2$ and derived similarly:

$$\frac{\partial \langle K \rangle}{\partial t} + \langle U_j \rangle \frac{\partial \langle K \rangle}{\partial x_j} = -\frac{1}{\rho}\frac{\partial \langle P \rangle \langle U_j \rangle}{\partial x_j} + \nu \, \nabla^2 \langle K \rangle$$
$$- \frac{\partial \tau_{ij}\langle U_i \rangle}{\partial x_j} - \nu \frac{\partial \langle U_i \rangle}{\partial x_j}\frac{\partial \langle U_i \rangle}{\partial x_j} + \tau_{ij}\langle S \rangle_{ij}. \tag{10.27}$$

The terms on the right-hand side account, respectively, for total pressure work, viscous diffusion, subgrid stress diffusion, viscous dissipation and in the last term, a gain (or loss) of resolved energy at the expense (or gain) of the unresolved energy. The expression $\tau_{ij}\langle S \rangle_{ij}$ has a role in (10.27) that is formally similar to that of an equivalent term, $R_{ij} \, \partial \overline{U}_i / \partial x_j$, which is present in the $\overline{K}$ equation, (2.28). In the latter case it is an average quantity accounting for the loss of $\overline{K}$ to production of K. This interpretation is fully compatible with (8.1) as long as $\nu_T > 0$, which is always assumed in RANS modeling.

In the context of a LES, $\tau_{ij} \langle S \rangle_{ij}$ is not an averaged quantity. Rather, it is part of the random field and varies strongly with local conditions. In fact, DNS calculations show that it can be both positive and negative. It is positive when energy is transferring from small to large scales, a phenomenon referred to as *backscatter*. If (10.25) is assumed to be true, it is readily seen that

$$\tau_{ij} \langle S \rangle_{ij} = -2\nu_T \langle S \rangle_{ij} \langle S \rangle_{ij}, \tag{10.28}$$

which is strictly negative as long as $\nu_T > 0$. In other words, there can be no backscatter unless ν_T is allowed to be negative. It is thus seen that the use of (10.25) with the restriction that ν_T be positive represents a potentially serious distortion of the physics of the subgrid scales.

If the last term in (10.27) is viewed globally (e.g., as an average over the flow domain), there is no problem with ν_T being strictly positive, since the domain-wide backscatter is virtually always subordinate to the energy flux to small scales. This explains why models that have $\nu_T > 0$ for stability reasons can still yield reasonable results in a mean sense. In practical terms, the best one can hope to do is reduce ν_T in regions of backscatter while still keeping it positive. Another option is to add other mathematical forms to (10.25) which do not affect numerical stability, yet do allow for backscatter.

One significant example of the latter course of action is referred to as the *scale similarity model*. This is derived by first substituting (10.3) into τ_{ij}, giving

$$\tau_{ij} = \left[\langle \langle U_i \rangle \langle U_j \rangle \rangle - \langle U_i \rangle \langle U_j \rangle \right] + \left[\langle u_i' \langle U_j \rangle \rangle + \langle u_j' \langle U_i \rangle \rangle \right] + \langle u_i' u_j' \rangle. \tag{10.29}$$

The first term on the right-hand side, which may be evaluated from $\langle U_i \rangle$, is the *Leonard stress* [24]. It may be thought of as representing the creation of small-scale motions from the resolved scales, as is easily visualized in the case of the sharp Fourier cutoff filter since in this case the Leonard stress can only be composed of terms at wavenumbers above the cutoff. The second term in (10.29), the *subgrid-scale cross-stress*, is imagined to connect resolved and unresolved scales directly and thus may be a source of backscatter. The final term, the *subgrid-scale Reynolds stress*, is expected to account for the influence of unresolved scales on resolved scales and thus may often be associated with backscatter.

The scale similarity model is derived from (10.29) after making some assumptions. First, it assumes that there is value in keeping the Leonard stress as an independent entity [i.e., as against adopting a model such as (10.25) in which the decomposition in (10.29) is not introduced]. Second, the cross-stress term in (10.29) is modeled [2] according to

$$\langle u_i' \langle U_j \rangle + u_j' \langle U_i \rangle \rangle = \langle U_i \rangle \langle U_j \rangle - \langle \langle U_i \rangle \rangle \langle \langle U_j \rangle \rangle, \tag{10.30}$$

in which case the sum of the Leonard and cross-term stresses becomes

$$\langle\langle U_i\rangle\langle U_j\rangle\rangle - \langle\langle U_i\rangle\rangle\langle\langle U_j\rangle\rangle. \tag{10.31}$$

Finally, the diffusion model, (10.25), which had previously been applied to all of τ_{ij} is now used to model just the subgrid Reynolds stress. The end result of these modeling assumptions is

$$\tau_{ij} - \tfrac{1}{3}\tau_{kk}\,\delta_{ij} = \left[\langle\langle U_i\rangle\langle U_j\rangle\rangle - \langle\langle U_i\rangle\rangle\langle\langle U_j\rangle\rangle\right]$$
$$- \tfrac{1}{3}\delta_{ij}\left[\langle\langle U_k\rangle\langle U_k\rangle\rangle - \langle\langle U_k\rangle\rangle\langle\langle U_k\rangle\rangle\right] - 2\nu_T\langle S\rangle_{ij}. \tag{10.32}$$

In this, ν_T can be taken to be any of the forms normally used in (10.25), although usually with adjusted model constants. It may be checked that (10.32) is self-consistent in the sense that both sides satisfy Galilean invariance [43]. In fact, a main motivation for the model (10.30) is that with it, Galilean invariance of (10.32) is assured. Computations show that the scale-invariant part of (10.32) (i.e., the first term on the right-hand side) makes substantial contributions to backscatter [19]. Moreover, because of the presence of this term, the last term in (10.32) has less of a role in simulations than it does in a model such as (10.25) [48]. On the other hand, the inclusion of the diffusive term is considered essential since without it, the scale similarity form does not allow for sufficient energy dissipation. It is generally accepted that "mixed" models such as (10.32) are superior to either the Smagorinsky model or the scale similarity models by themselves. Theoretical explanations for this conclusion are as yet rudimentary, and it is a popular topic for further study [29].

10.4.1 Smagorinsky Model

The first subgrid model to be widely used is that due to Smagorinsky [39], in which it is assumed that

$$\nu_T = (C_S h)^2 |\langle S\rangle|, \tag{10.33}$$

where C_S is referred to as the *Smagorinsky constant*, h is a length scale, and $|\langle S\rangle| \equiv (2\langle S\rangle_{ij}\langle S\rangle_{ij})^{1/2}$. In practice, h usually is defined to be a function of the local gridding with the convention that for a uniform mesh, h is taken to be twice the grid spacing. For grids whose spacing varies from one direction to another, h may be formed from the geometric mean of the grid spacings or by other formulas.

Equation (10.33) is reminiscent of the family of mixing-length models in that it assumes a velocity scale $\mathcal{U} = h|\langle S\rangle|$ as most appropriate to go along with the length scale h. Another route to the same result is the recognition that this choice of $\mathcal{U}$ is what is necessary to equilibrate the production rate of subgrid energy from the resolved field, namely $\tau_{ij}\langle S\rangle_{ij}$, with the subgrid energy dissipation rate estimate, $\mathcal{U}^3/h$, a scaling that was previously considered with regard to (2.97). Note that this argument is ultimately a global one, since it ignores the regions of backscatter.

As mentioned previously, it is necessary to reduce the length scale near boundaries artificially in order to conform to the reduced scale of the dominant vortical motions in this region. A common technique is to introduce *van Driest damping*, which has h replaced near the boundary by the empirical relation $h(1 - e^{-y^+/25})$. The need for such ad hoc modification is a significant liability of the Smagorinsky model, since it complicates the attempt to apply it to flows in complex geometries. This is a problem similar to that faced by the mixing-length model. It should also be remarked that applications of the Smagorinsky model are often done using wall function boundary conditions. This practice has many of the same problems and advantages as it has in the case of RANS modeling (see Section 8.9.1). In the present situation, it leads to a significant reduction in the number of mesh points needed to resolve the wall layer and thus speeds up the simulation significantly. On the other hand, the loss of resolution is not generally compensated for within the model itself, so valuable physics is lost.

It may be noticed that when used with (10.25), the Smagorinsky model is not dependent on a particular choice of filter[2] (i.e., the filter operation does not enter into the determination of $\langle U_i \rangle$). This is not true if the Smagorinsky model is used as part of the scale similarity model, since clearly (10.32) requires evaluation of the filtered product of the resolved velocity components. Numerical experiments have shown [35] that the Gaussian filter is a particularly good choice to use with the scale similarity model. It is believed that this is because both model and filter have a similar propensity to accommodate the influence of a relatively broad range of length scales. In contrast, a priori testing reveals that if the Smagorinsky model is used to represent τ_{ij} directly [i.e., as in (10.25)], the best correlation between τ_{ij} and the model occurs if the sharp Fourier cutoff filter is used. Evidently, this reflects the fact that both filter and model in this instance reflect the action of a relatively tight band of resolved scales.

In those instances when it is necessary to evaluate a filtering operation numerically, the best way to proceed depends on which filter is to be applied, the particular flow field, its symmetries, and the properties of the numerical mesh and algorithm. The sharp cutoff filter is readily done in spectral space, while the tophat filter is easy to implement in real space (i.e., by taking an average over nearby values in the mesh). After truncation so that it has finite support, the Gaussian filter can be applied by quadrature in either physical or spectral space. Many LES implementations use different filters in different directions, so a combination of techniques for evaluating filters is often necessary.

A classic analysis [26] of the physics of forced, stationary, homogeneous, isotropic turbulence has been used to determine a value of C_S. In this, assume that $0 < k < k_c$ is the range of resolved wavenumbers where k_c lies in the inertial range. In fact, in view of (10.15) it can be assumed that $k_c = \pi/h$. The energy dissipated from the resolved field, assuming that transport is governed by an eddy viscosity ν_T is $\nu_T \int_o^{k_c} k^2 E(k)\, dk$, which should be equal to the rate of viscous dissipation in the flow, ϵ, so that

---

[2]Of course, in a priori tests the choice of filter is important.

$$\epsilon = \nu_T \int_o^{k_c} k^2 E(k)\, dk. \tag{10.34}$$

Since dissipation is minimal at small wavenumbers, and k_c extends to the inertial range, it may be assumed that (2.105) can be used to evaluate the right-hand side of (10.34), yielding

$$\epsilon = \tfrac{3}{4}\nu_T C_K \epsilon^{2/3} k_c^{4/3}. \tag{10.35}$$

A second approximation of ϵ can be made from the equilibrium assumption to the effect that ϵ is balanced by an average of the production term, say $\overline{-\tau_{ij}\langle S_{ij}\rangle}$. Finally, ν_T is assumed to have the Smagorinsky form and the ratio $\overline{\langle S\rangle}^{3/2}/\langle S\rangle^{3/2}$ is taken to be unity. A calculation then gives $C_S \approx 0.16$.

This value of C_S proves to be overly dissipative in flows containing mean shear, such as a channel flow. In the latter case, a more suitable value of C_S is in the neighborhood of 0.065. In view of the fact that it is actually C_S^2 that appears in the eddy viscosity, an order-of-magnitude reduction in the coefficient of ν_T is realized by this change in coefficient.

The need to specify a constant value of C_S detracts from the desirability of the Smagorinsky model, since it is evident that different values are required in different regions. In typical applications, it is found to be overly diffusive and tends to smooth turbulence fluctuations, even to the extent of eradicating boundary layer structures. It is also known to prevent transition, in very much the same way that this happens for equivalent RANS models.

10.4.2 Alternative Eddy Viscosity Subgrid-Scale Models

Very much in the same way that one- and two-equation models have appeared as alternatives to the mixing-length model, similar kinds of generalizations have been developed with the goal of improving on the Smagorinsky model. One popular approach [37] uses the local subgrid kinetic energy,

$$K_{\text{sgs}} \equiv \tfrac{1}{2}\tau_{kk}, \tag{10.36}$$

as the basis for the velocity scale of the eddy viscosity, so that

$$\nu_T = C_k h \sqrt{K_{\text{sgs}}}. \tag{10.37}$$

A separate modeled equation for K_{sgs} is thus necessary in this approach. This usually is similar to the equivalent relations for K used in one-equation RANS models. For example, in one formulation [40] the K_{sgs} equation takes the form

$$\frac{\partial K_{\text{sgs}}}{\partial t} + \langle U_j\rangle \frac{\partial K_{\text{sgs}}}{\partial x_j} = -\tau_{ij}\langle S_{ij}\rangle - C_\epsilon \frac{K_{\text{sgs}}^{3/2}}{h} + \frac{\partial}{\partial x_j}\left(\overline{C_k h}\sqrt{K_{\text{sgs}}}\frac{\partial K_{\text{sgs}}}{\partial x_j}\right), \tag{10.38}$$

where the terms on the right-hand side represent production, dissipation, and transport, respectively. Note that since this is a one-equation model, the dissipation takes the characteristic form (8.17), with C_ϵ acting as a model constant. In this particular formulation an average coefficient, $\overline{C_k}$, replaces C_k in the transport term for reasons that are given in the next section.

Another approach toward eddy viscosity modeling [25,31] has its origins in spectral theory as it was employed in deriving (10.34) and (10.35). Specifically, (10.35) can be rewritten as

$$
\nu_T = \frac{4}{3} C_K^{-3/2} \left(\frac{E(k_c, t)}{k_c} \right)^{1/2}
\tag{10.39}
$$

after using (2.105). In effect, this is an assumption about the velocity and length scales upon which the eddy viscosity should depend [i.e., $\sqrt{E(k_c, t)k_c}$ and k_c^{-1}, respectively].

Introducing the structure function $S_2(\mathbf{x}, \mathbf{r}, t)$, where the nth-order structure function is defined via

$$
S_n(\mathbf{x}, \mathbf{r}, t) \equiv \overline{|\mathbf{u}(\mathbf{x} + \mathbf{r}, t) - \mathbf{u}(\mathbf{x}, t)|^n},
\tag{10.40}
$$

(10.39) can be expressed in terms of S_2, and in this form it is readily applied in physical space calculations. The connection is established by borrowing from the identity [3]

$$
\overline{|\mathbf{u}(\mathbf{x} + \mathbf{r}, t) - \mathbf{u}(\mathbf{x}, t)|^2} = 4 \int_0^\infty E(k, t) \left(1 - \frac{\sin kr}{kr} \right) dk,
\tag{10.41}
$$

which may be derived readily from (7.130) in the case of isotropic turbulence. At high Reynolds numbers, it is reasonable to evaluate this integral using (2.105) for the energy spectrum. In fact, for small k the term in parentheses in the integrand is small, while for large k the energy spectrum is in the dissipation range and is also small. The integration then yields

$$
\overline{|\mathbf{u}(\mathbf{x} + \mathbf{r}, t) - \mathbf{u}(\mathbf{x}, t)|^2} = 4.82 C_K (\epsilon r)^{2/3}.
\tag{10.42}
$$

Eliminating ϵ via (2.105) and once again assuming that $k_c = \pi/h$ yields the eddy viscosity formula

$$
\nu_T = 0.066 \, C_K^{-3/2} \, h \sqrt{\overline{\left(|\langle U_i(\mathbf{x} + \mathbf{r}, t) \rangle - \langle U_i(\mathbf{x}, t) \rangle|^2 \right)}_{|\mathbf{r}|=h}},
\tag{10.43}
$$

where an average over the values of $|\mathbf{r}| = h$ is intended in this expression. A number of variants of this approach have been formulated that attempt to add additional physics into the determination of the eddy viscosity.

10.4.3 Dynamic Models

Since local adjustment of the parameter C_S enhances the performance of the Smagorinsky model (10.33), it is natural to seek means by which C_S can be modified automatically throughout a flow domain so as to optimize its response to local conditions. This is the philosophy behind the *dynamic procedure* [12], which uses information available during the calculation to dynamically change parameters appearing in whatever subgrid model is chosen, so that it adapts to the local flow conditions as they vary in space and time. The dynamic procedure is most often used in the context of (10.25) incorporating (10.33) as well as with the scale similarity model (10.32), although it can, in principle, be applied to other models. For simplicity the present discussion is framed in terms of the specific example of the Smagorinsky model.

Dynamic models attempt to tune a subgrid model to local flow conditions by leveraging the information contained in the smallest resolved scales of the velocity field. Specifically, it is imagined that the smallest scales in the *resolved* computation, for a particular filter width, look like the largest *unresolved* scales in a computation done with a coarser filter. Then, while a calculation is in progress, the smallest resolved scales are used to estimate what the subgrid stress would look like for a coarser mesh calculation. This concept yields an estimate of the Smagorinsky constant if it can be assumed that C_S is the same for filtering done at both levels. In some recent treatments this condition is relaxed in favor of allowing a scale dependence to C_S [1]. The following discussion considers the formal implementation of these ideas.

In the dynamic model the coarse filter is referred to as the *test filter*, which differs from the standard or *grid filter* as it is commonly called, by its smoothing over a larger region. Thus, if h_t is the scale of the test filter, by definition $h_t > h$. For the purpose of the following discussion, test filtering is denoted by braces, as in $\{U\}$, for the test filtered velocity field. Frequently in what follows both test and grid filters are applied, first the grid, then the test filter, leading to the notation $\{\langle U_i \rangle\}$. For the sharp Fourier cutoff filter, it is clear that $\{\langle U_i \rangle\} = \{U_i\}$, but this does not necessarily hold for other filters.

If the test and grid filters are applied consecutively to the Navier–Stokes equation, the subgrid stress term

$$T_{ij} \equiv \{\langle U_i U_j \rangle\} - \{\langle U_i \rangle\}\{\langle U_j \rangle\} \tag{10.44}$$

appears in very much the same way that τ_{ij} appeared in (10.16). In fact, it is not hard to show that τ_{ij} and T_{ij} are related to each other through the identity [12]

$$T_{ij} = \{\tau_{ij}\} + \mathcal{L}_{ij}, \tag{10.45}$$

where

$$\mathcal{L}_{ij} \equiv \{\langle U_i \rangle \langle U_j \rangle\} - \{\langle U_i \rangle\}\{\langle U_j \rangle\} \tag{10.46}$$

closely resembles the Leonard stress.

The dynamic form of the Smagorinsky model is derived by assuming that both τ_{ij} and T_{ij} satisfy (10.25), in which case it can be asserted that

$$\tau_{ij} - \tfrac{1}{3}\tau_{kk}\,\delta_{ij} = -2Ch^2|\langle S\rangle|\langle S_{ij}\rangle \tag{10.47}$$

and

$$T_{ij} - \tfrac{1}{3}T_{kk}\,\delta_{ij} = -2Ch_t^2|\{\langle S\rangle\}|\{\langle S_{ij}\rangle\}, \tag{10.48}$$

where C has been introduced in place of C_s^2. In fact, (10.47) and (10.48) are meant to be viewed as a generalization of the Smagorinsky model which allows for the possibility of $C < 0$. C is assumed to be a function of position and time but independent of h or any other flow variables. Depending on the range of scales in a particular LES calculation, it is not hard to envision cases where it is unlikely for C to be the same in both formulas. For example, near a solid boundary, the test and grid filters could be experiencing very different flow phenomena with the expectation that C is different for each. For the present discussion no distinction in C is made, although, as mentioned previously, there is growing interest in incorporating some scale dependence in the coefficients [1].

The goal of the dynamic analysis formulated here is to come up with a means for predicting C. A relation for it can be obtained from (10.45) after (10.47) and (10.48) are substituted into it, yielding

$$\mathcal{L}_{ij} - \tfrac{1}{3}\delta_{ij}\mathcal{L}_{kk} = 2C M_{ij}, \tag{10.49}$$

where

$$M_{ij} \equiv h^2\{|\langle S\rangle|\langle S_{ij}\rangle\} - h_t^2|\{\langle S\rangle\}|\{\langle S_{ij}\rangle\}. \tag{10.50}$$

The appearance of C in front of the quantity M_{ij} in (10.49) is a result of the inconsistent assumption that C is a constant in space so that it can be taken outside the coarse filtering that is applied to τ_{ij} in (10.45). If this assumption is not made, (10.45) leads to an integral equation for $C(\mathbf{x}, t)$ instead of the simpler algebraic form in (10.49). Models that pursue the formally correct route are referred to as *dynamic localization models* [14]. They are not widely used today because the substantial extra cost of solving the integral equation has not been found to be balanced by an equivalent increase in performance.

Equation (10.49) is an overdetermined system for C (i.e., it consists of five independent relations, each of which could be solved separately for C). Since not all of these conditions can be met at the same time, it is reasonable to pursue a strategy in which (10.49) is satisfied only in a least-squares sense [27], that is, by minimizing the squared error

$$\mathcal{E} \equiv \left(\mathcal{L}_{ij} - \tfrac{1}{3}\delta_{ij}\mathcal{L}_{kk} - 2C M_{ij}\right)^2 \tag{10.51}$$

with respect to C. Thus, setting $\partial \mathcal{E}/\partial C = 0$ yields the requirement that

$$C = \frac{1}{2} \frac{\mathcal{L}_{ij} M_{ij}}{M_{ij}^2}, \qquad (10.52)$$

which is a minimum since $\partial^2 E/\partial C^2 = 8M_{ij}^2 > 0$. The ratio of length scales, h_t/h, remains as a free parameter in (10.52) which must be determined. Since h is given, the question is really how much coarser than h one should choose the test filter. Numerical experiments suggest that an optimal value is $h_t/h = 2$.

In practice it is found that C determined from (10.52) has an unacceptably rapid spatial variation that can lead to instability in the numerical computations of the re-solved field. Evidently, the ratio in (10.52) is a sensitive function of the numerator and denominator, which themselves vary rapidly throughout the flow. The numerator can, and does in practice, change sign. In and of itself, this is a theoretical advantage since it opens up the door to modeling backscatter. However, it is a distinct obstacle for achieving stable numerical solutions to the governing equations, since it is tanta-mount to imposing negative viscosity. To make the dynamic model work, it turns out to be necessary to prevent the occurrence of negative C as well as curtail the rapidity of its variation throughout the flow. Where possible, the normal practice is to smooth the local behavior of C by averaging it over a convenient region. Formally, this can be achieved by choosing C to minimize the average error in a region, that is, select C so as to minimize

$$\mathcal{E} \equiv \int_{\mathcal{V}} \left(\mathcal{L}_{ij} - \frac{1}{3} \delta_{ij} \mathcal{L}_{kk} - 2C M_{ij} \right)^2 d\mathcal{V}. \qquad (10.53)$$

The domain, $\mathcal{V}$, is normally chosen to include the homogeneous flow directions, so that C ends up being just a function of the inhomogeneous directions. For example, in a channel flow, $\mathcal{V}$ usually represents the planes at fixed y, in which case minimization of (10.53) with respect to C yields

$$C(y, t) = \frac{1}{2} \frac{\int_{\mathcal{V}} \mathcal{L}_{ij} M_{ij} \, d\mathcal{V}}{\int_{\mathcal{V}} M_{ij}^2 \, d\mathcal{V}}. \qquad (10.54)$$

This procedure not only smoothes out rapid variations in C but also prevents C from becoming negative. A particularly attractive property of (10.54) is that it eliminates the need to make ad hoc modifications to the subgrid length scale next to solid boundaries [i.e., there is no need to use the van Driest damping formula if (10.54) is used].

The mechanism by which the dynamic model is able to reflect the presence of small scales next to a boundary has to do with the occurrence of relatively large regions where C is negative. Thus integration of C over a region parallel to the wall surface leads to significant cancelation of positive and negative C values, with the result that its average value is small. To some extent this is fortuitous because it occurs regardless of whether or not the LES has sufficient mesh points to resolve the boundary layer structure. It is not, however, guaranteed that the dynamic model reacts equally well in other circumstances.

In complex flow geometries where no symmetries are present, the averaging in (10.54) cannot be done over a plane such as has proven to be so effective in channel flow. Alternatively, local averaging can be done (e.g., in a small volume around each point), but this cannot be expected to smooth C as much as one would like. For example, despite such averaging, C may still be negative. It is not uncommon for applications faced with this dilemma [48] to accept negative C only as long as $v + v_T > 0$ (i.e., the total viscous plus turbulent viscosity is positive), and in so doing guarantee the stability of the numerical calculation. If C is so negative as to violate this condition, it is forceably increased to prevent its occurrence. Clearly, such "clipping" procedures are undesirable. Fortunately, it is usually the case that this kind of ad hoc intervention has to be imposed at only a relatively small number of locations in the course of a typical calculation.

Another approach to smoothing the dynamic coefficient for nonhomogeneous flows is based on a weighted average over fluid particle paths [30]. In this, for a given point $\mathbf{x}$ in the flow at time t the spatial averaging in (10.53) is replaced by a weighted integration over the particle paths arriving at $\mathbf{x}$ at time t. Specifically, the error to minimize is

$$\mathcal{E} \equiv \int_{-\infty}^{t} (\mathcal{L}_{ij}(\mathbf{X}(s), s) - \frac{1}{3}\delta_{ij}\mathcal{L}_{kk}(\mathbf{X}(s), s) \tag{10.55}$$
$$- 2C(\mathbf{x}, t)M_{ij}(\mathbf{X}(s), s))^2 W(t - s) \, ds,$$

where $\mathbf{X}(s)$ is the path of a fluid particle satisfying $\mathbf{X}(t) = \mathbf{x}$ and $W(t)$ is a weighting function designed so as to force greater contributions from the immediate past than the distant past. By the same steps as led to (10.54), it follows that

$$C(\mathbf{x}, t) = \frac{1}{2} \frac{\int_{-\infty}^{t} \mathcal{L}_{ij}(\mathbf{X}(s), s)M_{ij}(\mathbf{X}(s), s)W(t - s) \, ds}{\int_{-\infty}^{t} M_{ij}(\mathbf{X}(s), s)^2 W(t - s) \, ds}. \tag{10.56}$$

Although this appears to be a formidable relation to evaluate in the course of a LES simulation, a significant simplification occurs if the weighting function is chosen to be an exponential of the form

$$W(t) = \frac{1}{T}e^{-t/T}, \tag{10.57}$$

where T is a time scale that has been selected after some empirical study to be

$$T = 1.5h(ND)^{-1/8}, \tag{10.58}$$

where $N \equiv \int_{-\infty}^{t} \mathcal{L}_{ij}(\mathbf{X}(s), s)M_{ij}(\mathbf{X}(s), s)W(t - s) \, ds$ is the numerator of (10.56) and $D \equiv \int_{-\infty}^{t} M_{ij}(\mathbf{X}(s), s)^2 W(t - s) \, ds$, is the denominator. In the case of (10.58) it is not hard to show that N and D satisfy relaxation-transport equations, specifically

$$\frac{\partial N}{\partial t} + \langle \mathbf{U} \rangle \cdot \nabla N = \frac{1}{T}(\mathcal{L}_{ij}M_{ij} - N), \tag{10.59}$$

and

$$\frac{\partial D}{\partial t} + \langle \mathbf{U} \rangle \cdot \nabla D = \frac{1}{T}(M_{ij}^2 - D), \tag{10.60}$$

respectively. Equations (10.59) and (10.60) are more readily approximated in the course of a LES than is (10.56). Overall, the additional expense of the path averaging is not prohibitive.

Some justification for the Lagrangian model comes from test calculations which show that the effect of path averaging is to make C relatively scale independent. This helps justify use of the same C for the test and grid filters. Calculations have also shown that the Lagrangian model responds well to the presence of large-scale structures. In fact, the Lagrangian model has the benefit of being responsive to events in the homogeneous directions, which would ordinarily be averaged out in the typical dynamic model formulation.

Another means [40] of allaying the problems associated with a rapidly varying and sometimes negative v_T produced by (10.52) is illustrated by the subgrid model incorporating (10.36) and (10.38). Here, C is used strictly in the production term in (10.38), where it cannot lead to instability, while a smooth global average of C is used in the transport terms. This method allows the dynamics of the K_{sgs} equation to ameliorate the variations in C naturally rather than through the somewhat artificial step of averaging.

A growing body of evidence [29] shows that use of the dynamic procedure generally improves the performance of subgrid models. A similar conclusion holds in regard to the similarity forms. Thus, dynamical variants of the similarity model appear to have the best overall performance of subgrid models. They are not, however, superior in all circumstances, so their extra expense may sometimes not be justified.

The art of subgrid modeling continues to receive considerable attention, with many new variations of the familiar approaches being developed as well as new means of including dynamic behavior. It can therefore be hoped that significant improvements beyond current models will appear in the future.

10.5 APPLICATIONS OF LES

With sufficient grid refinement, any LES will behave like a DNS because the contribution made by the subgrid stresses diminishes along with the grid spacing. It follows that LES calculations can be made to succeed merely by increasing the mesh resolution to the DNS range. The true test of the usefulness of a LES is to have it produce accurate predictions of flows at a density of mesh points for which it is known that a DNS will fail. Experience to date suggests that LES does offer some benefit of this kind, although it is often not as dramatic as might be hoped. In this regard it is useful to distinguish between free shear and bounded flows. In the former case the effect of increases in the Reynolds number is mainly to extend the inertial range to encompass smaller scales so that even with a fixed grid the resolution of the energy containing scales is not adversely affected as the Reynold number is increased. Subgrid-scale models which accurately account for scales at or below the inertial range are likely to still do so for larger Reynolds numbers and accuracy of the simulation will not be lost.

For bounded flows, if a grid is used that is of sufficient size to render an acceptable DNS at one Reynolds number, the use of a LES formulation can extend the range of the computation to only modestly higher Reynolds numbers. The problem in this case is that with much larger Reynolds numbers, significant energy begins to populate scales smaller than that of the unchanging mesh and the accuracy of the typical LES degrades significantly.

In fact, if the subgrid scale is larger than that of dynamically important vortical structures in the wall region, it is up to the subgrid-scale model to represent this dynamical contribution to the boundary layer. Evidently, in such cases more is being demanded of the subgrid modeling than it is able to deliver. When the mesh is fine enough to resolve the important structures then the demands on the subgrid model are more realistic. Although it is not often mentioned, another possible reason for the loss of accuracy at high Reynolds numbers is that the unresolved scales in such cases may make such significant contributions to mean quantities that relations such as (10.20) are untenable.

In light of these comments, attention is focused here on illustrating the capabilities of LES at Reynolds numbers clearly beyond the DNS range. Such calculations reported in the literature incorporate a mesh that might be adequate for a DNS at a lower Reynolds number but is not adequate for the Reynolds number of interest. Extremely coarse meshes are rarely employed since in such cases the loss of grid points is likely to be felt immediately in the resolution of the wall region, so that neither a reasonable DNS or LES is possible.

A calculation [47] of a three-dimensional turbulent boundary layer using DNS on a fine mesh and LES on a coarse mesh illustrates how LES can extend the range of DNS. The boundary layer flow in this example is generated over a flat plate by forcing the free stream to undergo a circular, time-dependent motion $U_\infty = U_0 \cos ft$ and $W_\infty = U_0 \sin ft$ in the directions parallel to the surface. Here, f is the frequency of rotation, and the outer flow moves as a whole in a circular fashion with radius U_0/f. For the results described here, the Reynolds number $R_\delta = 767$, where R_δ is made from the velocity scale U_0 and the depth of viscous penetration, $\delta \equiv \sqrt{2\nu/f}$. Periodic conditions are imposed in the lateral directions. A mesh with 65^3 points is used for the LES and $256 \times 80 \times 256$ for the DNS. Thus there is a 95% reduction in the number of mesh points between the two.

Both the dynamic model [12] and the dynamic similarity model [45] with Lagrangian averaging of the dynamic coefficients [30] are considered. The latter model gives substantially better predictions than the former, as is evident in Fig. 10.2, comparing the mean velocities (projected into a special coordinate system; see [42,47]) with the equivalent DNS result. In particular, the dynamic similarity model shows log-law behavior in good agreement to that predicted by the DNS.

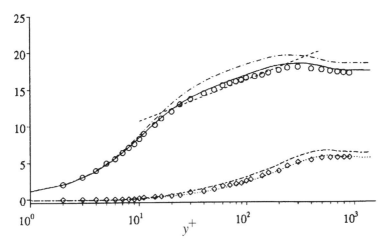

Fig. 10.2 *Mean velocities in a three-dimensional turbulent boundary layer at* $R_\delta \equiv 767$. *Dynamic similarity model [30,45]:* —, $\overline{\langle U \rangle}$; · · ·, $\overline{\langle W \rangle}$. *Dynamic model [12]:* — · —, $\overline{\langle U \rangle}$; — · —, $\overline{\langle W \rangle}$. *DNS [42]:* ○, $\overline{\langle U \rangle}$; ◇, $\overline{\langle W \rangle}$. *The log law,* $1/0.41 \log y^+ + 5$, *is included as a dashed line. (From [47]. Copyright © 1997 by The American Institute of Aeronautics and Astronautics. Reprinted with permission.)*

Examination of the normal Reynolds stresses predicted by the dynamic similarity model, in Fig. 10.3a, shows a relatively minor degradation near the wall, although the trend across the entire boundary layer, in Fig. 10.3b, appears to be excellent. The indication of this study is that there is significant importance as to which subgrid model is used, and at least for the relatively small Reynolds numbers of this study, LES can deliver accuracies similar to DNS with a great savings in cost.

A more demanding test of the capabilities of LES is brought out in computations of the flow past a circular cylinder at Reynolds number $\mathrm{Re}_d = 140{,}000$. Here, transition to turbulence occurs in thin separating shear layers and fills out a turbulent wake with embedded large-scale vortices of alternating sign. An illustration of this flow computed by a LES [6] is shown in Fig. 10.4. This study compares the performance of the Smagorinsky model incorporating van Driest damping with that of the dynamic Smagorinsky model. Several different meshes are employed so as to investigate grid dependence, and the spanwise extent of the flow is varied to see what effect this might have. The meshes are purposefully kept fine enough so that there is no need for wall functions. An important observation of this study is that the use of a subgrid model is essential: calculations done without it are found to lead to grossly inaccurate or non-convergent results. It is also observed in the case of the Smagorinsky model that the results show a significant effect of the Smagorinsky constant. For example, changing C_S from 0.1 to 0.065 causes a severe degradation in the quality of the predictions. Evidently, the amount of diffusion is insufficient in the latter case.

The time-averaged results are generally sensitive to both the mesh and the span-wise extent of the computational domain. Such dependencies appear to be an in-escapable part of the LES methodology. The Strouhal number, $\mathrm{St} \equiv \omega D / U_\infty$, defined as the dimensionless frequency at which the vortices shown in Fig. 10.4 are shed off

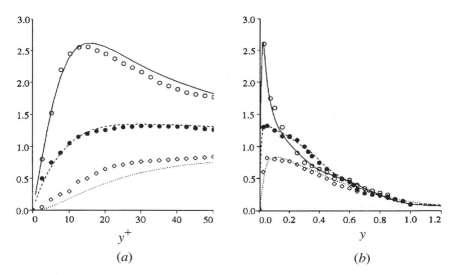

Fig. 10.3 *rms normal Reynolds stresses in a three-dimensional turbulent boundary layer at $R_\delta \equiv 767$; (a) near-wall; (b) complete boundary layer. Dynamic similarity model [30,45]: —, $\sqrt{\overline{(u')^2}}$; $\cdots$, $\sqrt{\overline{(v')^2}}$; $--$, $\sqrt{\overline{(w')^2}}$. DNS [42]: ○, $\sqrt{\overline{(u')^2}}$; ◇, $\sqrt{\overline{(v')^2}}$; ●, $\sqrt{\overline{(w')^2}}$. (From [47]. Copyright © 1997 by The American Institute of Aeronautics and Astronautics. Reprinted with permission.)*

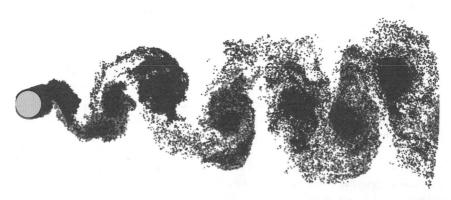

Fig. 10.4 *Visualization of tracers in the flow past a cylinder at $R_e = 140,000$. (Reprinted from [6], Copyright (2000) with permission from Elsevier Science.)*

the cylinder, is in the neighborhood of 0.2 for the various calculations, a value that is typical of those seen in experiments. More significant changes show up in other statistics, such as the drag coefficient, mean velocities, and Reynolds stresses as the mesh or domain size is varied. For example, the drag coefficient C_D for the dynamic model prediction changes from 1.239 to 1.454 as the mesh is enhanced, while the experimental value is 1.237. For the Smagorinsky model, the two runs are, respectively, 1.218 and 1.286.

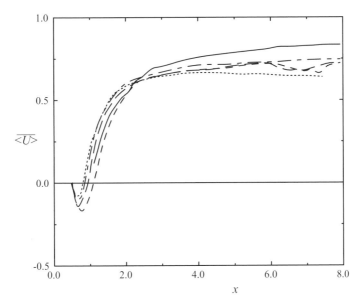

Fig. 10.5 $\langle \overline{U} \rangle$ *on the line* $y = 0.$ *— —, Smagorinsky model coarse mesh;* *— · —, Smagorinsky model fine mesh;* *— —, dynamic model coarse mesh;* *· · ·, dynamic model fine mesh;* *—,   experiment [7]. (Reprinted from [6], Copyright (2000) with permission from Elsevier Science.)*

The behavior of the mean streamwise velocity on the symmetry axis $y = 0$ is illustrated in Fig. 10.5. Here, results for both a coarse grid, with $165 \times 165 \times 64$ points, and a fine grid, with $325 \times 325 \times 64$ points, are shown for the two models. The mean velocity is increasingly underpredicted with downstream distance in the wake. Evidently, this is due to a corresponding drop-off in grid density which prevents adequate resolution of the energetic vortices governing the physics of the wake flow. Close to the cylinder, where the resolution is best, the results are better. In particular, the simulations show only relatively small errors in the predicted size of the recirculation zone, which is indicated by the region of negative mean velocity.

Figures 10.6 and 10.7 show the mean streamwise and normal velocities, respectively, at the station $x = 1$ downstream of the cylinder. Although the overall agreement is good, particularly for $\langle \overline{V} \rangle$, it is evident that the mesh refinement has not improved the solution; in fact, it is somewhat less accurate than the coarse mesh.

Predictions of the Reynolds stress component $\overline{v'^2}$ along the symmetry axis are given in Fig. 10.8, while the Reynolds shear stress $\overline{u'v'}$ on the cross plane $x = 1$ are given in Fig. 10.9. The simulations are less successful in capturing these experimental trends than was the case for the mean velocities. $\overline{v'^2}$ is very much overpredicted, and its peak is closer to the cylinder than the experimental data. The dynamic model on the fine grid provides the least accurate prediction of $\overline{v'^2}$ and $\overline{u'v'}$. Clearly there is an element of serendipity in this unexpected effect of mesh size.

Like the situation in RANS modeling, the evidence for judging the superiority of one model over another is not as clearcut as one might like. For example, in

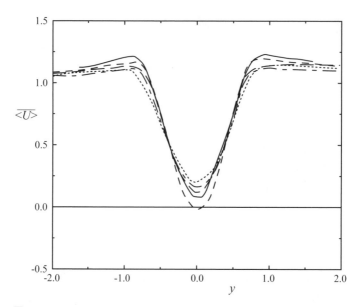

Fig. 10.6 $\langle \overline{U} \rangle$ *on the line* $x = 1$. *——, Smagorinsky model coarse mesh; —·—, Smagorinsky model fine mesh; – –, dynamic model coarse mesh; ···, dynamic model fine mesh; —, experiment [7]. (Reprinted from [6], Copyright (2000) with permission from Elsevier Science.)*

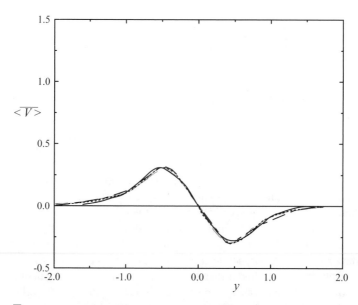

Fig. 10.7 $\langle \overline{V} \rangle$ *on the line* $x = 1$. *——, Smagorinsky model coarse mesh; —·—, Smagorinsky model fine mesh; – –, dynamic model coarse mesh; ···, dynamic model fine mesh; —, experiment [7]. (Reprinted from [6], Copyright (2000) with permission from Elsevier Science.)*

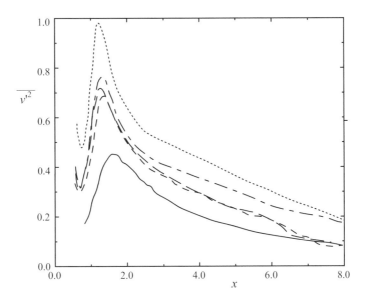

Fig. 10.8 $\overline{v'^2}$ *on the line* $y = 0$: — —, *Smagorinsky model coarse mesh;* — · —, *Smagorinsky model fine mesh;* — —, *dynamic model coarse mesh;* · · ·, *dynamic model fine mesh;* —, *experiment [7]. (Reprinted from [6], Copyright (2000) with permission from Elsevier Science.)*

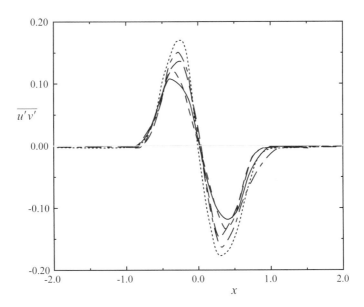

Fig. 10.9 $\overline{u'v'}$ *on the line* $x = 1$. — —, *Smagorinsky model coarse mesh;* — · —, *Smagorinsky model fine mesh;* — —, *dynamic model coarse mesh;* · · ·, *dynamic model fine mesh;* —, *experiment [7]. (Reprinted from [6], Copyright (2000) with permission from Elsevier Science.)*

the cylinder flow application, the dynamic model had the distinct advantage of not requiring a special wall treatment, but then the use of a finer mesh seemed to reduce its performance. Although one may speculate as to why this occurs—perhaps there is an unforeseen cancelation of errors in the coarse mesh—it remains an open question which model ought to be used in this and related applications.

10.6 VORTEX METHODS

It is clear from the foregoing that large eddy simulations are distinguished from DNS by the purposeful use of an underresolved mesh, combined with a model of how the unresolved flow scales affect the resolved ones. This same basic idea can be applied to the class of computational schemes known as vortex methods [36], in which convecting, gridfree vortex elements are used to represent flow fields rather than the more commonly employed fixed meshes. Evolution of the vortex elements is governed by rules designed to approximate the three-dimensional vorticity equation. A vortex method calculation has the characteristics of a LES when it sets a lower limit to the scale of the vortex elements and prevents finer-scale motions from being resolved. As in the case of traditional LES, provision is made for the effect of the small scales on the resolved scales. The potential advantages of vortex methods in regards to LES lies in the opportunity they provide to model vortex dynamics more efficiently and directly than is possible in an Eulerian grid-based setting. As we have noted in many previous contexts, vorticity dynamics offers a perspective on turbulent flow that is often the most natural one to take. The Lagrangian character of vortex methods is also an advantage since it allows for resolution of strong internal shear layers which would otherwise be smoothed in a grid-based method. Grid-free methods are also less subject than grid-based methods to numerical diffusion, and perhaps most significantly, they provide an opportunity to apply new ideas about subgrid modeling without the same concerns for numerical instability as in grid-based approaches.

Implementations of vortex methods differ according to what vortex elements are used and how diffusion and vortex stretching are modeled. They have in common the use of the Biot–Savart law to find velocities from given fields of vortical elements. The interest of this presentation is in describing some general aspects of vortex methods followed by a discussion of the special features that have been incorporated into one particular vortex method formulation which has been developed for use as a LES.

10.6.1 Biot–Savart Law

Consider an incompressible flow in $\mathfrak{R}^3$ with velocity field $\mathbf{U}$ and vorticity field $\mathbf{\Omega} = \nabla \times \mathbf{U}$. As in Section 3.6.1, assume a Helmholtz decomposition of the velocity field:

$$\mathbf{U} = \mathbf{U}_1 + \mathbf{U}_2, \tag{10.61}$$

where U_1 is solenoidal (i.e., $\nabla \cdot U_1 = 0$) and U_2 is irrotational (i.e., $\nabla \times U_2 = 0$). Since U_1 is solenoidal, there exists a vector field B, called the *vector potential*, satisfying $\nabla \cdot B = 0$ and such that

$$U_1 = \nabla \times B. \tag{10.62}$$

In this case, by virtue of the irrotationality of U_2, the definition of Ω, and the vector identity, $\nabla \times \nabla \times B = \nabla(\nabla \cdot B) - \nabla^2 B$, it follows that

$$\Omega = -\nabla^2 B. \tag{10.63}$$

The solution to this Poisson equation in unbounded space is

$$B(x, t) = -\int_{\Re^3} \frac{\Omega(y, t)}{|x - y|} \, dy, \tag{10.64}$$

and a calculation then gives

$$U_1(x, t) = -\int_{\Re^3} \frac{(x - y) \times \Omega(y, t)}{|x - y|^3} \, dy. \tag{10.65}$$

Equation (10.65) is referred to as the *Biot–Savart law*. It gives a velocity field U_1 whose curl is Ω. An alternative way of writing (10.65) follows by defining a matrix K according to

$$K_{ik}(x) = -\frac{\epsilon_{ijk}}{4\pi} \frac{x_j}{|x|^3}, \tag{10.66}$$

in which case it is not hard to show that

$$U_1(x, t) = \int_{\Re^3} K(x - y)\Omega(y, t) \, dy. \tag{10.67}$$

Without loss of generality the integration domain in (10.67) may be reduced to that of the support of Ω.

U_1 is the rotational part of U and must be supplemented by an irrotational flow U_2 used to satisfy boundary conditions. In fact, $\nabla \times U_2 = 0$ implies that U_2 can be written as the gradient of a potential, ϕ, so that $U_2 = \nabla\phi$. Taking the divergence of (10.61) then gives

$$\nabla^2 \phi = 0 \tag{10.68}$$

if U is incompressible. ϕ can be determined as the unique solution of the Laplace equation (10.68) compatible with a specified wall-normal velocity at the boundaries. For fixed solid walls the nonpenetration boundary condition applies. In general, at any boundary point with normal n, one asserts that

$$\frac{\partial \phi}{\partial n} = \nabla \phi \cdot \mathbf{n} = \mathbf{n} \cdot (\mathbf{U} - \mathbf{U}_1). \tag{10.69}$$

In this equation, $\mathbf{U} \cdot \mathbf{n}$ at the surface is a given quantity and $\mathbf{U}_1 \cdot \mathbf{n}$ is computed from (10.67). Thus, ϕ is determined by solving the Neumann problem formed by (10.68) and (10.69).

It is well known that the solution to the Neumman problem for ϕ can be determined in the form of an integral over a surface source distribution [18]. An equation from which the required surface source strength can be found is derived by enforcing (10.69). This kind of boundary element method is ideally suited to the needs of vortex methods since it means that the potential flow $\mathbf{U}_2$ can be determined without the need for a mesh covering the flow domain. Thus neither $\mathbf{U}_1$ nor $\mathbf{U}_2$ requires the creation of a mesh.

To summarize, the velocity field associated with a given vorticity field is computed by first determining $\mathbf{U}_1$ from (10.67), then obtaining ϕ in the form of a surface integral using (10.69) to predict the boundary source strength, and finally, evaluating $\mathbf{U}_2$ from the boundary source field. The way that individual vortex elements in a calculation enter this scheme is through their contributions to $\boldsymbol{\Omega}$ in (10.67). The form that the vortex elements may assume is now considered.

10.6.2 Vortex Elements

Implementations of the vortex method are distinguished according to which computational elements they use to represent the vorticity field, $\boldsymbol{\Omega}$. For the flow region away from boundaries there are two main choices: vortex blobs and vortex tubes. A *vortex blob* is an amorphous local volume of fluid with vorticity field of the general form $\boldsymbol{\Omega}(t) f_h(\mathbf{x} - \mathbf{x}')$, where $\boldsymbol{\Omega}(t)$ is the strength of the blob; $\mathbf{x}'$ is a central position within it; $f_h(\mathbf{r}) \equiv f(\mathbf{r}/h)/h^3$, where f is a smooth function with unit volume integral; and h is a length scale which essentially establishes the size of the vortex blob. A *vortex tube* is a special case of a blob, consisting of a short, straight, cylindrical volume with vorticity aligned in the axial direction. It is usually assumed that the vorticity in such a tube has a radially symmetric distribution. A fundamental difference between tubes and blobs is that the former tend to be connected together end to end, forming long filamentlike structures. These appear naturally during the development of a fluid flow. Blobs, on the other hand, are virtually always regarded as individual vortical particles with no connection to neighboring blobs.

The vorticity due to a collection of N tubes or blobs is just the sum of the individual contributions to vorticity, so that in the case of blobs, for example,

$$\boldsymbol{\Omega}(\mathbf{x}, t) = \sum_{i=1}^{N} \boldsymbol{\Omega}_i(t) f_h(\mathbf{x} - \mathbf{x}_i), \tag{10.70}$$

where $\mathbf{x}_i$ is the center of the ith blob and $\boldsymbol{\Omega}_i(t)$ is its strength. The smoothing function, f_h, is usually chosen with a view toward establishing the convergence of the vortex method to an exact solution of the Navier–Stokes equation as the number of elements is increased and $h \rightarrow 0$ [36].

Although it may not be obvious at first sight, the choice of f_h can be in potential conflict with the expectation that the vorticity field should be divergence free via the vector identity $\nabla \cdot \mathbf{\Omega} = \nabla \cdot (\nabla \times \mathbf{U}) \equiv 0$. In fact, for any individual blob, the condition $\nabla \cdot [\mathbf{\Omega}_i(t) f_h(\mathbf{x} - \mathbf{x}_i)] = 0$ is generally not satisfied. However, it can be shown that if one projects $\mathbf{\Omega}_i(t) f_h(\mathbf{x} - \mathbf{x}_i)$ onto the set of divergence-free vectors, the velocity field produced by this new vorticity, say $\mathbf{\Omega}_i^*(\mathbf{x}, t)$, is identical to that associated with the original blob:

$$\mathbf{U}_i(\mathbf{x}, t) = \int \mathbf{K}(\mathbf{x} - \mathbf{y})\mathbf{\Omega}_i(t) f_h(\mathbf{y} - \mathbf{x}_i) \, d\mathbf{y} = \int \mathbf{K}(\mathbf{x} - \mathbf{y})\mathbf{\Omega}_i^*(\mathbf{y} - \mathbf{x}_i, t) \, d\mathbf{y}. \quad (10.71)$$

Thus, there is no difference in the predicted velocity whether the vorticity of an element is divergence-free or not. Of course, other aspects of the calculation may be adversely affected, and if one wants to, it is possible to make modifications to $\mathbf{\Omega}_i(t) f_h(\mathbf{x} - \mathbf{x}_i)$ so that it is divergence free. In practice, it is not generally considered a high priority to make the necessary change. It may be noted that vortex tubes formed into closed loops do not violate the divergence-free condition of the vorticity field. However, for tubes that are not closed, the condition is violated at each end.

Although blobs and tubes offer an appropriate vehicle with which to represent the vorticity field in general circumstances, they are not well suited to describing the approximately two-dimensional region of very high vorticity in the viscous sublayer of wall-bounded flows. For example, a great concentration of vorticity near the boundary of even a low-Reynolds-number channel flow is evident in Fig. 4.4. At higher Reynolds numbers the relative magnitude of the wall and outer flow vorticity is much greater. The representation of $\mathbf{\Omega}$ in the boundary region through vortex tubes or blobs is inefficient: it is difficult to cover a largely two-dimensional region with tubelike or bloblike objects. However, vortex sheet elements are ideally suited for such conditions and are commonly used in this capacity in vortex methods. In some implementations they are idealized as sheets of zero thickness, while in others they have finite width. In some schemes the sheets convect with the flow; in others, they are held in a permanent fixed grid covering the viscous sublayer.

Whatever types of vortical elements are used in a vortex method, their collective contribution to the velocity field is found by summing their individual contributions as determined from (10.67). In the case of N vortex blobs as in (10.70), this gives

$$\mathbf{U}(\mathbf{x}, t) = \sum_{i=1}^{N} \mathbf{K}_h(\mathbf{x} - \mathbf{x}_i)\mathbf{\Omega}_i(t) + \mathbf{U}_2, \quad (10.72)$$

where $\mathbf{K}_h(\mathbf{x}) \equiv \mathbf{K} * f_h(\mathbf{x}) = \int \mathbf{K}(\mathbf{x} - \mathbf{y}) f_h(\mathbf{y}) \, d\mathbf{y}$ is a convolution integral. Unlike $\mathbf{K}$, $\mathbf{K}_h$ has no singularity: a property that is of considerable convenience if (10.72) is to be used in a numerical scheme.

With the use of the Biot–Savart law in computing velocities comes the need to solve an N-body problem; that is, the velocities at the positions of N vortex elements have to be computed, and each of these receives a contribution from N vortices, so there are $O(N^2)$ operations involved. Such calculations, if done by direct evaluation of (10.72), are prohibitively expensive when $N \approx 100,000$, even if the most powerful computers and full parallelization is employed. This number of vortices is far short of what is needed to do justice to the range of scales appearing in a typical turbulent flow. However, the advent of the fast multipole method (FMM) [16] enables a numerical solution to the N-body problem with just $O(N \log N)$ or even $O(N)$ labor and thus provides vortex methods with the means for becoming a feasible alternative to grid-based methods. In fact, modern vortex methods using a parallelized FMM can accommodate in excess of 1 million vortices, and at this level of information, they can treat turbulent flows.

10.6.3   Dynamical Equations

In a vortex method, the three-dimensional vorticity equation (2.108) is used as the basis for advancing the vortex elements in time. Often, a fractional step or splitting method is used in which the processes of advection, stretching, and diffusion are accounted for sequentially. How these procedures are carried out depends on which particular elements are incorporated in the method. Generally, in all approaches, advection requires moving the elements by their velocity as computed from (10.72), and this is what establishes vortex methods as Lagrangian particle methods.

In a vortex blob method the vortex stretching terms can be accounted for in a variety of ways. One technique is to substitute (10.70) and (10.72) into the stretching term in (2.108), yielding

$$(\nabla \mathbf{U})\mathbf{\Omega}(\mathbf{x}_i, t) = \sum_{m=1}^{N} \sum_{n=1}^{N} (\nabla[\mathbf{K}_h(\mathbf{x}_i - \mathbf{x}_n)\mathbf{\Omega}_n(t) + \mathbf{U}_2])\mathbf{\Omega}_m(t) f_h(\mathbf{x}_i - \mathbf{x}_m), \qquad (10.73)$$

where $\nabla \mathbf{U}$ is the tensor with components $\partial U_i / \partial x_j$. In view of the small support of f_h, the amount of labor in implementing (10.73) is much closer to $O(N)$ than the $O(N^2)$ suggested by the double sum. Another possibility is to reconstruct locally smooth velocity and vorticity fields via local least-squares modeling, in which case the stretching term can be calculated directly at any point [28]. In either case, when used in the context of a fractional step method, formulas such as (10.73) approximate the amount that $\mathbf{\Omega}_i(t)$ changes by stretching at each time step. Analysis of the numerical properties of blob methods shows that the accuracy of (10.73) depends on maintaining a dense, overlapping coverage of vortex blobs in those regions of the flow where there is significant vorticity. This requirement is more attainable in high-Reynolds-number laminar flow than when turbulence is present, and blobs are needed to resolve the evolution of the vorticity field to fine scales.

In the case of vortex tube methods, the stretching term is accounted for merely by translating the end of each tube segment (i.e., the relative motion of the endpoints models the stretching and reorientation effects). This is justified in high-Reynolds-number flow by appealing to Kelvin's theorem to the effect that vortex tubes convect with the fluid in inviscid motion. Whenever vortex tube segments stretch beyond a fixed length, they are subdivided. In this model of the stretching process the circulation of the tubes is unchanged despite how much they might stretch.

How best to take into account the effects of viscous vorticity diffusion has been a somewhat contentious issue, as evidenced by the number and variety of methods proposed. The first to see widespread use was the random vortex method (RVM), in which diffusion is modeled via a Monte Carlo scheme [36] in which vortex elements are made to undergo a random walk. Despite the simplicity of this approach and its easy implementation, the act of randomly moving vortex elements adds randomness to the velocity field even when simulating laminar flow. In the case of turbulence, the noise of this scheme can overwhelm the magnitude of the physical turbulence, and so an alternative method is required.

With the aim of circumventing the limitations of the RVM, a number of *deterministic* schemes have been formulated. Among the latter is the diffusion velocity method [33], in which an effective velocity is determined so that convection at this rate mimics the real effect of diffusion. In the particle strength exchange method [10], the differential form of the diffusion term is rewritten in an equivalent integral form allowing for easier evaluation of diffusion over a field of vortex elements. Other approaches include direct differentiation of the kernel functions [11], differentiation of least-squares fits to the vorticity [28], and a vortex redistribution scheme [38], among others. Each of these approaches has advantages and disadvantages; the decision as to which ought to be used in a vortex method depends on the application one has in mind and the kind of vortex elements. Moreover, entirely different approaches may be required if turbulence is to be modeled.

The deterministic diffusion models that have been derived tend to apply exclusively to vortex blob methods and not to tube methods. In fact, it is less clear how to accommodate diffusion of tubes. On the other hand, tube methods are better positioned than blob methods to remain stable under turbulent flow conditions. It is thus not entirely obvious how best to formulate a vortex method in the context of turbulent flow. However, it will be shown in the next section that by considering the physics of turbulence and most significantly, where and in what ways viscosity plays a crucial role in the dynamics, an effective vortex method for turbulence can be devised.

10.6.4 Vortex Method for Turbulent Flow

A useful guidepost in deciding how best to configure a vortex method for turbulent flow prediction is a consideration of the role of viscosity in turbulent physics. According to the discussion of earlier chapters, this is limited primarily to the viscous sublayer of bounded flows where vorticity diffusion from the surface is a major dynamic and to viscous energy dissipation in spatially intermittent regions throughout the flow. The latter are characterized by highly stretched small-scale vortices.

To capture these kinds of viscous effects in a vortex method, two important attributes must be imparted to the numerical scheme. First, the viscous sublayer adjacent to solid surfaces must be resolved well enough to capture viscous diffusion of vorticity from the boundary surface. It was remarked previously that this region is covered most efficiently by a sheet structure, so the requirement here boils down to an accurate calculation of viscous diffusion through the sheets. One relatively accurate scheme for accomplishing this is to use a fixed sheet mesh and accompanying finite-volume formulas for the diffusion rate as part of a finite-volume solution of the vorticity equation on the fixed sheet mesh. If the resolution is fine enough, viscous vorticity diffusion with its attendant vorticity generation at the solid surface can be determined with acceptable accuracy. In other words, numerical diffusion can be kept within bounds so that the desired Reynolds number is indeed modeled. Experience shows that it is exceedingly difficult to achieve similar accuracy by alternative deterministic techniques applied to convecting sheet elements.

With regard to the effect of viscosity on energy dissipation, first note that the principal mechanism causing energy cascade to small scales is vortex stretching. Moreover, as mentioned with regard to Fig. 1.3, the fine-scale structure of turbulence is tubelike, and it is therefore the stretching and folding of vortex tubes that takes energy to small scales. It is thus natural to chose vortex tubes as the primary grid-free element of the simulation. Such objects are unconditionally stable in tracking the stretching process: They are able to fold and stretch without loss or gain in circulation, thus avoiding the difficulties attendant in maintaining bounded vorticity amplitude in blob methods. On the other hand, it is prohibitively expensive to run a vortex tube calculation in which the tubes are permitted to fold and stretch until the fine, dissipative scales are reached. Moreover, there is no clearcut means for accurately accommodating viscous diffusion once energy arrives at the dissipation scales.

One answer to these problems lies in the hairpin removal algorithm developed by Chorin [8,9], in which folded or kinked vortex tubes at a sufficiently acute angle are removed from the computation. The rationale for this step is that once it is assumed that the vortex stretching process has taken energy to the inertial range scales, there is no need to further track the folding process to the viscous scales. In fact, folded vortex tubes contribute primarily to local energy—since the far-field velocities cancel—and removal of such hairpins then mimics the local dissipation process.

An additional benefit of tubes is that they are the principal dynamical feature of the near-wall region in bounded turbulent flows. Particularly in the form of quasi-streamwise vortices, they control the momentum exchange near boundaries that leads to the turbulent Reynolds shear stress. A tube method is a natural means for representing these structures. New tubes are created at the outer plane of sheets when the vorticity exceeds a threshold. This limits the number of new vortices to what may be imagined to be significant ejection events. The smallest resolved scales in the simulation are controlled by the minimum length of a tube segment and the density of the sheet mesh. Under ideal circumstances, for a given Reynolds number, these would be chosen to enable resolution of the physical wall region structures as well as the inertial range scales.

10.6.5   Sample Results from a Vortex Method LES

The application of vortex methods to LES is in its infancy, but judging from the results of preliminary calculations, it is likely to find increasing use in the future. One flow field that has been studied with a view toward establishing the effectiveness of the methodology is that of a tripped, zero-pressure-gradient turbulent boundary layer past a flat plate. Some results that illustrate the unique advantages of vortex methods are presented here.

Figure 10.10 shows the vortex elements in a boundary layer calculation as viewed from above. The flow is tripped by a bump placed at $x = 0.1$ with height 0.003. The plate used in the simulation has a total length of 2 with a Reynolds number $\text{Re}_x = U_\infty x/\nu = 400,000$ at the end of the plate. Upstream of the trip, the vorticity is purely spanwise and the flow is laminar. Immediately downstream of the bump the flow remains laminar for a short distance, but soon transitions into a highly perturbed state. This behavior is reminiscent of the physical transition process in which two-dimensional Tollmein–Schlichting waves first appear, followed by a focusing of their spanwise vorticity, which subsequently undergoes instability, causing the appearance of streamwise vorticity and finally, turbulence [44]. A close-up view of the vortex elements in the latter part of the transition region shown in Fig. 10.11, indicates that reorientation of spanwise vorticity into the streamwise direction is associated with the presence of alternating regions of faster and slower streamwise motions (i.e., the beginnings of the streaky structure that characterizes turbulent boundary layers and which is found to underlie the fully turbulent regions of the simulation). A side view of the vortices in Fig. 10.12 reveals a thickening of the boundary layer in the transition region and the generation of wall-normal vorticity. Farther downstream the flow develops many structural features in the form of coherent vortices. The side images show the characteristic crumpled outer-edge pattern of the boundary layer formed from individual large-scale structures. A detail of one such mushroom-shaped object is shown in Fig. 10.13. Its appearance has much in common with similar objects seen in smoke visualizations of the turbulent boundary layer (see Fig. 4.25 and [17,44]).

Some preliminary statistics at $x = 1.1$ corresponding to $\text{Re}_x = 200,000$ are shown in Fig. 10.14. $\overline{U}(y)$ is plotted versus the standard log-law $\overline{U}^+ = (1/0.41)\log y^+ + 5$, the sublayer relation $\overline{U}^+ = y^+$, and a DNS prediction [41] at $R_\theta = 670$. The latter is an appropriate comparison because $R_\theta = 575$ for the vortex method simulation. It

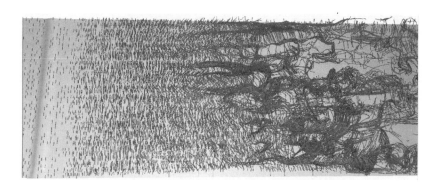

Fig. 10.10 *View from above of vortex tubes in a zero-pressure-gradient boundary layer. (From [4].)*

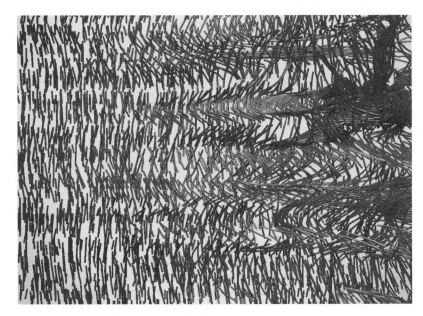

Fig. 10.11   *Detail of Fig. 10.10 in late transition showing the reorientation of spanwise vortices. (From [4].)*

Fig. 10.12   *Side view of vortex tubes in a zero-pressure-gradient boundary layer. (From [4].)*

is evident that apart from a small overprediction in the log layer, the physical trend is well duplicated.

Some results for the Reynolds stresses are presented in Fig. 10.15. The magnitude of the individual normal stresses are well accounted for in the simulation as is the Reynolds shear stress. In fact, this simulation is somewhat underresolved and it is anticipated that greater quantitative accuracy will follow when resolution is enhanced. It is particularly interesting that the structure predicted by the vortex method generates a significant Reynolds shear stress. This is a strong indication of the physical soundness of the methodology, that is, the Reynolds stress and its attendant momentum exchange are not likely to be correctly predicted without capturing the physical structures and their dynamics as discussed in Chapter 4.

10.7   SUMMARY

LES is a method for turbulent flow simulation that has grown in popularity in recent years due both to the diminishing expectation that RANS modeling will be able to

Fig. 10.13 *Mushroom-shaped object in boundary layer simulation. (From [4].)*

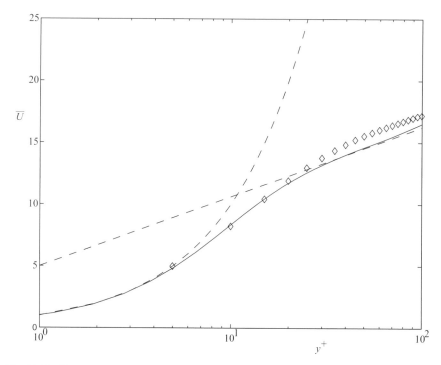

Fig. 10.14 $\overline{U}$ *at* $x = 1.1$. —, *DNS [41];* $\diamond$, *vortex method. Dashed lines show linear and log law profiles.* *(From [4].)*

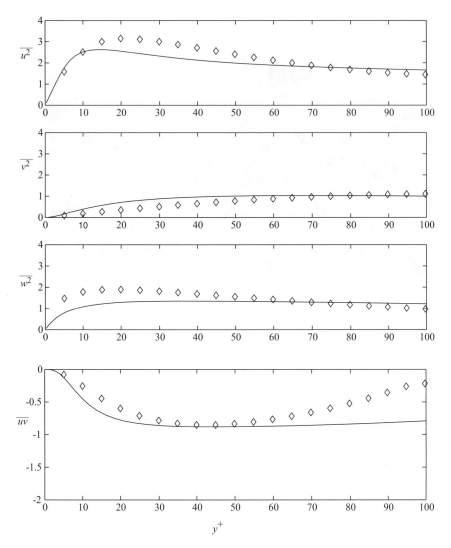

Fig. 10.15 *Reynolds stresses at x = 1.1. Solid lines are from a DNS [41]; vortex method [4].*

deliver accurate flow predictions and the rising availability of ever-more-powerful computers to support turbulent flow simulations at high Reynolds numbers. The question of how best to capture subgrid scale effects in LES has received a great deal of attention, and significant progress beyond the classical Smagorinsky model has been achieved. It is often the case that the increased responsiveness to local conditions offered by the dynamic procedure is worth the additional computational expense. This model is also particularly advantageous near boundaries, where it obviates the need for ad hoc modifications to model coefficients, such as the practice of applying van

Driest damping. Adding a similarity term to the subgrid model is also beneficial, presumably because it allows for some inclusion of backscatter.

On the whole, the search for better subgrid models continues, and many other aspects of the science remain unsettled, including how best to manage grid size and the errors associated with numerical schemes. Finally, alternatives to grid-based LES via vortex methods are just beginning to be applied to the solution of engineering flows. Preliminary results suggest that the approach holds out much promise of maturing into an effective means for implementing LES.

REFERENCES

1. Agel, F., Meneveau, C. and Parlange, M. B. (2000) "A scale-dependent dynamic model for large-eddy simulation: application to a neutral atmospheric boundary layer," *J. Fluid Mech.* **415**, 261–284.

2. Bardina, J. Ferziger, J. H. and Reynolds, W. C. (1980), "Improved subgrid scale models for large-eddy simulation," *AIAA Pap. 80-1357.*

3. Batchelor, G. K. (1960) *The Theory of Homogeneous Turbulence*, Cambridge University Press, Cambridge.

4. Bernard, P. S. and Dimas, A. (2001) "Vortex method modeling of complex, turbulent engineering flows," *Proc. 2nd International Conference on Vortex Methods*, Istanbul, Turkey, pp. 41–54.

5. Boris, J. P., Grinstein, F. F., Oran, E. S. and Kolbe, R. L. (1992) "New insights into large eddy simulation," *Fluid Dyn. Res.* **10**, 199–228.

6. Breuer, M. (2000) "A challenging test case for large eddy simulation: high Reynolds number circular cylinder flow," *Int. J. Heat Fluid Flow* **21**, 648–654.

7. Cantwell, B. and Coles, D. (1983) "An experimental study on entrainment and transport in the turbulent near wake of a circular cylinder," *J. Fluid Mech.* **136**, 321–374.

8. Chorin, A. J. (1993) "Hairpin removal in vortex interactions II," *J. Comput. Phys.* **107**, 1–9.

9. Chorin, A. J. (1994) *Vorticity and Turbulence*, Springer-Verlag, New York.

10. Cottet, G. H. and Mas-Gallic, S. (1990) "A particle method to solve the Navier–Stokes system," *Numer. Math.* **57**, 805–827.

11. Fishelov, D. (1990) "A new vortex scheme for viscous flows," *J. Comput. Phys.* **86**, 211–224.

12. Germano, M., Piomelli, U., Moin, P. and Cabot, W. H. (1991) "A dynamic subgrid-scale eddy viscosity model," *Phys. Fluids A*, **3**, 1760–1765.

13. Ghosal, S. (1996) "An analysis of numerical errors in large eddy simulation of turbulence," *J. Comput. Phys.* **125**, 187–206.

14. Ghosal, S., Lund, T. Moin, P. and Akselvoll, K. (1995) "A dynamic localization model for large-eddy simulation of turbulent flow," *J. Fluid Mech.* **286**, 229–255.

15. Ghosal, S. and Moin, P. (1995) "The basic equations for the large eddy simulation of turbulent flows in complex geometry," *J. Comput. Phys.* **118**, 24–37.

16. Greengard, L. and Rokhlin, V. (1987) "A fast algorithm for particle simulations," *J. Comput. Phys.*, **73**, 325–348.

17. Head, M. R. and Bandyopadhyay, P. (1981) "New aspects of turbulent boundary layer structure," *J. Fluid Mech.* **107**, 297–338.

18. Hess, J. L. and Smith, A. M. O. (1967) "Calculation of potential flow about arbitrary bodies," *Prog. Aerosp. Sci.* **8**, 1–138.

19. Horiuti, K. (1989) "The role of the Bardina model in large eddy simulation of turbulent channel flow," *Phys. Fluids A* **1**, 426–428.

20. Kawamura, T. and Kuwahara, K. (1984) "Computation of high Reynolds number flow around a circular cylinder with surface roughness," *AIAA Pap. 84-0340.*

21. Kim, J., Moin, P. and Moser, R. (1987) "Turbulence statistics in fully developed channel flow at low Reynolds number," *J. Fluid Mech* **177**, 133–166.

22. Kravchenko, A. G. and Moin, P. (1997) "On the effect of numerical errors in large eddy simulations of turbulent flows," *J. Comput. Phys.* **131**, 310–322.

23. Kravchenko, A. G., Moin, P. and Moser, R. (1996) "Zonal embedded grids for numerical simulations of wall-bounded turbulent flows," *J. Comput. Phys.* **127**, 412–423.

24. Leonard, A. (1974) "Energy cascade in large-eddy simulations of turbulent fluid flows," *Adv. Geophys.* **18A**, 137–148.

25. Lesieur, M. and Métais, O. (1996) "New trends in large-eddy simulations of turbulence," *Annu. Rev. Fluid Mech.* **28**, 45–82.

26. Lilly, D. K. (1967) "The representation of small-scale turbulence in numerical simulation experiments," in *Proc. IBM Scientific Computing Symposium on Environmental Sciences,* p. 195.

27. Lilly, D. K. (1992) "A proposed modification of the Germano subgrid scale closure method," *Phys. Fluids A* **4**, 633–635.

28. Marshall, J. S., Grant, J. R., Gossler, A. A. and Huyer, S. A. (2000) "Vorticity transport on a Lagrangian tetrahedral mesh," *J. Comput. Phys.* **161**, 85–113.

29. Meneveau, C. and Katz, J. (2000) "Scale-invariance and turbulence models for large-eddy simulation," *Annu. Rev. Fluid Mech.* **32**, 1–32.

30. Meneveau, C., Lund, T. S. and Cabot, W. H. (1996) "A Lagrangian dynamic subgrid-scale model of turbulence," *J. Fluid Mech.* **319**, 353–385.

31. Métais, O. and Lesieur, M. (1992) "Spectral large-eddy simulation of isotropic and stably stratified turbulence," *J. Fluid Mech.* **239**, 157–194.

32. Moin, P. (2000) "Numerical issues in large eddy simulation of complex turbulent flows and application to aeroacoustics," in *Advanced Turbulent Flow Computations* (R. Peyret and E. Krause, Eds.), Springer-Verlag, New York, pp. 131–154.

33. Ogami, Y. and Akamatsu, T. (1990) "Viscous flow simulation using the discrete vortex model—the diffusion velocity method," *Comput. Fluids* **19**, 433–441.

34. Piomelli, U. (1993) "High Reynolds number calculations using the dynamic subgrid-scale stress model," *Phys. Fluids A* **5**, 1484–1490.

35. Piomelli, U., Moin, P. and Ferziger, J. (1988) "Model consistency in LES of turbulent channel flows," *Phys. Fluids* **31**, 1884–1891.

36. Puckett, E. G. (1993)"Vortex methods: an introduction and survey of selected research topics," in *Incompressible Computational Fluid Dynamics: Trends and Advances* (M. D. Gunzburger and R. A. Nicolaides, Eds.), Cambridge University Press, Cambridge, pp. 335–407.

37. Schumann, U. (1975) "Subgrid scale model for finite difference simulations of turbulent flows in plane channel and annuli," *J. Comput. Phys.* **18**, 376–404.

38. Shankar, S. and van Dommelen, L. (1996) "A new diffusion procedure for vortex methods," *J. Comput. Phys.* **127**, 88–109.

39. Smagorinsky, J. (1963) "General circulation experiments with the primitive equations," *Mon. Weather Rev.* **91**, 99–165.

40. Sohankar, A., Davidson, L. and Norberg, C. (2000) "Large eddy simulation of flow past a square cylinder: comparison of different subgrid scale models," *ASME J. Fluids Eng.* **122**, 39–47.

41. Spalart, P. R. (1988) "Direct simulation of a turbulent boundary layer up to $Re_\theta = 1410$," *J. Fluid Mech.* **187**, 61–98.

42. Spalart, P. R. (1989) "Theoretical and numerical study of a three-dimensional turbulent boundary layer," *J. Fluid Mech.* **205**, 319–340.

43. Speziale, C. G. (1985) "Galilean invariance of subgrid scale models in the large-eddy simulation of turbulence," *J. Fluid Mech.* **156**, 55–62.

44. Van Dyke, M. (1982) *An Album of Fluid Motion*, Parabolic Press, Stanford, Calif.

45. Vreman, B., Geurts, B. and Kuerten, J. (1994) "On the formulation of the dynamic mixed subgrid-scale model," *Phys. Fluids* **6**, 4057–4059.

46. Vreman, B., Geurts, B. and Kuerten, J. (1996) "Comparison of numerical schemes in large-eddy simulation of the temporal mixing layer," *Int. J. Numer. Methods Fluids* **22**, 297–311.

47. Wu, X. and Squires, K. D. (1997) "Large eddy simulation of an equilibrium three-dimensional turbulent boundary layer," *AIAA J.* **35**, 67–74.

48. Zang, Y., Street, R. L. and Koseff, J. R. (1993) "A dynamic mixed subgrid-scale model and its application to turbulent recirculating flows," *Phys. Fluids A* **5**, 3186–3196.

11

Analysis of Turbulent Scalar Fields

In a great many engineering flows, interest in the statistical properties of the velocity field is subsidiary to that of the temperature or scalar concentration fields. For example, in the analysis of the cooling of electronic components, there is little interest in fluid forces but a great concern about what the effect of turbulent motion has on heat dissipation. Similar remarks apply to plumes formed by the release of contaminants into the atmosphere or rivers or lakes. The goal of this chapter is to survey some of the main directions of investigation into predicting the behavior of contaminants in turbulent flow. This presentation builds on the brief introductory material in Section 2.8.

It was previously established through Eq. (2.134) that there is a fundamental connection between the Eulerian and Lagrangian views of the evolution of the scalar concentration field. This connection will be put to good use here in the case of a turbulent scalar plume evolving within a uniform mean flow. Not only is this plume fundamental to the subject and relatively easy to analyze from a mathematical perspective, but it is a prototype of many of the most common practical problems involving scalar transport.

The following analysis yields a fairly complete analytical description of a plume in a uniform mean field. Some gaps in the argument still exist, but enough is known to be valuable in practical contexts. In particular, the insights gained for the simple plume sets the stage for investigating the turbulent scalar flux rate, $\overline{u_i c}$, which plays a key role in determining the evolution of general mean scalar fields in turbulent flow. Further clarification of the physics surrounding $\overline{u_i c}$ is given using the backward analysis of Lagrangian fluid particle trajectories which was previously involved in the study of momentum and vorticity transport in Chapter 6. The final section is devoted to a discussion of techniques for predicting the mean scalar field. This includes methods approaching the problem from both the Eulerian and Lagrangian perspectives.

11.1 PLUMES

A plume results from the continuous release of contaminant into a moving fluid. If q is the rate of release of contaminant per volume per second into a flow, a plume originating from a small volume dV at a point $\mathbf{x}$ can be approximated as being the end result of a continuous series of discrete puffs every dt time units that release $q\,dt\,dV$ contaminant into the flow. Looked at this way, it is evident that a useful strategy for describing plumes is first to predict the attributes of a solitary puff and then consider the result of stringing many puffs together to form a plume. This is the strategy typically pursued, and it ties in closely with the observations leading to (2.134). Thus the object now is to seek an expression for the pdf of the positions of particles released into a turbulent flow, namely $\mathcal{P}(\mathbf{x}, t)$, which, through (2.134), is equivalent to giving the concentration of a puff. Before considering this problem, however, it is helpful to consider briefly the case of a puff of contaminant in a laminar flow.

11.1.1 Laminar Puff

A puff is formally the solution of (2.128) with q given by (2.130). This is a standard Green's function problem whose solution is well known to be

$$C(\mathbf{x}, t) = \frac{Q}{(4\pi\,\mathcal{D}t)^{3/2}}\,e^{-[(x-Ut)^2+y^2+z^2]/4\mathcal{D}t}. \tag{11.1}$$

Defining

$$\sigma^2 = 2\mathcal{D}t, \tag{11.2}$$

then according to (11.1) and (11.2), it follows that

$$\mathcal{P}(\mathbf{x}, t) = \frac{1}{(2\pi\sigma^2)^{3/2}}e^{-[(x-Ut)^2+y^2+z^2]/2\sigma^2}, \tag{11.3}$$

which is the pdf of a Gaussian field with variance given by (11.2). Furthermore, σ^2 is related to the mean-squared spread of the puff through the identities

$$\sigma^2 = \int (x - Ut)^2 \mathcal{P}(\mathbf{x}, t)\,dV = \int y^2 \mathcal{P}(\mathbf{x}, t)\,dV = \int z^2 \mathcal{P}(\mathbf{x}, t)\,dV, \tag{11.4}$$

since $\mathcal{P}(\mathbf{x}, t)$ describes the locations of the particles comprising the puff at time t. From (11.2) it is clear that $\sqrt{2\mathcal{D}t}$ is a measure of the size of the puff as it grows by molecular diffusion while translating through the fluid.

The Gaussian pdf in (11.3) can also be arrived at [16] by modeling the diffusion process as if it evolved according to a Brownian motion process in which marked molecules leaving from the initial point travel through the fluid in a series of statistically independent jumps, each occurring over time dt. For each of the three directions, the distance of a jump is sampled from a probability distribution with standard deviation $\sqrt{2D\,dt}$. By the central limit theorem the pdf resulting from such a sum of independent random variables is Gaussian with variance given by $(2D\,dt) \times (t/dt) = 2Dt$, where t/dt is the number of steps in the random walk.

Equations (11.1) and (11.3) tie together the Eulerian viewpoint incorporating a Fickian diffusion law with attendant diffusion coefficient D, with the probabilistic path model with spreading characterized by σ^2. As will be seen, generalization of these ideas to the turbulent case provides a means to usefully analyze some important aspects of turbulent scalar transport.

11.1.2 Turbulent Puff

Turbulent scalar transport treated through the Eulerian point of view, as expressed in (2.129), is ultimately limited by the essential difficulties of modeling the scalar flux rate. Thus, unlike the laminar puff, it is not possible to solve a differential equation to obtain the concentration of the puff. The Lagrangian framework, on the other hand, since it does not depend on modeling $\overline{u_i c}$, potentially offers a way around this impasse. The particular case of a puff of contaminant released into a homogeneous turbulent flow with uniform mean velocity provides an important example where this strategy turns out to be effective.

The ultimate goal here, although this proves to be too ambitious, is to determine the exact functional form of $\mathcal{P}(\mathbf{x}, t)$ for a turbulent puff. A more attainable goal is determining the second moments of $\mathcal{P}$, and through them to gain some approximate knowledge of $\mathcal{P}$ itself. This strategy is part of Taylor's famous paper on diffusion by continuous movements [11]. To illustrate the argument, consider a puff released in a homogeneous turbulent flow with uniform mean velocity $(\overline{U}, 0, 0)$. If $\mathbf{X}(s)$ denotes a fluid particle trajectory, the lateral position of a fluid particle leaving from $\mathbf{x} = 0$ at $t = 0$ satisfies

$$Y(t) = \int_0^t V(s)\,ds \tag{11.5}$$

since $Y(0) = 0$ and where $V(s) = v(\mathbf{X}(s), s)$ is the Lagrangian velocity of the fluid particle in the y direction. Squaring and averaging (11.5) yields

$$\overline{Y^2}(t) = \int_0^t \int_0^t \overline{V(s)V(s')}\,ds\,ds'. \tag{11.6}$$

Since the turbulence is assumed to be homogeneous, it makes sense to define a velocity autocorrelation function $R(s)$ via

$$\overline{v^2}R(s' - s) = \overline{V(s)V(s')} \tag{11.7}$$

(i.e., as is written here, the functional dependence of R is only on the difference between s and s' since absolute time is immaterial in this context). Note as well that $R(0) = 1$ since $\overline{v^2} = \overline{V(s)^2}$ is independent of s. According to (11.6) and (11.7),

$$\overline{Y^2}(t) = \overline{v^2} \int_0^t ds \int_0^t ds' \, R(s' - s). \tag{11.8}$$

Breaking up the inner integration into two parts, one from $0 \rightarrow s$ and the other from $s \rightarrow t$, and making a change of variables in the second of the two integrals, yields

$$\overline{Y^2}(t) = \overline{v^2} \int_0^t ds \int_0^s ds' \, R(s' - s) + \overline{v^2} \int_0^t ds \int_0^s ds' \, R(s - s'). \tag{11.9}$$

By virtue of homogeneity, $R(s' - s) = R(s - s')$, so that

$$\overline{Y^2}(t) = 2\overline{v^2} \int_0^t ds \int_0^s ds' \, R(s' - s). \tag{11.10}$$

After switching the order of integration in (11.10), it may be deduced that

$$\overline{Y^2}(t) = 2\overline{v^2} \int_0^t (t - s) R(s) \, ds. \tag{11.11}$$

Note that similar results can be derived for $\overline{(X - \overline{U}t)^2}$ and $\overline{Z^2}$.

For small values of t, the range of s in the integral in (11.11) is close to zero, in which case $R(s) \approx 1$, so that

$$\overline{Y^2}(t) \approx \overline{v^2} t^2. \tag{11.12}$$

On the other hand, when t is large, (11.11) converges to

$$\overline{Y^2}(t) \approx 2\overline{v^2} \mathcal{T}(t - t_1), \tag{11.13}$$

where

$$\mathcal{T} = \int_0^\infty R(s) \, ds \tag{11.14}$$

is the Lagrangian integral time scale, and t_1 is the constant $(1/\mathcal{T}) \int_0^\infty s R(s) \, ds$. For large enough t,

$$\overline{Y^2}(t) \approx 2\overline{v^2} \mathcal{T} t. \tag{11.15}$$

It may be concluded from this that the mean-squared particle displacement grows quadratically in time in the early going, finally slowing to linear growth at large times. The exact details of how $\overline{Y^2}(t)$ changes between (11.12) and (11.15) depends on knowing $R(t)$, a quantity that is not known from theory but has been measured in experiments [9] and computed in DNS [15]. These suggest that a reasonable approximation to $R(t)$ is the exponential form,

$$R(t) = e^{-t/\mathcal{T}}, \tag{11.16}$$

which is self-consistent in the sense that (11.14) is satisifed. Substitution of (11.16) into (11.11) gives

$$\frac{\overline{Y^2}(t)}{2\mathcal{T}^2 \overline{v^2}} = \left(\frac{t}{\mathcal{T}} - 1 + e^{-t/\mathcal{T}}\right). \tag{11.17}$$

Although not exact, this is a good approximation since (11.11) is not particularly sensitive to the exact form of $R(t)$. Moreover, (11.17) is supported by experimental results for plume spread shown in Section 11.1.3. It may be concluded that the mean-squared particle displacement of a puff is fairly well known from (11.17) if $\mathcal{T}$ and $\overline{v^2}$ are given.

Knowledge of the second-order moments of particle displacement is particularly valuable if $\mathcal{P}(\mathbf{x}, t)$ is Gaussian, since in that case it would be fully determined in the form

$$\mathcal{P}(\mathbf{x}, t) = \frac{1}{(2\pi\sigma^2)^{3/2}} e^{-\left[(x-\overline{U}t)^2+y^2+z^2\right]/2\sigma^2} \tag{11.18}$$

where the variance according to (11.11) satisfies

$$\sigma^2(t) = \overline{\left[X(t) - \overline{U}t\right]^2} = \overline{Y^2}(t) = \overline{Z^2}(t) = 2\overline{v^2} \int_0^t (t - s)R(s)\,ds, \tag{11.19}$$

since $\overline{u^2} = \overline{v^2} = \overline{w^2}$ in homogeneous turbulence and $R(s)$ is the same for all directions. As it turns out, there is considerable heuristic and experimental evidence, suggesting that $\mathcal{P}$ is for the most part Gaussian. The thrust of the argument is as follows:

1. At small times, fluid particles disperse in directions and speeds directly proportional to the local velocity. Moreover, experimental measurements of the pdf of the velocity field at a fixed point in homogeneous turbulence show it to be very nearly Gaussian [1]. Thus the small time particle displacements fall into a Gaussian distribution and (11.18) should be valid for small times.

2. At large times, the paths of the fluid particles traveling through a turbulent field may be considered to be the sum of a large number of (apparently) random independent movements. For example, for each unit of time proportional to $\mathcal{T}$ it is expected that complete decorrelation of the velocity field has been achieved, so that the subsequent motion is independent of the prior motion. If this characterization of particle paths is legitimate, it suggests that the large time displacement of fluid particles in a turbulent flow is similar to that of the laminar puff considered in Section 11.1.1. Thus a Gaussian pdf should also be expected for large times.

3. At intermediate times, there is no guarantee of Gaussianity in $\mathcal{P}$, although as a practical matter it is reasonable to assume that Gaussianity holds at least in an approximate sense, since a drastic departure from Gaussianity is unlikely.

If Gaussianity of the mean turbulent puff can be accepted as true, then (11.18) together with (11.19) must be a solution to (2.129). It is not hard to show by direct substitution that this will be the case as long as

$$\overline{u_i c} = -\frac{1}{2}\frac{d\sigma^2}{dt}\frac{\partial \overline{C}}{\partial x_i}; \tag{11.20}$$

that is, a gradient diffusion law is implied in which, according to (11.19), the turbulent eddy diffusivity $\mathcal{D}_t$ is given by

$$\mathcal{D}_t = \frac{1}{2}\frac{d\sigma^2}{dt} = \overline{v^2}\int_0^t R(s)\,ds. \tag{11.21}$$

For small t, $\mathcal{D}_t$ is time dependent, while for large t it asymptotes to the constant value $\mathcal{T}\,\overline{v^2}$. The latter result is not unexpected, since the linear spreading rate of the mean-squared displacement of the turbulent puff—which is implied by (11.15)—looks formally equivalent to (11.2), so that, by analogy, the turbulent diffusivity should be given by $\mathcal{T}\,\overline{v^2}$. This form, with $\mathcal{T}$ replaced by $\mathcal{T}_{22}$, is equivalent to that derived previously for momentum and vorticity transport in (6.22) and (6.31), respectively, and will be encountered again in Section 11.2.

The time dependence of the turbulent diffusivity during the early period of the puff contradicts the expectation that it should be a property of the turbulent flow and not of the puff itself. This suggests that (11.20) is only coincidentally in the form of a gradient transport law (i.e., these results should not be construed as implying that the underlying transport physics is that of gradient transport as it is usually understood). This point will be pursued further in Section 11.2, where the Lagrangian transport analysis is used to explore the physics of the transport correlation in the vicinity of the plume origin.

The laminar puff was shown to be derivable from a model of a Brownian motion process in which fluid particles proceed through the flow in a series of random hops. It was also mentioned that the same ideas apply to the turbulent puff at large times [i.e., times large enough for many time steps of duration $O(\mathcal{T})$ to exist]. For short times, this model is inappropriate, as is clear from the differences between (11.2) and (11.12), in which the short time spread of the turbulent puff is quadratic and not linear in time. For short time intervals the particles in the turbulent puff feel a persistent correlation in the velocity which dies off only after a duration of $\mathcal{T}$. Consequently, the particle movements cannot be viewed as the sum of independent steps as they can at large times.

The question naturally arises as to whether there is a probabilistic model that can yield the characteristics of a turbulent puff. This should produce the variance (11.17) and converge to a Gaussian process at large times. In fact, the Langevin equations

[8] represent a random process which has exactly these characteristics and thus has played a significant role in the modeling of turbulent scalar transport. Without going into detail on the motivation and justification, it suffices to say that for the lateral diffusion in a puff, the Langevin equations consist of

$$\frac{dV(t)}{dt} = -\frac{1}{T}V(t) + \xi(t) \tag{11.22}$$

and

$$\frac{dY(t)}{dt} = V(t), \tag{11.23}$$

which are, in essence, an expression of the dynamics of a particle moving in a fluid. The first term on the right-hand side of (11.22) represents a frictional retarding force, while $\xi(t)$ is a white noise process, that is, a random field with the property, in this instance, that

$$\overline{\xi(t)\xi(s)} = \frac{2\overline{v^2}}{T}\delta(t - s). \tag{11.24}$$

Solution of (11.22) and (11.23) for $\overline{Y^2}(t)$ using (11.24) yields (11.17). Moreover, for large times, it can be shown that the stochastic process for $Y(t)$ is Gaussian with variance $2T\overline{v^2}t$.

Equations (11.22) and (11.23) can be made the basis for numerical schemes that predict the behavior of a puff in homogeneous turbulence. As in the case of the laminar puff, the contaminant placed into the flow is broken up into N parts, each of which follows trajectories determined by a numerical approximation to the Langevin equations. Random flight methods discussed in Section 11.3.2 build on these ideas to create an approximate means of solving for the scalar field in general circumstances.

11.1.3 Continuous Point-Source Plumes

As suggested previously, a scalar plume emanating from a small volume dV in a turbulent flow, such as the effluent from an industrial smokestack, may be modeled as a continuous sequence of turbulent puffs, one each time interval dt bringing $q\,dt\,dV$ contaminant into the flow. By summing the scalar fields produced by the sequence of puffs starting from an initial instant, say $t = 0$, until t is very large, while simultaneously taking the limit as $dt \to 0$, an expression is arrived at for the mean concentration field of the plume. Proceeding formally, let $P(\mathbf{x}, t)$ given by (11.18) denote the expected positions at time t of the contaminant in a puff released at $t = 0$ from the neighborhood of the origin $\mathbf{x} = 0$. The scalar field at time t contributed by a puff released at time s is $q^*\,ds\,P(\mathbf{x}, t - s)$ for $t > s$ and zero for $t < s$, where $q^* = q\,dV$ is the amount of scalar per second released into the flow. Summing over discrete time intervals and taking the appropriate limit as $ds \to 0$ yields

$$\overline{C}(\mathbf{x}, t) = \int_0^t q^* \mathcal{P}(\mathbf{x}, t - s) \, ds. \tag{11.25}$$

The concentration field of a steady plume is obtained by taking $\lim_{t \to \infty}$ in (11.25), after first changing the variable of integration to $\tau = t - s$. The result is

$$\overline{C}(\mathbf{x}) = \int_0^\infty q^* \mathcal{P}(\mathbf{x}, \tau) \, d\tau. \tag{11.26}$$

The integral on the right-hand side is convergent, since for any fixed $\mathbf{x}$, $\mathcal{P}(\mathbf{x}, \tau) = 0$ once τ is large enough (i.e., puffs traveling long enough in time will have passed the position $\mathbf{x}$).

Details of the mean plume concentration in a homogeneous turbulent flow can be obtained from (11.18) and (11.26) once $R(\tau)$ is given. Even without information about $R(\tau)$, one can still gain insight into the plume behavior in the near and far fields by using the general results (11.12) and (11.15). In particular, near the plume origin, $\overline{C}(\mathbf{x})$ is influenced only by puffs released in the immediate past which have yet to convect far downstream. This means that only the part of the integral in (11.26) that is for short times will influence the plume near its origin. For these times, say $t < t^*$, one can assume that (11.12) is applicable. Moreover, it does no harm—as long as attention is focused on just the near field of the plume, say $x < \overline{U}t^*$—to perform the integration in (11.26) as if (11.12) holds for all time. In fact, the part of the integration where $t > t^*$ is of no consequence for the region near the plume origin since the puffs being included in this case have all convected downstream beyond the position $\overline{U}t^*$. The result of the integration is [6]

$$\overline{C}(\mathbf{x}) = \frac{q^*}{(2\pi)^{3/2} r^2 v_{\text{rms}}} e^{-1/2\alpha^2} \left[1 + \sqrt{\frac{\pi}{2}} \frac{x}{\alpha r} e^{x^2/2\alpha^2 r^2} \operatorname{erfc}\left(-\frac{1}{\sqrt{2}} \frac{x}{\alpha r} \right) \right], \tag{11.27}$$

where $v_{\text{rms}} \equiv \sqrt{\overline{v^2}}$, $\alpha = v_{\text{rms}}/\overline{U}$ is the relative turbulence intensity, and $r \equiv \sqrt{x^2 + y^2 + z^2}$.

Typically, values of $\alpha \approx 0.1$ can be expected to occur in real turbulent flows such as the atmosphere. In this case, the rate of lateral plume spread is much slower than convection downstream, so the plume appears slender; that is, most of the scalar is concentrated around the central axis, where

$$r \approx x. \tag{11.28}$$

In this case, significant simplification to (11.27) is possible. In particular, the argument of the complementary error function will be large, so that this term is near its asymptotic limit of 2. In addition, the second term in the brackets is much larger than unity. Putting these approximations together yields

$$\overline{C}(\mathbf{x}) = \frac{q^*}{2\pi \alpha x^2 v_{\text{rms}}} e^{-(1/2\alpha^2)(1 - x^2/r^2)}. \tag{11.29}$$

Finally, consistent with (11.12) and (11.19), one may define

$$\sigma^2(x) = \overline{v^2}x^2/\overline{U}^2 = \alpha^2 x^2, \tag{11.30}$$

so that (11.29) becomes

$$\overline{C}(\mathbf{x}) = \frac{q^*}{2\pi\sigma^2\overline{U}}e^{-(y^2+z^2)/2\sigma^2} \tag{11.31}$$

for the mean concentration field in the near field of the plume.

At large distances from the source, the plume is not influenced by recently emitted puffs, only by those that have been diffusing long enough for (11.15) to be applicable. In this case it is legitimate to integrate (11.26) assuming that (11.15) holds for all time. This calculation is similar to the classical result for diffusion from a point source in a uniformly translating medium since the turbulent diffusivity, $\mathcal{T}\overline{v^2}$, is constant. The result is the formula

$$\overline{C}(\mathbf{x}) = \frac{q^*}{(4\pi)\mathcal{T}\overline{v^2}r}e^{-\left(\overline{U}/2\mathcal{T}\overline{v^2}\right)(r-x)} \tag{11.32}$$

which is valid at appropriately far distances from the source. For slender plumes it can be asserted that

$$r = x\left(1 + \frac{y^2+z^2}{x^2}\right)^{1/2} \approx x\left(1 + \frac{y^2+z^2}{2x^2}\right), \tag{11.33}$$

which leads, after substitution into (11.32), to exactly (11.31) again, where in this case

$$\sigma^2(x) = 2\mathcal{T}\alpha^2\overline{U}x. \tag{11.34}$$

Thus, it is seen that (11.31) with (11.30) and (11.34) may be regarded as a good approximation to the point-source plume both near and far from the source.

Finally, it is reasonable to hypothesize that (11.31) also describes the plume for distances intermediate to those for which (11.30) and (11.34) apply. In this case (11.11) needs to be transformed into the spatial domain via $x = \overline{U}t$, yielding

$$\sigma^2(x) = 2\alpha^2\int_0^x (x-x')R(x'/\overline{U})\, dx'. \tag{11.35}$$

For small x, (11.35) gives (11.30), and for large x, it converges to (11.34).

The validity of (11.31) as a description of slender plumes is supported by experiments such as that shown in Fig. 11.1, giving the measured lateral spread of a plume with downstream distance. In the figure, the dimensionless ratio $\overline{U}\sigma^2/(2\mathcal{T}\overline{v^2}x^*)$ is plotted versus x/x^*, where

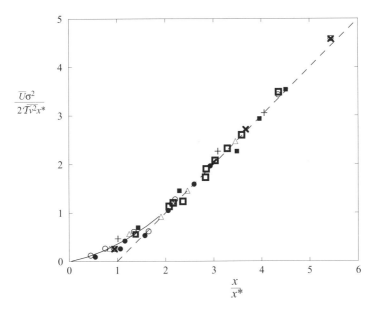

Fig. 11.1 *Mean-squared lateral diffusion in a plume of benzene–carbon tetrachloride droplets in a turbulent water flow. Dashed line is (11.37) and solid line is (11.38). (Reprinted with permission from [7]. Copyright 1944 American Chemical Society.)*

$$x^* \equiv \frac{\int_0^\infty x' R(x'/\overline{U})\, dx'}{\overline{U}\, \mathcal{T}} \tag{11.36}$$

is the spatial equivalent of the time t_1 introduced in (11.13). It is not hard to show using (11.35) and (11.36) that as $x \to \infty$,

$$\frac{\overline{U}\sigma^2}{2\mathcal{T}\,\overline{v^2}x^*} \to \left(\frac{x}{x^*} - 1\right) \to \frac{x}{x^*}, \tag{11.37}$$

which is consistent with (11.34). This trend is evident in the figure. Moreover, for small x, the x^2 behavior predicted by (11.30) is captured.

Assuming that (11.16) holds in the form $R(x/\overline{U}) = e^{-x/x^*}$, then analogously to (11.17), it follows that

$$\frac{\overline{U}\sigma^2}{2\mathcal{T}\,\overline{v^2}x^*} = \frac{x}{x^*} - 1 + e^{-x/x^*}, \tag{11.38}$$

in which $x^* = \overline{U}\, \mathcal{T}$. This curve is plotted as a solid line in Fig. 11.1 and is seen to well represent the entire range of experimental data, including the transition from x^2 to x behavior.

In the same way that the concentration formula for a puff was used in Section 11.1.2 to determine the scalar flux correlation in a time evolving flow, (11.31) can be used to determine the scalar flux in a steady slender plume. First note that in a steady, uniform, mean flow, $\overline{C}$ satisfies

$$\overline{U}\frac{\partial \overline{C}}{\partial x} = -\frac{\partial \overline{vc}}{\partial y} - \frac{\partial \overline{wc}}{\partial z} + \mathcal{D} \nabla^2 \overline{C} + q, \tag{11.39}$$

where the streamwise flux term $\partial \overline{uc}/\partial x$ is neglected in comparison to the lateral flux terms by virtue of the slender plume assumption. Without loss of generality in the following argument, the molecular transport term may also be neglected. Then a calculation shows that (11.31) satisfies (11.39) as long as

$$\overline{vc} = -\frac{\overline{U}}{2}\frac{d\sigma^2(x)}{dx}\frac{\partial \overline{C}}{\partial y} \tag{11.40}$$

and a similar result for $\overline{wc}$. The eddy diffusivity implied by (11.40), after substituting (11.35), is given by

$$\mathcal{D}_t = \frac{\overline{U}}{2}\frac{d\sigma^2(x)}{dx} = \overline{v^2}\int_0^x R(x'/\overline{U})\,dx', \tag{11.41}$$

which varies with distance from the source until asymptoting to $\mathcal{T}\,\overline{v^2}$ in the far field.

Just as the physical inappropriateness of a time-dependent diffusivity was noted in the case of a puff, here a plume-dependent, spatially varying diffusivity is encountered. Such behavior cannot be reconciled to the homogeneous turbulent conditions of this problem, in which the diffusivity is the same everywhere and certainly not plume dependent [4,12]. Thus, as in the case of a puff, despite the formal legitimacy of (11.40), it cannot be taken to imply that the physics of gradient transport is satisfied in a plume, at least not until $\mathcal{D}_t$ become constant in the far field.

Modeling similar to that entailed in deriving (11.31) has been developed and applied to the prediction of plumes in more general contexts (e.g., with a nonuniform mean velocity field and Reynolds stresses). The validity of the resulting Gaussian models in such circumstances is questionable and is reflected in the accuracy of the scalar fields predicted. An alternative strategy for predicting plume behavior in non-homogeneous conditions is to use plume-dependent diffusivities based on (11.41). However, besides the questionable accuracy of the diffusivity formulas for these new circumstances, the practical consequences of plume-dependent diffusivities is significant, since it complicates the prediction of multipoint or distributed source plumes. In effect, the scalar field originating from each plume or subplume has to be held separately, at least until the far field.

In the next section we use a backward particle path analysis of the physics of scalar transport with the goal of clarifying the transport mechanisms near the plume origin. In this it will be seen that the transport physics can be understood without the need to introduce a plume-dependent diffusivity.

11.2   SCALAR TRANSPORT

The discussion of slender scalar plumes showed that a gradient transport law with plume-dependent diffusivity is compatible with a Gaussian plume structure. At the same time, the turbulent eddy diffusivity should be a local property of turbulent motion independent of the particular diffusing quantity. To help resolve this conflict, the origin of the scalar flux correlation is now examined with the help of the backward Lagrangian particle path analysis first described in Chapter 6. The starting point is to write $c = C - \overline{C}$ at a given point $(\mathbf{x}, t)$ in terms of the identity

$$c = c^b + (\overline{C}^b - \overline{C}) + (C - C^b), \qquad (11.42)$$

where b denotes the random points at time $t - \tau$ from which fluid particles originate that arrive at $\mathbf{x}$ at time t. Then it follows that

$$\overline{u_i c} = \overline{u_i c^b} + \overline{u_i(\overline{C}^b - \overline{C})} + \overline{u_i(C - C^b)}. \qquad (11.43)$$

As before, the time τ at which $\overline{u_i c^b}$ is negligible may be taken as a mixing time. It is important to note that mixing time defined this way does not have to be equivalent to that defined previously in the case of momentum transport (i.e., the time when $\overline{u_i u_j^b} = 0$). By way of illustration of this point, consider the plume depicted in Fig. 11.2 with the backward particle paths arriving at two locations: $(x^+, y^+) = (20, 30)$ near the source and $(x^+, y^+) = (160, 30)$ far from the source. The $\overline{C}$ contours are from a DNS of a plume developing from a Gaussian line source of small support centered at $x^+ = 0$, $y^+ = 15$ in a turbulent channel flow [2]. In this calculation the Schmidt number is 0.71, as it would be for water vapor diffusing within air.

It is clear that for relatively small values of τ, the fluid particles arriving at $x^+ = 20$ are upstream of the plume at $t - \tau$, and hence $C^b = \overline{C}^b = c^b = 0$. In this case, $\overline{vc^b} = 0$ identically, so that the mixing time is relatively small. For the downstream point the fluid particles remain within the plume for a longer time, and here it can be expected that $\overline{vc^b}$ is first zero, due to the physics of random mixing, for τ less than the time when the particles are all upstream of the plume. In this case it is expected that τ is closer in magnitude to the mixing time for momentum, for example.

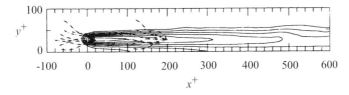

Fig. 11.2 $\overline{C}$ *contours for a plume centered at* $x^+ = 0$, $y^+ = 15$ *in a channel flow DNS. Backward paths arriving at two locations are indicated. (From [2].)*

The consequences of these behaviors is illustrated in Figs. 11.3 and 11.4, showing the result of an evaluation of (11.43) at the upstream and downstream point, respectively, using paths computed in a DNS of the same line source plume as in Fig. 11.2. At the upstream point, mixing is achieved rapidly at $\tau^+ \approx 4$, and at this time the flux is seen to be caused by $u_i(C - C^b)$. In contrast, at $x^+ = 160$, mixing is achieved when $\tau^+ \approx 12$, and in this instance the transport is explained by the displacement term $u_i(\overline{C^b} - \overline{C})$. Note that in both instances, before mixing occurs, all terms contribute to the decomposition in (11.43), but it is only at mixing that a relatively clear physical explanation is offered for transport.

These results show that the transport in a plume transitions from a mechanism related to the scalar change term to the displacement term with downstream distance. The rise of influence of the latter accompanies a reduction in importance of the former. Consideration of the displacement term in more detail provides some clues with which to explain the plume-dependent diffusivity contained in (11.40).

Mimicking the analysis of Chapter 6, a first-order approximation to the displacement term is the gradient model

$$\overline{u_i(\overline{C^b} - \overline{C})} = -\overline{u_i L_j}\frac{d\overline{C}}{dx_j}, \tag{11.44}$$

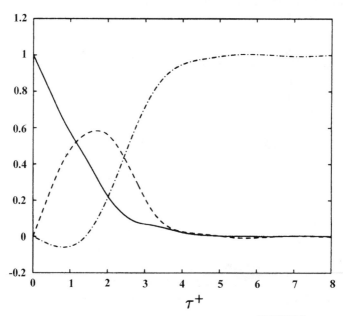

Fig. 11.3 $\overline{vc}$ *decomposition* (11.43) *at* $x^+ = 20$, $y^+ = 30$. —, $\overline{vc^b}$; – –, $\overline{v(\overline{C^b} - \overline{C})}$; – · –, $\overline{v(C - C^b)}$. (*From [2].*)

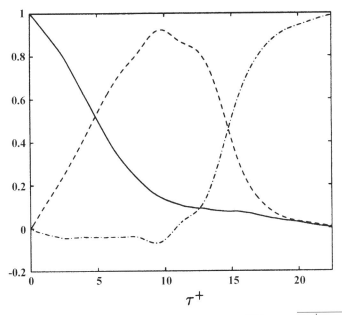

Fig. 11.4 $\overline{vc}$ *decomposition (11.43) at* $x^+ = 160$, $y^+ = 30$. —, $\overline{vc^b}$; – –, $\overline{v(\overline{C}^b - \overline{C})}$; – · –, $\overline{v(C - C^b)}$. *(From [2].)*

where $L_j = \int_{t-\tau}^{t} U_j(s) \, ds$ and the eddy diffusivity $\overline{u_i L_j}$ is the same as it was previously. If a gradient law for scalar transport is valid, (11.44) suggests that it should have the form

$$\overline{u_\alpha c} = -\mathcal{T}_{\alpha j}\overline{u_\alpha u_j}\frac{d\overline{C}}{dx_j}. \tag{11.45}$$

For the lateral flux in a plume, (11.45) gives an eddy diffusivity $\mathcal{T}_{22}\overline{v^2}$, in agreement with the far-field result $\mathcal{T}\overline{v^2}$ in (11.40). This implies that the higher-order terms that are truncated to get (11.44) might very well not be important in the far field of the plume.

The gradient model represents a one-point approximation to a correlation that is inherently nonlocal (i.e., depends on events covering a finite region of the flow field). This is evident in the structure of the displacement correlation if it is modeled in a probabilistic sense through the introduction of $\mathcal{P}(\mathbf{y}, \mathbf{x})$ denoting the pdf of initial particle position, say $\mathbf{y}$, which arrives at $\mathbf{x}$ after a time τ has elapsed. A straightforward model of the displacement transport correlation then is

$$\overline{v(\overline{C}^b - \overline{C})}(\mathbf{x}) = \int_{V_y} \frac{y_2 - x_2}{\tau} \left[\overline{C}(\mathbf{y}) - \overline{C}(\mathbf{x})\right] \mathcal{P}(\mathbf{y}, \mathbf{x}) \, d\mathbf{y}, \tag{11.46}$$

where $\mathcal{V}_y$ is the support of $\mathcal{P}$ in the variable **y**. To have (11.46) apply everywhere in the plume, τ can be chosen to be equal to the mixing time far downstream of the source (i.e., where it is a maximum). To approximate v, this expression incorporates a simple model of particle paths as traveling in straight lines. Equation (11.46) will be identically zero when **x** is near the source, since $\mathcal{V}_y$ will be upstream of the plume in this case. The integral, however, increases with downstream distance from the source until it recovers the gradient law. Note that it takes no special knowledge of the plume origin to achieve this result: It is inherent in the distribution of $\overline{C}$ only.

In light of these results, it appears that the rise of the eddy diffusivity in (11.40) to a constant plume-independent value mirrors the rise in legitimacy of a gradient transport law given by the Lagrangian analysis. Moreover, the implausibility of a plume-dependent diffusivity can be explained by recognizing that (11.40), in fact, models the rise in dominance of the displacement transport mechanism, which is itself fully independent of the plume origin. The conclusion is reached that the plume dependency of (11.40) derives from its being a one-point model of a phenomenon that is intrinsically nonlocal in character.

Whether transport in the immediate vicinity of the source is important or not depends on the details of the interaction of the turbulent flow with the physical device from which the contaminant emerges. In the numerical simulation yielding the plume in Fig. 11.2, the source is a region in the flow, and it is relatively simple to see in this case from whence a transport correlation arises. It is helpful first to note the identity

$$\overline{u_i(C - C^b)} = \mathcal{D} \int_{t-\tau}^{t} \overline{u_i \, \nabla^2 C(s)} \, ds + \int_{t-\tau}^{t} \overline{u_i q(s)} \, ds, \qquad (11.47)$$

where $\nabla^2 C(s)$ and $q(s)$ are evaluated on fluid particle paths. Assuming that the molecular diffusion effect in the first term is small, it is the last term that explains the transport at the plume source. This expression depends on a correlation between velocity and the amount of scalar acquired by fluid particles passing through the source region, $\int q(s) \, ds$. A net scalar flux is implied when the scalar gained by fluid particles depends on the direction and speed of travel. Figure 11.5 illustrates how a flux can originate by this mechanism. At point A, $\overline{vc} > 0$ since fluid particles for which $v > 0$ are more likely to cross the source than those for which $v < 0$. By the same token, $\overline{vc} < 0$ at point B.

Although the transport analysis in this section helps shed some light on the physics of plumes, it does not readily lead to practical transport models. For example, there is no obvious means for acquiring information about quantities such as the pdf of initial particle displacement. Instead, models used in engineering studies must follow along the lines of the models used in solving the RANS equations. These are discussed in the next section, together with a distinctly different approach to transport modeling which is motivated by the possibility of the Lagrangian description of plumes through the random movement of fluid particles.

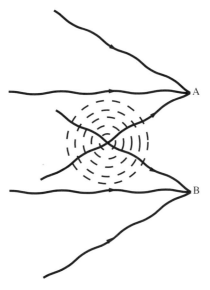

Fig. 11.5   *Origin of scalar flux due to the source term in (11.47).*

11.3   SCALAR TRANSPORT MODELING

The prediction of scalar fields such as concentration and temperature in turbulent flow is most often done by methods which are strongly reminiscent of those discussed in Chapter 8 in regard to the mean velocity field. In fact, many schemes for predicting mean scalars are direct extensions of RANS models discussed previously. These close the $\overline{C}$ equation and, depending on the model complexity, may also close equations for scalar variance $\overline{c^2}$, the scalar dissipation rate ϵ_c, and the flux vector $\overline{u_i c}$. As an alternative to grid-based methods, a family of grid-free methods, referred to as random flight models (RFMs), have been developed as extensions of the previously mentioned Langevin model of turbulent diffusion from a point source. In this, scalar fields are represented by tracer particles whose trajectories through the flow are calculated in such a way as to mimic the diffusion of contaminants by turbulent eddies. RFMs often tend to be applied in meteorological applications where it may be exceedingly difficult to impose a mesh or set up a closed system of equations with appropriate boundary conditions. Another advantage of RFMs is that unlike typical gradient transport models, which are affected adversely by the large concentration gradients near the plume source, they do not unduly distort the near field of plumes [2].

It goes without saying that regardless of the technique used in predicting the scalar field, the outcome cannot be more accurate than the mean velocity field and turbulence statistics upon which the solution depends. In fact, Chapter 9 suggests that there will be many flows for which inadequate turbulence predictions pose a major obstacle to the successful computation of scalar fields, without ever taking into account the quality of the scalar modeling itself. This means that it is necessary to be doubly cautious in interpreting the results of predictions of scalar fields via closure methods.

Some of the basic approaches toward modeling scalar fields in turbulent flow are considered next. For simplicity, the discussion is confined to a passive scalar, C, which will be taken to be a concentration field but could apply equally well to the temperature field. In later discussions, the Schmidt number, $Sc \equiv \nu/\mathcal{D}$, will be referred to with the understanding that it should be replaced by the Prandtl number, $Pr \equiv \nu/\alpha$, where α is the thermal diffusivity, if the scalar in question is the temperature.

11.3.1 Closure Schemes for the Scalar Transport Equations

Closure models for the mean scalar field have a hierarchy which is similar to that for the average momentum equation. Thus, at the simplest level, the eddy diffusivity formulation

$$\overline{u_i c} = -\mathcal{D}_t \frac{\partial \overline{C}}{\partial x_i} \tag{11.48}$$

is used in closing (2.129). The most basic model of this type assumes that $\mathcal{D}_t$ is related to ν_t through a constant, turbulent Schmidt number Sc_t, defined by

$$Sc_t \equiv \frac{\nu_t}{\mathcal{D}_t} \tag{11.49}$$

(i.e., $\mathcal{D}_t$ is directly proportional to ν_t). Thus the model used for ν_t has a direct bearing on the diffusivity of the scalar field. For example, if ν_t has the form used in the K–ϵ closure, $\mathcal{D}_t = (C_\mu/Sc_t)K^2/\epsilon$.

The point was made earlier that if gradient transport is a legitimate expression of both the momentum and scalar fluxes in turbulent flow, it must be the case that $Sc_t = 1$. This is also the substance of the Reynolds analogy used in the study of heat and mass transfer [3]. Typically, however, the value $Sc_t = 0.9$ is used in practice. This can be thought of as being a consequence of the fact that gradient transport modeling is being used to represent all the transport physics, including that part which should not be represented by the gradient form.

More advanced scalar transport closures attempt to enhance the effectiveness of (11.48) and (11.49) by using information about $\overline{c^2}$ and the scalar molecular dissipation rate,

$$\epsilon_c \equiv \mathcal{D} \overline{\left(\frac{\partial c}{\partial x_j} \right)^2}. \tag{11.50}$$

This is similar to the way in which K and ϵ are enlisted in the attempt to make a model of the eddy viscosity which reflects local turbulent flow conditions. An opening for the use of $\overline{c^2}$ and ϵ_c comes from assuming that the turbulent Schmidt number is related to the ratio of time scales K/ϵ and $\overline{c^2}/\epsilon_c$ by the formula

$$Sc_t = \left(\frac{\overline{c^2}/2\epsilon_c}{K/\epsilon} \right)^{1/2} . \tag{11.51}$$

Models using this assumption with (11.48) and (11.49) are referred to as four-equation models, in view of the number of extra equations that must be solved together with the mean velocity and scalar equations.

Closures to the $\overline{c^2}$ and ϵ_c equations are developed by generalizing the modeling assumptions used in closing the K and ϵ equations. First consider the exact equation for scalar variance, which is derived from (2.128):

$$\frac{\partial \overline{c^2}}{\partial t} + \overline{U}_j \frac{\partial \overline{c^2}}{\partial x_j} = \mathcal{D} \nabla^2 \overline{c^2} - \frac{\partial}{\partial x_j} \left(\overline{u_j c^2} \right) - \overline{c u_j} \frac{\partial \overline{C}}{\partial x_j} - \epsilon_c, \tag{11.52}$$

where the terms on the right-hand side represent, respectively, molecular and turbulent transport, production from the mean scalar field, and molecular dissipation. Closure for (11.52) requires a model for the transport correlation $\overline{u_j c^2}$ which is typically given using the classic formulation of Daly and Harlow [5]. The resulting equation is then

$$\frac{\partial \overline{c^2}}{\partial t} + \overline{U}_j \frac{\partial \overline{c^2}}{\partial x_j} = \mathcal{D} \nabla^2 \overline{c^2} + \frac{\partial}{\partial x_j} \left(C_{c_1} R_{jk} \frac{K}{\epsilon} \frac{\partial \overline{c^2}}{\partial x_k} \right)$$

$$+ \mathcal{D}_t \left(\frac{\partial \overline{C}}{\partial x_j} \right)^2 - \epsilon_c. \tag{11.53}$$

Accompanying this is a closure to the exact ϵ_c equation. Without presenting details of the derivation, which pursues a very similar strategy as that of the ϵ equation, this is often of the form

$$\frac{\partial \epsilon_c}{\partial t} + \overline{U}_j \frac{\partial \epsilon_c}{\partial x_j} = \mathcal{D} \nabla^2 \epsilon_c + \frac{\partial}{\partial x_j} \left(C_{c_2} R_{jk} \frac{K}{\epsilon} \frac{\partial \epsilon_c}{\partial x_k} \right)$$

$$+ \left(C_{c_3} \frac{\epsilon_c}{\overline{c^2}} + C_{c_4} \frac{\epsilon}{K} \right) \mathcal{D}_t \left(\frac{\partial \overline{C}}{\partial x_j} \right)^2 - C_{c_5} \frac{\epsilon_c}{K} R_{ij} \frac{\partial \overline{U}_i}{\partial x_j} \tag{11.54}$$

$$- \left(C_{c_6} \frac{\epsilon_c}{\overline{c^2}} + C_{c_7} \frac{\epsilon}{K} \right) \epsilon_c.$$

Equations (11.53) and (11.54) are sufficiently complicated that it is a formidable challenge to find a calibration scheme which can clearly establish what the many constants C_{c_1} through C_{c_7} should be. Thus there are quite a few different proposals in this regard [10]. Further compounding the problem is the fact that the closure in (11.53) and (11.54) is restricted to high-Reynolds-number flow so that near boundaries, modifications of the $\overline{c^2}$ and ϵ_c equations are often required.

The models that have been described here are built up from the eddy diffusivity model (11.48). The search for better models with greater generality has followed along the same lines as the RANS techniques mentioned in Chapter 8, and the reader is referred to the literature to see the details. Algebraic models, nonlinear scalar transport models, and closures to the exact equation for the scalar flux $\overline{u_i c}$ are encountered. Suffice it to say that the best approach to take in scalar field modeling is far from settled at the present time. Moreover, as emphasized previously, all scalar predictions inherit the strengths and weaknesses of the RANS models upon which they depend, so the reliability of scalar prediction schemes is particularly uncertain.

11.3.2 Random Flight Models

As suggested previously, the analogy between Brownian motion and scalar diffusion in a laminar flow can be used to justify computing the concentration of laminar plumes via a Monte Carlo method. To be specific, consider the particular case of diffusion from a point source in a two-dimensional laminar velocity field (U, V). Assume that mass is released into the flow at the rate q (mass/sec). For each time step of length dt, $q\, dt/N$ mass is placed into the flow on each of N particles. The ith particle follows a trajectory with coordinates (x_i^n, y_i^n), where n refers to the discrete time $t_n = n\, dt$. The particles march forward in time according to a numerical scheme in which, for example,

$$x_i^{n+1} = x_i^n + U(x_i^n, y_i^n, t_n)\, dt + \xi_i$$
$$y_i^{n+1} = y_i^n + V(x_i^n, y_i^n, t_n)\, dt + \eta_i,$$

(11.55)

where ξ_i and η_i are independent Gaussian random variables drawn from a distribution with mean zero and variance $2\mathcal{D}\, dt$. According to (11.55), the new position, (x_i^{n+1}, y_i^{n+1}), is determined from the old position, (x_i^n, y_i^n), by adding a convective step caused by the velocity field and a random hop whose purpose is to model molecular diffusion.

An estimate of the concentration field C_{pq} near a point (x_p, y_q) can be had by counting how many particles fall into a small square of area dA surrounding the point, say N_{pq}, and then computing

$$C_{pq} \approx \frac{N_{pq} q\, dt}{N\, dA}.$$

(11.56)

For constant diffusivity, as in this case, the evolving concentration field determined by this procedure is rigorously justified. In fact, as $dt \to 0$, $dA \to 0$, and $N \to \infty$, $C_{pq} \to C(x_p, y_q)$. It is interesting to note that this method is a fully legitimate means of modeling diffusion in a turbulent flow as long as the velocity field comes from a DNS. However, in view of the difficulty of performing DNS, such applications are rare.

The main application of Monte Carlo schemes in turbulent flow uses the random movements of scalar particles to represent the turbulent motion, not molecular diffusion. These RFMs require the mean velocity field and other statistics to be available from empirical data and/or RANS calculations. They are formulated as generalizations of the Langevin model of a plume in a uniform mean flow of homogeneous turbulence that was introduced in Section 11.1.2. A typical implementation of such a scheme has the particle paths advancing from their initial position at the origin according to the algorithm (in the y direction with similar relations in the other directions)

$$V_i^{n+1} = V_i^n \left(1 - \frac{dt}{\mathcal{T}}\right) + \xi$$

$$Y_i^{n+1} = Y_i^n + dt\, V_i^n,$$

(11.57)

where ξ is taken from a Gaussian distribution with zero mean and variance

$$\sigma^2 = dt\, 2\overline{v^2}/\mathcal{T}.$$

(11.58)

The question then is: how can this methodology be generalized to inhomogeneous flows with mean shear? This has been the object of considerable study (e.g., [13,14]), and conditions have been derived that must be satisfied for the Monte Carlo simulation to be successful. For example, in the case of a nonhomogeneous flow in which $\overline{U} = \overline{U}(y)$ and $\overline{v^2} = \overline{v^2}(y)$, the naive assumption that (11.57) could be applied with the obvious modification that ξ is evaluated using the local value of the variance in (11.58) leads to an unphysical gathering of concentration in regions where $\overline{v^2}$ has a local minimum. In the physical problem a uniform concentration field should result. Preventing this in (11.57) is the fact that particles will have a reduced tendency to move away from regions where σ^2 is small and an increased likelihood of moving away from locations where $\overline{v^2}$ has a maximum. Thus concentration collects in the low-$\overline{v^2}$ regions.

According to the well-mixed criterion [13], a correct determination of $\overline{C}$ in inhomogeneous flow with the y dependencies $\overline{U}(y)$ and $\overline{v^2}(y)$ requires that ξ be selected with mean $dt\,(d\overline{v^2}/dy)$, variance $dt\,(2\overline{v^2}/\mathcal{T} + d\overline{v^3}/dy)$, and skewness $dt\,(3\overline{v^3}/\mathcal{T} + d\overline{v^4}/dy - 3\overline{v^2}\,d\overline{v^2}/dy)$, as well as conditions on higher-order moments. In a numerical scheme, the number of moments to be included in a model of ξ has to be limited. If just the mean and variance are used, a Gaussian random variable with these properties can be used as a model. If the skewness is also included, a random variable with the three given moments can be created (nonuniquely) as a sum of two Gaussian variables. For more complicated problems (e.g., where diffusion is planar or three-dimensional) the necessary conditions to be satisfied in shear flow become difficult to implement in practice. Despite these challenges, the RFM incorporating simplifying assumptions are used to model complex inhomogeneous flows. In many cases it is not obvious that there is a better alternative.

11.4 SUMMARY

It has been seen that the possibility of determining the fate of contaminants in turbulent flow depends in part on the extent and accuracy to which the underlying turbulent field is known. At one extreme, puffs and plumes released into a uniformly translating homogeneous field are well understood. In fact, theory provides an accurate quantitative description of their behavior, and a considerable amount is known about the physics of transport under such circumstances.

The presence of shear and, with it, anisotropy, has a substantial effect on the possibility of predicting the dispersal of scalars accurately. Closure schemes that depend on models for the flux vector, as well as more complex algebraic, nonlinear, and flux equation models, inherit all the limitations of RANS models with the additional uncertainty raised by the extra layer of scalar-velocity moment modeling. The random flight models, which are advantageous in predicting homogeneous plumes, need considerably more development before they can achieve quantitative accuracy in nonhomogeneous flow conditions.

With the availability of DNS, accurate solutions to scalar transport problems are readily obtained through either grid-based or Monte Carlo methods. This suggests that to the extent that LES calculations are accurate, they offer a potentially useful route for modeling scalar transport. This direction is being actively pursued at this time.

REFERENCES

1. Batchelor, G. K. (1960) *The Theory of Homogeneous Turbulence*, Cambridge University Press, Cambridge.

2. Bernard, P. S. and Rovelstad, A. L. (1994) "On the physical accuracy of scalar transport models in inhomogeneous turbulence," *Phys. Fluids* **6**, 3093–3108.

3. Burmeister, L. C. (1983) *Convective Heat Transfer*, Wiley, New York.

4. Csanady, G. T. (1973) *Turbulent Diffusion in the Environment*, D. Reidel, Dordrecht, The Netherlands.

5. Daly, B. J. and Harlow, F. H. (1970) "Transport equations in turbulence," *Phys. Fluids* **13**, 2634–2649.

6. Fleishman, B. A. and Frenkiel, F. N. (1955) "Diffusion of matter emitted from a line source in a non-isotropic turbulent flow," *J. Meteorol.* **12**, 141–145.

7. Kalinske, A. A. and Pien, C. L. (1944) "Eddy diffusion," *Ind. Eng. Chem.* **36**, 220–222.

8. Reichl, L. E. (1998) *A Modern Course in Statistical Physics*, Wiley, New York.

9. Sato, Y. and Yamamoto, K. (1985) "Mixing and diffusion mechanisms of fluids in isotropic turbulence—based on measured data for Lagrangian velocity correlation and turbulent diffusion," *Kagaku Kogaku Ronbunshu* **11**, 555–562.

10. Speziale, C. G. and So, R. M. C. (1998) "Turbulence modeling and simulation," in *The Handbook of Fluid Dynamics* (R. W. Johnson, Ed.), CRC Press, Boca Raton, Fla., Chap. 14.

11. Taylor, G. I. (1921) "Diffusion by continuous movements," *Proc. London Math. Soc.* **XX**, 196–211.

12. Taylor, G. I. (1959) "The present position in the theory of turbulent diffusion," *Adv. Geophys.* **6**, 101–112.

13. Thomson, D. J. (1984) "Random walk modeling of diffusion in inhomogeneous turbulence," *Q. J. R. Meteorol. Soc.* **110**, 1107–1120.

14. Thomson, D. J. (1987) "Criteria for the selection of stochastic models of particle trajectories in turbulent flows," *J. Fluid Mech.* **180**, 529–556.

15. Yeung, P. K. and Pope, S. B. (1989) "Lagrangian statistics from direct numerical simulations of isotropic turbulence," *J. Fluid Mech.* **207**, 531–586.

16. Zauderer, E. (1989) *Partial Differential Equations of Applied Mathematics*, 2nd ed., Wiley, New York.

12

Turbulence Theory

12.1 INTRODUCTION

This final chapter is concerned with theories that use the methods of mathematical and statistical physics to probe the fundamental nature of turbulent flow. Just exactly what is meant by fundamental and how it is studied varies from theory to theory. In one case it may mean trying to explain how the observed behavior of turbulent flow is intrinsic to solutions of the Navier-Stokes equation. In another it may mean treating turbulence as an example of a random system such as are routinely investigated in the field of statistical mechanics. Often, theoretical analyses make no pretense of providing a means for predicting practical turbulent flows, although they can help in developing methods that are useful in this regard. Turbulence theories are also filled with important ideas that help one to think about turbulent flow in novel and useful ways.

What is classified here as *theories of turbulence*, is a vast literature of many competing ideas and methodologies which cannot begin to be well surveyed in this limited context. Moreover, many turbulence theories deserve an entire book or more for a complete discussion [4,13,14,16]. Here we adopt the point of view that there is value in providing at least an entranceway into some of the major examples of turbulence theories, even if the penetration is not particularly deep. Even so, one can gain a reasonable idea of what these approaches consist of, and then pursue them further in more specialized references. To be specific, four different theories are discussed here. These are chosen primarily because they are highly visible in the folklore of turbulence and cover a significant range of possibilities for theoretical analysis.

It should be abundantly clear from the considerations of previous chapters (e.g., the general transport question in Chapter 6) that the behavior of the large energy-containing scales of turbulence is not a fertile ground for a turbulence theory. These kinds of motions are strongly affected by the flow geometry and other unique aspects of particular problems, and it is not likely that a general theory can account for them.

On the other hand, the small-scale motions, which it is believed can adjust relatively rapidly to changes in the flow at larger scales, are exactly the sort of turbulence phenomena to which a theory may supply a succinct and satisfying explanation. For example, a theory to explain the appearance of the $-5/3$ energy spectrum in the inertial subrange or the properties of homogeneous turbulence, described in Chapter 7, is a reasonable goal to pursue. As modest as this may seem, it nonetheless represents a formidable intellectual challenge.

The first approach considered, EDQNM (eddy-damped quasi-normal Markovian) [22,23], is largely a closure scheme based on ideas about the fundamental statistical behavior of turbulence as manifested in the multipoint correlations. Next is the DIA (direct interaction approximation) [12], which is the progenitor of a very large field of theoretical inquiry into turbulence motivated by mathematical techniques that have been applied successfully by physicists studying quantum physics and other realms. Considered third is a technique that is loosely based on a renormalization group analysis such as that used in statistical physics, popularly known in turbulence modeling as the RNG (renormalization group) closure [29]. Finally, some recent investigations into the basic properties of turbulence through the statistical mechanics of three-dimensional vortex systems are described [4]. This is the the latest chapter in a line of inquiry extending back to early work on two-dimensional vortex systems [21]. Before going into these analyses, it is helpful to provide some background on the mathematical properties of Gaussian random fields since many turbulence theories view turbulence as a perturbation away from strict Gaussian behavior. Some general remarks about the different theories covered here are then provided before giving details of each.

12.2 GAUSSIAN RANDOM FIELDS

The velocity field in turbulent flow is but one example of a random field, a mathematical object with the property that it is a random variable for each value of its arguments. The velocities at a collection of positions and times have a joint pdf to describe their statistical dependencies. For a *stationary* random field the joint pdfs depend on at most the relative time intervals of when the velocities are evaluated. This implies that the pdf of the velocity at a given point does not change in time. Similar ideas apply to *homogeneous* random fields. In this instance the joint pdfs of velocities at several spatial points at the same time depend on just the relative distances between them.

Consider a random field, $u(\mathbf{x})$, where for notational convenience, time is not indicated explicitly. $u(\mathbf{x})$ is referred to as a Gaussian random field if the following property holds: For any finite number, N, of points $\mathbf{x}_1, \mathbf{x}_2, ..., \mathbf{x}_N$, the joint pdf of the N random variables $u(\mathbf{x}_1), u(\mathbf{x}_2), \ldots, u(\mathbf{x}_N)$ (all evaluated at the same time) is of the form

$$\frac{1}{(2\pi)^N |\Lambda|} e^{-(1/2)(\mathbf{u}-\mathbf{m})^t \Lambda^{-1}(\mathbf{u}-\mathbf{m})} \tag{12.1}$$

where $\mathbf{u}$ is the N vector with components $u_i = u(\mathbf{x}_i)$, $i = 1, \ldots, N$,

$$m_i = \overline{u(\mathbf{x}_i)} \tag{12.2}$$

is the vector of means,

$$\Lambda_{ij} = \overline{(u(\mathbf{x}_i) - m_i)(u(\mathbf{x}_j) - m_j)} \tag{12.3}$$

is the $N \times N$ covariance tensor, and $|\Lambda|$ is the determinant of Λ.

The Mth-order moment of a random field is defined as the quantity

$$\overline{[u(\mathbf{x}_1) - m_1]^{\mu_1} [u(\mathbf{x}_2) - m_2]^{\mu_2} \cdots [u(\mathbf{x}_N) - m_N]^{\mu_N}}, \tag{12.4}$$

where μ_i, $i = 1, \ldots, N$, are integers such that $M = \mu_1 + \mu_2 + \cdots + \mu_N$. For the particular case of a Gaussian random field, it follows that

$$\overline{[u(\mathbf{x}_1) - m_1]^{\mu_1} [u(\mathbf{x}_2) - m_2]^{\mu_2} \cdots [u(\mathbf{x}_N) - m_N]^{\mu_N}}$$
$$= \frac{1}{(2\pi)^N |\Lambda|} \int (u_1 - m_1)^{\mu_1} \cdots (u_N - m_N)^{\mu_N} e^{-(1/2)(\mathbf{u}-\mathbf{m})^t \Lambda^{-1}(\mathbf{u}-\mathbf{m})} \, d\mathbf{u}. \tag{12.5}$$

It is not hard to show from (12.5) that all odd-order moments of a Gaussian process are zero. Moreover, it can also be shown that all even moments of order four or higher can be expressed in terms of the second-order moments. For example, in the particular case when $M = 4$,

$$\overline{(u(\mathbf{x}_1) - m_1)(u(\mathbf{x}_2) - m_2)(u(\mathbf{x}_3) - m_3)(u(\mathbf{x}_4) - m_4)}$$
$$= \Lambda_{12}\Lambda_{34} + \Lambda_{13}\Lambda_{24} + \Lambda_{14}\Lambda_{23}. \tag{12.6}$$

Thus the fourth-order moments of a Gaussian process can be expressed as quadratic products of second-order moments.

For any random process, not necessarily Gaussian, the Mth-order *cumulant* is defined as the difference between the Mth-order moment and the equivalent Gaussian form, such as given on the right-hand side of (12.6) in the case of fourth-order moments. By this reckoning it is evident that the fourth-order cumulant of a general random field $u(\mathbf{x})$ is the difference between the left- and right-hand sides of (12.6). Of course, if the random field is Gaussian, all of its cumulants are zero, and in particular, (12.6) holds exactly.

According to (12.1), Gaussian random fields have the property that $u(\mathbf{x})$ is a Gaussian random variable at each fixed point $\mathbf{x}$. This is also approximately true for the velocity field at fixed points in a homogeneous turbulent flow, as mentioned previously. But the velocity field in such a turbulent flow is not a Gaussian random field [i.e., it does not obey (12.1)]. Nonetheless, it appears plausible at first sight that turbulent flow is in some sense not too far removed from Gaussianity. If so, there might be some ways to use information about Gaussian random fields to aid in the development of a theory of turbulent fluid motion (e.g., as a perturbation away from a Gaussian state). This idea in different manifestations forms an integral part of some of the theories presented here.

12.3   OVERVIEW OF TURBULENCE THEORIES

One of the first ways in which the special properties of Gaussian random fields were used to create a turbulence theory was to imagine that the velocity field has vanishing fourth-order cumulants, as if it were a Gaussian process in this sense, but not necessarily in any other sense. This assumption has the effect of allowing fourth-order correlations to be computed as products of second-order correlations, thereby creating a closure. Since the velocity field is not Gaussian, this quasi-normal hypothesis cannot be exactly true. Moreover, the likelihood of it even being a good approximation is called into question by the fact that the third-order moments themselves are distinctly non-Gaussian (i.e., they cannot be zero since they are necessary for the transfer of energy to small scales). In fact, (7.184) shows that S_{ijk} would be zero and then so too would T_{ijn}, from (7.186). Without third-order moments there would be no turbulence.

Early attempts at studying the properties of the QN closure found that it is associated with an unphysical growth in the third-order moments, which subsequently causes the appearance of a negative energy spectrum (i.e., a violation of realizability [20]). Further hypotheses beyond quasi-normality are needed to prevent third-order moments from getting too large. A popular approach of this sort is to use eddy damping in coordination with a Markovinization assumption, as described below [22,23]. The resulting closure is referred to as EDQNM, which has the virtue of being applicable to high-Reynolds-number flows and is much less expensive to use than DNS. On the other hand, the steps taken to remedy the problems brought on by the QN assumption depend on functions with little rigorous basis for their determination. Moreover, the mathematical complexity of the approach is such as to limit its usefulness to relatively simple canonical flows, such as decaying isotropic turbulence.

Since quasi-normality conflicts with realizability, other means for taking into account departures from Gaussianity that do guarantee realizability have been sought. Many of these fall into the category referred to as renormalized perturbation theories (RPTs), of which the most influential and long-standing is the direct interaction approximation (DIA) [12]. Modern-day variants of DIA continue to attract considerable attention [16]. In a RPT the velocity field is expanded in an infinite series around a Gaussian base state, and this is used to develop an expansion of the nonlinear transfer term in the spectrum equation. Closure to the energy spectrum equation is achieved by suitably renormalizing it into a closed form. Generally, this entails summing the effects of higher-order terms in an approximate way that leads to a useful means of including their dynamical effects. There are many technical difficulties with such procedures, and the DIA itself leads to a $-3/2$ inertial range spectrum instead of a $-5/3$ spectrum. Later modifications, most notably the Lagrangian history DIA, were able to remedy this defect. The present discussion is limited to giving the main themes of the derivation of the DIA both for its intrinsic interest and so that it can be contrasted with the other models.

Both EDQNM and DIA make essential use of multiple space and/or time correlations. As such, they are examples of two-point closures. In fact, they lead to closure of the energy spectrum equation (7.190), which is just the Fourier transform of the two-point correlation equation. A third approach to turbulence theory is the renormalization group (RNG) theory [28,29,33]. Unlike EDQNM or DIA, the goal in this case is not to develop a two-point closure but rather a RANS or LES model through a formal analysis based on statistical assumptions. First, RNG assumes the validity of a correspondence principle to the effect that the Navier–Stokes equation with appropriate forcing can be made to reproduce an inertial range spectrum. A dynamical equation for the low-wavenumber part of the velocity field is then determined by a renormalization procedure in which the contributions of high-wavenumber motions on the low-wavenumber motions is redirected into an effective viscosity. The approach is fraught with assumptions and has generated considerable controversy since its inception.

The final subject of this chapter concerns the thermodynamics of vortex systems [4]. The idea of this approach is to establish a framework for understanding the fine-scale features of turbulence as a collection of interacting vortices, and from this explore its connection with the macroscopic properties of turbulent flows. This type of analysis attempts to get at the heart of such questions as the composition and mechanics of the energy cascade in the inertial range and how it is that it adheres to a $-5/3$ spectrum. Many open questions remain in this and other turbulence theories whose answers are being pursued actively today.

12.4   EDQNM

The obvious advantage of analyzing Navier–Stokes turbulence through the perspective of Fourier space is that the correlations that appear in the Fourier-transformed equations, such as (7.166) and (7.190) and equations for higher-order moments, are only between the undifferentiated products of Fourier coefficients. Thus the numerous correlations between velocities and their derivatives occurring in the physical space equations are avoided. The hierarchy of coupled moment equations in Fourier space that can be derived are in a form suitable for modeling through such statistical assumptions as quasi-normality. This is the point of view adopted by EDQNM, whose goal is to derive a closed equation for the energy spectrum. Since the present discussion is limited to isotropic turbulence, the route to the desired relation is through (7.120), namely,

$$E_{ij}(\mathbf{k}) = P_{ij}(\mathbf{k})\frac{E(k,t)}{4\pi k^2}, \tag{12.7}$$

where $P_{ij}(\mathbf{k})$ is defined in (7.167). Consequently, it is necessary first to consider closure to the energy spectrum tensor equation, (7.190), which means finding a way of evaluating the third-order moments originating in the stretching terms. In the EDQNM model the latter are computed by solving their own governing equation, which is closed via use of the quasi-normality hypothesis.

An equation for third moments is derived by using (7.168) to carry out the operations summarized by

$$\overline{\left(\frac{\partial \widehat{u}_i}{\partial t}(\mathbf{k}) + \cdots\right) \widehat{u}_p(\mathbf{a})\widehat{u}_q(\mathbf{b})} + \overline{\widehat{u}_i(\mathbf{k})\left(\frac{\partial \widehat{u}_p}{\partial t}(\mathbf{a}) + \cdots\right)\widehat{u}_q(\mathbf{b})}$$

$$+ \overline{\widehat{u}_i(\mathbf{k})\widehat{u}_p(\mathbf{a})\left(\frac{\partial \widehat{u}_q}{\partial t}(\mathbf{b}) + \cdots\right)}. \tag{12.8}$$

For notational simplicity the time dependence of the Fourier coefficients is not written out explicitly. Multiplying (12.8) by $(L/2\pi)^6$ and taking the limit as $L \to \infty$, as in the steps leading to (7.186), gives

$$\left(\frac{\partial}{\partial t} + \nu\left(k^2 + a^2 + b^2\right)\right) T_{ipq}(\mathbf{k}, \mathbf{a}, t)$$

$$= \lim_{L \to \infty}\left(\frac{L}{2\pi}\right)^6\left[M_{ijm}(\mathbf{k})\sum_{\mathbf{l}}\overline{\widehat{u}_j(\mathbf{l})\widehat{u}_m(\mathbf{k} - \mathbf{l})\widehat{u}_p(\mathbf{a})\widehat{u}_q(-\mathbf{k} - \mathbf{a})}\right.$$

$$+ M_{pjm}(\mathbf{a})\sum_{\mathbf{l}}\overline{\widehat{u}_j(\mathbf{l})\widehat{u}_m(\mathbf{a} - \mathbf{l})\widehat{u}_i(\mathbf{k})\widehat{u}_q(-\mathbf{k} - \mathbf{a})}$$

$$\left.+ M_{qjm}(-\mathbf{k} - \mathbf{a})\sum_{\mathbf{l}}\overline{\widehat{u}_j(\mathbf{l})\widehat{u}_m(-\mathbf{k} - \mathbf{a} - \mathbf{l})\widehat{u}_i(\mathbf{k})\widehat{u}_p(\mathbf{a})}\right], \tag{12.9}$$

where (7.173) has been used to assert that $\mathbf{b} = -\mathbf{k} - \mathbf{a}$ and $M_{ijm}(\mathbf{k})$ is defined in (7.169). Equation (12.9) is actually much simpler than it looks. In fact, consider the first of the quartic correlations on the right-hand side after application of the quasi-normal hypothesis:

$$\overline{\widehat{u}_j(\mathbf{l})\widehat{u}_m(\mathbf{k} - \mathbf{l})\widehat{u}_p(\mathbf{a})\widehat{u}_q(-\mathbf{k} - \mathbf{a})}$$

$$= \overline{\widehat{u}_j(\mathbf{l})\widehat{u}_m(\mathbf{k} - \mathbf{l})}\ \overline{\widehat{u}_p(\mathbf{a})\widehat{u}_q(-\mathbf{k} - \mathbf{a})}$$

$$+ \overline{\widehat{u}_j(\mathbf{l})\widehat{u}_p(\mathbf{a})}\ \overline{\widehat{u}_m(\mathbf{k} - \mathbf{l})\widehat{u}_q(-\mathbf{k} - \mathbf{a})} \tag{12.10}$$

$$+ \overline{\widehat{u}_j(\mathbf{l})\widehat{u}_q(-\mathbf{k} - \mathbf{a})}\ \overline{\widehat{u}_m(\mathbf{k} - \mathbf{l})\widehat{u}_p(\mathbf{a})}.$$

From (7.171) it follows that the first of the three products of second-order terms on the right-hand side is nonzero only when $\mathbf{k} = 0$, but then no contribution is made to (12.9) in this case since $M_{ijm}(0) = 0$. The second quadratic product in (12.10) contributes only when $\mathbf{l} = -\mathbf{a}$, and the last term only when $\mathbf{l} = \mathbf{a} + \mathbf{k}$. Taking advantage of these results, grouping a factor $(L/2\pi)^3$ with each of the quadratic correlations, taking the limit as $L \to \infty$, and using (7.182), the first term on the right-hand side of (12.9) yields

$$\lim_{L\to\infty} \left(\frac{L}{2\pi}\right)^6 \left[M_{ijm}(\mathbf{k}) \sum_{\mathbf{l}} \overline{\hat{u}_j(\mathbf{l})\hat{u}_m(\mathbf{k}-\mathbf{l})\hat{u}_p(\mathbf{a})\hat{u}_q(-\mathbf{k}-\mathbf{a})} \right]$$
$$= M_{ijm}(\mathbf{k}) \left[E_{jp}(\mathbf{a})E_{mq}(\mathbf{k}+\mathbf{a}) + E_{jq}(\mathbf{k}+\mathbf{a})E_{mp}(\mathbf{a}) \right] \qquad (12.11)$$
$$= 2M_{ijm}(\mathbf{k})E_{jp}(\mathbf{a})E_{mq}(\mathbf{k}+\mathbf{a}),$$

where the last equality is a result of switching j and m in the second term and noting the identity

$$M_{ijm}(\mathbf{k}) = M_{imj}(\mathbf{k}). \qquad (12.12)$$

Similar results follow for the remaining terms on the right-hand side of (12.9). Collecting these results together and substituting into (12.9) gives the following equation for triple moments:

$$\left(\frac{\partial}{\partial t} + \nu(k^2 + a^2 + b^2) \right) T_{ipq}(\mathbf{k}, \mathbf{a}, t) = H_{ipq}(\mathbf{k}, \mathbf{a}, t), \qquad (12.13)$$

where

$$H_{ipq}(\mathbf{k}, \mathbf{a}, t)$$
$$= 2M_{ijm}(\mathbf{k})E_{jp}(\mathbf{a})E_{mq}(\mathbf{k}+\mathbf{a})$$
$$+ 2M_{pjm}(\mathbf{a})E_{ji}(-\mathbf{k})E_{mq}(\mathbf{k}+\mathbf{a}) \qquad (12.14)$$
$$+ 2M_{qjm}(-\mathbf{k}-\mathbf{a})E_{ji}(-\mathbf{k})E_{mp}(\mathbf{a}).$$

Equation (12.13) has the structure of a first-order nonhomogeneous ODE and may be solved easily, yielding

$$T_{ipq}(\mathbf{k}, \mathbf{a}, t) = \int_0^t e^{-\nu(k^2+a^2+b^2)(t-\tau)} H_{ipq}(\mathbf{k}, \mathbf{a}, \tau)\, d\tau, \qquad (12.15)$$

where for simplicity the initial condition $T_{ipq}(\mathbf{k}, \mathbf{a}, 0) = 0$ is assumed.

Equation (12.15) is what is needed to effect the closure of (7.190). In this, (12.7) is first substituted into (12.14), which is then substituted into (12.15). After a rather extensive calculation (details may be found, e.g., in [16]), the following closed equation for the energy spectrum results:

$$\frac{\partial E(k,t)}{\partial t} + 2\nu k^2 E(k,t)$$

$$= \int d\mathbf{a} \int_0^t d\tau \, e^{-\nu(k^2+a^2+b^2)(t-\tau)} L(\mathbf{k},\mathbf{a}) E(b,\tau) \left[a^2 E(k,\tau) - k^2 E(a,\tau) \right],$$

(12.16)

where

$$L(\mathbf{k},\mathbf{a}) \equiv \frac{1}{2\pi} \frac{M_{jpq}(\mathbf{k}) M_{pjm}(\mathbf{a}) P_{mq}(\mathbf{k}+\mathbf{a})}{a^2 b^2},$$

(12.17)

and remember that $b = |\mathbf{a}+\mathbf{k}|$. The volume integral in (12.16) may be further reduced to a planar integral over the wavenumber amplitudes a and b (see [13]). This is helpful in obtaining numerical solutions of this nonlinear integrodifferential equation.

The derivation of (12.16) has entailed only one assumption, that of quasi-normality. As suggested previously, this is an unacceptable closure hypothesis since numerical solutions to (12.16) are found to develop negative spectra, thus violating the realizability condition to the effect that $E(k,t) \geq 0$. Subsequent investigation [22] traced the source of the problem to a failure of quasi-normality to provide for sufficient damping of the growth of the triple correlations. In other words, the QN hypothesis causes the right-hand side of (12.13) to be too large, so that T_{ipq}, which ultimately lies behind the right-hand side of (12.16), is overestimated. It is also the case that the skewness factor, which is related to the triple correlations, does not show a tendency toward equilibration during the decay process, as is seen in experimental measurements.

The EDQNM formalism is crafted with a view toward curing the specific weaknesses attendant on the QN hypothesis. Thus an eddy-damping term is added to the right-hand side of (12.13) as a practical means for limiting the growth in triple correlations. In this case, the model (12.13) is replaced by

$$\left[\frac{\partial}{\partial t} + \nu(k^2 + a^2 + b^2) \right] T_{ipq}(\mathbf{k},\mathbf{a},t)$$

$$= H_{ipq}(\mathbf{k},\mathbf{a},t) - \mu_{kab} T_{ipq}(\mathbf{k},\mathbf{a},t),$$

(12.18)

where μ_{kab} has the units of inverse time and is referred to as the *eddy-damping rate* of the third-order moments. $1/\mu_{kab}$ is a characteristic time scale of the damping. The larger μ_{kab} is, the more the growth of the triple correlations will be damped. There is no formal means for predicting μ_{kab}, so it must be modeled. A typical approach is to assume that $\mu_{kab} = \mu_k + \mu_a + \mu_b$ and then use dimensional arguments to suggest appropriate forms for μ_k, μ_a, and μ_b. One general principle which is also adhered to is to assume that the damping time should decrease with increasing wavenumber (i.e., $1/\mu_k \to 0$ as $k \to \infty$). This reflects the belief that the amount of damping should increase with increasing k rather than decrease, since motions at higher wavenumbers are expected to equilibrate faster than those at smaller wavenumbers. If one may assume that μ_k depends on k and $E(k,t)$, one of many possible forms of μ_k which is monotonically increasing is

$$\mu_k = C \left(\int_0^k k^2 E(k, t) \, dk \right)^{1/2},$$ (12.19)

where C is a constant. Equation (12.19) is often adopted in practice.

The consequence of the addition of eddy damping is to modify (12.16) so that now the closed energy spectrum equation has the form

$$\frac{\partial E(k, t)}{\partial t} + 2\nu k^2 E(k, t)$$

$$= \int d\mathbf{a} \int_0^t d\tau \, e^{-[\nu(k^2 + a^2 + b^2) + \mu_{kab}](t-\tau)}$$ (12.20)

$$\times L(\mathbf{k}, \mathbf{a}) E(b, \tau) \left[a^2 E(k, \tau) - k^2 E(a, \tau) \right].$$

This is the eddy-damped quasi-normal closure (EDQN), which, as it turns out, is still not entirely free of the occurrence of negative energy spectra. One additional hypothesis, referred to as *Markovinization*, does achieve realizability. In this, it is assumed that the total time scale in the exponential term in (12.20), namely, $[\nu(k^2 + a^2 + b^2) + \mu_{kab}]^{-1}$, is sufficiently small so that the exponential term will be significant only in a narrow time region around the point $t = \tau$. If it may be further assumed that the remainder of the integrand in (12.20) does not vary rapidly near $t = \tau$, then under these hypotheses, it may be taken outside the integral. Clearly, Markovinization has more justification for larger wavenumbers than lower ones, since the time scale is smaller in the former case, but it is nonetheless applied as an approximation at all wavenumbers. The result is that (12.20) simplifies further to the eddy-damped quasi-normal Markovian (EDQNM) form

$$\frac{\partial E(k, t)}{\partial t} + 2\nu k^2 E(k, t)$$

$$= \int \theta_{kab}(t) L(\mathbf{k}, \mathbf{a}) E(b, t) \left[a^2 E(k, t) - k^2 E(a, t) \right] d\mathbf{a},$$ (12.21)

where

$$\theta_{kab}(t) = \int_0^t e^{-[\nu(k^2 + a^2 + b^2) + \mu_{kab}](t-\tau)} \, d\tau.$$ (12.22)

In practice it is often the case that μ_{kab} is taken to be time independent, so that the integration in (12.22) can be done explicitly, yielding

$$\theta_{kab}(t) = \frac{1 - e^{-[\nu(k^2 + a^2 + b^2) + \mu_{kab}]t}}{\nu(k^2 + a^2 + b^2) + \mu_{kab}}.$$ (12.23)

The properties of the EDQNM theory have been explored over many years (e.g., [2,11]). Among its attractive features is full compatibility with Kolmogorov's $-5/3$

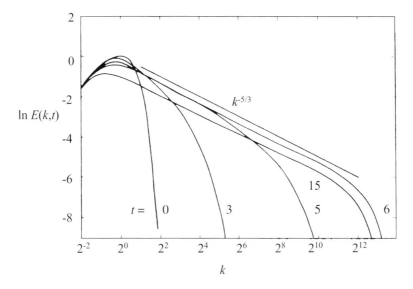

Fig. 12.1   *Time history of energy spectrum in isotropic decay computed as a solution to the EDQNM equations. (From [1]. Reprinted with the permission of Cambridge University Press.)*

law, which is not too surprising in view of the scaling used in setting μ_{kab}. An illustration of the kinds of information that EDQNM gives is shown in Fig. 12.1, where the computed time history of $E(k, t)$ during isotropic decay is shown. It is seen that from an initial state with energy concentrated at large scales, an inertial range $-5/3$ spectrum develops as energy spreads out to higher wavenumbers. During the ensuing decay, the $-5/3$ law is maintained while the total energy dissipates.

12.5   DIRECT INTERACTION APPROXIMATION

If a perturbation is applied to a steady-state turbulent system and then removed, after a period of time the system will return to a steady state. For example, if one of the Fourier modes associated with the velocity field is increased by the addition of energy, in time the excess energy will spread among the other modes until the system rearrives at a steady state. The direct interaction approximation[1] develops closure to the energy spectrum equation by coupling it to an equation for a response function which carries fundamental information about how turbulent systems respond to perturbations. An

---

[1]Due to Kraichnan [12]. A comprehensive discussion of DIA and many related and subsequent theories may be found, for example, in [14,16].

essential aspect of this analysis is to regard the DIA solution as a (not necessarily small) perturbation from a Gaussian ground state.

The starting point for deriving the DIA is (7.168) for turbulence in a periodic box of size L, rewritten in the form

$$\left(\frac{\partial}{\partial t} + \nu k^2\right) \widehat{u}_i(\mathbf{k}, t) = \lambda M_{ijm}(\mathbf{k}) \sum_{\mathbf{l}} \widehat{u}_j(\mathbf{l}, t)\widehat{u}_m(\mathbf{k} - \mathbf{l}, t) + P_{ij}(\mathbf{k})\widehat{f}_j(\mathbf{k}, t), \quad (12.24)$$

in which a parameter λ, referred to as the *ordering parameter*, is placed in front of the nonlinear term. Clearly, $\lambda = 1$, but it is kept explicitly in order to later develop a perturbation expansion. The last term on the right-hand side of (12.24) appears as a result of including a forcing term, $\mathbf{f}(\mathbf{x}, t)$, in the Navier–Stokes equation. $\widehat{f}_i(\mathbf{k}, t)$ is the Fourier coefficient corresponding to $\mathbf{f}(\mathbf{x}, t)$, and P_{ij} is the projection operator (7.167). The projection of the Fourier coefficient, namely, $P_{ij}\widehat{f}_j$, appears in (12.24) as a consequence of the steps used in eliminating pressure from (7.162), leading to (7.166). In fact, it must be the case that the forcing vector field has zero dilatation, so that $k_i\widehat{f}_i = 0$. If this condition is not satisfied, it cannot be guaranteed that the velocity field remains incompressible past its initial state. In view of the definition of P_{ij}, it follows that $P_{ij}\widehat{f}_j = \widehat{f}_i$, so the last term in (12.24) can, in fact, be replaced by $\widehat{f}_i$.

For the DIA, the forcing function $\widehat{f}_i(\mathbf{k}, t)$ is taken to be a Gaussian random field with homogeneous, isotropic, and stationary correlation tensor given by

$$\lim_{L \to \infty} \left(\frac{L}{2\pi}\right)^3 \overline{\widehat{f}_i(\mathbf{k}, t)\widehat{f}_j(-\mathbf{k}, t')} = P_{ij}(\mathbf{k})w(k, t - t'), \quad (12.25)$$

where

$$w(k, t - t') = W(k)\delta(t - t') \quad (12.26)$$

for some function $W(k)$. Equation (12.25) anticipates the later goal of deriving an equation for the energy spectrum equation by taking into account the limit as $L \to \infty$. The appearance of the delta function in (12.26) means that the forcing at any instant is independent of previous times. Consequently, whatever turbulent correlations develop in time due to the forcing originate purely from the character of the equations of motion and not the forcing field itself.

To bring into play the physics associated with the response of the turbulent system to small perturbations, the DIA makes essential and intimate use of a quantity, $\widehat{G}_{ij}(\mathbf{k}, t, t')$, referred to as the *infinitesimal response tensor*. This tensor function captures the way in which the Navier–Stokes equation responds to perturbations. In particular, suppose that $\widehat{u}_i$ and $\widehat{f}_i$ satisfy (12.24) and $\widehat{f}_i$ is perturbed according to $\widehat{f}_i + \delta\widehat{f}_i$, leading to a solution $\widehat{u}_i + \delta\widehat{u}_i$. Subtracting the versions of (12.24) written for the perturbed and nonperturbed solutions from each other, and dropping terms containing factors such as $\delta\widehat{u}_j(\mathbf{l}, t)\delta\widehat{u}_m(\mathbf{k} - \mathbf{l}, t)$—which are assumed to be infinitesimal—results in the following linearized equation for the velocity perturbation:

$$\left(\frac{\partial}{\partial t} + \nu k^2\right) \delta \widehat{u}_i(\mathbf{k}, t)$$

$$= 2\lambda M_{ijm}(\mathbf{k}) \sum_{\mathbf{l}} \widehat{u}_j(\mathbf{l}, t) \delta \widehat{u}_m(\mathbf{k} - \mathbf{l}, t) + P_{ij}(\mathbf{k}) \delta \widehat{f}_j(\mathbf{k}, t), \tag{12.27}$$

in which use has been made of (12.12), and what were initially two terms in the sum have been combined into a single term by using a transformation of the dummy index $\mathbf{l}$.

The infinitesimal response tensor is defined mathematically as a Green's tensor associated with (12.27). Specifically, it is defined as the solution of

$$\left(\frac{\partial}{\partial t} + \nu k^2\right) \widehat{G}_{in}(\mathbf{k}, t, t')$$

$$= 2\lambda M_{ijm}(\mathbf{k}) \sum_{\mathbf{l}} \widehat{u}_j(\mathbf{l}, t) \widehat{G}_{mn}(\mathbf{k} - \mathbf{l}, t, t') + P_{in}(\mathbf{k}) \delta(t - t'). \tag{12.28}$$

$\widehat{G}_{in}$ may be interpreted as capturing the effect at time t of an earlier unit disturbance at time t' created by the forcing term on the right-hand side. According to this point of view, $\widehat{G}_{in}(\mathbf{k}, t, t') = 0$ for $t' > t$, since disturbances do not propagate backward in time. If (12.28) could be solved for $\widehat{G}_{in}(\mathbf{k}, t, t')$, (in fact, this is not a likely prospect, due to the presence of the random Fourier coefficients of the turbulent velocity field), the solution to (12.27) would be given by

$$\delta \widehat{u}_i(\mathbf{k}, t) = \int_{-\infty}^{t} \widehat{G}_{in}(\mathbf{k}, t, t') \delta \widehat{f}_n(\mathbf{k}, t') \, dt' \tag{12.29}$$

(i.e., the weighted integral of the forcing $\delta \widehat{f}_n$ occurring over earlier times $t' \leq t$). It is implicit in (12.29) that the system has existed for a long enough time so that its initial state has no bearing on its current state.

The immediate goal now is to use (12.24) and (12.28) as the basis for deriving a closed system of coupled equations for the space-time energy spectrum tensor, $E_{in}(\mathbf{k}, t, t')$, and the (ensemble) averaged response tensor, $G_{in}(\mathbf{k}, t, t') \equiv \overline{\widehat{G}_{in}}$ $(\mathbf{k}, t, t')$. $E_{ij}(\mathbf{k}, t, t')$ is the Fourier transform of the space-time velocity correlation tensor, $\mathcal{R}_{ij}(\mathbf{r}, t, t') = \overline{u_i(\mathbf{x}, t) u_j(\mathbf{x} + \mathbf{r}, t')}$, and as such it generalizes the previously introduced single time energy spectrum tensor, $E_{ij}(\mathbf{k}, t)$, defined in (1.16).

An exact equation for G_{in} is determined by averaging (12.28) and is given by

$$\left(\frac{\partial}{\partial t} + \nu k^2\right) G_{in}(\mathbf{k}, t, t')$$

$$= 2\lambda M_{ijm}(\mathbf{k}) \sum_{\mathbf{l}} \overline{\widehat{u}_j(\mathbf{l}, t) \widehat{G}_{mn}(\mathbf{k} - \mathbf{l}, t, t')} + P_{in}(\mathbf{k}) \delta(t - t'). \tag{12.30}$$

The equation governing $E_{ij}(\mathbf{k}, t, t')$ is derived from (12.24) by multiplying through by $(L/2\pi)^3 \widehat{u}_n(-\mathbf{k}, t')$, averaging, and taking the infinite space limit. The result is

$$\left(\frac{\partial}{\partial t} + \nu k^2\right) E_{in}(\mathbf{k}, t, t')$$

$$= I_{in}(\mathbf{k}, t, t') + \lambda M_{ijm}(\mathbf{k}) \lim_{L \to \infty} \left(\frac{L}{2\pi}\right)^3 \sum_{\mathbf{l}} \overline{\widehat{u}_j(\mathbf{l}, t)\widehat{u}_m(\mathbf{k} - \mathbf{l}, t)\widehat{u}_n(-\mathbf{k}, t')}, \tag{12.31}$$

where

$$I_{in}(\mathbf{k}, t, t') \equiv P_{ij}(\mathbf{k}) \lim_{L \to \infty} \left(\frac{L}{2\pi}\right)^3 \overline{\widehat{f}_j(\mathbf{k}, t)\widehat{u}_n(-\mathbf{k}, t')} \tag{12.32}$$

reflects the correlation between the forcing and velocity fields at different times.

Preventing closure to the system of equations (12.30) and (12.31) are the nonlinear terms on the right-hand side of each equation. The DIA strategy for closing these transfer terms involves expanding them in infinite series whose terms represent increasing levels of interaction between scales. At the lowest level is the linear viscous response to a Gaussian forcing or perturbation. Each successive term in the expansion represents the effect of progressively greater levels of coupling between modes created by the action of the nonlinear transfer on the forced, viscous modes. After the formal expansions are set up, truncation of the nonlinear expansions at second order is achieved after invoking several key hypotheses.

Formally, one proceeds by first defining $\widehat{u}_i^0$ and $\widehat{G}_{in}^0$ as representing the Gaussian base state. This is the solution to the linear equations formed when $\lambda = 0$ in (12.24) and (12.28), namely,

$$\left(\frac{\partial}{\partial t} + \nu k^2\right) \widehat{u}_i^0(\mathbf{k}, t) = \widehat{f}_i(\mathbf{k}, t) \tag{12.33}$$

and

$$\left(\frac{\partial}{\partial t} + \nu k^2\right) \widehat{G}_{in}^0(\mathbf{k}, t, t') = P_{in}(\mathbf{k})\delta(t - t'). \tag{12.34}$$

$\widehat{u}_i^0$ and $\widehat{G}_{in}^0$ capture the viscous response to a Gaussian forcing; in particular, vortex stretching has no effect on these quantities. Now, perturbations of $\widehat{u}_i$ and $\widehat{G}_{in}$ away from the Gaussian state are assumed to take the form

$$\widehat{u}_i = \widehat{u}_i^0 + \lambda \widehat{u}_i^1 + O(\lambda^2) \tag{12.35}$$

and

$$\widehat{G}_{in} = \widehat{G}_{in}^0 + \lambda \widehat{G}_{in}^1 + O(\lambda^2). \tag{12.36}$$

$\widehat{u}_i^1$ and $\widehat{G}_{in}^1$ are the first terms in an infinite sequence of higher-order functions that capture the coupling effect between modes. Equations for these quantities are derived by substituting (12.35) and (12.36) into (12.24) and (12.28), and collecting terms in like powers of λ. This yields for $\widehat{u}_i^1$ and $\widehat{G}_{in}^1$, respectively, the equations

$$\left(\frac{\partial}{\partial t} + vk^2\right) \widehat{u}_i^1(\mathbf{k}, t) = M_{ijm}(\mathbf{k}) \sum_{\mathbf{l}} \widehat{u}_j^0(\mathbf{l}, t)\widehat{u}_m^0(\mathbf{k} - \mathbf{l}, t) \qquad (12.37)$$

and

$$\left(\frac{\partial}{\partial t} + vk^2\right) \widehat{G}_{in}^1(\mathbf{k}, t, t') = 2M_{ijm}(\mathbf{k}) \sum_{\mathbf{l}} \widehat{u}_j^0(\mathbf{l}, t)\widehat{G}_{mn}^0(\mathbf{k} - \mathbf{l}, t, t') \qquad (12.38)$$

and continues on with equations corresponding to higher-order powers of λ. Note that the solution for $\widehat{u}_i^1$ and $\widehat{G}_{in}^1$ will depend on the base state. Similarly, high-order terms can also ultimately be expressed as functionals of the base state.

Equation (12.34) is fully deterministic (i.e., $\widehat{G}_{in}^0$ is not random and may be solved for directly, e.g., using a Fourier transform in time). It is also evident that $\widehat{G}_{in}^0$ is the tensor Green's function corresponding to the differential equation (12.33) satisfied by $\widehat{u}_i^0(\mathbf{k}, t)$, so that a solution for the latter can be expressed as

$$\widehat{u}_i^0(\mathbf{k}, t) = \int_{-\infty}^{t} \widehat{G}_{in}^0(\mathbf{k}, t, s)\widehat{f}_n(\mathbf{k}, s) \, ds. \qquad (12.39)$$

The validity of this relation is easily confirmed by back substitution into (12.33). Note that (12.39) implies that $\widehat{u}_i^0(\mathbf{k}, t)$ is Gaussian since $\widehat{f}_n(\mathbf{k}, s)$ is assumed to have this property. By taking advantage of the same Green's function machinery, (12.37) and (12.38) can be solved, yielding

$$\widehat{u}_i^1(\mathbf{k}, t) = \int_{-\infty}^{t} \widehat{G}_{in}^0(\mathbf{k}, t, s)M_{njm}(\mathbf{k}) \sum_{\mathbf{l}} \widehat{u}_j^0(\mathbf{l}, s)\widehat{u}_m^0(\mathbf{k} - \mathbf{l}, s) \, ds \qquad (12.40)$$

and

$$\widehat{G}_{in}^1(\mathbf{k}, t, t') = 2 \int_{t'}^{t} \widehat{G}_{ip}^0(\mathbf{k}, t, s)M_{pjm}(\mathbf{k}) \sum_{\mathbf{l}} \widehat{u}_j^0(\mathbf{l}, s)\widehat{G}_{mn}^0(\mathbf{k} - \mathbf{l}, s, t') \, ds. \qquad (12.41)$$

The integration in (12.41) begins at t' since $\widehat{G}_{mn}^0(\mathbf{k} - \mathbf{l}, s, t') = 0$ for $s < t'$, as is evident from (12.34). Once again, it is easy to see that (12.40) and (12.41) solve (12.37) and (12.38), respectively.

The expressions for $\widehat{u}_i^1$ and $\widehat{G}_{in}^1$ in (12.40) and (12.41) are in terms of zeroth-order quantities. Similar relations can be derived for all the higher-order terms in (12.35) or (12.36), so that the expansion is ultimately in a form depending on just the ground states $\widehat{u}_i^0$ and $\widehat{G}_{in}^0$. At this point the expressions are ready for their intended purpose of closing the transfer terms in (12.30) and (12.31). After making these substitutions, formal expansions of the nonlinear terms in powers of λ result. These have coefficients depending on various complicated integrals of the ground state. This effort is much simplified by taking advantage of the Gaussianity of $\widehat{u}_i^0$. For example, its odd-order moments are zero and the zero cumulant conditions, such as in (12.6), are satisfied. The end result of these steps—through second order in λ—are the following equations:

$$
\left(\frac{\partial}{\partial t} + \nu k^2\right) G_{in}(\mathbf{k}, t, t')
$$

$$
= 4\lambda^2 M_{ijm}(\mathbf{k}) \int d\mathbf{l} \int_{t'}^{t} \widehat{G}_{jp}^0(\mathbf{l}, t, s) M_{pqr}(\mathbf{l}) \widehat{G}_{rn}^0(\mathbf{k}, s, t') E_{mq}^0(\mathbf{k} - \mathbf{l}, t, s) \, ds
$$

$$
+ O(\lambda^3) + P_{in}(\mathbf{k})\delta(t - t') \tag{12.42}
$$

and

$$
\left(\frac{\partial}{\partial t} + \nu k^2\right) E_{in}(\mathbf{k}, t, t')
$$

$$
= \lambda^2 M_{ijm}(\mathbf{k}) \int d\mathbf{l} \left[\int_{-\infty}^{t'} \widehat{G}_{np}^0(-\mathbf{k}, t', s) M_{pqr}(-\mathbf{k}) 2 E_{jq}^0(\mathbf{l}, t, s) E_{rm}^0(\mathbf{k} - \mathbf{l}, t, s) \, ds \right.
$$

$$
\left. + \int_{-\infty}^{t} \widehat{G}_{jp}^0(\mathbf{l}, t, s) M_{pqr}(\mathbf{l}) 4 E_{qm}^0(\mathbf{k} - \mathbf{l}, t, s) E_{rn}^0(-\mathbf{k}, t', s) \, ds \right]
$$

$$
+ O(\lambda^3) + I_{in}(\mathbf{k}, t, t'), \tag{12.43}
$$

where, in the spirit of (7.179) and (7.181),

$$
E_{in}^0(\mathbf{k}, t, t') \equiv \lim_{L \to \infty} \left(\frac{L}{2\pi}\right)^3 \overline{\widehat{u}_i^0(\mathbf{k}, t)\widehat{u}_j^0(-\mathbf{k}, t')} \tag{12.44}
$$

and use is made of (12.26). Equations (12.42) and (12.43) are coupled via their mutual dependence on the ground states E_{in}^0 and $\widehat{G}_{in}^0$.

If terms of all powers of λ were included in (12.42) and (12.43), these would be exact equations in which the nonlinear transfer terms appear as perturbation series. The crux of DIA lies in the way in which a practical closure is derived from these infinite series. Three steps are taken to effect a closure. First, a renormalization is carried out in which the quantities E_{ij}^0 and $\widehat{G}_{ij}^0$ appearing in (12.42) and (12.43) are replaced by E_{ij} and G_{ij}, respectively. Second, the perturbation series in each equation is truncated to its lowest-order term [i.e., the terms now indicated by $O(\lambda^3)$ are omitted]. Finally, the parameter λ is set equal to unity. The second of these steps is referred to as the *direct interaction approximation*, from whence the method gets its name. It refers to the fact that the $O(\lambda^2)$ terms in (12.42) and (12.43) involve interacting triads of wavenumbers $\mathbf{k}, \mathbf{l}$, and $\mathbf{k} - \mathbf{l}$ forming a triangle just as was noted previously in the context of analyzing the energy exchange mechanism in Section 7.7. The neglected higher-order terms account for energy transfer between wavenumber vectors in which other wavenumbers are dynamically involved in the role of an intermediary. These type of interactions are referred to as "indirect" and are omitted from the DIA.

The renormalization step may be viewed as carrying out a summation over some of the terms in the λ expansion. Confidence in the validity of this step comes through indirect evidence in the fact that it proves to be an effective means of solving a number of model systems sharing some, but not all, of the characteristics of Navier–Stokes turbulence.

The interest of the present discussion is on isotropic turbulence, so that the equations can be simplified via the isotropic formulas [similar to (12.7)]

$$E_{ij}(\mathbf{k}, t, t') = P_{ij}(\mathbf{k})E(k, t, t') \tag{12.45}$$

and

$$G_{ij}(\mathbf{k}, t, t') = P_{ij}(\mathbf{k})G(k, t, t'). \tag{12.46}$$

Substituting these into (12.42) and (12.43), contracting indices, and implementing the DIA renormalization assumptions yields the following closed system of equations for $G(k, t, t')$ and $E(k, t, t')$:

$$
\left(\frac{\partial}{\partial t} + \nu k^2\right) G(k, t, t')
$$

$$
= -\int d\mathbf{l}\, L(\mathbf{k}, \mathbf{l}) \int_{t'}^{t} G(l, t, s)G(k, s, t')E(|\mathbf{k} - \mathbf{l}|, t, s)\, ds + \delta(t - t'), \tag{12.47}
$$

where $L(\mathbf{k}, \mathbf{l})$ was defined in (12.17) and

$$
\left(\frac{\partial}{\partial t} + \nu k^2\right) E(k, t, t')
$$

$$
= \int_{-\infty}^{t'} G(k, t', s)w(k, t - s)\, ds + \int d\mathbf{l}\, L(\mathbf{k}, \mathbf{l}) \left[\int_{-\infty}^{t'} G(k, t', s) \right. \tag{12.48}
$$

$$
\left. \times E(l, t, s)E(|\mathbf{k} - \mathbf{l}|, t, s)\, ds - \int_{-\infty}^{t} G(l, t, s)E(|\mathbf{k} - \mathbf{l}|, t, s)E(k, t', s)\, ds \right].
$$

The first term on the right-hand side of (12.48) has its origin in the velocity/forcing correlation, $I_{in}(\mathbf{k}, t, t')$, defined in (12.32) after invoking the assumption of maximal randomness [12], which is described briefly next.

$I_{in}(\mathbf{k}, t, t')$ depends on the correlation between $\widehat{f}_j(\mathbf{k}, t)$ and $\widehat{u}_n(-\mathbf{k}, t')$. While the latter is a result of the flow evolution and all that that entails, for the purpose of modeling I_{in} it may be assumed that the main source of correlation between these quantities originates in that part of $\widehat{u}_n(-\mathbf{k}, t')$ which is directly attributable to the recent action of $\widehat{f}_j(-\mathbf{k}, s)$. This is quantized in the expression (12.29), which for the present application may be expressed as

$$\widehat{u}_n(-\mathbf{k}, t') = \int_{-\infty}^{t'} \widehat{G}_{nm}(-\mathbf{k}, t', s) \widehat{f}_m(-\mathbf{k}, s) \, ds. \tag{12.49}$$

Consequently, substituting (12.49) into (12.32) gives

$$I_{in}(\mathbf{k}, t, t') = P_{ij}(\mathbf{k}) \int_{-\infty}^{t'} \widehat{G}_{nm}(-\mathbf{k}, t', s) \lim_{L \to \infty} \left(\frac{L}{2\pi}\right)^3 \left[\overline{\widehat{f}_i(\mathbf{k}, t) \widehat{f}_m(-\mathbf{k}, s)}\right] ds \tag{12.50}$$

and after using (12.25) and (12.46) and renormalizing, this becomes

$$I_{ij}(\mathbf{k}, t, t') = P_{ij}(\mathbf{k}) \int_{-\infty}^{t'} G(k, t', s) w(k, t - s) \, ds. \tag{12.51}$$

This then leads to the expression containing w in (12.48).

Returning to the main derivation, note that the objective here is an equation for the energy density $E(k, t)$, which, according to (7.120) and (12.45), satisfies

$$E(k, t) = 4\pi k^2 E(k, t, t). \tag{12.52}$$

Thus, setting $t' = t$ in (12.48) and using (12.52), it is found that

$$\left(\frac{\partial}{\partial t} + vk^2\right) E(k, t) = T(k, t) + 4\pi k^2 W(k, t), \tag{12.53}$$

where (12.26) has been used and the transfer term in this equation is given by

$$T(k, t)$$

$$= 8\pi k^2 \int d\mathbf{l} \, L(\mathbf{k}, \mathbf{l}) \int_{-\infty}^{t} E(|\mathbf{k} - \mathbf{l}|, t, s) \left[G(k, t, s) E(l, t, s)\right. \tag{12.54}$$

$$\left. - G(l, t, s) E(k, t, s)\right] ds.$$

Equation (12.47), together with (12.52) and (12.53), form a closed system with which to solve for the energy spectrum.

Some interesting results from a simulation of decaying isotropic turbulence via DIA [17] are shown in Figs. 12.2 and 12.3. In the first of these the computed transfer spectrum is shown at several times during the decay. The initial state underrepresents the dissipation scales and there is a significant movement of energy to large k balanced by a corresponding loss of energy from the larger scales. Subsequently, the peaks in the loss and gain of energy reduce rapidly as the transfer spectrum relaxes toward a zero state. A look at the scaled one-dimensional dissipation spectrum $k_1^2 E_{11}(k_1, t)$ at some time after the initial transients have dissipated is shown in Fig. 12.3. This is

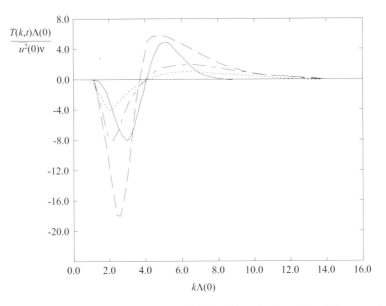

$$\frac{T(k,t)\Lambda(0)}{\overline{u^2}(0)\nu}$$

Fig. 12.2 *Time history of the transfer spectrum, $T(k,t)$, in decaying isotropic turbulence predicted by DIA. —, $t = 0$; – –, $t = 0.5$; — · —, $t = 1.0$; · · ·, $t = 1.6$, where time is scaled by the initial integral scale, $\Lambda(0)$, and rms velocity $[\overline{u^2}(0)]^{1/2}$. (From [17]. Reprinted with the permission of Cambridge University Press.)*

a low-Reynolds-number calculation, and it compares well with a number of experimental results included in the figure. Also shown in the figure is a calculation based on the local energy-transfer (LET) theory, which has a number of similarities with DIA [16].

12.6 RENORMALIZATION GROUP

In high-Reynolds-number flow, the inertial subrange of the energy spectrum with its $-5/3$ power law may extend over a considerable wavenumber range (see Fig. 2.3). The fluid motions corresponding to the inertial range occupy a range of physical scales, and it is not hard to imagine that throughout this range the turbulence evolves under a common physical process. This means that two observers of different size, but both within the inertial range scales, see the same dynamics. For example, if they are witnessing the transfer of energy to small scales via vortex stretching, it will appear the same to both of them. Physical phenomena of this type may be appropriate for a description through the mathematics of the renormalization group. Such ideas have had numerous applications in theoretical physics, including most notably the theory of critical phenomena [32]. In the case of Navier–Stokes turbulence, several avenues for applying RG methodology have been pursued [8,9,29,33]. The goal in this case is

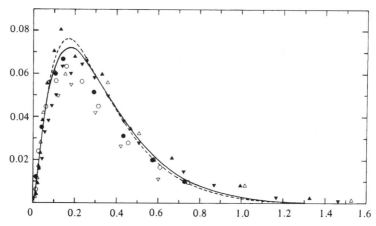

Fig. 12.3 *Comparison of scaled one-dimensional dissipation spectrum,* $k_1^2 E_{11}(k_1, t)/k_s^2 \sqrt{\overline{u^2}(t)}$, *where* $k_s \equiv (15 R_\lambda)^{1/3} \lambda^{-1}$, *predicted by:* $--$, *DIA;* $-$, *LET; and experiments:* $\triangledown$, $R_\lambda = 39.4$ [31]; $\circ$, $R_\lambda = 49.0$ *and* $\bullet$, $R_\lambda = 35.0$ [3]; $\triangle$, $R_\lambda = 38.1$ *and* $\blacktriangle$, $R_\lambda = 36.6$ [5]; $\blacktriangledown$, $R_\lambda = 45.2$ [10]. *(From [17]. Reprinted with the permission of Cambridge University Press.)*

to be able to model the entire effect of the inertial range succinctly without calculating its dynamics in detail.

One particular implementation of the RG analysis [33] has been used to provide a heuristic derivation of a closure scheme, which is referred to as the RNG version of the $K-\epsilon$ closure. The route to this end includes many assumptions whose basis in physics is not transparent, and considerable controversy surrounds the legitimacy of this formalism. Moreover, putting aside such questions, use of the RNG model, although appearing in some commercial codes, does not guarantee significant improvement over alternative closure formulations. Be that as it may, there is some value in considering the general ideas behind the approach so that they may be contrasted to those in the EDQNM and DIA theories, besides providing some insight into the derivation of the RNG model.

The RNG approach employs the space-time Fourier transform

$$\widehat{u}_i(\mathbf{k}, \omega) = \int dt \int u_i(\mathbf{x}, t) e^{-\iota(\mathbf{k} \cdot \mathbf{x} - \omega t)} \, d\mathbf{x}, \tag{12.55}$$

with inverse transform

$$u_i(\mathbf{x}, t) = \frac{1}{(2\pi)^4} \int d\omega \int_{k < \Lambda} \widehat{u}_i(\mathbf{k}, \omega) e^{\iota(\mathbf{k} \cdot \mathbf{x} - \omega t)} \, d\mathbf{k}, \tag{12.56}$$

where it is understood that the time integral in (12.55) is over $(-\infty, \infty)$ and the space integral is over $\Re^3$. The wavenumber domain is taken to be the sphere $k < \Lambda$ in $\mathbf{k}$ space, where Λ bounds the magnitude of wavenumbers with nonzero amplitude. It is expected that such a bound exists due to the effectiveness of viscous dissipation in removing energy from small scales. Λ is referred to as an *ultraviolet cutoff*, in recognition of the fact that it bounds the active wavenumbers from above, just as ultraviolet light is positioned just above the largest wavenumbers in the visible spectrum. To guarantee that the integration in (12.55) is convergent, it is imagined that $\mathbf{u}(\mathbf{x}, t) \rightarrow 0$ in the distant far field but in such a way that the analysis of the flow in a neighborhood of the origin is unaffected by this condition. Alternatively, the analysis can be pursued for periodic flow in a cube of dimension L and then the limit taken as $L \rightarrow \infty$, as in previous sections.

Applying a space-time Fourier transform to the Navier–Stokes equation with forcing term shows that $\widehat{u}_i(\mathbf{k}, \omega)$ obeys the dynamical equation

$$
\begin{aligned}
\widehat{u}_i(\mathbf{k}, \omega) \\
&= G^0(\mathbf{k}, \omega) f_i(\mathbf{k}, \omega) + \lambda_0 G^0(\mathbf{k}, \omega) M_{imn}(\mathbf{k}) \int_{l \leq \Lambda} d\mathbf{l} \\
&\times \int d\omega' \, \widehat{u}_n(\mathbf{l}, \omega') \widehat{u}_m(\mathbf{k} - \mathbf{l}, \omega - \omega'),
\end{aligned}
\tag{12.57}
$$

where $M_{ijm}(\mathbf{k})$ is defined in (7.169) and

$$
G^0(\mathbf{k}, \omega) \equiv [-\iota\omega + \nu k^2]^{-1}
\tag{12.58}
$$

is referred to as the *bare propagator*. The parameter $\lambda_0 = 1$ is to be used similarly to the way that it was used in the case of DIA. Clearly, (12.57) is similar to (12.24) but with the further step of taking the transform of the time variable.

While (12.57) is valid for all $\mathbf{k}$, the goal of the RNG approach is to develop a renormalized version of this relation valid only for small $\mathbf{k}$. In fact, renormalization is the act of systematically removing the presence of large $\mathbf{k}$ (small-scale) terms in (12.57), yet maintaining their influence in the remaining terms. Generally speaking, it is terms down through the lower end of the inertial range which are eliminated, although formally the procedure continues until the limit $\mathbf{k} \rightarrow 0$ is taken. The similarity with LES should be noted and, indeed, the RNG approach can be used to generate subgrid scale models.

From the Fourier perspective given by (12.57), the effect of high wavenumbers on lower wavenumbers comes about through the nonlinear term. In the RNG procedure these contributions are taken into account through an effective or renormalized viscosity (essentially, a turbulent eddy viscosity).

An important aspect of the RNG theory centers on the use and meaning of the forcing term in (12.57). Generally speaking it is used to stir the fluid in such a way as to create motion with the characteristics of the inertial range. In particular, it is assumed that the forcing autocorrelation function has the general stationary, homogeneous isotropic form

$$
\overline{f_i(\mathbf{k}, \omega) f_j(\mathbf{l}, \omega')} = 2W(k)(2\pi)^4 P_{ij}(\mathbf{k})\delta(\mathbf{k} + \mathbf{l})\delta(\omega + \omega'),
\tag{12.59}
$$

where the choice of $W(k)$ determines what sort of spectrum and fluid behavior will occur. Dimensional arguments show that for the choice $W(k) = W_0 k^{-3}$, W_0 a constant, $E(k, t)$ will have a $-5/3$ spectrum, so this is the form of W that is generally used [33]. However, for technical reasons that will surface later, it is assumed that

$$W(k) = W_0 k^{1-\epsilon^*} \tag{12.60}$$

and then the condition $\epsilon^* = 4$ is imposed. It should be cautioned that the issues involved in these statements are complex and the cause of much lingering controversy. For example, arguments both for and against the physicality of forcing with $W \sim k^{-3}$ have been offered [16]. It is beyond the scope of this presentation to consider such questions, other than to reiterate that many of the steps taken here are open to criticism since they are as yet without rigorous justification.

The way in which high-wavenumber effects are to be eliminated from (12.57) is via an iterative process in which, starting at the largest values of k and working toward the smallest, layer by layer of high-wavenumber contributions to the transfer term are removed. As suggested above, it is assumed that all excited wavenumbers satisfy $k < \Lambda$. The set of wavenumbers in this sphere may be divided into a region $0 < k \leq \Lambda - \Delta\Lambda$ and a thin shell of largest wavenumbers defined by $\Lambda - \Delta\Lambda < k \leq \Lambda$, where $\Delta\Lambda$ is small. The effect of the scales in the thin shell on the remainder is to be absorbed into an equation for just the latter scales. To accomplish this, it is convenient to decompose the Fourier components via the division

$$\widehat{u}_i(\mathbf{k}, \omega) = \widehat{u}_i^{<}(\mathbf{k}, \omega) + \widehat{u}_i^{>}(\mathbf{k}, \omega), \tag{12.61}$$

where

$$\widehat{u}_i^{<}(\mathbf{k}, \omega) \equiv \begin{cases} \widehat{u}_i(\mathbf{k}, \omega) & 0 < k \leq \Lambda - \Delta\Lambda \\ 0 & k > \Lambda - \Delta\Lambda \end{cases} \tag{12.62}$$

$$\widehat{u}_i^{>}(\mathbf{k}, \omega) \equiv \begin{cases} 0 & 0 < k \leq \Lambda - \Delta\Lambda \\ \widehat{u}_i(\mathbf{k}, \omega) & k > \Lambda - \Delta\Lambda \end{cases} \tag{12.63}$$

and a similar decomposition for $f_i = f_i^{<} + f_i^{>}$. Thus $\widehat{u}_i^{>}(\mathbf{k}, \omega)$ and $f_i^{>}(\mathbf{k}, \omega)$ represent the Fourier modes with smallest length scales.

The object now is to get an equation for $\widehat{u}_i^{<}(\mathbf{k}, \omega)$ that depends exclusively on scales in the range $k \leq \Lambda - \Delta\Lambda$. Evaluating (12.57) for $\mathbf{k}$ in the desired range and using (12.61) gives

$$\widehat{u}_i^{<}(\mathbf{k}, \omega)$$

$$= G^0(\mathbf{k}, \omega) f_i^{<}(\mathbf{k}, \omega)$$

$$+ \lambda_0 G^0(\mathbf{k}, \omega) M_{imn}(\mathbf{k}) \int_{l \leq \Lambda} d\mathbf{l} \int d\omega' [\widehat{u}_n^{<}(\mathbf{l}, \omega') \widehat{u}_m^{<}(\mathbf{k} - \mathbf{l}, \omega - \omega')$$

$$+ 2\widehat{u}_n^{>}(\mathbf{l}, \omega') \widehat{u}_m^{<}(\mathbf{k} - \mathbf{l}, \omega - \omega') + \widehat{u}_n^{>}(\mathbf{l}, \omega') \widehat{u}_m^{>}(\mathbf{k} - \mathbf{l}, \omega - \omega')]. \tag{12.64}$$

It is to be understood that the range of wavenumber integration in (12.64) will vary for each of the terms in the integrand, depending on where they are identically zero, due to the definitions in (12.62) and (12.63). To proceed in the use of (12.64) it is necessary to eliminate the dependence on $\widehat{u}_i^>(\mathbf{k}, \omega)$ in the last two terms inside the integral. In the same way that (12.57) was used as the basis for (12.64), so it also can be used as the first step in getting a relation for $\widehat{u}_i^>(\mathbf{k}, \omega)$. Thus from (12.57) it follows that

$$
\begin{aligned}
&\widehat{u}_i^>(\mathbf{k}, \omega) \\
&= G^0(\mathbf{k}, \omega) f_i^>(\mathbf{k}, \omega) \\
&\quad + \lambda_0 G^0(\mathbf{k}, \omega) M_{imn}(\mathbf{k}) \int_{l \le \Lambda} d\mathbf{l} \int d\omega' [\widehat{u}_n^<(\mathbf{l}, \omega') \widehat{u}_m^<(\mathbf{k} - \mathbf{l}, \omega - \omega') \\
&\quad + 2\widehat{u}_n^>(\mathbf{l}, \omega') \widehat{u}_m^<(\mathbf{k} - \mathbf{l}, \omega - \omega') + \widehat{u}_n^>(\mathbf{l}, \omega') \widehat{u}_m^>(\mathbf{k} - \mathbf{l}, \omega - \omega')].
\end{aligned}
\tag{12.65}
$$

One way to proceed in the analysis is to substitute (12.65) repeatedly into itself and by so doing create a representation of $\widehat{u}_i^>$ solely in terms of an infinite expansion in $\widehat{u}_i^<$. An equivalent means to the same end is first to expand $\widehat{u}_i^>(\mathbf{k}, \omega)$ in a perturbation series:

$$
\widehat{u}_i^> = \widehat{u}_i^{(0)} + \lambda_0 \widehat{u}_i^{(1)} + \lambda_0^2 \widehat{u}_i^{(2)} + \cdots,
\tag{12.66}
$$

substitute this into (12.65), and equate coefficients of like powers of λ_0 to determine $\widehat{u}_i^{(0)}, \widehat{u}_i^{(1)}$ and so on. Following the latter approach yields

$$
\widehat{u}_i^{(0)}(\mathbf{k}, \omega) = G^0(\mathbf{k}, \omega) f_i^>(\mathbf{k}, \omega)
\tag{12.67}
$$

as the ground state, which is Gaussian, and

$$
\begin{aligned}
&\widehat{u}_i^{(1)}(\mathbf{k}, \omega) \\
&= G^0(\mathbf{k}, \omega) M_{imn}(\mathbf{k}) \int_{l \le \Lambda} d\mathbf{l} \int d\omega' [\widehat{u}_n^<(\mathbf{l}, \omega') \widehat{u}_m^<(\mathbf{k} - \mathbf{l}, \omega - \omega') \\
&\quad + 2\widehat{u}_n^<(\mathbf{l}, \omega') \widehat{u}_m^{(0)}(\mathbf{k} - \mathbf{l}, \omega - \omega') + \widehat{u}_n^{(0)}(\mathbf{l}, \omega') \widehat{u}_m^{(0)}(\mathbf{k} - \mathbf{l}, \omega - \omega')],
\end{aligned}
\tag{12.68}
$$

and so on, for the remaining terms in (12.66). Ultimately, all of the terms in this expansion depend, insofar as the velocity field is concerned, on just the ground state. Formally, the analysis thus far is very similar to that used in the development of DIA.

Substitution of (12.68) and other higher-order relations into (12.66) gives $\widehat{u}_i^>$ in terms of its Gaussian ground state only. This may be then substituted into (12.64) so that the only dependence on wavenumbers above $\Lambda - \Delta\Lambda$ is via the ground state and the goal of the renormalization is to remove this dependence. Since $\widehat{u}_i^{(0)}$ and f_i, are Gaussian random fields, their moments are known [e.g., the second-order moments of $\widehat{u}_i^{(0)}$ are known through (12.59) and (12.67)]. Elimination of the high-k band of wavenumbers is now achieved by averaging over the ensemble of states corresponding to just the different realizations of the high-wavenumber part of the velocity field. In other words, the idea is to keep the lower-wavenumber terms fixed while averaging over motions at the larger wavenumbers. Denoting this kind of averaging by $\langle \cdots \rangle$, it follows by definition that $\langle \widehat{u}_i^< \rangle = \widehat{u}_i^<$, and similarly, $\langle f_i^< \rangle = f_i^<$. To proceed in the analysis it is necessary to assume, without formal justification, that $\widehat{u}_i^{(0)}$ retains the same stochastic properties under the subensemble of states (i.e., those that vary with high k) as it does under the full ensemble. The result of averaging, then, is to eliminate terms of odd order in $u_i^{(0)}$ or f_i. Moreover, zero cumulants of even orders are assumed.

The resulting equations are formidable and must undergo a further simplification in which unimportant terms are omitted and important terms are kept. The ensuing analysis is particularly subtle. First, the observation is made that the transfer term in (12.57) contains both "local" and "nonlocal" triadic contributions to the wavenumber integral. The former refers to the situation when k, l and $|\mathbf{k} - \mathbf{l}|$ are roughly the same magnitude, and the nonlocal case is when one of these is much larger than another. The significance of this distinction is that it is believed that the forcing term (which is used to stir the fluid to create a $-5/3$ law), essentially replicates the effect of the local part of the transfer term. Thus, in order not to double count the influence of such local interactions, this part of the transfer term must be omitted from the convolution integral during the renormalization process. This is accomplished formally via the *distant interaction approximation*, which is to say that the part of the transfer term which needs to be kept is the part that remains after taking a limit in which k becomes very small in comparison to l.

The equation resulting from these operations is

$$(-\iota\omega + \nu k^2)\widehat{u}_i^<(\mathbf{k}, \omega)$$

$$= f_i^<(\mathbf{k}, \omega) + \lambda_0 M_{imn}(\mathbf{k}) \int_{l \leq \Lambda - \Delta\Lambda} d\mathbf{l} \int d\omega' [\widehat{u}_n^<(\mathbf{l}, \omega')\widehat{u}_m^<(\mathbf{k} - \mathbf{l}, \omega - \omega')]$$

$$+ \widehat{u}_p^<(\mathbf{k}, \omega)8\lambda_0^2 M_{imn}(\mathbf{k}) \int_{\Lambda > l > \Lambda - \Delta\Lambda} d\mathbf{l}$$

$$\times \int d\omega' [G_0(|\mathbf{k} - \mathbf{l}|, \omega - \omega')|G_0(\mathbf{l}, \omega)|^2 M_{npq}(\mathbf{k} - \mathbf{l})P_{mq}(\mathbf{l})W(l)],$$

(12.69)

where terms of higher order in λ_0 have been dropped. Whether or not this last step can be justified depends on properties of the solution in the neighborhood of the point $\epsilon^* = 0$, which turns out to have considerable significance for the RNG approach [6], as shown below.

A significant observation about (12.69), which is essential for the RNG analysis, is that the last term on the right-hand side depends linearly on velocity just as do the terms on the left-hand side. This suggest the following rearrangement of (12.69):

$$(-\iota\omega + (\nu + \Delta\nu)k^2)\widehat{u}_i^<(\mathbf{k}, \omega)$$

(12.70)

$$= f_i^<(\mathbf{k}, \omega) + \lambda_0 M_{imn}(\mathbf{k}) \int_{l \leq \Lambda - \Delta\Lambda} d\mathbf{l} \int d\omega' [\widehat{u}_n^<(\mathbf{l}, \omega')\widehat{u}_m^<(\mathbf{k} - \mathbf{l}, \omega - \omega')],$$

where

$$\widehat{u}_i^<(\mathbf{k}, \omega)k^2 \, \Delta v$$

$$\equiv -\widehat{u}_p^<(\mathbf{k}, \omega)8\lambda_0^2 M_{imn}(\mathbf{k}) \int_{\Lambda>l>\Lambda-\Delta\Lambda} d\mathbf{l} \int d\omega'[G_0(|\mathbf{k}-\mathbf{l}|, \omega-\omega') \quad (12.71)$$

$$\times |G_0(\mathbf{l}, \omega)|^2 M_{npq}(\mathbf{k}-\mathbf{l}) P_{mq}(\mathbf{l}) W(l)].$$

Thus each step of the renormalization can be thought of as modifying the net viscosity affecting transport. A rescaling of $\Lambda - \Delta\Lambda$ back to Λ in (12.70) shows that the end result of one step of the renormalization is to produce an equation of the same form as (12.57), the only difference being in the viscosity. Thus the entire sequence of steps can now be repeated to strip away another part of the high-wavenumber spectrum. The complete renormalization repeats this process a large number of times until a limiting state of the eddy viscosity is reached. In fact, since $\Delta\Lambda$ is small and $\Delta v(\Lambda)$ is known from (12.71), the ratio $\Delta v/\Delta\Lambda$ can be computed. In the limit as $\Delta\Lambda \to 0$, and after evaluating the integral in (12.71) using (12.60), it is found that v satisfies the differential equation

$$\frac{dv(\Lambda)}{d\Lambda} = -A(\epsilon^*)\frac{v(\Lambda)\overline{\lambda}^2(\Lambda)}{\Lambda}, \quad (12.72)$$

where $A(\epsilon^*) \equiv (6 - \epsilon^*)/60\pi$ and

$$\overline{\lambda}(\Lambda) \equiv \lambda_0 \left(\frac{W_0}{v^3(\Lambda)\Lambda^{\epsilon^*}}\right)^{1/2}. \quad (12.73)$$

Equation (12.72) is exactly integrable and after integration using the initial conditions $\Lambda = \Lambda_0$, $v(\Lambda_0) = v_0$, and $\lambda_0 = 1$ gives

$$v(\Lambda) = v_0 \left(1 + \frac{3A(\epsilon^*)W_0}{v_0^3}\frac{\Lambda^{-\epsilon^*} - \Lambda_0^{-\epsilon^*}}{\epsilon^*}\right)^{1/3}. \quad (12.74)$$

For large scales, when $\Lambda \ll \Lambda_0$, this simplifies to

$$v(\Lambda) \sim \left(\frac{3W_0 A(\epsilon^*)}{\epsilon^*}\right)^{1/3} \Lambda^{-\epsilon^*/3}, \quad (12.75)$$

and after substituting this into (12.73) it is found that

$$\overline{\lambda} \sim (\epsilon^*)^{1/2}[3A(\epsilon^*)]^{-1/2}. \quad (12.76)$$

Note that $v(\Lambda) \to \infty$ as $\Lambda \to 0$. Also, $v(\Lambda)$ is independent of Λ_0 and v_0, which are associated with the dissipation range.

At this point the significance of the parameter ϵ^* must be considered. An important observation is that while the preceding development was based on a λ_0 expansion, in view of (12.73) it could just as well have been based on an expansion in $\bar{\lambda}$. In fact, $\bar{\lambda}$ appears naturally in place of λ_0 in the governing equations if they are nondimensionalized with scales appropriate to the wavenumber shell being eliminated. In this capacity $\bar{\lambda}$ has the role of a Reynolds number. It is also clear from (12.76) that $\bar{\lambda}^{-2} \sim \epsilon^*$, which means that ϵ^* can be used as a parameter in the scaling, with the implication that it must be small to justify the truncations used in carrying out the renormalization analysis. As it stands, the lowest-order terms in ϵ^* are included exactly. Higher order terms only affect other higher-order terms in the expansion. Thus, to maintain self-consistency, $A(\epsilon^*)$ in (12.75) should be replaced by $A(0) = 1/10\pi$ (i.e., the first term in an expansion in powers of ϵ^*), and (12.75) becomes

$$v(\Lambda) \sim \left(\frac{3W_0}{10\pi\epsilon^*} \right)^{1/3} \Lambda^{-\epsilon^*/3}. \tag{12.77}$$

It may be recalled that it is necessary to take $\epsilon^* = 4$ in order to have inertial range forcing. In fact, it is not obvious how to legitimize this step without violating the implicit assumption that ϵ^* is small.

Despite the many questions raised by these and other assumptions in the RNG analysis, the method has had some significant use in developing turbulent flow models. For example, it provides a subgrid model for LES [29,33] and RANS models such as used in the RNG form of the K–ϵ closure [34]. It has also been applied to developing SMC [26] and scalar transport modeling [27], to cite a few applications. RNG analysis also provides a means of making a prediction about the energy spectrum. This last point will now be discussed.

At the end of the renormalization process the equation for $\widehat{u}_i^<$ which remains accounts for the flow with wavenumbers up to a value $k << \Lambda_0$. This equation may be extracted from the previous formulas by setting $k = \Lambda$ so that all wavenumbers in the energy-containing range are included. Of course, the highest of these is next to the inertial range and violates the spirit of the distant-interaction approximation wherein k is supposed to be far removed from the shells eliminated in the renormalization. In any event, the following results are deduced:

$$v(k) \sim \left(\frac{3W_0}{40\pi} \right)^{1/3} k^{-4/3}, \tag{12.78}$$

where the renormalized velocity field satisfies the equation

$$\widehat{u}_i(\mathbf{k}, \omega) = G(\mathbf{k}, \omega) M_{imn}(\mathbf{k}) \int_{l \leq \Lambda} d\mathbf{l} \int d\omega' \, \widehat{u}_j(\mathbf{l}, \omega') \, \widehat{u}_m(\mathbf{k} - \mathbf{l}, \omega - \omega'), \tag{12.79}$$

where

$$G(\mathbf{k}, \omega) = [-\imath\omega + v(k)k^2]^{-1}. \tag{12.80}$$

In (12.79) the local part of the integration is restored and the forcing term is discarded since its effect after renormalization is accommodated in the turbulent viscosity. The parameter W_0 survives from the force term and can be related to the energy dissipation rate ϵ, since the input energy needs to balance the energy loss. Using (12.79) as the basis for deriving the energy spectrum equation and then equating the total energy transfer to dissipation yields

$$1.59 \, \epsilon = \frac{W_0}{\pi}, \tag{12.81}$$

so that from (12.78) it follows that

$$\nu(k) = 0.49\epsilon^{1/3}k^{-4/3}. \tag{12.82}$$

Finally, using (12.82) with (12.79) and assuming steady state, it is possible to derive the result

$$E(k) = 1.61\epsilon^{2/3}k^{-5/3}, \tag{12.83}$$

where the numerical value of the coefficient agrees well with the data in Fig. 2.3.

12.7   THERMODYNAMICS OF VORTEX SYSTEMS

The previous theories considered in this chapter depend on formal mathematical analyses of the statistical properties of the Navier–Stokes equation. A contrasting point of view is now considered which begins with a physical model of turbulent flow as comprised of a collection of vortices and proceeds from this vantagepoint to construct what is tantamount to a statistical mechanics of vortex systems. The hope is to extract useful properties of the aggregate of vortices as well as insight into their collective behavior (e.g., why they must support a $-5/3$ inertial range spectrum). The insight provided by this work also lends support to the vortex method analysis described in Section 10.6.

Initial work in the statistical mechanics of vortex systems was limited to an analysis of two-dimensional turbulence through collections of point vortices [21]. Minus the physics of vortex stretching, the two-dimensional case is reasonably tractable and its theory is now highly developed [7,18,19,25]. The equivalent study of three-dimensional vortex systems is much more difficult and remains the object of recent investigations. Here, a brief overview of this approach is presented, including some of its principal results. Many details and additional thoughts on this methodology may be found in the volume by Chorin [4].

As in previous sections of this chapter, the focus is on understanding the universal properties of turbulence, not those specially created by geometric or other constraints. Thus, considered here is turbulent flow in a large region containing the origin and whose velocity decays to zero in the far field (i.e., $|\mathbf{u}| \rightarrow 0$ as $|\mathbf{x}| \rightarrow \infty$). Assume that the total kinetic energy of this flow field, E, is constant or slowly varying in time. Consistent with previous observations (e.g., Figs. 1.3 and 4.37), it may be hypothesized that the turbulent field may be represented as a system of organized vortical motions that are responsible for the principal dynamics of the flow. The vortices are imagined to be most often tubelike, although the possibility of other forms, such as sheetlike vortices, is not excluded. From the perspective of statistical mechanics, the collection of vortices that define the fluid at one instant of time represent a "state" of the turbulent field out of an ensemble of allowable states: the *microcanonical ensemble* [24]. This is philosophically similar to taking a particular collection of molecular positions and momenta as defining the state of a gas. Now, the vortices in a particular state cannot take on arbitrary forms; for example, the energy of any acceptable configuration must equal E. Moreover, the vortex tubes are known to stretch as time goes on, creating a distribution of scales from some relatively large energy-containing scale down to a small scale in the dissipation range. The acceptable states of the vortices fill out this range of scales and are consistent with the kinematics and dynamics of the stretching process.

It is clear that the Reynolds number has a strong influence on setting the scale of the smallest vortices. However, the interest here is with the limiting case as $\nu \to 0$, so that the inertial range extends indefinitely to large wavenumbers. Thus there will be no need to bring a detailed consideration of dissipation into the subsequent discussion in this section.

The immediate task now is to characterize the allowable form of the vortical states. Two important observations are useful in illuminating how the vortex structures must appear in order to be consistent with the stretching process. In the first instance throughout the inertial range the spectrum remains in the $-5/3$ form: this suggests that the underlying vortical objects which support it are in some sense self-similar. Second, as known from experiments, the energy dissipation in turbulent flow is spatially intermittent (i.e., it occurs at spatially separated regions as against a uniform distribution everywhere), and this must also be reflected in the makeup of the vortex structure of the flow.

As noted in previous contexts, the fact that the flow is self-similar over the wavenumber range of the $-5/3$ spectrum suggests that the vortical structure should look the same to all observers in the inertial range. In particular, consider two observers of different sizes. Whatever they see the structure as, if blown up or shrunk so as to match what the other observer sees, they should "look" identical (i.e., have the same basic properties). The consequence of this is significant. It means that small observers see a structure inside the structure seen by large observers, who cannot see such details. If this sequence of viewers is imagined to continue on indefinitely to smaller and smaller observers, it is clear that the vorticity structure exists on an abstract set that has a more complicated structure than merely a section of three-dimensional space. In fact, the limiting sets obtained via this process are described as being *fractal* [15], and associated with them is a parameter called the *fractal dimension*, which in this case is a number less than 3. It is beyond the scope of this discussion to consider the precise mathematics of these statements, but the essential idea is clear. The vortex systems in the abstract phase space are composed of fractal objects that contain the support of the vorticity field (see also [30]).

Coinciding with the primarily intermittent locations where the turbulent energy is being dissipated are the locations where energy is being brought to dissipation scales by vortex stretching. This suggests that a further property of the vortices that needs to be recognized is that their fractal structure must be such as to support intermittency (i.e., it must lead to dispersed regions where the vortices are concentrated in small scales).

With these ideas as to the properties of the states of the turbulent fluid, it is now of interest to develop a thermodynamics of three-dimensional vortex systems that can answer such questions as why intermittency appears and most significantly, why a $-5/3$ energy spectrum is a natural companion of a turbulent field composed of such vortices. These two issues are considered in turn.

The question of intermittency may be attacked from the point of view of showing that it is an inevitable consequence of the constraint of fixed E on the stretching process. In other words, with E fixed and the vortices stretching, they must bunch up, creating intermittent disjoint zones. If not, the energy will become unbounded. To demonstrate this, it is first necessary to derive a means of determining the energy of a given vortex configuration.

The total kinetic energy/mass is

$$E = \frac{1}{2} \int \mathbf{u} \cdot \mathbf{u} \, d\mathbf{x}, \tag{12.84}$$

which is finite by the assumption made about $\mathbf{u}$ in the far field. The object now is to convert this to a formula involving the vorticity field. Taking note of (10.62), it follows from a vector identity that

$$\mathbf{u} \cdot \mathbf{u} = \mathbf{u} \cdot (\nabla \times \mathbf{B}) = \nabla \cdot (\mathbf{B} \times \mathbf{u}) + \mathbf{B} \cdot \mathbf{\Omega}, \tag{12.85}$$

which may then be substituted into (12.84). It is readily shown through the divergence theorem that the integral of the expression $\nabla \cdot (\mathbf{B} \times \mathbf{u})$ is zero in the limit as $|\mathbf{x}| \to \infty$. Thus it follows that

$$E = \frac{1}{2} \int \mathbf{B} \cdot \mathbf{\Omega} \, d\mathbf{x} \tag{12.86}$$

and after using (10.64),

$$E = \frac{1}{8\pi} \int d\mathbf{x} \int d\mathbf{y} \frac{\mathbf{\Omega}(\mathbf{x}) \cdot \mathbf{\Omega}(\mathbf{y})}{|\mathbf{x} - \mathbf{y}|}. \tag{12.87}$$

Given a particular member of the ensemble of flows with a definite vorticity field, formula (12.87) allows for a calculation of the energy.

The integrations in (12.87) are over only the supports of the vorticity field. For a vortex tube model each integral follows separately along the vortices, with the factor $1/|\mathbf{x} - \mathbf{y}|$ being large when the same position on the same vortex is reached in both integrals. It is convenient to isolate this part of the integration from the remainder since it involves a sensitive integration of the details of the vortex structure and its interaction with the large factor. This contribution to energy is referred to as the *self-energy* and is denoted here as E_2 in the decomposition $E = E_1 + E_2$. E_1 is an *interaction energy* accounting for the interactions between more distant parts of a vortex with itself and others. In the evaluation of E for a given vortex state it is necessary to take E_1 and E_2 into account separately.

One strategy for showing why intermittency develops depends on using simple estimates of how E_1 and E_2 vary under a change in scale. The idea is to begin with a straight vortex with given energy and then reevaluate the energy after stretching the vortex. The energy is then reevaluated again after another stretching of the previously stretched vortex, and so on, until all the sequence of stretchings have been carried out and the vortex achieves its fractal shape. The end result of this process must still yield a vortex with given energy E, and it is in enforcing this that the necessity of intermittency becomes evident. In fact, without giving the details of the scaling (see [4]), it is shown that the behavior of E_2 during this calculation depends on the relative magnitude of the distance between points on the vortices and the lengths of the vortices between these points. It is found that if the vortices are allowed to fill up all the volume available to them, this ratio is unity and the self-energy becomes unbounded. To prevent growth of E_2 it is necessary for the distance stretched by vortex segments to exceed the distance between points on the vortices, a condition that can be met by the folding of vortices. In essence, there is a cancelation of velocity in the folded vortex segments that prevents energy growth. If folding accompanies stretching, some regions will be occupied by greater concentrations of vortices, and intermittency is explained.

Additional insight and justification for this viewpoint can be had by developing a relatively simple model of the vortex structure, in which the vortices are represented as occupying the grid lines of a three-dimensional lattice. Sample vortices can be created through a Markov process that creates a self-avoiding random walk on the lattice. This model of the physical system can be used to perform numerical experiments that give clues as to how the actual system will behave. For the present case, a vortex can be laid out in the lattice, its energy evaluated, and then a refinement of the lattice made and the energy reevaluated on the refinement. A computation of how energy changes in this case shows that to prevent unbounded growth in the self-energy the volume available to the vortices must shrink by excluding volume without vorticity or where the vorticity is not stretching. The end result of applying the shrinkage over an unending sequence of volumes is to create a fractal volume occupied by the vorticity.

The collection of fractilized vortex tubes and filaments are assumed to form the microcanonical ensemble of states which is appropriate for analysis via the standard machinery of statistical thermodynamics, including the mathematical framework associated with the canonical ensemble [24]. This is appropriate for a system whose subsets interact with the remainder as if it was a large heat bath in equilibrium. For the vortex model, the complete system is taken to be a large collection of vortex tubes in a lattice arrangement with N segments. These can be generated via the self-avoiding random walk so that the vortices do not pass through the same node twice and the singularity in (12.87) is avoided. Each such vortex forms a subset of the whole, which is assumed to be interacting only weakly with the rest of the field, is denoted as C, and has an energy $E(C)$ that can be computed from (12.87).

By a standard argument, it can be shown that the probability of occurrence of C is given by

$$P(C) = \frac{1}{Z} e^{-\beta E}, \tag{12.88}$$

where $\beta \equiv 1/T$, T is the temperature, and Z is the partition function,

$$Z \equiv \sum_C e^{-\beta E}, \tag{12.89}$$

whose definition ensures that the sum of $P(C)$ over all configurations is unity. The average energy of the members of the ensemble is

$$\overline{E} = \sum_C E(C) P(C). \tag{12.90}$$

The entropy of the total system is defined by

$$S = -\sum_C P(C) \log P(C), \tag{12.91}$$

and a calculation using (12.88) and (12.90) confirms that

$$\frac{dS}{d\overline{E}} = \frac{1}{T}. \tag{12.92}$$

The temperature in use here is an abstract temperature and should not be identified with the usual thermodynamic temperature, although a connection can be made if the thermodynamic system in question were a gas, for example. In the present situation, there is no restriction on the sign of T; it may be positive or negative. In fact, when T increases without bound through positive values and becomes infinite ($\beta : \infty \to 0^+$), further increase in T can be shown to be into negative temperatures ($\beta : 0^- \to -\infty$). The $|T| = \infty$, $\beta = 0$ point is the boundary between positive and negative temperatures.

The lattice model of the vortex system is sufficiently simple that numerical studies of the behavior of the thermodynamic variables can be carried out. This shows that the maximum of the entropy occurs at the point of infinite temperature separating the negative and positive temperature states. This result is consistent with (12.92), which shows that entropy can have a maximum with respect to energy when $|T| = \infty$. Since thermodynamic systems tend toward the maximum entropy state, the implication is that three-dimensional vortex systems will tend to establish an equilibrium state at infinite temperature. In this state, (12.88) implies that $P(C)$ is a constant independent of C, so all configurations are equally likely.

Assuming that this is a reasonably accurate picture of the kind of equilibrium to be encountered in turbulent flow in the inertial range, it is valuable to investigate the properties of this equilibrium as well as to see how it can be maintained via the dynamics of vortex systems. The chief question to be asked concerning this model is whether it is associated with the $-5/3$ energy spectrum. Helpful in answering this is to borrow from techniques used in the analysis of polymer physics, whose systems are also made up of long folded objects. In fact, the label *polymer* fits the kinds of objects assumed in the lattice model of vortices, namely, equally likely three-dimensional paths in the lattice which may be constructed via self-avoiding random walks.

Without giving details, an estimate of the energy spectrum associated with vortices in the form of polymers may be calculated with the help of several previous results. In the first place, using relations equivalent to (2.87) and (2.89), it may be shown that for isotropic turbulence the Fourier transform of the two-point vorticity correlation function $\overline{\Omega_i(\mathbf{x})\Omega_i(\mathbf{x}+\mathbf{r})}$ is proportional to $k^2 E(k)$. Moreover, under similar circumstances and assumptions used in the derivation of (10.42), it can be shown that a power law spectrum in $E(k)$, say $E(k) \sim k^{-\gamma}$, corresponds to a midrange of r, in which

$$\overline{\Omega_i(\mathbf{x})\Omega_i(\mathbf{x}+\mathbf{r})} \sim r^{\gamma-3}. \tag{12.93}$$

Thus the possibility of finding an explanation for Kolmogorov's inertial range law is tied to the nature of the two-point vorticity correlation tensor as determined from a field of vortical objects residing in an equilibrium state at infinite temperature. An evaluation of (12.93) is relatively easy to come by using a numerical model of the vortices, except when r is small. In this case, a careful evaluation of the local structure of the vortices is necessary.

An argument has been made that the decomposition

$$\gamma = D - D_c + \tilde{D} \tag{12.94}$$

can be justified. In this, $D \approx 3$ is the dimension of the support of the vorticity field, D_c is the fractal dimension of the centerline of the vortices, and $\tilde{D}$ is the exponent that would appear if only the nonlocal contributions were included. A numerical calculation incorporating a direct evaluation of the two-point vorticity correlation has revealed that $\tilde{D} \approx 0.37$. Moreover, the estimate $D_c \approx 1.70$ can be computed from the behavior of computer-generated self-avoiding Brownian motion paths. Taken together, this gives $\gamma \approx 1.67 \approx 5/3$ (i.e., the Kolmogorov law).

It has to be cautioned that there are many qualifications surrounding this favorable result, and it is certain that considerably more theoretical and numerical investigation are necessary before great confidence can be placed in it. Nonetheless, it is evident that the approach to understanding turbulent flow through the thermodynamics of vortex systems is steeped in the physics of the phenomenon and offers many practical and conceptual advantages.

12.8 SUMMARY

The theoretical approaches to the turbulence problem described in this chapter underlie many current investigations into the most fundamental nature of turbulent flow at the universal scales of motion. That there is a plurality of methods with the same goal in mind suggests that no single approach yet enjoys a clear consensus of opinion as being most useful either for modeling turbulence or elucidating its physics. Differences of opinion include basic questions such as whether or not turbulence should be viewed as an equilibrium or nonequilibrium process. In fact, the thermodynamic theory in Section 12.7 challenges the prevailing view that motion at inertial and dissipation range scales is part of a nonequilibrium process. In particular, the equilibrium state compatible with theories such as DIA and EDQNM is one of thermal equilibrium, for which there is no energy flux between scales. Thus the presence of energy transfer in a cascade between regions of energy production and energy dissipation means that the turbulent field must be in a state far from equilibrium. The contrary view, summarized in part in Section 12.7, holds that the tools of equilibrium thermodynamic systems can legitimately be applied to describing inertial range turbulence.

It is not yet clear where and when these and other basic questions about turbulence will be answered. Nonetheless, there is reason to hope that future years will see substantial progress in elucidating the fundamental nature of turbulence, if for no other reason than that the continuing advancement in computational tools allows for the simulation of turbulence at a wider range of conditions and with better means for extracting information useful to testing and analyzing theoretical predictions. Whatever new truths emerge from such sources should have collateral benefit in the development of predictive tools such as those discussed elsewhere in this volume.

REFERENCES

1. Andre, J. C. and Lesieur, M. (1977) "Influence of helicity on high Reynolds number isotropic turbulence," *J. Fluid Mech.* **81**, 187–207.

2. Cambon, C. and Scott, J. F. (1999) "Linear and nonlinear models of anisotropic turbulence," *Annu. Rev. Fluid Mech.* **31**, 1–53.

3. Chen, W. Y. (1968) "Spectral energy transfer and higher-order correlations in grid turbulence" Ph.D. dissertation, University of California, San Diego.

4. Chorin, A. J. (1994) *Vorticity and Turbulence*, Springer-Verlag, New York.

5. Comte-Bellot, G. and Corrsin, S. (1971) "Simple Eulerian time correlation of full and narrow band velocity signals in grid generated, isotropic turbulence," *J. Fluid Mech.* **48**, 273–337.

6. Eyink, G. L. (1994) "The renormalization group method in statistical hydrodynamics," *Phys. Fluids* **6**, 3063–3078.

7. Eyink, G. L. and Spohn, H. (1993) "Negative temperature states and large scale, long lived vortices in 2-dimensional turbulence," *J. Stat. Phys.* **70**, 833–886.

8. Forster, D., Nelson, D. R. and Stephen, M. J. (1976) "Long time tails and large eddy behavior of a randomly stirred fluid," *Phys. Rev. Lett.* **36**, 867–870.

9. Forster, D., Nelson, D. R. and Stephen, M. J. (1977) "Large distance and long time properties of a randomly stirred fluid," *Phys. Rev. A* **16**, 732–749.

10. Frenkiel, N. F. and Klebanoff, P. S. (1971) "Statistical properties of velocity derivatives in a turbulent field," *J. Fluid Mech.* **48**, 183–208.

11. Godeferd, F. S., Cambon, C. and Scott, J. F. (2001) "Two-point closures and their applications: report on a workshop," *J. Fluid Mech.* **436**, 393–407.

12. Kraichnan, R. (1959) "The structure of isotropic turbulence at very high Reynolds numbers," *J. Fluid Mech.* **5**, 497–543.

13. Lesieur, M. (1997) *Turbulence in Fluids*, 3rd Ed., Kluwer Academic, Dordrecht, The Netherlands.

14. Leslie, D. C. (1973) *Developments in the Theory of Turbulence*, Clarendon Press, Oxford.

15. Mandelbrot, B. (1977) *Fractals, Form, Chance and Dimension*, W. H. Freeman, San Francisco.

16. McComb, W. D. (1990) *The Physics of Fluid Turbulence*, Clarendon Press, Oxford.

17. McComb, W. D. and Shanmugasundaram, V. (1984) "Numerical calculation of decaying isotropic turbulence using the LET theory," *J. Fluid Mech.* **143**, 95–123.

18. Miller, J. (1990) "Statistical mechanics of Euler equations in two dimensions," *Phys. Rev. Lett.* **65**, 2137–2140.

19. Montgomery, D. and Joyce, G. (1974) "Statistical mechanics of negative temperature states," *Phys. Fluids* **17**, 1139–1145.

20. Ogura, Y. (1963) "A consequence of the zero fourth cumulant approximation in the decay of isotropic turbulence," *J. Fluid Mech.* **16**, 33–40.

21. Onsager, L. (1948) "Statistical hydrodynamics," *Nuovo Cimento Suppl.* **6**, 279–287.

22. Orszag, S. A. (1970) "Analytical theories of turbulence," *J. Fluid Mech.* **41**, 363–386.

23. Orszag, S. A. (1977) "Statistical theories of turbulence," in *Fluid Dynamics 1973*, Les Houches Summer School of Theoretical Physics (R. Balian and J. L. Peube, Eds.), Gordon and Breach, New York, pp. 237–374.

24. Reichl, L. E. (1998) *A Modern Course in Statistical Physics*, Wiley, New York.

25. Robert, R. (1991) "A maximum entropy principle for two-dimensional perfect fluid dynamics," *J. Stat. Phys.* **65**, 531–554.

26. Rubinstein, R. and Barton, J. M. (1990) "Nonlinear Reynolds stress models and the renormalization-group," *Phys. Fluids A* **2**, 1472–1476.

27. Rubinstein, R. and Barton, J. M. (1992) "Renormalization-group analysis of the Reynolds stress transport equation," *Phys. Fluids A* **4**, 1759–1766.

28. Smith, L. M. and Reynolds, W. C. (1992) "On the Yakhot–Orszag renormalization group method for deriving turbulence statistics and models," *Phys. Fluids A* **4**, 364–390.

29. Smith, L. M. and Woodruff, S. L. (1998) "Renormalization-group analysis of turbulence," *Annu. Rev. Fluid Mech.* **30**, 275–310.

30. Sreenivasan, K. R. (1991) "Fractals and multi-fractals in fluid turbulence," *Annu. Rev. Fluid Mech.* **23**, 435–472.

31. Stewart, R. W. and Townsend, A. A. (1951) "Similarity and self-preservation in isotropic turbulence," *Philos. Trans. R. Soc. London Ser. A* **243**, 359–386.

32. Wilson, K. G. (1975) "Renormalization group methods," *Adv. Math.* **16**, 170–186.

33. Yakhot, V. and Orszag, S. A. (1986) "Renormalization group analysis of turbulence. I. Basic theory," *J. Sci. Comput.* **1**, 3–51.

34. Yakhot, V., Orszag, S. A., Thangam, S., Gatski, T. B. and Speziale, C. G. (1992) "Development of turbulence models for shear flows by a double expansion technique," *Phys. Fluids A* **4**, 1510–1520.

Author Index

Subject Index